D. Feng, W.C. Siu, H. Zhang (Eds.)

Multimedia Information Retrieval and Management

Springer

Berlin
Heidelberg
New York
Hong Kong
London
Milan
Paris
Tokyo

Engineering  **ONLINE LIBRARY**

http://www.springer.de/engine/

David Dagan Feng
Wan Chi Siu
Hong-Jiang Zhang (Eds.)

Multimedia Information Retrieval and Management

Technological Fundamentals and Applications

With 168 Figures

 Springer

D. Feng, PhD
Centre for Multimedia Signal Processing
Hong Kong Polytechnic University
and
School of Information Technologies
The University of Sydney

W. C. Siu, PhD
Centre for Multimedia Signal Processing
Hong Kong Polytechnic University

H. Zhang, PhD
Microsoft Research, Asia

Multimedia Signal Processing Book Series
Book Series Sponsored by: Centre for Multimedia Signal Processing,
The Hong Kong Polytechnic University
Series Editor: Prof. W.C. Siu

ISBN 3-540-00244-8 Springer-Verlag Berlin Heidelberg New York

Library of Congress Cataloging-in-Publication-Data applied for
A catalog record for this book is available from the Library of Congress.

Bibliographic information published by Die Deutsche Bibliothek Die Deutsche Bibliothek lists this publication in the Deutsche Nationalbibliographie; detailed bibliographic data is available in the Internet at http://dnb.ddb.de

Springer-Verlag Berlin Heidelberg New York
a member of BertelsmannSpringer Science+Business Media GmbH
http://www.springer.de

© Springer-Verlag Berlin Heidelberg 2003
Printed in Germany

Typesetting: print-data delivered by authors
Cover design: design & production, Heidelberg
Printed on acid free paper   62/3020/M - 5 4 3 2 1 0

Preface

Multimedia information technologies, which provide comprehensive and intuitive information for a broad range of applications, have a strong impact on modern life, and have changed our way of learning and thinking. Over the past two decades, there has been an explosive growth in the use of digital multimedia (including audio, video, images and graphics) over the Internet and wireless communication. As the use of digital multimedia increases, effective data storage and management become increasingly important. In fields which use large quantities of data (e.g. audio, video, image and digital libraries; geographical and medical image databases; etc), we need to minimize the volume of data stored while meeting the often conflicting demand for accurate data representation. In addition, the data need to be managed such that it facilitates efficient searching, browsing and cooperative work. This area has been a very active research area in recent years. This book will provide readers with an up-to-date and comprehensive picture of cutting edge technologies in multimedia information retrieval and management, which directly affect our industry, economy and social life

The book is divided into two major parts: Technological Fundamentals which covers the core theories of the area; and Applications which describes the broad range of practical uses for this technology. The book includes:
- a complete set of techniques for digital audio, music, images and video content-based retrieval;
- multimedia low-level feature extraction and high level semantic description, as well as the most up-to-date MPEG-7 standard;
- multimedia authentication and watermarking; and
- a broad range of practical applications (e.g., digital libraries, medical images, and biometrics, etc).

It is suitable for use as a textbook for senior undergraduate and junior postgraduate students, and as a handbook for multimedia researchers, developers, and practitioners.

The contributors of this book are outstanding researchers in their areas and include: Professor Shih-Fu Chang, Columbia University, the IEEE Distinguished Lecturer for the Year of 2002; Professor H. K. Huang, University of Southern California, the author of "Picture Archiving and Communication Systems" (PACS) and a pioneer of the PACS systems; and Professor Edward A. Fox, Director of the Internet Technology Innovation Center at Virginia Tech and Co-Editor-in-Chief for the ACM Journal of Education Resources in Computing. The names and affiliations of all of the contributors are listed in the next few pages. We would like to thank the contributors for their excellent team work which enabled us to meet our tight deadline, which in turn has ensured that the information included is up-to-date. We would also like to thank Mr Z.Y. Wang and Ms W. Ho for their professional assistance in editing this book and Springer-Verlag for their support and efficiency throughout the process.

David Dagan Feng, Wan-Chi Siu and Hong-Jiang Zhang (Eds.)

About the Editors

David Dagan Feng received his PhD degree in Computer Science from UCLA in 1988. After briefly working as assistant professor in USA, he joined the University of Sydney, Australia, as lecturer, senior lecturer, reader, and then professor. He is the former Head of Department of Computer Science/School of Information Technologies. He is currently Associate Dean of the Faculty of Science and Director of the Biomedical & Multmedia Information (BMIT) Research Group at the University of Sydney, and Professor and Deputy Director, Center for Multimedia Signal Processing, Department of Electronic and Information Engineering, Hong Kong Polytechnic University. He has published over 200 scholarly research papers, pioneered several new research directions, made a number of landmark contributions in his field, and received the Crump Prize for Excellence in Medical Engineering. More important, however, is that many of his research results have been translated into solutions to real-life problems worldwide and have made tremendous improvements to the quality of life. He is currently special area editor of IEEE Transactions on Information Technology in Biomedicine, advisor for the International Journal of Image and Graphics, chairman of IFAC-TC-BIOMED, and chairman of the International Programme and National Organizing Committees for the IFAC 2003 Symposium. He is a Fellow of ACS, HKIE, IEE and IEEE.

Wan-Chi Siu received PhD degree from Imperial College of Science, Technology & Medicine, London, in 1984. He was with the Chinese University of Hong Kong between 1975 and 1980. He joined the Hong Kong Polytechnic University as a lecturer in 1980 and has been Chair Professor since 1992. He took up administrative duties as Associate Dean of the Engineering Faculty 1992-94, Head of the Department of Electronic and Information Engineering 1994-2000, and Dean of the Engineering Faculty 2000-2002. Professor Siu has been Director of Centre for Multimedia Signal Processing since September 1998. He has published over 200 research papers, and his research interests include digital signal processing, fast computational algorithms, transforms, image and video coding, and computational aspects of pattern recognition and neural networks. Professor Siu is a member of the editorial board of the Journal of VLSI Signal Processing Systems for Signal, Image and Video Technology and the EURASIP Journal on Applied Signal Processing. He was a guest editor of a Special Issue of the IEEE Transactions on Circuits and Systems, Pt.II published in May 1998, and associate editor of the IEEE Transactions on Circuits and Systems, Pt.II between 1995-97. Professor Siu has held the position of general chair or technical program chair of many international conferences. In particular, he was Technical Program chair of the IEEE International Symposium on Circuits and Systems (ISCAS'97), and is now the general chair of the 2003 IEEE International Conference on Acoustics, Speech and Signal Processing (ICASSP'2003) which will be held in Hong Kong. Between 1991 and 1995, Prof. Siu was a member of the Physical Sciences and Engineering Panel of the Research Grants Council (RGC), Hong Kong Government, and in 1994 he chaired the first Engineering and Information Technology Panel to assess the research quality of 19 Cost Centers (departments) from all universities in Hong Kong.

Hong-Jiang Zhang received his PhD from the Technical University of Denmark, Lyngby, Denmark, in 1991. He has held the position of research manager at the Institute of Systems Science, National University of Singapore, and HP Labs, USA. Dr. Zhang joined Microsoft Research Asia (MRA) in 1999, where he is currently senior researcher and assistant managing director of MRA. Dr. Zhang is well known in the multimedia research community for his pioneering work in video and image content analysis, representation, retrieval and browsing. He has authored 3 books, 200 referred papers and book chapters, and numerous special issues of professional journals in multimedia processing and content-based media retrieval. He holds over 30 US patents and pending applications. "Image and Video Processing in Multimedia Systems" which was co-authored by Zhang, was the first book to address content-based image and video retrieval research. Since its publication by Kluwer in 1995, the book has become a classical reference text and the research contained therein has formed the technology basis for several start-up companies. Zhang currently serves on the editorial boards of five professional journals and a dozen committees of international conferences, including being the Program Committee Co-chair of the ACM 1999 Multimedia Conference.

Contributors

Dr. Ramazan Savaş Aygün
Department of Computer Science
and Engineering
The State University of New York
at Buffalo
Buffalo, USA

Dr. Michael Brown
Department of Computer Science
The Hong Kong University of
Science and Technology
Hong Kong

Dr. Weidong Cai
School of Information
Technologies
The University of Sydney
NSW, Australia

Prof. Shih-Fu Chang
Department of Electrical
Engineering
The Columbia University
New York, NY, USA

Dr. David Cheung
Department of Computer Science
and Information Systems
The University of Hong Kong
Hong Kong

Dr. Zheru Chi
Centre for Multimedia Signal
Processing
Department of Electronic and
Information Engineering
The Hong Kong Polytechnic
University
Hong Kong

Dr. Schahram Dustdar
Institute for Information Systems
Vienna University of Technology
Austria

Prof. David Dagan Feng
Centre for Multimedia Signal
Processing
Department of Electronic and
Information Engineering
The Hong Kong Polytechnic
University
Hong Kong
and
School of Information
Technologies
The University of Sydney
NSW, Australia

Prof. Edward A. Fox
Department of Computer Science
Virginia Polytechnic Institute and
State University
Blacksburg, VA, USA

Dr. Roger Fulton
Departments. of PET and Nuclear
Medicine
The Royal Prince Alfred Hospital
NSW, Australia

Prof. H. K. Bernie Huang
Children's Hospital Los
  Angeles/University of Southern
  California
Los Angeles, CA, USA,
and
Department of Optometry and
  Radiography
The Hong Kong Polytechnic
  University
Hong Kong

Dr. Jesse Jin
School of Information
  Technologies
The University of Sydney
NSW, Australia
and
School of Computer Science and
  Engineering
The University of New South
  Wales
NSW, Australia

Dr. Benjamin Kao
Department of Computer Science
  and Information Systems
The University of Hong Kong
Hong Kong

Dr. K. M. Lam
Centre for Multimedia Signal
  Processing
Department of Electronic and
  Information Engineering
The Hong Kong Polytechnic
  University
Hong Kong

Dr. C. K. Li
Department of Electronic and
  Information Engineering
The Hong Kong Polytechnic
  University
Hong Kong

Dr. Qing Li
Department of Computer
  Engineering and Information
  Technology
The City University of Hong
  Kong
Hong Kong

Dr. Fuhui Long
Medical Center
Duke University
Durham, USA

Dr. Paul Mather
Department of Computer Science
Virginia Polytechnic Institute and
  State University
Blacksburg, VA, USA

Prof. W. C. Siu
Centre for Multimedia Signal
  Processing
Department of Electronic and
  Information Engineering
The Hong Kong Polytechnic
  University
Hong Kong

Prof. John Smith
IBM T.J. Watson Research Center
Hawthorne, NY, USA

Prof. Yuqing Song
Department of Computer and
   Information Science
The University of Michigan at
   Dearborn
Dearborn, Michigan, USA

Prof. Hari Sundaram
Department of Computer Science
   and the Institute for Studies in
   the Arts
Arizona State University
Tempe, AZ, USA

Dr. Mitchell D. Swanson
Department of Electrical
   Engineering
The University of Minnesota
Minneapolis, USA

Dr. Qi Tian
Laboratories for Information
   Technology
Singapore

Dr. Jimmy To
Department of Electronic and
   Information Engineering
The Hong Kong Polytechnic
   University
Hong Kong

Dr. Changsheng Xu
Laboratories for Information
   Technology
Singapore

Dr. Jane You
Department of Computing
The Hong Kong Polytechnic
   University
Hong Kong

Prof. Aidong Zhang
Department of Computer Science
   and Engineering
The State University of New York
   at Buffalo
Buffalo, USA

Prof. David Zhang
Department of Computing
The Hong Kong Polytechnic
   University
Hong Kong

Dr. Hongjiang Zhang
Microsoft Research Asia
Beijing, China

Dr. Bin Zhu
Microsoft Research Asia
Beijing, China

Prof. Yueting Zhuang
Department of Computer Science
Zhejiang University
Hangzhou, China

Table of Contents

Part I: Technological Fundamentals

Part II: Applications

Part I
Technological
Fundamentals

1 FUNDAMENTALS OF CONTENT-BASED IMAGE RETRIEVAL

Dr. Fuhui Long, Dr. Hongjiang Zhang and Prof. David Dagan Feng

We introduce in this chapter some fundamental theories for content-based image retrieval. Section 1.1 looks at the development of content-based image retrieval techniques. Then, as the emphasis of this chapter, we introduce in detail in Section 1.2 some widely used methods for visual content descriptions. After that, we briefly address similarity/distance measures between visual features, the indexing schemes, query formation, relevance feedback, and system performance evaluation in Sections 1.3, 1.4 and 1.5. Details of these techniques are discussed in subsequent chapters. Finally, we draw a conclusion in Section 1.6.

1.1 Introduction

Content-based image retrieval, a technique which uses visual contents to search images from large scale image databases according to users' interests, has been an active and fast advancing research area since the 1990s. During the past decade, remarkable progress has been made in both theoretical research and system development. However, there remain many challenging research problems that continue to attract researchers from multiple disciplines.

Before introducing the fundamental theory of content-based retrieval, we will take a brief look at its development. Early work on image retrieval can be traced back to the late 1970s. In 1979, a conference on Database Techniques for Pictorial Applications [6] was held in Florence. Since then, the application potential of image database management techniques has attracted the attention of researchers [12, 13, 16, 18]. Early techniques were not generally based on visual features but on the textual annotation of images. In other words, images were first annotated with text and then searched using a text-based approach from traditional database management systems. Comprehensive surveys of early *text-based image retrieval* methods can be found in [14, 93]. Text-based image retrieval uses traditional database techniques to manage images. Through text descriptions, images can be organized by topical or semantic hierarchies to facilitate easy navigation and browsing based on standard Boolean queries. However, since automatically generating descriptive texts for a wide spectrum of images is not feasible, most text-based image retrieval systems require manual annotation of images. Obviously, annotating images manually is a cumbersome and expensive task for large image databases, and is often subjective, context-sensitive and incomplete. As a result, it is difficult for the traditional text-based methods to support a variety of task-dependent queries.

In the early 1990s, as a result of advances in the Internet and new digital image sensor technologies, the volume of digital images produced by scientific, educational, medical, industrial, and other applications available to users increased dramatically. The difficulties faced by text-based retrieval became more and more severe. The efficient management of the rapidly expanding visual information became an urgent problem. This need formed the driving force behind the emergence of content-based image retrieval techniques. In 1992, the National Science Foundation of the United States organized a workshop on visual information management systems [49] to identify new directions in image database management systems. It was widely recognized that a more efficient and intuitive way to represent and index visual information would be based on properties that are inherent in the images themselves. Researchers from the communities of computer vision, database management, human-computer interface, and information retrieval were attracted to this field. Since then, research on content-based image retrieval has developed rapidly [11, 23, 24, 35, 49, 50, 103]. Since 1997, the number of research publications on the techniques of visual information extraction, organization, indexing, user query and interaction, and database management has increased enormoulsy. Similarly, a large number of academic and commercial retrieval systems have been developed by universities, government organizations, companies, and hospitals. Comprehensive surveys of these techniques and systems can be found in [31, 77, 87].

Content-based image retrieval uses the visual contents of an image such as *color*, *shape*, *texture*, and *spatial layout* to represent and index the image. In typical content-based image retrieval systems (Figure 1-1), the visual contents of the images in the database are extracted and described by multi-dimensional feature vectors. The feature vectors of the images in the database form a feature database. To retrieve images, users provide the retrieval system with example images or sketched figures. The system then represents these examples or sketch figures by internal feature vectors. The similarities /distances between the feature vectors of the query example or sketch and those of the images in the database are then calculated and retrieval is performed with the aid of an indexing scheme. The indexing scheme provides an efficient way to search for the image database. Recent retrieval systems have incorporated users' relevance feedback to modify the retrieval process in order to generate perceptually and semantically more meaningful retrieval results. In this chapter, we introduce these fundamental techniques for content-based image retrieval.

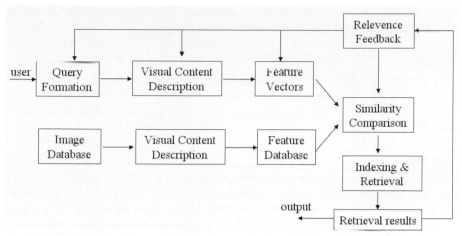

Figure 1-1 Diagram for content-based image retrieval system.

1.2 Image Content Descriptors

Generally speaking, image content may include both visual and semantic content. Visual content can be very general or domain specific. *General visual content* include color, texture, shape, spatial relationship, etc. *Domain specific visual content*, like human faces, is application dependent and may involve domain knowledge. *Semantic content* is obtained either by textual annotation or by complex inference procedures based on visual content. This chapter concentrates on general visual contents descriptions. Later chapters discuss domain specific and semantic contents.

A good visual content descriptor should be invariant to the accidental variance introduced by the imaging process (e.g., the variation of the illuminant of the scene). However, there is a tradeoff between the invariance and the discriminative power of visual features, since a very wide class of invariance loses the ability to discriminate between essential differences. Invariant description has been largely investigated in computer vision (like object recognition), but is relatively new in image retrieval [8].

A visual content descriptor can be either global or local. A global descriptor uses the visual features of the whole image, whereas a local descriptor uses the visual features of *regions* or *objects* to describe the image content. To obtain the local visual descriptors, an image is often divided into parts first. The simplest way of dividing an image is to use a *partition*, which cuts the image into tiles of equal size and shape. A simple partition does not generate perceptually meaningful regions but is a way of representing the global features of the image at a finer resolution. A better method is to divide the image into homogenous regions according to some criterion using *region segmentation* algorithms that have been extensively investigated in computer vision. A more complex way of dividing an image is to undertake a complete *object segmentation* to obtain semantically meaningful objects (like ball, car, horse). Currently, automatic object segmentation for broad domains of general images is unlikely to succeed.

In this section, we will introduce some widely used techniques for extracting color, texture, shape and spatial relationship from images.

COLOR

Color is the most extensively used visual content for image retrieval [43, 44, 45, 65, 71, 89, 91, 104]. Its three-dimensional values make its discrimination potentiality superior to the single dimensional gray values of images. Before selecting an appropriate color description, color space must be determined first.

Color Space

Each pixel of the image can be represented as a point in a 3D color space. Commonly used color space for image retrieval include *RGB*, *Munsell*, *CIE L*a*b**, *CIE L*u*v**, *HSV* (or *HSL*, *HSB*), and *opponent color* space. There is no agreement on which is the best. However, one of the desirable characteristics of an appropriate color space for image retrieval is its perceptual *uniformity* [65]. Perceptual uniformity means that two color pairs that are equal in similarity distance in a color space are perceived as equal by viewers. In other words, the measured proximity among colors must be directly related to the psychological similarity among them.

RGB space is a widely used color space for image display. It is composed of three color components *red*, *green*, and *blue*. These components are called "*additive primaries*" since a color in RGB space is produced by adding them together. In contrast, CMY space is a color space primarily used for printing. The three color components are *cyan*, *magenta*, and *yellow*. These three components are called "*subtractive primaries*" since a color in CMY space is produced through light absorption. Both RGB and CMY space are device-dependent and perceptually non-uniform.

The CIE L*a*b* and CIE L*u*v* spaces are device independent and considered to be perceptually uniform. They consist of a luminance or *lightness* component (*L*) and two *chromatic* components *a* and *b* or *u* and *v*. CIE L*a*b* is designed to deal with subtractive colorant mixtures, while CIE L*u*v* is designed to deal with additive colorant mixtures. The transformation of RGB space to CIE L*u*v* or CIE L*a*b* space can be found in [47].

HSV (or HSL, or HSB) space is widely used in computer graphics and is a more intuitive way of describing color. The three color components are *hue*, *saturation* and *value* (or lightness, *brightness*). The hue is invariant to the changes in illumination and camera direction and hence more suited to object retrieval. RGB coordinates can be easily translated to the HSV (or HLS, or HSB) coordinates by a simple formula [27].

The opponent color space uses the opponent color axes (*R-G, 2B-R-G, R+G+B*). This representation has the advantage of isolating the brightness information on the third axis. With this solution, the first two chromaticity axes, which are invariant to the changes in illumination intensity and shadows, can be down-sampled since humans are more sensitive to brightness than they are to chromatic information.

In the following sections, we will introduce some commonly used color descriptors: the color moments, color histogram, color coherence vector, and color correlogram.

Color Moments

Color moments have been successfully used in many retrieval systems (like *QBIC* [26, 67]), especially when the image contains only   objects. The *first order* (*mean*), the *second* (*variance*) and the *third order* (*skewness*) color moments have been proved to be efficient and effective in representing color distributions of images [89]. Mathematically, the first three moments are defined as:

$$\mu_i = \frac{1}{N}\sum_{j=1}^{N} f_{ij} \qquad (1\text{-}1)$$

$$\sigma_i = \left(\frac{1}{N}\sum_{j=1}^{N}(f_{ij}-\mu_i)^2\right)^{\frac{1}{2}} \qquad (1\text{-}2)$$

$$s_i = \left(\frac{1}{N}\sum_{j=1}^{N}(f_{ij}-\mu_i)^3\right)^{\frac{1}{3}} \qquad (1\text{-}3)$$

where f_{ij} is the value of the i-th color component of the image pixel j, and N is the number of pixels in the image.

Usually, the color moments defined on L*u*v* and L*a*b* color spaces performs better than those on HSV space.   Using the additional third-order moment improves the overall retrieval performance compared to using only the first and second order moments. However, this third-order moment sometimes makes the feature representation more sensitive to scene changes and thus may decrease the performance.

Since only 9 (three moments for each of the three color components) numbers are used to represent the color content of each image, color moments are very compact representations compared to other color features. Due to this compactness, they may also lower the discrimination power. Usually, color moments can be used as the first pass to narrow down the search space before other sophisticated color features are used for retrieval.

Color Histogram

The color histogram serves as an effective representation of the color content of an image if the color pattern is unique compared with the rest of the data set. The color histogram is easy to compute and effective in characterizing both the global and local distributions of colors in an image. In addition, it is robust to translation and rotation about the viewing axis and changes only slowly with the scale, occlusion and viewing angle.

Since any pixel in the image can be described by three components in a certain color space (for instance, red, green, and blue components in RGB space, or hue, saturation, and value in HSV space), a *histogram*, i.e., the distribution of the number of pixels for each quantized bin, can be defined for each component. Clearly, the more bins a color histogram contains, the more discrimination power it has. However, a histogram with a large number of bins will not only increase the computational cost, but will also be inappropriate for building efficient indexes for image databases.

Furthermore, a very fine bin quantization does not necessarily improve the retrieval performance in many applications. One way to reduce the number of bins is to use the opponent color space which enables the brightness of the histogram to be down sampled. Another way is to use clustering methods to determine the K best colors in a given space for a given set of images. Each of these best colors will be taken as a histogram bin. Since that clustering process takes the color distribution of images over the entire database into consideration, the likelihood of histogram bins in which no or very few pixels fall will be minimized. Another option is to use the bins that have the largest pixel numbers since a small number of histogram bins capture the majority of pixels of an image [35] Such a reduction does not degrade the performance of histogram matching, but may even enhance it since small histogram bins are likely to be noisy.

When an image database contains a large number of images, histogram comparison will saturate the discrimination. To solve this problem, the *joint histogram* technique is introduced [71]. In addition, color histogram does not take the spatial information of pixels into consideration, thus very different images can have similar color distributions. This problem becomes especially acute for large scale databases. To increase discrimination power, several improvements have been proposed to incorporate spatial information. A simple approach is to divide an image into sub-areas and calculate a histogram for each of those sub-areas. As introduced above, the division can be as simple as a rectangular partition, or as complex as a region or even object segmentation. Increasing the number of sub-areas increases the information about location, but also increases the memory and computational time.

Color Coherence Vector

In [72] a different way of incorporating spatial information into the color histogram, *color coherence vectors* (*CCV*), was proposed. Each histogram bin is partitioned into two types, i.e., coherent, if it belongs to a large uniformly-colored region, or incoherent, if it does not. Let α_i denote the number of coherent pixels of the ith color bin in an image and β_i denote the number of incoherent pixels. Then, the CCV of the image is defined as the vector $<(\alpha_1, \beta_1), (\alpha_2, \beta_2), \ldots, (\alpha_N, \beta_N)>$. Note that $<\alpha_1+\beta_1, \alpha_2+\beta_2, \ldots, \alpha_N+\beta_N>$ is the color histogram of the image.

Due to its additional spatial information, it has been shown that CCV provides better retrieval results than the color histogram, especially for those images which have either mostly uniform color or mostly texture regions. In addition, for both the color histogram and color coherence vector representation, the HSV color space provides better results than CIE L*u*v* and CIE L*a*b* space.

Color Correlogram

The *color correlogram* [44] was proposed to characterize not only the color distributions of pixels, but also the spatial correlation of pairs of colors. The first and the second dimension of the three-dimensional histogram are the colors of any pixel pair and the third dimension is their spatial distance. A color correlogram is a table indexed by color pairs, where the k-th entry for (i, j) specifies the probability of finding a pixel of color j at a distance k from a pixel of color i in the image. Let I

represent the entire set of image pixels and $I_{c(i)}$ represent the set of pixels whose colors are *c(i)*. Then, the color correlogram is defined as:

$$\gamma_{i,j}^{(k)} = \Pr_{p_1 \in I_{c(i)}, p_2 \in I} \left[p_2 \in I_{c(j)} \mid \mid p_1 - p_2 \mid = k \right] \tag{1-4}$$

where $i, j \in \{1, 2, ..., N\}$, $k \in \{1, 2, ..., d\}$, and $\mid p_1 - p_2 \mid$ is the distance between pixels p_1 and p_2. If we consider all the possible combinations of color pairs the size of the color correlogram will be very large ($O(N^2 d)$), therefore a simplified version called the *color autocorrelogram* is often used instead. The color autocorrelogram only captures the spatial correlation between identical colors and thus reduces the dimension to $O(Nd)$.

Compared to the color histogram and CCV, the color autocorrelogram provides the best retrieval results, but is also the most computational expensive due to its high dimensionality.

Invariant Color Features

Color not only reflectsthe material of surface, but also varies considerably with the change of illumination, the orientation of the surface, and the viewing geometry of the camera. This variability must be taken into account. However, invariance to these environmental factors is not considered in most of the color features introduced above.

Invariant color representation has been introduced to content-based image retrieval recently. In [33], a set of color invariants for object retrieval was derived based on the Schafer model of object reflection. In [25], specular reflection, shape and illumination invariant representation based on blue ratio vector (r/b, g/b,1) is given. In [34], a surface geometry invariant color feature is provided.

These invariant color features, when applied to image retrieval, may yield illumination, scene geometry and viewing geometry independent representation of color contents of images, but may also lead to some loss in discrimination power among images.

TEXTURE

Texture is another important property of images. Various texture representations have been investigated in pattern recognition and computer vision. Basically, texture representation methods can be classified into two categories: *structural* and *statistical*. Structural methods, including *morphological operator* and *adjacency graph*, describe texture by identifying structural primitives and their placement rules. They tend to be most effective when applied to textures that are very regular. Statistical methods, including *Fourier power spectra, co-occurrence matrices, shift-invariant principal component analysis (SPCA), Tamura feature, Wold decomposition, Markov random field, fractal model*, and *multi-resolution filtering* techniques such as *Gabor and wavelet transform*, characterize texture by the statistical distribution of the image intensity. In this section, we introduce a number of texture representations [7, 19, 21, 28, 29, 30, 48, 51, 53, 54, 57, 58, 62, 63, 64, 70, 75, 92, 99], which have been used frequently and have proved to be effective in content-based image retrieval systems.

Tamura Features

The Tamura features [92], including *coarseness, contrast, directionality, linelikeness, regularity,* and *roughness,* are designed in accordance with psychological studies on the human perception of texture. The first three components of Tamura features have been used in some early well-known image retrieval systems, such as *QBIC* [26, 67] and *Photobook* [73]. The computations of these three features are given as follows.

Coarseness

Coarseness is a measure of the granularity of the texture. To calculate the coarseness, moving averages $A_k(x,y)$ are computed first using $2^k \times 2^k$ ($k = 0, 1, \ldots, 5$) size windows at each pixel (x, y), i.e.,

$$A_k(x, y) = \sum_{i=x-2^{k-1}}^{x+2^{k-1}-1} \sum_{j=y-2^{k-1}}^{y+2^{k-1}-1} g(i, j)/2^{2k} \tag{1-5}$$

where $g(i, j)$ is the pixel intensity at (i, j).

Then, the differences between pairs of non-overlapping moving averages in the horizontal and vertical directions for each pixel are computed, i.e.,

$$E_{k,h}(x, y) = \left| A_k(x + 2^{k-1}, y) - A_k(x - 2^{k-1}, y) \right|$$

$$E_{k,v}(x, y) = \left| A_k(x, y + 2^{k-1}) - A_k(x, y - 2^{k-1}) \right| \tag{1-6}$$

After that, the value of k that maximizes E in either direction is used to set the best size for each pixel, i.e.,

$$S_{best}(x, y) = 2^k \tag{1-7}$$

The coarseness is then computed by averaging S_{best} over the entire image, i.e.,

$$F_{crs} = \frac{1}{m \times n} \sum_{i=1}^{m} \sum_{j=1}^{n} S_{best}(i, j) \tag{1-8}$$

Instead of taking the average of S_{best}, an improved version of the coarseness feature can be obtained by using a histogram to characterize the distribution of S_{best}. Compared with using a single value to represent coarseness, using histogram-based coarseness representation can greatly increase the retrieval performance. This modification makes the feature capable of dealing with an image or region which has multiple texture properties, and thus is more useful to image retrieval applications.

Contrast

The formula for the contrast is as follows:

$$F_{con} = \frac{\sigma}{\alpha_4^{1/4}} \tag{1-9}$$

where the kurtosis $\alpha_4 = \mu_4/\sigma^4$, μ_4 is the fourth moment about the mean, and σ^2 is the variance. This formula can be used for both the entire image and a region of the image.

Directionality

To compute the directionality, image is convoluted with two 3x3 arrays (i.e.,

$$\begin{matrix} -1 & 0 & 1 \\ -1 & 0 & 1 \\ -1 & 0 & 1 \end{matrix} \quad \text{and} \quad \begin{matrix} 1 & 1 & 1 \\ 0 & 0 & 0 \\ -1 & -1 & -1 \end{matrix}$$) and a gradient vector at each pixel is computed.

The magnitude and angle of this vector are defined as:

$$|\Delta G| = (|\Delta_H| + |\Delta_V|)/2$$
$$\theta = \tan^{-1}(\Delta_V/\Delta_H) + \pi/2$$

(1-10)

where Δ_H and Δ_V are the horizontal and vertical differences of the convolution.

Then, by quantizing θ and counting the pixels with the corresponding magnitude $|\Delta G|$ larger than a threshold, a histogram of θ, denoted as H_D, can be constructed. This histogram will exhibit strong peaks for highly directional images and will be relatively flat for images without strong orientation. The entire histogram is then summarized to obtain an overall directionality measure based on the sharpness of the peaks:

$$F_{dir} = \sum_{p}^{n_p} \sum_{\phi \in w_p} (\phi - \phi_p)^2 H_D(\phi)$$

(1-11)

In this sum p ranges over n_p peaks; and for each peak p, w_p is the set of bins distributed over it; while ϕ_p is the bin that takes the peak value.

Wold Features

Wold decomposition [28, 57] provides another approach to describing textures in terms of perceptual properties. The three Wold components, *harmonic, evanescent,* and *indeterministic,* correspond to *periodicity, directionality,* and *randomness* of texture respectively. Periodic textures have a strong harmonic component, highly directional textures have a strong evanescent component, and less structured textures tend to have a stronger indeterministic component.

For a homogeneous regular random field $\{y(m,n), (m,n) \in Z^2\}$, 2D Wold decomposition allows the field to be decomposed into three mutually orthogonal components:

$$y(m,n) = u(m,n) + d(m,n) = u(m,n) + h(m,n) + e(m,n)$$

(1-12)

where $u(m,n)$ is the indeterministic component; and $d(m,n)$ is the deterministic component which can be further decomposed into the harmonic component $h(m,n)$ and evanescent component $e(m,n)$. In the frequency domain, a similar expression exists:

$$F_y(\xi,\eta) = F_u(\xi,\eta) + F_d(\xi,\eta) = F_u(\xi,\eta) + F_h(\xi,\eta) + F_e(\xi,\eta)$$

(1-13)

where $F_y(\xi,\eta), F_u(\xi,\eta), F_d(\xi,\eta), F_h(\xi,\eta), F_e(\xi,\eta)$ are the spectral distribution functions (SDF) of $\{y(m,n)\}$, $\{u(m,n)\}$, $\{d(m,n)\}$, $\{h(m,n)\}$ and $\{e(m,n)\}$ respectively.

In the spatial domain, the three orthogonal components can be obtained by the maximum likelihood estimation (MLE), which involves fitting a high-order AR process, minimizing a cost function, and solving a set of linear equations. In the frequency domain, Wold components can be obtained by global thresholding of Fourier spectral magnitudes of the image. In [57], a method using harmonic peak extraction and MRSAR modeling without an actual decomposition of the image is presented. This method is designed to tolerate a variety of inhomogeneities in natural texture patterns.

Simultaneous Auto-Regressive (SAR) Model

The *SAR model* is an instance of *Markov random field* (*MRF*) models, which have been very successful in texture modeling in the past decades. Compared with other MRF models, SAR uses fewer parameters. In the SAR model, pixel intensities are taken as random variables. The intensity $g(x,y)$ at pixel (x,y) can be estimated as a linear combination of the neighboring pixel values $g(x',y')$ and an additive noise term $\varepsilon(x,y)$, i.e.,

$$g(x, y) = \mu + \sum_{(x',y')\in D} \theta(x', y')g(x', y') + \varepsilon(x, y) \qquad (1\text{-}14)$$

where μ is a bias value determined by the mean of the entire image; D is the neighbor set of (x, y); $\theta(x',y')$ is a set of weights associated with each of the neighboring pixels; $\varepsilon(x,y)$ is an independent Gaussian random variable with zero mean and variance σ^2. The parameters θ and σ are used to measure texture. For instance, a higher σ value implies a finer granularity or less coarseness; a higher $\theta(x, y+1)$ and $\theta(x, y-1)$ values indicate that the texture is vertically oriented. The least square error (LSE) technique or the maximum likelihood estimation (MLE) method is usually used to estimate the parameters of the SAR model.

The SAR model is not rotation invariant. To derive a *rotation-invariant SAR model* (*RISAR*), pixels lying on circles of different radii centered at each pixel (x,y) serve as its neighbor set D. Thus the intensity $g(x,y)$ at pixel (x,y) can be estimated as

$$g(x, y) = \mu + \sum_{i=1}^{p} \theta_i(x, y)l_i(x, y) + \varepsilon(x, y) \qquad (1\text{-}15)$$

where p is the number of circular neighborhood. To make the computational cost inexpensive and to achieve rotation invariance at the same time, p can neither be too large nor too small. Usually $p = 2$. $l(x,y)$ can be computed by:

$$l_i(x, y) = \frac{1}{8i} \sum_{(x',y')\in N_i} w_i(x', y')g(x', y') \qquad (1\text{-}16)$$

where N_i is the ith circular neighborhood of (x,y); $w_i(x',y')$ is a set of pre-computed weights indicating the contribution of the pixel (x',y') in the ith circle.

To describe textures of different granularities, the *multi-resolution simultaneous auto-regressive model* (*MRSAR*) [64] has been proposed to enable multi-scale texture analysis. An image is represented by a multi-resolution Gaussian pyramid with low-pass filtering and sub-sampling applied at several successive levels. Either the SAR or RISAR model may then be applied to each level of the pyramid.

MRSAR has been proved [63, 75]to have better performance on the *Brodatz texture database* [7] than many other texture features, such as principal component analysis, Wold decomposition, and wavelet transform.

Gabor Filter Features

The *Gabor filter* has been widely used to extract image features, especially texture features [22, 48]. It is optimal in terms of minimizing the joint uncertainty in space and frequency, and is often used as an orientation and scale tunable edge and line (bar) detector. There have been many approaches proposed to characterize textures of images based on Gabor filters. The basic idea of using Gabor filters to extract texture features is as follows.

A two dimensional Gabor function $g(x, y)$ is defined as:

$$g(x, y) = \frac{1}{2\pi\sigma_x\sigma_y} \exp\left[-\frac{1}{2}\left(\frac{x^2}{\sigma_x^2} + \frac{y^2}{\sigma_y^2}\right) + 2\pi j W x\right] \qquad (1\text{-}17)$$

where, σ_x and σ_y are the standard deviations of the Guassian envelopes along the x and y direction.

Then a set of Gabor filters can be obtained by appropriate dilations and rotations of $g(x, y)$:

$$g_{mn}(x, y) = a^{-m} g(x', y')$$
$$x' = a^{-m}(x\cos\theta + y\sin\theta) \qquad (1\text{-}18)$$
$$y' = a^{-m}(x\sin\theta + y\cos\theta)$$

where $a > 1$, $\theta = n\pi/K$, $n = 0, 1, ..., K\text{-}1$, and $m = 0, 1, ..., S\text{-}1$. K and S are the number of orientations and scales. The scale factor a^{-m} is to ensure that energy is independent of m.

Given an image $I(x, y)$, its Gabor transform is defined as:

$$W_{mn}(x, y) = \int I(x, y) g_{mn}^{*}(x - x_1, y - y_1) dx_1 dy_1 \qquad (1\text{-}19)$$

where * indicates the complex conjugate. Then the mean μ_{mn} and the standard deviation σ_{mn} of the magnitude of $W_{mn}(x, y)$, i.e., $f = [\mu_{00}, \sigma_{00}, ..., \mu_{mn}, \sigma_{mn}, ..., \mu_{S-1k-1}, \sigma_{S-1K-1}]$ can be used to represent the texture feature of a homogenous texture region.

Wavelet Transform Features

Similar to the Gabor filtering, the *wavelet transform* [21, 62] provides a multi-resolution approach to texture analysis and classification [19, 54]. Wavelet transforms decompose a signal with a family of basis functions $\psi_{mn}(x)$ obtained through translation and dilation of a mother wavelet $\psi(x)$, i.e.,

$$\psi_{mn}(x) = 2^{-m/2} \psi(2^{-m} x - n) \qquad (1\text{-}20)$$

where *m* and *n* are dilation and translation parameters. A signal *f*(*x*) can be represented as:

$$f(x) = \sum_{m,n} c_{mn} \psi_{mn}(x) \tag{1-21}$$

The computation of the wavelet transforms of a 2D signal involves recursive filtering and sub-sampling. At each level, the signal is decomposed into four frequency sub-bands, LL, LH, HL, and HH, where L denotes low frequency and H denotes high frequency. Two major types of wavelet transforms used for texture analysis are the *pyramid-structured wavelet transform* (*PWT*) and the *tree-structured wavelet transform* (*TWT*). The PWT recursively decomposes the LL band. However, for some textures the most important information often appears in the middle frequency channels. To overcome this drawback, the TWT decomposes other bands such as LH, HL or HH when needed.

After the decomposition, feature vectors can be constructed using the mean and standard deviation of the energy distribution of each sub-band at each level. For three-level decomposition, PWT results in a feature vector of 3x4x2 components. For TWT, the feature will depend on how sub-bands at each level are decomposed. A fixed decomposition tree can be obtained by sequentially decomposing the LL, LH, and HL bands, and thus results in a feature vector of 52x2 components. Note that in this example, the feature obtained by PWT can be considered as a subset of the feature obtained by TWT. Furthermore, according to the comparison of different wavelet transform features [58], the particular choice of wavelet filter is not critical for texture analysis.

SHAPE

Shape features of objects or regions have been used in many content-based image retrieval systems [32, 36, 46, 94]. Compared with color and texture features, shape features are usually described after images have been segmented into regions or objects. Since robust and accurate image segmentation is difficult to achieve, the use of shape features for image retrieval has been limited to special applications where objects or regions are readily available. The state-of-art methods for shape description can be categorized into either *boundary-based* (e.g. rectilinear shapes [46], polygonal approximation [2], finite element models [84], and Fourier-based shape descriptors [1, 52, 74]) or *region-based* methods (e.g. statistical moments [41, 102]). A good shape representation feature for an object should be invariant to translation, rotation and scaling. In this section, we briefly describe some of these shape features that have been commonly used in image retrieval applications. For a comprehensive overview of the shape matching techniques, see [97].

Moment Invariants

Classical shape representation uses a set of *moment invariants*. If the object *R* is represented as a binary image, then the central moments of order *p+q* for the shape of object *R* are defined as:

$$\mu_{p,q} = \sum_{(x,y)\in R} (x - x_c)^p (y - y_c)^q \tag{1-22}$$

where (x_c, y_c) is the center of object. This central moment can be normalized to be scale invariant [47]:

$$\eta_{p,q} = \frac{\mu_{p,q}}{\mu_{0,0}^{\gamma}}, \quad \gamma = \frac{p+q+2}{2} \tag{1-23}$$

Based on these moments, a set of moment invariants to translation, rotation, and scale can be derived [41, 102]:

$$\phi_1 = \mu_{2,0} + \mu_{0,2}$$

$$\phi_2 = (\mu_{2,0} - \mu_{0,2})^2 + 4\mu_{1,1}^2$$

$$\phi_3 = (\mu_{3,0} - 3\mu_{1,2})^2 + (\mu_{0,3} - 3\mu_{2,1})^2$$

$$\phi_4 = (\mu_{3,0} + \mu_{1,2})^2 + (\mu_{0,3} + \mu_{2,1})^2$$

$$\phi_5 = (\mu_{3,0} - 3\mu_{1,2})(\mu_{3,0} + \mu_{1,2})\left[(\mu_{3,0} + \mu_{1,2})^2 - 3(\mu_{0,3} + \mu_{2,1})^2\right]$$
$$+ (\mu_{0,3} - 3\mu_{2,1})(\mu_{0,3} + \mu_{2,1})\left[(\mu_{0,3} + \mu_{2,1})^2 - 3(\mu_{3,0} + \mu_{1,2})^2\right]$$

$$\phi_6 = (\mu_{2,0} - \mu_{0,2})[(\mu_{3,0} + \mu_{1,2})^2 - (\mu_{0,3} + \mu_{2,1})^2] + 4\mu_{1,1}(\mu_{3,0} + \mu_{1,2})(\mu_{0,3} + \mu_{2,1})$$

$$\phi_7 = (3\mu_{2,1} - \mu_{0,3})(\mu_{3,0} + \mu_{1,2})\left[(\mu_{3,0} + \mu_{1,2})^2 - 3(\mu_{0,3} + \mu_{2,1})^2\right]$$

$$\tag{1-24}$$

An expanded set of moment invariants can also be found in [102], which gives a wider choice for users to possibly achieve higher recognition rate and faster speed of realization.

Turning Angles
The contour of a 2D object can be represented as a closed sequence of successive boundary pixels (x_s, y_s), where $0 \le s \le N\text{-}1$ and N is the total number of pixels on the boundary. The *turning function* or *turning angle* $\theta(s)$, which measures the angle of the counterclockwise tangents as a function of the arc-length s according to a reference point on the object's contour, can be defined as:

$$\theta(s) = \tan^{-1}\left(\frac{y_s'}{x_s'}\right)$$

$$y_s' = \frac{dy_s}{ds} \tag{1-25}$$

$$x_s' = \frac{dx_s}{ds}$$

One major problem with this representation is that it is variant to the rotation of object and the choice of the reference point. If we shift the reference point along the boundary of the object by an amount t, then the new turning function becomes $\theta(s+t)$. If we rotate the object by angle ω, then the new function becomes $\theta(s)+\omega$.

Therefore, to compare the shape similarity between objects A and B with their turning functions, the minimum distance needs to be calculated over all possible shifts t and rotations ω, i.e.,

$$d_p(A,B) = \left(\min_{\omega \in R, t \in [0,1]} \int_0^1 |\theta_A(s+t) - \theta_B(s) + \omega|^P \, ds \right)^{\frac{1}{p}} \tag{1-26}$$

Here we assume that each object has been re-scaled so that the total perimeter length is 1. This measure is invariant under translation, rotation, and change of scale.

Fourier Descriptors

Fourier descriptors describe the shape of an object with the Fourier transform of its boundary. Again, consider the contour of a 2D object as a closed sequence of successive boundary pixels (x_s, y_s), where $0 \leq s \leq N\text{-}1$ and N is the total number of pixels on the boundary. Then three types of contour representations, i.e., *curvature*, *centroid distance*, and *complex coordinate function*, can be defined.

The curvature $K(s)$ at a point s along the contour is defined as the rate of change in tangent direction of the contour, i.e.,

$$K(s) = \frac{d}{ds}\theta(s) \tag{1-27}$$

where $\theta(s)$ is the turning function of the contour, defined as (1-25).

The centroid distance is defined as the distance function between boundary pixels and the centroid (x_c, y_c) of the object:

$$R(s) = \sqrt{(x_s - x_c)^2 + (y_s - y_c)^2} \tag{1-28}$$

The complex coordinate is obtained by simply representing the coordinates of the boundary pixels as complex numbers:

$$Z(s) = (x_s - x_c) + j(y_s - y_c) \tag{1-29}$$

The Fourier transforms of these three types of contour representations generate three sets of complex coefficients, representing the shape of an object in the frequency domain. Lower frequency coefficients describe the general shape property, while higher frequency coefficients reflect shape details. To achieve rotation invariance (i.e., contour encoding is irrelevant to the choice of the reference point), only the amplitudes of the complex coefficients are used and the phase components are discarded. To achieve scale invariance, the amplitudes of the coefficients are divided by the amplitude of DC component or the first non-zero coefficient. The translation invariance is obtained directly from the contour representation.

The Fourier descriptor of the curvature is:

$$f_K = \left[|F_1|, |F_2|, ..., |F_{M/2}| \right] \tag{1-30}$$

The Fourier descriptor of the centroid distance is:

$$f_R = \left[\frac{|F_1|}{|F_0|}, \frac{|F_2|}{|F_0|}, ..., \frac{|F_{M/2}|}{|F_0|} \right] \tag{1-31}$$

where F_i in (1-30) and (1-31) denotes the ith component of Fourier transform coefficients. Here only the positive frequency axes are considered because the curvature and centroid distance functions are real and, therefore, their Fourier transforms exhibit symmetry, i.e., $|F_{-i}| = |F_i|$.

The Fourier descriptor of the complex coordinate is:

$$f_Z = \left[\frac{|F_{-(M/2-1)}|}{|F_1|}, \dots, \frac{|F_{-1}|}{|F_1|}, \frac{|F_2|}{|F_1|}, \dots, \frac{|F_{M/2}|}{|F_1|} \right] \tag{1-32}$$

where F_1 is the first non-zero frequency component used for normalizing the transform coefficients. Here both negative and positive frequency components are considered. The DC coefficient is dependent on the position of a shape, and therefore, is discarded.

To ensure the resulting shape features of all objects in a database have the same length, the boundary $((x_s, y_s), 0 \le s \le N\text{-}1)$ of each object is re-sampled to M samples before performing the Fourier transform. For example, M can be set to $2^m = 64$ so that the transformation can be conducted efficiently using the fast Fourier transform.

Circularity, Eccentricity, and Major Axis Orientation

Circularity is computed as:

$$\alpha = \frac{4\pi S}{P^2} \tag{1-33}$$

where S is the size and P is the perimeter of an object. This value ranges between 0 (corresponding to a perfect line segment) and 1 (corresponding to a perfect circle).

The *major axis orientation* can be defined as the direction of the largest eigenvector of the second order covariance matrix of a region or an object. The eccentricity can be defined as the ratio of the smallest eigenvalue to the largest eigenvalue.

SPATIAL INFORMATION

Regions or objects with similar color and texture properties can be easily distinguished by imposing spatial constraints. For instance, regions of blue sky and ocean may have similar color histograms, but their spatial locations in images are different. Therefore, the spatial location of regions (or objects) or the spatial relationship between multiple regions (or objects) in an image is very useful for searching images.

The most widely used representation of spatial relationship is the *2D strings* proposed by Chang *et al* [17]. It is constructed by projecting images along the x and y directions. Two sets of symbols, V and A, are defined on the projection. Each symbol in V represents an object in the image. Each symbol in A represents a type of spatial relationship between objects. As its variant, the *2D G-string* [15] cuts all the objects along their minimum bounding box and extends the spatial relationships into two sets of spatial operators. One defines local spatial relationships. The other defines the global spatial relationships, indicating that the projection of two objects are disjoin, adjoin or located at the same position. In addition, *2D C-string* [55] is proposed to minimize the number of cutting objects. *2D-B string* [56] represents an object by two symbols, standing for the beginning and ending boundary of the object. All these methods can facilitate three types of query. Type 0 query finds all images containing object O_1, O_2, ..., O_n. Type 1 finds all images containing objects that have certain relationship between each other, but the distance between them is insignificant. Type 2 finds all images that have certain distance relationship with each other.

In addition to the 2D string, *spatial quad-tree* [82], and *symbolic image* [37] are also used for spatial information representation. However, searching images based on spatial relationships of regions remains a difficult research problem in content-based image retrieval, because reliable segmentation of objects or regions is often not feasible except in very limited applications. Although some systems simply divide the images into regular sub-blocks [90], only limited success has been achieved with such spatial division schemes since most natural images are not spatially constrained to regular sub-blocks. To solve this problem, a method based on the *radon transform*, which exploits the spatial distribution of visual features without a sophisticated segmentation is proposed in   [38, 100].

1.3 Similarity Measures and Indexing Schemes

SIMILARITY/DISTANCE MEASURES

Instead of exact matching, content-based image retrieval calculates visual similarities between a query image and images in a database. Accordingly, the retrieval result is not a single image but a list of images ranked by their similarities with the query image. Many similarity measures have been developed for image retrieval based on empirical estimates of the distribution of features in recent years. Different *similarity/distance measures* will affect retrieval performances of an image retrieval system significantly. In this section, we will introduce some commonly used similarity measures. We denote $D(I, J)$ as the distance measure between the query image I and the image J in the database; and $f_i(I)$ as the number of pixels in bin i of I.

Minkowski-Form Distance

If each dimension of image feature vector is independent of each other and is of equal importance, the *Minkowski-form distance* L_p is appropriate for calculating the distance between two images. This distance is defined as:

$$D(I,J) = (\sum_i | f_i(I) - f_i(J)|^p)^{1/p} \qquad (1\text{-}34)$$

when p=1, 2, and ∞, $D(I, J)$ is the L_1, L_2 (also called Euclidean distance), and L_∞ distance respectively. Minkowski-form distance is the most widely used metric for image retrieval. For instance, MARS system [78] used Euclidean distance to compute the similarity between texture features; Netra   [61, 60] used Euclidean distance for color and shape feature, and L_1 distance for texture feature; Blobworld [9] used Euclidean distance for texture and shape feature. In addition, Voorhees and Poggio [99] used L_∞ distance to compute the similarity between texture images.

The *Histogram intersection* can be taken as a special case of L_1 distance, which is used by Swain and Ballard [91] to compute the similarity between color images. The intersection of the two histograms of I and J is defined as:

$$S(I,J) = \frac{\sum_{i=1}^{N} \min(f_i(I), f_i(J))}{\sum_{i=1}^{N} f_i(J)} \qquad (1\text{-}35)$$

It has been shown that histogram intersection is fairly insensitive to changes in image resolution, histogram size, occlusion, depth, and viewing point.

Quadratic Form (QF) Distance

The Minkowski distance treats all bins of the feature histogram entirely independently and does not account for the fact that certain pairs of bins correspond to features which are perceptually more similar than other pairs. To solve this problem, *quadratic form distance* is introduced:

$$D(I,J) = \sqrt{(\mathbf{F}_I - \mathbf{F}_J)^T \mathbf{A} (\mathbf{F}_I - \mathbf{F}_J)} \qquad (1\text{-}36)$$

where $A=[a_{ij}]$ is a similarity matrix, and a_{ij} denotes the similarity between bin i and j. $\mathbf{F}_I$ and $\mathbf{F}_J$ are vectors that list all the entries in $f_i(I)$ and $f_i(J)$.

Quadratic form distance has been used in many retrieval systems [40, 67] for color histogram-based image retrieval. It has been shown that quadratic form distance can lead to perceptually more desirable results than Euclidean distance and histogram intersection method as it considers the cross similarity between colors.

Mahalanobis Distance

The *Mahalanobis distance* metric is appropriate when each dimension of image feature vector is dependent of each other and is of different importance. It is defined as:

$$D(I,J) = \sqrt{(\mathbf{F}_I - \mathbf{F}_J)^T C^{-1} (\mathbf{F}_I - \mathbf{F}_J)} \qquad (1\text{-}37)$$

where C is the covariance matrix of the feature vectors.

The Mahalanobis distance can be simplified if feature dimensions are independent. In this case, only a variance of each feature component, c_i, is needed.

$$D(I,J) = \sum_{i=1}^{N} (\mathbf{F}_I - \mathbf{F}_J)^2 / c_i \qquad (1\text{-}38)$$

Kullback-Leibler (KL) Divergence and Jeffrey-Divergence (JD)

The *Kullback-Leibler (KL) divergence* measures how compact one feature distribution can be coded using the other one as the codebook. The KL divergence between two images I and J is defined as:

$$D(I,J) = \sum_{i} f_i(I) \log \frac{f_i(I)}{f_i(J)} \qquad (1\text{-}39)$$

The KL divergence is used in [66] as the similarity measure for texture.

The *Jeffrey-divergence (JD)* is defined by:

$$D(I,J) = \sum_i f_i(I) \log \frac{f_i(I)}{\hat{f}_i} + f_i(J) \log \frac{f_i(J)}{\hat{f}_i} \qquad (1\text{-}40)$$

where $\hat{f}_i = [f_i(I) + f_i(J)]/2$. In contrast to KL-divergence, JD is symmetric and numerically more stable when comparing two empirical distributions.

INDEXING SCHEME

Another important issue in content-based image retrieval is effective indexing and fast searching of images based on visual features. Because the feature vectors of images tend to have high dimensionality and therefore are not well suited to traditional indexing structures, *dimension reduction* is usually used before setting up an efficient indexing scheme.

One of the techniques commonly used for dimension reduction is *principal component analysis (PCA)*. It is an optimal technique that linearly maps input data to a coordinate space such that the axes are aligned to maximumly reflect the variations in the data. The QBIC system uses PCA to reduce a 20-dimensional shape feature vector to two or three dimensions [26, 27]. In addition to PCA, many researchers have used *Karhunen-Loeve (KL) transform* to reduce the dimensions of the feature space. Although the KL transform has some useful properties such as the ability to locate the most important sub-space, the feature properties that are important for identifying the pattern similarity may be destroyed during blind dimensionality reduction [53]. Apart from PCA and KL transformation, *neural network* has also been demonstrated to be a useful tool for dimension reduction of features [10].

After dimension reduction, the multi-dimensional data are indexed. A number of approaches have been proposed for this purpose, including *R-tree* (particularly, *R*-tree* [5]), *linear quad-trees* [98], *K-d-B tree* [76] and *grid files* [68]. Most of these multi-dimensional indexing methods have reasonable performance for a small number of dimensions (up to 20), but explore exponentially with the increasing of the dimensionality and eventually reduce to sequential searching. Furthermore, these indexing schemes assume that the underlying feature comparison is based on the Euclidean distance, which is not necessarily true for many image retrieval applications. One attempt to solve the indexing problems is to use hierarchical indexing scheme based on the *Self-Organization Map (SOM)* proposed in [103]. In addition to benefiting indexing, SOM provides users a useful tool to browse the representative images of each type. Details of indexing techniques are given in Chapter 8.

1.4 User Interaction

For content-based image retrieval, user interaction with the retrieval system is crucial since flexible formation and modification of queries can only be obtained by involving the user in the retrieval procedure. User interfaces in image retrieval systems typically consist of a query formulation part and a result presentation part.

QUERY SPECIFICATION

Specifying what kind of images a user wishes to retrieve from the database can be done in many ways. Commonly used query formations are: *category browsing, query by concept, query by sketch*, and *query by example*. Category browsing is to browse through the database according to the category of the image. For this purpose, images in the database are classified into different categories according to their semantic or visual content [95]. Query by concept is to retrieve images according to the conceptual description associated with each image in the database. Query by sketch and query by example [3] is to draw a sketch or provide an example image from which images with similar visual features will be extracted from the database. The first two types of queries are related to the semantic description of images which will be introduced in the following chapters.

Query by sketch allows user to draw a sketch of an image with a graphic editing tool provided either by the retrieval system or by some other software. Queries may be formed by drawing several objects with certain properties like color, texture, shape, sizes and locations. In most cases, a coarse sketch is sufficient, as the query can be refined based on retrieval results.

Query by example allows the user to formulate a query by providing an example image. The system converts the example image into an internal representation of features. Images stored in the database with similar features are then searched. Query by example can be further classified into query by external image example, if the query image is not in the database, and query by internal image example, if otherwise. For query by internal image, all relationships between images can be pre-computed. The main advantage of query by example is that the user is not required to provide an explicit description of the target, which is instead computed by the system. It is suitable for applications where the target is an image of the same object or set of objects under different viewing conditions. Most of the current systems provide this form of querying.

Query by group example allows user to select multiple images. The system will then find the images that best match the common characteristics of the group of examples. In this way, a target can be defined more precisely by specifying the relevant feature variations and removing irrelevant variations in the query. In addition, group properties can be refined by adding negative examples. Many recently developed systems provide both query by positive and negative examples.

RELEVANCE FEEDBACK

Human perception of image similarity is subjective, semantic, and task-dependent. Although content-based methods provide promising directions for image retrieval, generally, the retrieval results based on the similarities of pure visual features are not necessarily perceptually and semantically meaningful. In addition, each type of visual feature tends to capture only one aspect of image property and it is usually hard for a user to specify clearly how different aspects are combined. To address these problems, interactive *relevance feedback*, a technique in traditional text-based information retrieval systems, was introduced. With relevance feedback [79, 66, 80, 42], it is possible to establish the link between high-level concepts and low-level features.

Relevance feedback is a supervised active learning technique used to improve the effectiveness of information systems. The main idea is to use positive and negative examples from the user to improve system performance. For a given query, the system first retrieves a list of ranked images according to a predefined similarity metrics. Then, the user marks the retrieved images as relevant (positive examples) to the query or not relevant (negative examples). The system will refine the retrieval results based on the feedback and present a new list of images to the user. Hence, the key issue in relevance feedback is how to incorporate positive and negative examples to refine the query and/or to adjust the similarity measure. Detail discussions on various feedback approaches can be found in chapter 3.

1.5 Performance Evaluation

To evaluate the performance of retrieval system, two measurements, namely, *recall* and *precision* [87], are borrowed from traditional information retrieval. For a query q, the data set of images in the database that are relevant to the query q is denoted as $R(q)$, and the retrieval result of the query q is denoted as $Q(q)$. The precision of the retrieval is defined as the fraction of the retrieved images that are indeed relevant for the query:

$$precision = \frac{|Q(q) \cap R(q)|}{|Q(q)|} \tag{1-41}$$

The recall is the fraction of relevant images that is returned by the query:

$$recall = \frac{|Q(q) \cap R(q)|}{|R(q)|} \tag{1-42}$$

Usually, a tradeoff must be made between these two measures since improving one will sacrifice the other. In typical retrieval systems, recall tends to increase as the number of retrieved items increases; while at the same time the precision is likely to decrease. In addition, selecting a relevant data set $R(q)$ is much less stable due to various interpretations of the images. Further, when the number of relevant images is greater than the number of the retrieved images, recall is meaningless. As a result, precision and recall are only rough descriptions of the performance of the retrieval system.

Recently MPEG7 recommend a new retrieval performance evaluation measure, the *average normalized modified retrieval rank* (*ANMRR*) [105]. It combines the precision and recall to obtain a single objective measure. Denote the number of ground truth images for a given query q as $N(q)$ and the maximum number of ground truth images for all Q queries, i.e., max($N(q_1)$, $N(q_2)$, …, $N(q_Q)$), as M. Then for a given query q, each ground truth image k is assigned a rank value $rank(k)$ that is equivalent to its rank in the ground truth images if it is in the first K (where $K=min[4N(q), 2M]$) query results; or a rank value $K+1$ if it is not. The *average rank* $AVR(q)$ for query q is computed as:

$$AVR(q) = \sum_{k=1}^{N(q)} \frac{rank(k)}{N(q)} \tag{1-43}$$

The modified retrieval rank MRR(q) is computed as:

$$MRR(q) = AVR(q) - 0.5 - 0.5 * N(q) \tag{1-44}$$

$MRR(q)$ takes value 0 when all the ground truth images are within the first K retrieval results.

The *normalized modified retrieval rank* $NMRR(q)$, which ranges from 0 to 1, is computed as:

$$NMRR(q) = \frac{MRR(q)}{K + 0.5 - 0.5 * N(q)} \tag{1-45}$$

Then the average normalized modified retrieval rank ANMRR over all Q queries is computed as:

$$ANMRR = \frac{1}{Q} \sum_{q=1}^{Q} NMRR(q) \tag{1-46}$$

1.6 Conclusion

In this chapter, we introduced some fundamental techniques for *content-based image retrieval*, including *visual content description, similarity/distance measures, indexing scheme, user interaction* and *system performance evaluation*. Our emphasis is on visual feature description techniques. Details of indexing of high-dimensional features, user relevance feedback, and semantic description of visual contents will be addressed in chapters 3, 4, 8 and 9.

General visual features most widely used in content-based image retrieval are color, texture, shape, and spatial information. Color is usually represented by the color histogram, color correlogram, color coherence vector, and color moment under a certain color space. Texture can be represented by Tamura feature, Wold decomposition, SAR model, Gabor and Wavelet transformation. Shape can be represented by moment invariants, turning angles, Fourier descriptors, circularity, eccentricity, and major axis orientation and radon transform. The spatial relationship between regions or objects is usually represented by a 2D string. In addition, the general visual features on each pixel can be used to segment each image into homogenous regions or objects. Local features of these regions or objects can be extracted to facilitate region-based image retrieval.

There are various ways to calculate the similarity distances between visual features. This chapter introduced some basic metrics, including the Minkowski-form distance, quadratic form distance, Mahalanobis distance, Kullback-Leibler divergence and Jeffrey divergence. Up to now, the Minkowski and quadratic form distance are the most commonly used distances for image retrieval.

Efficient indexing of visual feature vectors is important for image retrieval. To set up an indexing scheme, dimension reduction is usually performed first to reduce the dimensionality of the visual feature vector. Commonly used dimension reduction methods are PCA, ICA, Karhunen-Loeve (KL) transform, and neural network methods. After dimension reduction, an indexing tree is built up. The most commonly used tree structures are R-tree, R*-tree, quad-tree, K-d-B tree, etc. Details of indexing techniques will be introduced in Chapter 8.

Image retrieval systems rely heavily on user interaction. On the one hand, images to be retrieved are determined by the user's specification of the query. On the other hand, query results can be refined to include more relevant candidates through the relevance feedback of users. Updating the retrieval results based on the user's feedback can be achieved by updating the images, the feature models, the weights of features in similarity distance, and select different similarity measures. Details will be introduced in Chapter 3.

Although content-based retrieval provides an intelligent and automatic solution for efficient searching of images, the majority of current techniques are based on low level features. In general, each of these low level features tends to capture only one aspect of an image property. Neither a single feature nor a combination of multiple features has explicit semantic meaning. In addition, the similarity measures between visual features do not necessarily match human perception. Users are interested in are semantically and perceptually similar images, the retrieval results of low-level feature based retrieval approaches are generally unsatisfactory and often unpredicable. Although relevance feedback provides a way of filling the gap between semantic searching and low-level data processing, this problem remains unsolved and more research is required. New techniques in semantic descriptions of visual contents will be addressed in chapters 4 and 9.

References

[1]   K. Arbter, W. E. Snyder, H. Burkhardt, and G. Hirzinger, "Application of affine-invariant Fourier descriptors to recognition of 3D objects," *IEEE Trans. Pattern Analysis and Machine Intelligence*, vol. 12, pp. 640-647, 1990.

[2]   E. M. Arkin, L.P. Chew, D..P. Huttenlocher, K. Kedem, and J.S.B. Mitchell, "An efficiently computable metric for comparing polygonal shapes," *IEEE Trans. Pattern Analysis and Machine Intelligence*, vol. 13, no. 3, pp. 209-226, 1991.

[3]   J. Assfalg, A. D. Bimbo, and P. Pala, "Using multiple examples for content-based retrieval," *Proc. Int'l Conf. Multimedia and Expo*, 2000.

[4]   J. R. Bach, C. Fuller, A. Gupta, A. Hampapur, B. Horowitz, R. Humphrey, R. Jain, and C. F. Shu, "The virage image search engine: An open framework for image management," *In Proc. SPIC Storage and Retrieval for Image and Video Database*, Feb. 1996.

[5]   N. Beckmann, *et al*, "The R*-tree: An efficient robust access method for points and rectangles," *ACM SIGMOD Int. Conf. on Management of Data*, Atlantic City, May 1990.

[6]   A. Blaser, Database Techniques for Pictorial Applications, *Lecture Notes in Computer Science*, Vol.81, Springer Verlag GmbH, 1979.

[7]   P. Brodatz, "Textures: A photographic album for artists & designers," Dover, NY, 1966.

[8]   H. Burkhardt, and S. Siggelkow, "Invariant features for discriminating between equivalence classes," *Nonlinear Model-based Image Video Processing and Analysis*, John Wiley and Sons, 2000.

[9]   C. Carson, M. Thomas, S. Belongie, J. M. Hellerstein, and J. Malik, "Blobworld: A system for region-based image indexing and retrieval," In D. P. Huijsmans and A. W. M. Smeulders, ed. *Visual Information and Information System, Proceedings of the Third International Conference VISUAL'99*, Amsterdam, The Netherlands, June 1999, Lecture Notes in Computer Science 1614. Springer, 1999.

[10]   J.A. Catalan, and J.S. Jin, "Dimension reduction of texture features for image retrieval using hybrid associative neural networks," *IEEE International Conference on Multimedia and Expo*, Vol.2, pp. 1211 -1214, 2000.

[11]   A. E. Cawkill, "The British Library's Picture Research Projects: Image, Word, and Retrieval," *Advanced Imaging*,Vol.8, No.10, pp.38-40, October 1993.

[12]   N. S. Chang, and K. S. Fu, "A relational database system for images," *Technical Report TR-EE 79-82*, Purdue University, May 1979.

[13]   N. S. Chang, and K. S. Fu, "Query by pictorial example," *IEEE Trans. on Software Engineering*, Vol.6, No.6, pp. 519-524, Nov.1980.

[14]   S. K. Chang, and A. Hsu, "Image information systems: where do we go from here?" *IEEE Trans. on*

Knowledge and Data Engineering, Vol.5, No.5, pp. 431-442, Oct.1992.

[15] S. K. Chang, E. Jungert, and Y. Li, "Representation and retrieval of symbolic pictures using generalized 2D string", *Technical Report*, University of Pittsburgh, 1988.

[16] S. K. Chang, and T. L. Kunii, "Pictorial database systems," *IEEE Computer Magazine*, Vol. 14, No.11, pp.13-21, Nov.1981.

[17] S. K. Chang, Q. Y. Shi, and C. Y. Yan, "Iconic indexing by 2-D strings," *IEEE Trans. on Pattern Anal. Machine Intell.*, Vol.9, No.3, pp. 413-428, May 1987.

[18] S. K. Chang, C. W. Yan, D. C. Dimitroff, and T. Arndt, "An intelligent image database system," *IEEE Trans. on Software Engineering*, Vol.14, No.5, pp. 681-688, May 1988.

[19] T. Chang, and C.C.J. Kuo, "Texture analysis and classification with tree-structured wavelet transform," *IEEE Trans. on Image Processing*, vol. 2, no. 4, pp. 429-441, October 1993.

[20] I. J. Cox, M. L. Miller, T. P. Minka, T. Papathomas, and P. N. Yianilos, "The Bayesian image retrieval system, PicHunter: Theory, implementation, and psychophysical experiments," *IEEE Trans. on Image Processing*, Vol.9, No.1, pp. 20-37, Jan. 2000.

[21] I. Daubechies, "The wavelet transform, time-frequency localization and signal analysis," *IEEE Trans. on Information Theory*, Vol. 36, pp. 961-1005, Sept. 1990.

[22] J. G. Daugman, "Complete discrete 2D Gabor transforms by neural networks for image analysis and compression," *IEEE Trans. ASSP*, vol. 36, pp. 1169-1179, July 1998.

[23] J. Dowe, "Content-based retrieval in multimedia imaging," *In Proc. SPIE Storage and Retrieval for Image and Video Database*, 1993.

[24] C. Faloutsos et al, "Efficient and effective querying by image content," *Journal of intelligent information systems*, Vol.3, pp.231-262, 1994.

[25] G. D. Finlayson, "Color in perspective," *IEEE Trans on Pattern Analysis and Machine Intelligence*, Vol.8, No. 10, pp.1034-1038, Oct. 1996.

[26] M. Flickner, H. Sawhney, W. Niblack, J. Ashley, Q. Huang, B. Dom, M. Gorkani, J. Hafner, D. Lee, D. Petkovic, D. Steele, and P. Yanker, "Query by image and video content: The QBIC system." *IEEE Computer*, Vol.28, No.9, pp. 23-32, Sept. 1995.

[27] J. D. Foley, A. van Dam, S. K. Feiner, and J. F. Hughes, *Computer graphics: principles and practice*, 2nd ed., Reading, Mass, Addison-Wesley, 1990.

[28] J. M. Francos. "Orthogonal decompositions of 2D random fields and their applications in 2D spectral estimation," N. K. Bose and C. R. Rao, editors, *Signal Processing and its Application*, pp.20-227. North Holland, 1993.

[29] J. M. Francos, A. A. Meiri, and B. Porat, "A unified texture model based on a 2d Wold like decomposition," *IEEE Trans on Signal Processing*, pp.2665-2678, Aug. 1993.

[30] J. M. Francos, A. Narasimhan, and J. W. Woods, "Maximum likelihood parameter estimation of textures using a Wold-decomposition based model," *IEEE Trans. on Image Processing*, pp.1655-1666, Dec.1995.

[31] B. Furht, S. W. Smoliar, and H.J. Zhang, *Video and Image Processing in Multimedia Systems*, Kluwer Academic Publishers, 1995.

[32] J. E. Gary, and R. Mehrotra, "Shape similarity-based retrieval in image database systems," *Proc. of SPIE, Image Storage and Retrieval Systems*, Vol. 1662, pp. 2-8, 1992.

[33] T. Gevers, and A.W.M.Smeulders, "Pictoseek: Combining color and shape invariant features for image retrieval," *IEEE Trans. on image processing*, Vol.9, No.1, pp102-119, 2000.

[34] T. Gevers, and A. W. M. Smeulders, "Content-based image retrieval by viewpoint-invariant image indexing," *Image and Vision Computing*, Vol.17, No.7, pp.475-488, 1999.

[35] Y. Gong, H. J. Zhang, and T. C. Chua, "An image database system with content capturing and fast image indexing abilities", *Proc. IEEE International Conference on Multimedia Computing and Systems*, Boston, pp.121-130, 14-19 May 1994.

[36] W. I. Grosky, and R. Mehrotra, "Index based object recognition in pictorial data management," *CVGIP*, Vol. 52, No. 3, pp. 416-436, 1990.

[37] V. N. Gudivada, and V. V. Raghavan, "Design and evaluation of algorithms for image retrieval by spatial similarity," *ACM Trans. on Information Systems*, Vol. 13, No. 2, pp. 115-144, April 1995.

[38] F. Guo, J. Jin, and D. Feng, "Measuring image similarity using the geometrical distribution of image contents", *Proc. of ICSP*, pp.1108-1112, 1998.

[39] A. Gupta, and R. Jain, "Visual information retrieval," *Communication of the ACM*, Vol.40, No..5, pp.71-79, May, 1997.

[40] J. Hafner, *et al.*, "Efficient color histogram indexing for quadratic form distance functions," IEEE *Trans. on Pattern Analysis and Machine Intelligence*, Vol. 17, No. 7, pp. 729-736, July 1995.

[41]   M. K. Hu, "Visual pattern recognition by moment invariants," in J. K. Aggarwal, R. O. Duda, and A. Rosenfeld, *Computer Methods in Image Analysis*, IEEE computer Society, Los Angeles, CA, 1977.

[42]   J. Huang, S. R. Kumar, and M. Metra, "Combining supervised learning with color correlograms for content-based image retrieval," *Proc. of ACM Multimedia '95*, pp. 325-334, Nov. 1997.

[43]   J. Huang, S.R. Kumar, M. Metra, W. J., Zhu, and R. Zabith, "Spatial color indexing and applications," *Int'l J. Computer Vision*, Vol.35, No.3, pp. 245-268, 1999.

[44]   J. Huang, *et al.*, "Image indexing using color correlogram," *IEEE Int. Conf. on Computer Vision and Pattern Recognition*, pp. 762-768, Puerto Rico, June 1997.

[45]   M. Ioka, "A method of defining the similarity of images on the basis of color information," *Technical Report RT-0030*, IBM Tokyo Research Laboratory, Tokyo, Japan, Nov. 1989.

[46]   H. V. Jagadish, "A retrieval technique for similar shapes," *Proc. of Int. Conf. on Management of Data, SIGMOID '91*, Denver, CO, pp. 208-217, May 1991.

[47]   A. K. Jain, *Fundamental of Digital Image Processing*, Englewood Cliffs, Prentice Hall, 1989.

[48]   A. K. Jain, and F. Farroknia, "Unsupervised texture segmentation using Gabor filters," *Pattern Recognition*, Vo.24, No.12, pp. 1167-1186, 1991.

[49]   R. Jain, *Proc. US NSF Workshop Visual Information Management Systems*, 1992.

[50]   R. Jain, A. Pentland, and D. Petkovic, *Workshop Report: NSF-ARPA Workshop on Visual Information Management Systems*, Cambridge, Mass, USA, June 1995.

[51]   A. Kankanhalli, H. J. Zhang, and C. Y. Low, "Using texture for image retrieval," *Third Int. Conf. on Automation, Robotics and Computer Vision*, pp. 935-939, Singapore, Nov. 1994.

[52]   H. Kauppinen, T. Seppnäen, and M. Pietikäinen, "An experimental comparison of autoregressive and Fourier-based descriptors in 2D shape classification," *IEEE Trans. Pattern Anal. and Machine Intell.*, Vol. 17, No. 2, pp. 201-207, 1995.

[53]   W. J. Krzanowski, *Recent Advances in Descriptive Multivariate Analysis*, Chapter 2, Oxford science publications, 1995.

[54]   A. Laine, and J. Fan, "Texture classification by wavelet packet signatures," *IEEE Trans. Pattern Analysis and Machine Intelligence*, Vol. 15, No. 11, pp. 1186-1191, Nov. 1993.

[55]   S. Y. Lee, and F. H. Hsu, "2D C-string: a new spatial knowledge representation for image database systems," *Pattern Recognition*, Vol. 23, pp 1077-1087, 1990.

[56]   S. Y. Lee, M.C. Yang, and J. W. Chen, "2D B-string: a spatial knowledge representation for image database system," *Proc. ICSC'92 Second Int. computer Sci. Conf.*, pp.609-615, 1992.

[57]   F. Liu, and R. W. Picard, "Periodicity, directionality, and randomness: Wold features for image modeling and retrieval," *IEEE Trans. on Pattern Analysis and Machine Learning*, Vol. 18, No. 7, July 1996.

[58]   W. Y. Ma, and B. S. Manjunath, "A comparison of wavelet features for texture annotation," *Proc. of IEEE Int. Conf. on Image Processing*, Vol. II, pp. 256-259, Washington D.C., Oct. 1995.

[59]   W. Y. Ma, and B. S. Manjunath, "Image indexing using a texture dictionary," *Proc. of SPIE Conf. on Image Storage and Archiving System*, Vol. 2606, pp. 288-298, Philadelphia, Pennsylvania, Oct. 1995.

[60]   W. Y. Ma, and B. S. Manjunath, "Netra: A toolbox for navigating large image databases," *Multimedia Systems*, Vol.7, No.3, pp.:184-198, 1999.

[61]   W. Y. Ma, and B. S. Manjunath, "Edge flow: a framework of boundary detection and image segmentation," *IEEE Int. Conf. on Computer Vision and Pattern Recognition*, pp. 744-749, Puerto Rico, June 1997.

[62]   S. G. Mallat, "A theory for multiresolution signal decomposition: the wavelet representation," *IEEE Trans. Pattern Analysis and Machine Intelligence*, Vol. 11, pp. 674-693, July 1989.

[63]   B. S. Manjunath, and W. Y. Ma, "Texture features for browsing and retrieval of image data," *IEEE Trans. on Pattern Analysis and Machine Intelligence*, Vol. 18, No. 8, pp. 837-842, Aug. 1996.

[64]   J. Mao, and A. K. Jain, "Texture classification and segmentation using multiresolution simultaneous autoregressive models," *Pattern Recognition*, Vol. 25, No. 2, pp. 173-188, 1992.

[65]   E. Mathias, "Comparing the influence of color spaces and metrics in content-based image retrieval," *Proceedings of International Symposium on Computer Graphics, Image Processing, and Vision*, pp. 371 -378, 1998.

[66]   T. P. Minka, and R. W. Picard, "Interactive learning using a 'society of models', " *IEEE Int. Conf. on Computer Vision and Pattern Recognition*, pp. 447-452, 1996.

[67]   W. Niblack et al., "Querying images by content, using color, texture, and shape," *SPIE Conference on Storage and Retrieval for Image and Video Database*, Vol. 1908, pp.173-187, April 1993.

[68]   J. Nievergelt, H. Hinterberger, and K. C. Sevcik, "The grid file: an adaptable symmetric multikey

file structure," *ACM Trans. on Database Systems*, pp. 38-71, March 1984.

[69]   V. E. Ogle, and M. Stonebraker, "Chabot: Retrieval from a relational database of images," *IEEE Computer*, Vol.28, No.9, pp. 40-48, Sept. 1995.

[70]   T. Ojala, M. Pietikainen, and D. Harwood, "A comparative study of texture measures with classification based feature distributions," *Pattern Recognition*, Vol.29, No.1, pp.51-59, 1996.

[71]   G.Pass, and R. Zabith, "Comparing images using joint histograms," *Multimedia Systems*, Vol.7, pp.234-240, 1999.

[72]   G. Pass, and R. Zabith, "Histogram refinement for content-based image retrieval," *IEEE Workshop on Applications of Computer Vision*, pp. 96-102, 1996.

[73]   A. Pentland, R.W. Picard and S. Sclaroff, "Photobook: Content-Based Manipulation of Image Databases," *Proc. Storage and Retrieval for Image and Video Databases II*, Vol. 2185, San Jose, CA, USA February, 1994.

[74]   E. Persoon, and K. Fu, "Shape discrimination using Fourier descriptors," *IEEE Trans. Syst., Man, and Cybern.*, Vol. 7, pp. 170-179, 1977.

[75]   R. W. Picard, T. Kabir, and F. Liu, "Real-time recognition with the entire Brodatz texture database," *Proc. IEEE Int. Conf. on Computer Vision and Pattern Recognition*, pp. 638-639, New York, June 1993.

[76]   J. T. Robinson, "The k-d-B-tree: a search structure for large multidimensional dynamic indexes," *Proc. of SIGMOD Conference*, Ann Arbor, April 1981.

[77]   Y. Rui, T. S. Huang, and S. F. Chang, "Image retrieval: current techniques, promising directions and open issues, " *Journal of Visual Communication and Image Representation*, Vol.10, pp. 39-62, 1999.

[78]   Y. Rui, T.S.Huang, and S. Mehrotra, "Content-based image retrieval with relevance feedback in MARS," *Proceedings of International Conference on Image Processing*, Vol.2, pp. 815 -818, 1997.

[79]   Y. Rui, T. S. Huang, M. Ortega, and S. Mehrotra, "Relevance feedback: a power tool for interactive content-based image retrieval," *IEEE Trans. on Circuits and Systems for Video Technology*, 1998.

[80]   Y. Rui, *et al*, "A relevance feedback architecture in content-based multimedia information retrieval systems," *Proc of IEEE Workshop on Content-based Access of Image and Video Libraries*, 1997.

[81]   G. Salton, and M. McGill, *Introduction to Modern Information Retrieval*. McGraw-Hill, New York, NY, 1983.

[82]   H. Samet, "The quadtree and related hierarchical data structures," *ACM Computing Surveys*, Vol.16, No.2, pp.187-260, 1984.

[83]   H. Samet, *The Design and Analysis of Spatial Data Structures*, Addison-Wesley, 1989.

[84]   S. Sclaroff, and A. Pentland, "Modal matching for correspondence and recognition," *IEEE Trans. on Pattern Analysis and Machine Intelligence*, Vol. 17, No. 6, pp. 545-561, June 1995.

[85]   S. Sclaroff, L. Taycher, and M. L. Cascia, "ImageRover: a content-based image browser for the World Wide Web," Boston University CS Dept. *Technical Report* 97-005, 1997.

[86]   A. W. M. Smeulders, S. D. Olabariagga, R. van den Boomgaard, and M. Worring, "Interactive segmentation," *Proc. Visual'97: Information Systems*, pp.5-12, 1997.

[87]   A. M. W. Smeulders, M. Worring, S. Santini, A. Gupta, and R. Jain, "Content-based image retrieval at the end of the early years, " *IEEE Trans. on Pattern Analysis and Machine Intelligence*, Vol.22, No.12, pp. 1349-1380, Dec. 2000.

[88]   J. R. Smith, and S. F. Chang, "VisualSEEk: a fully automated content-based image query system," *ACM Multimedia 96*, Boston, MA, Nov. 1996.

[89]   M. Stricker, and M. Orengo, "Similarity of color images," *SPIE Storage and Retrieval for Image and Video Databases III*, vol. 2185, pp.381-392, Feb. 1995.

[90]   M. Stricker, and M. Orengo, "Color indexing with weak spatial constraint," *Proc. SPIE Conf. On Visual Communications, 1996*.

[91]   M. J. Swain, and D. H. Ballard, "Color indexing," *International Journal of Computer Vision*, Vol. 7, No. 1, pp.11-32, 1991.

[92]   H. Tamura, S. Mori, and T. Yamawaki, "Texture features corresponding to visual perception," *IEEE Trans. On Systems, Man, and Cybernetics*, vol. Smc-8, No. 6, June 1978.

[93]   H. Tamura, and N.Yokoya, "Image database systems: A survey, " *Pattern Recognition*, Vol.17, No.1, pp. 29-43, 1984.

[94]   D. Tegolo, "Shape analysis for image retrieval," *Proc. of SPIE, Storage and Retrieval for Image and Video Databases -II*, no. 2185, San Jose, CA, pp. 59-69, February 1994.

[95]   A. Vailaya, M. A. G. Figueiredo, A. K. Jain, and H. J. Zhang, "Image classification for content-based indexing," *IEEE Trans. on Image Processing*, Vol.10, No.1, Jan. 2001.

[96]   N. Vasoncelos, and A. Lippman, "A probabilistic architecture for content-based image retrieval," *Proc. Computer vision and pattern recognition*, pp. 216-221, 2000.

[97]   R. C. Veltkamp, and M. Hagedoorn, "State-of-the-art in shape matching," *Technical Report UU-CS-1999-27*, Utrecht University, Department of Computer Science, Sept. 1999.

[98]   J. Vendrig, M. Worring, and A. W. M. Smeulders, "Filter image browsing: exploiting interaction in retrieval," *Proc. Viusl'99: Information and Information System*, 1999.

[99]   H. Voorhees, and T. Poggio. "Computing texture boundaries from images," *Nature*, 333:364-367, 1988.

[100]   H. Wang, F. Guo, D. Feng, and J. Jin, "A signature for content-based image retrieval using a geometrical transform," *Proc. Of ACM MM'98*, Bristol, UK, 1998.

[101]   W. H. Wong, W. C. Siu, and K. M. Lam, "Generation of moment invariants and their uses for character recognition," *Pattern Recognition Letters*, Vol. 16, pp. 115-123, Feb. 1995.

[102]   L. Yang, and F. Algregtsen, "Fast computation of invariant geometric moments: A new method giving correct results," *Proc. IEEE Int. Conf. on Image Processing*, 1994.

[103]   H. J. Zhang, and D. Zhong, "A Scheme for visual feature-based image indexing," *Proc. of SPIE conf. on Storage and Retrieval for Image and Video Databases III*, pp. 36-46, San Jose, Feb. 1995.

[104]   H. J. Zhang, *et al*, "Image retrieval based on color features: An evaluation study," *SPIE Conf. on Digital Storage and Archival*, Pennsylvania, Oct. 25-27, 1995.

[105]   MPEG Video Group, Description of core experiments for MPEG-7 color/texture descriptors, *ISO/MPEGJTC1/SC29/WG11 MPEG98/M2819*, July 1999.

2 CONTENT-BASED VIDEO ANALYSIS, RETRIEVAL AND BROWSING

Dr. Hongjiang Zhang

This chapter presents video structure parsing, content representation, and content based video retrieval and browsing technologies. First, we introduce research issues in this active research area in Section2.1. Section 2.2 describes video parsing algorithms, namely, shot boundary detection, scene grouping and story segmentation. Video content representation schemes are presented in detail in Section 2.3, followed by discussion on technologies for video summarization in Section 2.4. Discussions on video similarity, clustering and content-based video retrieval and browsing technologies are presented in Section 2.5. Section 2.6 concludes the chapter by giving an overview of current research issues in video content analysis, retrieval and browsing.

2.1 Introduction

Considering the vast amount of video data, the development of a means for the quick relevance assessment of video documents is critical. The fundamental needs are similar to that of image databases: video data should be structured and indexed. Content-based image retrieval technologies presented in Chapter 1 can be extended to video retrieval. However, such an extension is not straightforward. A video clip a sequence of image frames therefore indexing each of them as a still image introduces extremely high redundancy and is impossible given the number of frames in a video for even one minute. Video is a structured medium in which actions and events, in a time and space, comprise stories or convey particular visual information. That is, a video program should be viewed as a document rather than a non-structured sequence of frames. The indexing of video should be seen as analogous to text document indexing, where a structural analysis is performed to decompose the document into paragraphs, sentences and words, before an index is built. Similarly, we need to identify the structure of the video, and decompose the video into basic components, then build indexes based on the structural information and individual image frames [64].

Moreover, content-based browsing is another problem when assessing the relevance of video source material, considering the large amount of data of video [21, 64]. By browsing, we mean an informal but quick access of content which may lack any specific goal or focus. To achieve content-based browsing, we need to re-present the information landscape or structure of the video in a more abstract and/or summarized manner.

Current video indexing schemes involve three primary processes: video parsing, content analysis, and abstraction, which are significantly different and more complicated than those involved in image content indexing [3, 21, 64]. Parsing is the process of the extracting temporal structure of a video that involves the detection of temporal boundaries and identification of story units. The content feature extraction process is similar to that of image feature extraction, but extended to the extraction of features that describe object motions, events and actions in video sequences. Abstraction is a process which extracts or constructs a subset of video data from the original video, such as key-frames or key-sequences as entries for shots, scenes or stories. The outcome of the abstraction process forms the basis for content-based video browsing: key-frames, for instance, enable browsing with a content-based hierarchical viewer.

Based on the content features or meta-data resulting from these three processes, indexes for the video can be built through a clustering process which classifies sequences or shots, into different visual categories or indexing structures. Again, similar to that in image database systems, schemes and tools are needed to utilize the indices and content meta-data to query, search and browse large video databases to locate desired video clips. Figure 2-1 illustrates these processes.

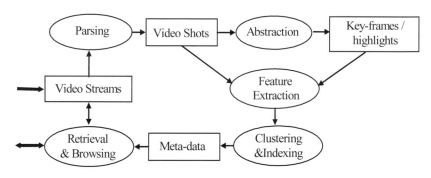

Figure 2-1 Process diagram for video content analysis and retrieval

2.2 Video Structure Analysis

Like text documents, video also has an underlying synthetic structure, although it is not as obvious as that of documents. A video sequence can be viewed as a well organized document and can be parsed into logical units at different levels [7]. The top-level consists of sequences or stories, which are composed of a set of scenes. Scenes are further partitioned into shots. Each shot contains a sequence of frames recorded contiguously and representing a continuous action in time or space. Decomposing a long video sequence into meaningful components is the task of video parsing or temporal structuring.

Video parsing is a topic that has been intensively studied, compared to other issues in video indexing. However, while the algorithms for shot detection have matured and been put to practical use, scene detection and story segmentation algorithms remain a challenging research topic [7, 71, 59]. Part of the problem is that detection of scenes and stories in a video relies on an understanding of semantics. In contrast, shots are a physical unit in video and their boundaries are determined by editing points or when the camera is switched on or off. Analogous to words or sentences in text documents, shots are a good choice as the basic unit for video content indexing. As illustrated in the next sections, shot based analysis, classification and grouping, form the basis for content-based video indexing, browsing and retrieval.

In this section, we focus on algorithms for partitioning video into shots and scene grouping. Algorithms for story segmentation will also be briefly discussed.

SHOT BOUNDARY DETECTION

Shot boundary detection is the process of detecting the boundaries between two consecutive shots, so that a sequence of frames belonging to a single shot will be grouped together. There are a number of different types of transitions or boundaries between shots. The simplest transition is a cut: an abrupt shot change which occurs between two consecutive frames. More sophisticated transitions include dissolve, fade-in, fade-out and wipes, etc. [7]. A robust partitioning algorithm should be able to detect all of these different boundaries with a high degree of accuracy.

The detection of shot boundaries is dependent on the fact that consecutive frames on either side of a boundary generally display a significant change in content. Therefore, what is required is a quantitative measure that can quantitatively determine the difference between such a pair of frames. If the difference exceeds a given threshold, it indicates a shot boundary. Hence, establishing suitable metrics is the key issue in automatic partitioning. The ideal metric for automatic video partitioning will be able to differentiate between the following three image changes: (1) shot change, either abrupt or gradual; motions, introduced by both camera operation and object motion; and (2) luminosity changes and noise.

The major differences among automatic video partitioning algorithms are the frame features and difference metrics, which are used to quantitatively measure changes between consecutive frames and schemes to apply the metrics. Difference metrics used in video partitioning can be divided into two major types: (1) those based local pixel feature comparison, such as pixel values [35, 61, 44] and edges [60], and (2) those based on global features such as histograms and statistic distributions of pixel-to-pixel change.

The most popular metric for cut detection is the difference between the histograms of two consecutive frames. The assumption is that two frames having an unchanging background and objects will show little difference in their respective histograms. The histogram comparison algorithm is less sensitive to object motion than the pairwise pixel comparison algorithm, since it ignores spatial changes in a frame [61]. This is similar to the case of histogram-based image retrieval where there are many ways to compare the histograms of two consecutive video frames [35].

More sophisticated shot boundaries (such as dissolve, wipe, fade-in, and fade-out) are much more difficult to detect because the change between consecutive frames is more gradual compared to a sharp cut. Furthermore, changes resulting from camera operations may be of the same order as those from gradual transitions, further complicating the detection.

The twin-comparison algorithm was the first published algorithm to detect gradual shot changes [61]. This algorithm uses two comparisons: (1) the difference between consecutive frames to detect a sharp cut, and (2) the accumulated difference over a sequence of frames to detect gradual transitions. This algorithm also applies a global motion analysis to filter out sequences of frames involving global or large moving objects, which may confuse detection of the transition. Experiments show the twin comparison algorithm is very effective and achieves a very high level of accuracy.

Hampapur et al., have studied the use of editing a model fitting algorithm for detecting different types of gradual transitions [23]. The algorithm detects the editing effect by fitting sequences of inter-frame changes into editing models, one for each given type of transition [2]. However, the potential problem with such model based algorithms is that as the number of different types of editing effects increases it becomes more difficult to model them. Furthermore, transition sequences may not follow any particular editing model, due to noises and/or a combination of editing effects. Although such problems may also be apparent in other detection algorithms, Hampapur's algorithm is more limited.

There are many variations of the algorithms discussed, as well as algorithms which use other image features such as edge distribution. Several studies compare shot change detection algorithms presented [8, 16]. It is interesting to observe that histogram based algorithms outperform other algorithms in both accuracy and speed, in the experiments carried out by Boreczky and Rowe [8].

There are a few interesting and recent works on gradual transition detection that are not based on the video production model. Bouthemy et al. [9] proposed a unified framework to detect cuts, wipes and dissolves based on a *2D* parametric motion model. Cuts can be detected by observing the size of the dominant motion support layers between two adjacent frames. A significant drop in the size in a frame clearly marks the presence of a cut. However, gradual transitions are not easily detected using this method. As a result, the Hinkley test is employed to detect wipes and dissolves by investigating the size of the support layer over a period of time. Although the proposed approach is computationally intensive and can not be used for real-time applications, the estimated dominant motion is a useful by-product that is ready to be utilized for motion characterization.

Ngo et al, proposed a novel detection algorithm based on the analysis of spatio-temporal slices [36]. Ngo presented various transitional patterns in slices generated by cut, wipe and dissolves. These transitions are detected by a proposed spatio-temporal energy model which takes the color and texture patterns of slices into account.

JPEG, MPEG and H.26X [21] have become industry standards, and an increasing amount of video data have been and will continue to be stored and distributed in one of these compressed formats. It is therefore advantageous for the analysis tools to operate directly on compressed representations, saving on the computational cost of decompression. More importantly, the compressed domain representation of video defined by these standards, namely DCT(Discrete Cosine Transform) coefficients and motion vectors, are more effective and readily available features for compressed video partitioning. In this sub-section, we discuss some basic algorithms developed to utilize such content information.

The first comparison metric based on DCT for partitioning JPEG compressed videos was developed by Arman et al. [6] and extended to MPEG by Zhang et al. [65]. In this algorithm, a subset of the blocks in each frame and a subset of the DCT coefficients for each block were used as a vector representation for each frame. The difference metric between frames is then defined by content correlation. For detecting shots boundaries, DC components of DCT coefficients of video frames provide sufficient information [56] and using DC sequences makes it easy to apply a histogram comparison. That is, instead of comparing histograms of pixel values, we compare histograms of DCT-DC coefficients of frames. This algorithm has proved to be very effective, achieving both high detection accuracy and speed in detecting sharp cuts [56].

In addition to the pixel value, motion resulting from either moving objects and/or camera works is another important visual content in video data. In general, the motion vectors should show continuity between frames within a camera shot and show discontinuity between frames across two shots. Thus, a continuity metric for a field of motion vectors is an alternative criterion for detecting segment boundaries. Combining the DCT based and motion based metrics into a hybrid algorithm will improve the detection accuracy and processing speed in partitioning MPEG compressed video [65]. Recently, Jun et al. proposed a fast and efficient dissolve detector by using the macroblock information of MPEG videos [28]. The authors found the typical patterns of macroblocks during dissolves. By investigating the ratio of forward macroblocks in B-frames, and the spatial distribution of forward and backward macroblocks, promising results were obtained when the proposed method was tested on news and sports videos.

A key issue in most of the shot boundary detection algorithm is how to determine the thresholds. Obviously, adaptive or statistical threshold setting, such as in [61], is desirable. The adaptive threshold setting method proposed by Vasconcelos and Lippman is based on the Bayesian formulation by modeling the shot duration and shot activities For instance, in [54], a better experimental result was obtained compared to the fixed threshold strategy commonly practiced for cut detection.

In summary, experimental evaluations have shown that the shot detection algorithms, based on compression domain features, perform with at least the same order of accuracy as those using video data in the original format, though the detection of sharp cuts are more reliable than the detection of gradual transitions. On the other hand, algorithms based on compressed features achieved a higher processing speed, which makes it possible to partition real time video using software alone in today PC's.

SCENE GROUPING

In media production, the level immediately higher than shots is called scene. From the narrative point of view, a scene consists of a series of consecutive shots (i.e. shots which have been shot in the same location or share some thematic visual content). The process of detecting these video scenes is analogous to paragraphing in text document parsing and requires a higher level content analysis.

Two different approaches have been proposed for the automatic recognition of sequences of programs: filming rule-based and a priori program model based. Aigrain et al., have used filming rules to detect local (in time) clues of macro-scope change [4]. These rules refer to transition effects, shot repetition, shot setting similarity, apparition of music in the soundtrack, editing rhythm and camera work. After detecting the local clues, their temporal organization is analyzed to produce the segmentation and to choose one or two representative shots for each sequence.

Yeung and Yeo have proposed a similar approach to scene or story detection, called time-constrained clustering [59, 71]. In this approach, both the visual similarity and temporal locality of shots are considered in each shot's grouping and event detection. This is based on the idea that content presented in video programs tends to be localized in time: two visually similar shots occurring next to each other may represent a continuing event; while shots separated by large time durations may represent different contents or belong to different scenes. Using information about special temporal events, namely, dialogues and fast action shots, is another essential part of this approach.

If scene detection is considered as a process of shot grouping, it is a key issue to define a measure of shot correlation. The shot correlation can be obtained by computing the similarity of key frames in consecutive shots. A pseudo-color-object is defined for a more accurate measure of shot correlation based on the dominant color grouping and tracking [30]. Moreover, an expanding window method is used to group shots into scenes. Experimental results indicate this method always outperforms other key-frame or average color histogram based methods. In summary, general scene detection requires content analysis at a higher level. In the near future, it is doubtful that this task can be fully automated using current image processing and computer vision techniques to analyse visual content.

STORY SEGMENTATION

Story segmentation needs more semantic understanding of video content. A priori model-based algorithms use certain structure models of special video programs such as news and particular sports [22, 46, 66]. For those special video programs, the temporal structures are normally very rigid. Thus, if one can identify a few key types of shots, such as the anchor-person in a news video, then, the rest of shots can be classified according to the structure models, without sophisticated semantic content analysis of the entire video. For example, Zhang et al. have developed an approach to automatically parse and index TV news [66]. They recognized specific types of shots, such as anchor-person shots using motion information, and then used the news program model to analyze the succession of shot types and produce a segmentation of news stories. However, the problem with such an approach is that it is often impossible to build a priori models which can be extended to other application domains. Even for the news video, this approach is not able to classify the segmented news stories into appropriate categories since it does not have the capability to extract semantic content of each shot. However, the fusion of information from image, audio and closed-caption or transcript text analysis is a feasible solution and successful examples include the *Informedia* project [26] and many others [10, 33, 43].

2.3 Video Content Representation

After temporal segmentation and motion analysis, another key step in video content analysis is to identify representation primitives of video content and extract visual features, based on which the content of shots can be classified, indexed and compared. These content primitives are often also referred to as the content meta-data of video. Here we restrict our discussion to visual representations. Ideally these primitives should be semantic. A user can easily employ the primitives to define interesting or significant events. However, automatic extraction of the primitives is often not feasible. In this section, we introduce a structured content representation scheme for video, namely, a set of motion and color based shot representations.

A video program or documents can be decomposed into three structural levels: stories, scenes and shots.The representation of a video document should have the same structural levels. The structured representation scheme in Figure 2-2 is a good example that meets the above criterion.

Sequence
    Sequence-ID: x (a unique index key of the sequence)
    Scenes: {Scene(1), Scene (2), ... Scene (L)}
    Summary: *A*
Scene
    Scene-ID: y_l (a unique index key of the scene l)
    Shots: {Shot(1), Shot(2), ... Shot(M)}
    Key-frames: **{KF$_l$(1), KF$_l$(2), ..., KF$_l$(I)}**
Shot
    Shot-ID: z_m (a unique index key of the shot m)
    Primitives: {$f_s(1), f_s(2), ..., f_s(N)$}
    Key-frames: **{KF$_m$(1), KF$_m$(2), ..., KF$_m$(J)}**

Figure 2-2 A structured video content representation scheme: three levels of content descriptors.

Note that the bolded items in Figure 2-2 are primitives based on visual abstractions of a sequence, which are not usually contained explicitly in conventional video representation schemes. These primitives are included to make the representation useful in content-based video browsing and in the indexing of video contents as discussed later in this section. The video summary A, as part of the representation, could be a set of pointers to highlight segments extracted by using video summarization approaches presented in Section 2.4. Similarly, key-frames are included in both the scene and shot representation primitives. However, the set of key-frames of a scene should be a sub-set of the key-frames of all shots comprising the scene; thus, $KF_l(i)$ is a set of pointers to $KF_m(j)$, i.e. $KF_l(j)$ is a more abstracted set extracted from $KF_m(j)$. If key-frames are not extracted in the level of scene, then, the key-frame item should be removed from the scene representation.

Give the representation structure as in Figure 2-2, the key issue in video content representation is determining which set of visual features should be used as content primitives. We concentrate our discussions on appropriate features for representing the content of shots and assume that a scene can represented by, and retrieved based on, the content primitives of all of its shots. The features are divided into two groups: those associated with key-frames, and remaining shot features. The former are only derived from single key-frames, which only capture representative information at the few sampling points where key-frames are located. The latter are primarily derived from the sequence of frames of a shot, including temporal variation of any given image feature or feature set, and motions associated with the shot or even some objects in the shot. We refer the first group as key-frames-based and the other shot-based. Therefore, these two groups constitute a complete set of content primitives for video shots and form the basis for shot based video indexing and retrieval.

KEY-FRAME-BASED REPRESENTATIONS
Key-frames-based representation of video content uses the same features as those for content-based still image retrievals discussed in the previous chapter. These features are extracted from each of the key-frames associated with a shot [64, 68]. Apart from the low-level primitives, one can also include high-level semantic features, such as objects, e.g. news-anchor, in a key-frame [66].

Therefore, we can add one more extension, for each of the key-frames in a shot, to the representation scheme in Figure 2-2. This as shown in Figure 2-3.

> *Key-frames:*
> *KF-ID:* $\mathbf{KF_m(j)}$
> *Primitives:* $\{f_{KF}(1), f_{KF}(2),..., f_{KF}(K)\}$.

Figure 2-3 Representation scheme for key-frame primitives – an extension to the video representation scheme shown in Figure 2-5

SHOT BASED REPRESENTATIONS
Key-frame-based features are insufficient to support event-based classification and retrieval since they do not capture motion and temporal changes in a shot, which are an essential and unique feature of video. Early work by Zhang et al., [64, 68] have proposed a set of shot-based features, including statistical motion measures, and the temporal mean and variance of color features over a shot, to provide information about activities, and motion complexity and distribution that might be useful in queries. In recent years, more effective shot based representations, including pseudo-color-object, perceived motion spectrum, motion texture, and slice motion characterization have been proposed [30, 31, 37]. A more detail description of this feature set is given in below.

Global motion feature

To enhance representation power, model parameters of global or dominant motion in a shot are used as a feature vector for representing motion content. This feature captures the general motions in the shot such as camera pan or zoom, and can be used to classify the video sequences into static and motion sequences. It can be obtained using the motion analysis algorithms presented above. Motion models can range from a simple translation to complex planar parallax motion. The affine motion model (only six model parameters) provides a good compromise between complexity and stability.

Statistic motion features were introduced to describe motion distributions in terms of directions and magnitude. They complement the global motion feature that is ineffective when there is no dominant motion in a shot. They include directional distribution of motion, directional average speed and local average speed, which are all derived from optical flow calculated between consecutive frames [33].

Both the global motion and this set of features may change over frames within a shot, these two feature sets may have to be a time series, with one set of values per frame. A simpler way to accommodate temporal changes is to use the temporal average for all frames in a shot, or to use values taken at a few sampling points, such as points where key-frames are extracted. [68, 72].

Temporal Slice based Representation

Based on the analysis of spatio-temporal image volumes, an effective motion-based content representation was proposed in [37]. In the spatio-temporal slices of image volumes, motion is depicted as oriented patterns. Using a tensor histogram computation algorithm, motion can be characterized efficiently. Not only can the camera motion in a video can be annotated as static, pan, tilt, zoom, etc., but the moving object can be segmented and tracked efficiently in the spatio-temporal images of video shots.

A video can be arranged as a volume with (x; y) representing image dimensions and t representing the temporal dimension. We can view the volume as formed by a set of 2D temporal slices each with dimension (x; t) or (y; t). Each spatio-temporal slice is then a collection of 1D scans taken at the same selected position of every frame, over time. The slice is used to extract an indicator to capture the motion coherency of the video. For convenience, we refer to H(x; t) as the horizontal slice and V(y; t) as the vertical slice.

Figure 2-4 shows various patterns in slices due to movement of the camera and object. The orientation of a slice reflects the type of motion. A static sequence exhibits horizontal lines across H(x; t) and V(y; t); while camera panning and tilting results in one slice indicating the speed and direction of the motion, and the other slice exploring the panoramic information [40]. For zooming, the lines in slices are either expanded in or out in a V-shape pattern. In a multiple motion case, more than one H(x; t) and one V(y; t) are, generally required for analysis. For instance, a sequence with object motion shows both static and panning patterns in different slices. A sequence which tracks an object over time manifests two motion patterns in a horizontal slice, one indicates camera panning and one shows object motion.

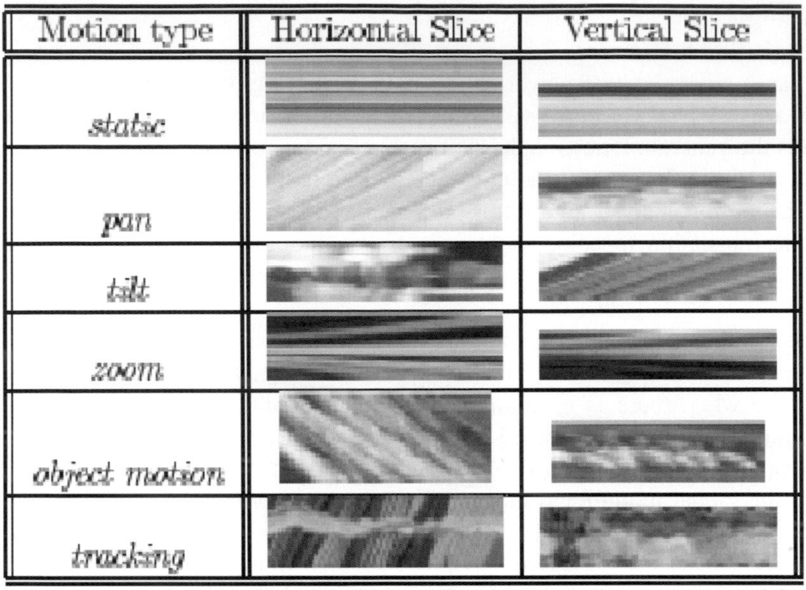

Motion type	Horizontal Slice	Vertical Slice
static		
pan		
tilt		
zoom		
object motion		
tracking		

Figure 2-4 Motion patterns in slices. The horizontal and vertical slices are extracted from the center of an image volume. The x-axis is in time dimension while the y-axis is in image dimension.

An approach based on the structure tensor computation was introduced in [37] to estimate the local orientations of a slice. With this method, we can classify motion-types as well as separate different motion-layers. This method is described in detail as follows.

The tensor Γ of slice H can be expressed as:

$$\Gamma = \begin{bmatrix} J_{xx} & J_{xt} \\ J_{xt} & J_{tt} \end{bmatrix} = \begin{bmatrix} \sum_w H_x^2 & \sum_w H_x H_t \\ \sum_w H_x H_t & \sum_w H_t^2 \end{bmatrix} \tag{2-1}$$

where H_x and H_t are partial derivatives along the spatial and temporal dimensions respectively. The rotation angle θ of Γ indicates the direction of a gray level change in w. We can rewrite Equation 2-1 as:

$$\Gamma = \begin{bmatrix} \lambda_x & 0 \\ 0 & \lambda_t \end{bmatrix} = R \begin{bmatrix} J_{xx} & J_{xt} \\ J_{xt} & J_{tt} \end{bmatrix} R^T \tag{2-2}$$

where

$$R = \begin{bmatrix} \cos\theta & \sin\theta \\ -\sin\theta & \cos\theta \end{bmatrix}$$

From Equation 2-2, since we have three equations with three unknowns; thus, θ can be solved and expressed as:

$$\theta = \frac{1}{2}\tan^{-1}\frac{2J_{xt}}{J_{xx}-J_{tt}} \tag{2-3}$$

The local orientation Φ of a w in slices is computed as:

$$\phi = \begin{cases} \theta - \dfrac{\pi}{2} & \theta > 0 \\ \theta + \dfrac{\pi}{2} & otherwise \end{cases} \qquad \phi - \left[-\dfrac{\pi}{2}, \dfrac{\pi}{2} \right] \tag{2-4}$$

It is useful to add in a certainty measure to describe how well Φ approximates the local orientation of w .The certainty c is estimated as:

$$c = \frac{\left(J_{xx} - J_{tt} \right)^2 + 4J_{xt}^2}{\left(J_{xx} + J_{tt} \right)^2} = \left(\frac{\lambda_x - \lambda_t}{\lambda_x + \lambda_t} \right)^2 \tag{2-5}$$

and c = [0; 1]. For an ideal local orientation, c = 1 when either $\lambda x = 0$ or $\lambda t = 0$. For an isotropic structure i.e., $\lambda x = \lambda t$, c = 0.

The distribution of local orientations across time inherently reflects the motion trajectories in an image volume. A 2D tensor histogram $M(\Phi; t)$ with the dimensions as a 1D orientation histogram and time, can be constructed to model the distribution. Mathematically, the histogram can be expressed as:

$$M(\phi, t) = \sum_{\Omega(\phi, t)} c(\Omega) \tag{2-6}$$

where $\Omega(\Phi; t) = \{ H(x; t) | \Gamma(x; t) = \Phi \}$ which means that each pixel (in slices) votes for the bin $(\Phi; t)$ with the certainty value c. The resulting histogram is associated with a confident measure of:

$$C = \frac{1}{T \times M \times N} \sum_{\phi} \sum_{t} M(\phi, t) \tag{2-7}$$

where T is the temporal duration and $M \times N$ is the image size. In principle, a histogram with low C should be rejected for further analysis.

Motion trajectories can be traced by tracking the histogram peaks over time. These trajectories can correspond to (i) object and/or camera motions; and (ii) motion parallax with respect to different depths. Figure 2-5 shows two examples, in (a) one trajectory indicates the non-stationary background, and one indicates the moving objects; in (b) the trajectories correspond to parallax motion sequence, Equation 2-8 is employed to detect the zoom.

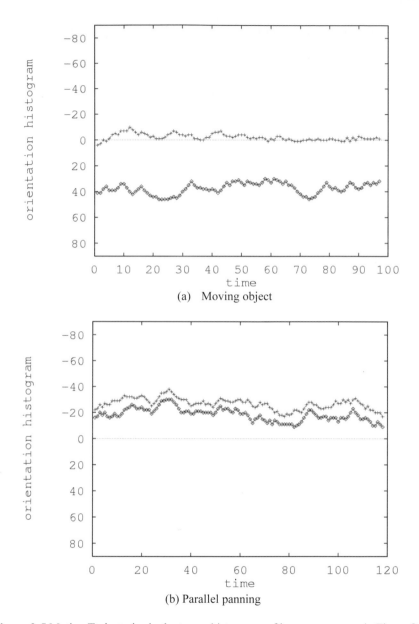

Figure 2-5 Motion Trajectories in the tensor histograms of image sequences in Figure 2-1

The tensor histogram offers useful information for characterizing dominant motions. A sequence with static or slight motion has a trajectory at $\Phi = [-\Phi_a, \Phi_a]$. Ideally, Φ_a should equal 0. The horizontal slices of a panning sequence form a trajectory at $\Phi > \Phi_a$ or $\Phi < -\Phi_a$. If $\Phi < -\Phi_a$, the camera pans to the right; if $\Phi > \Phi_a$, the camera pans to the left. A tilting sequence is similar to a panning sequence, except that the trajectory is traced in the tensor histogram generated by vertical slices. Throughout the experiments, the parameter Φ_a is empirically set to $\pi/36$. A zoom operation, instead of being modeled as a single trajectory, is detected by:

$$\frac{\sum_\phi \sum_{t>0} M(\phi,t)}{\sum_\phi \sum_{t<0} M(\phi,t)} \approx 1 \qquad (2\text{-}8)$$

where the tensor votes are approximately symmetric at $\Phi = 0$.

Pseudo-Color-Object based Representation

Pseudo-color-object based shot representation was introduced to calculate the semantic correlation of consecutive shots [30]. The representation is calculated by calculating the color histogram of each frame in a shot, and identifying the dominant colors of each frame and the shot. Compared to the temporal slice based representation, this representation focuses more on temporal changes reflected in dominant colors in a shot. Therefore, more information than just motion is embedded in this representation.

First, the color histogram for each frame in a shot is calculated using a given color space. *HSV* color space was used in [30] since it is natural and approximately perceptually uniform and one can define a quantization of *HSV* to produce a collection of colors that is compact and complete. In this method, the *HSV* color space is quantized by a 3D Cartesian coordinate system with 10 values for X and Y, 5 values for Z (the lightness) , as shown in Figure 2-6 because: (1) the *HSV* space is cylindrical and (2) in the cylindrical *HSV* color space, the similarity between two colors given by indices $(h1,s1,v1)$ and $(h2,s2,v2)$ is given by the Euclidean distance between the color points $(x1,y1,z1)$ and $(x2,y2,z2)$, respectively. The granularity of the color quantization will affect the extraction of dominant objects. A fine quantification will be able to discriminate between more objects, while it may also cause the extraction of dominant objects being sensitive to lighting dominant objects between frames which may result in loss of tracking of dominant objects .

To determine the dominant colors of a video shot, pixels from each frame of the shot are projected into the quantized HSV color space. The normalized distribution of these pixels in the 3-D color space thus forms a normalized 3D color histograms for the frame. All dominant local maximum points in the 3-D color histogram are identified; and a sphere surrounding each local maximum point within a small neighborhood (with diameter of 3 quatization units) in the color space is defined as a color object. These colors objects (top 20 in our implementation) with the largest numbers of pixels are identified as dominant objects. These dominant objects capture the most significant color information of a frame and are more resilient to noise. Then a 3-D *dominant color histogram, $hist_k(k, x, y, z)$* is formed for each frame by counting only the pixels included in dominant color objects, where k denotes the frame number, and (x, y, z) denotes a color bin. No object segmentation is performed in the spatial domain, though the segmentation in *HSV* color space could be mapped back to a frame image, leading to a spatial segmentation; rather, pixels falling into the dominant regions are considered an color object in the color space, which seldom represents a spatial object in a frame.

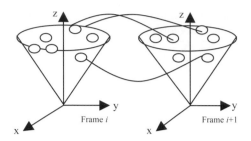

Figure 2-6 Color object segmentation and tracking

Then, color objects in consecutive frames are tracked in the color space to identify the dominant objects of a shot. If the centers of two color objects in two consecutive frames are sufficiently close, these two color objects are recognized as the same color object. Such a color tracking process will continue until all frames in the shot are tracked. After tracking, only the color objects that have longer duration in a shot are retained as dominant objects. In the words, we form an overall dominant color histogram for each shot, $hist_d^a$ (x, y, z) (a denotes a shot), consisting of only dominant color objects that are not only dominant in a frame, but are also dominant across the entire shot. To give more weight to color objects with a longer duration in a shot since they are more dominant, the histogram bins, corresponding to each dominant objects are weighted by its relative duration in a shot as:

$$hist_d^A(x,y,z) = hist_d^a(x,y,z) \times d_l / d_0 \qquad (2-9)$$

where d_0 is the duration of the shot, and d_l is duration of the dominant color object with color (x, y, z). Also, $hist_d^a$ (x, y, z) is normalized by normalizing the mean size of each dominant color object within the shot. Therefore, the dominant color histogram for a shot represents both structural content in a frame and temporal content in a shot.

Also, these dominant color objects often represent dominant objects or background in a shot. Hence, the correlation between the color objects in two shots is a good representation of the correlation between the two shots.

Perceived Motion based Representation

Usually, global motion and statistical motion representation can only characterize dominant motion or camera motion. However, human viewers pay little attention to camera operations, though camera operations are used to direct the viewers' attention. Viewers perceive video content more from object motion, hence a perceived motion representation was proposed in [31]. The perceived motion based representation of video content matches human perception. Also, it avoids object segmentation and global motion estimation in computing the representation which are both computationally expensive and error prone.

For an object in a shot, the more intensive the motion, and the longer the duration of its appearance, the easier it is for humans to perceive the object. The motion energy at each pixel or block position (i,j) of motion vector field can be represented by the average of the motion magnitude over the duration of a shot after removing atypical samples(see Figure 2-7). On the other hand, the spatio-temporal consistency of motion directions reflects the intensity of global motion: the more consistent the motion directions in a shot, the higher the intensity of global motion. The spatio-temporal motion consistency at each position (i,j) can be obtained by tracking the variation of the angle in a spatial window and along the temporal axis, as shown in Figure 2-7). Based on these two observations, a measure called *perceived motion energy* (PME) is designed to represent the motion content of a video that agrees with human perceptions.

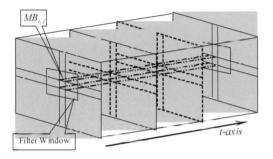

Figure 2-7 Computation of perceived motion energy: motion vectors around each block $MB_{i,j}$ are spatially filters within a window and temporally filtered over the duration of a shot to obtain motion intensity and consistency.

To compute the perceived motion energy of a shot, the motion vectors are spatially filtered to eliminate atypical vectors, which cause inaccurate energy accumulation. In [31], a modified median filter was used for this purpose. As illustrated in Figure 2-7, the magnitude of motion vector at block $MB_{i,j}$ is computed as:

$$Mag_{i,j} = \begin{cases} Mag_{i,j} & \left(if\ Mag_{i,j} \le Max\,4th(Mag_k) \right) \\ Max\,4th(Mag_{i,j}) & \left(if\ Mag_{i,j} > Max\,4th(Mag_k) \right) \end{cases} \qquad (2\text{-}10)$$

When the list is sorted in descending order, the function $Max4th(Mag_k)$ returns the fourth value of all magnitude values in the filter window.

The motion energy at each block position (i,j) is accumulated over the duration of a shot. This is achieved by a temporal energy filtering process. In [31], an alpha-trimmed filter is adopted to filter in a 3-D spatio-temporal tracking volume, as illustrated in Figure 2-7 with the spatial window size of W_t^2 and the temporal duration of L_t. The $Mag_{i,j}$ values in the tracking volume are sorted first. After the values at two ends of sorted list are trimmed, the remaining magnitudes are averaged to form motion energy as follows:

$$MixEn_{i,j} = \frac{1}{(M - 2 \cdot \lfloor \alpha M \rfloor) \cdot W_t^2} \sum_{m=\lfloor \alpha M \rfloor + 1}^{M - \lfloor \alpha M \rfloor} Mag_{i,j}(m) \qquad (2\text{-}11)$$

where M is the total number of motion vectors accumulated in tracking volume, and αM equals to the largest integer not greater than αM; and $Mag_{i,j}(m)$ is the magnitudes value in the sorted list of tracking volume. The trimming parameter α $(0 \le \alpha \le 0.5)$ controls the number of data samples excluded from the accumulating computation. $MixEn_{i,j}$ is referred to as mixture energy since energies of both object and background motions are mixed in this number. In order to form a motion energy spectrum, the mixture energy is normalized into range [0,1].

To extract object motion or perceived motion energy from the mixture energy $MixEn_{i,j}$, a global motion filter is employed. If we consider the angle value of a motion vector as a stochastic variable, and its probability distribution function is computable, the consistency of motion directions in a tracking volume can be measured by entropy. The normalized entropy reflects the ratio of camera motion to object motion: the higher the entropy, the poorer the consistency of motion directions.

The probability distribution function of motion angel variation can be obtained from the normalized motion direction histogram. The 2π range of directions is quantized into n angle ranges. Then, number of samples in each range is accumulated over the tracking volume at each block position (i,j) to form an direction histogram with n bins, denoted by $AH_{i,j}(t)$, $t \in [1, n]$. The probability distribution function $p(t)$ is defined as:

$$p(t) = \frac{AH_{i,j}(t)}{\sum_{k=1}^{n} AH_{i,j}(k)} \tag{2-12}$$

In Equation 2-12, the direction entropy, denoted by $AngEn_{i,j}$, can be computed as:

$$AngEn_{i,j} = -\sum_{t=1}^{n} p(t) \log p(t) \tag{2-13}$$

where the value range of $AngEn_{i,j}$ is $(0, \log n)$. When $p(t)=1/n$, $AngEn_{i,j}$ reaches the maximum value $\log n$, the normalized direction entropy can be considered as a ratio of global motion, denoted by $GMR_{i,j}$,

$$GMR_{i,j} = \frac{AngEn_{i,j}}{\log n} \tag{2-14}$$

where $GMR_{i,j}$ values rang in $(0,1)$. When $GMR_{i,j}$ approaches 0, it implies the camera motion is dominant in the mixture energy $MixEn_{i,j}$.

Finally, the perceived motion energy at each block position (i,j) of defined as:

$$PME_{i,j} = GMR_{i,j} \times MixEn_{i,j} \tag{2-15}$$

If all $PME_{i,j}$ values, at all block positions are quantized to 256 levels of gray, a *perceived motion energy spectrum* (PMES) image is generated for a shot. This PMES image is the motion content representation of the shot.

In a PMES image, light denotes high energy, namely, the lighter the region in image, the more intensive the motion in this region. Figure 2-8 shows some examples of PMES image. The regions in PMES images with high values usually correspond to moving objects. In other words, PMES highlights moving objects and distinguishes them from camera or background movement. It is for this reason that PMES values represent human perceived motion energy. As studied in [31], PMES is a powerful method for representing the motion content of video shots. PMES is effective when used in motion-based video retrieval and activity-based video abstraction and summarization.

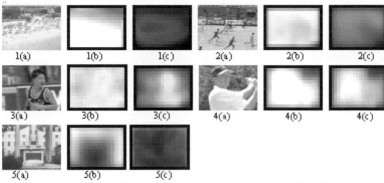

Figure 2-8 Examples of PMES images. (a) Key-frames of shot, (b) MinEn; (c) PMES; 1. Pure panning, 2. Panning with small active objects, 3. Tracking, 4. Zooming with intensive object motion, and 5. Pure zooming.

Object Based Representation

If a query to a video database involves objects in a video clip, the video frame-based representation described above does not provide sufficient resolution to support such queries. Therefore, incorporating object-based primitives into video content representation is advantageous when dealing with queries about objects and their motions.

In the object-augmented representation, the main attributes attached to key or dominant objects are motion, shape and life cycle, as well as other image features such as color and texture. The life cycle of an object is the duration from the time (or relative frame number) when the object appears into a shot (birth) to the time when the object disappears (death). The object motions can be modeled as affine. Furthermore, the object motion can be decomposed into a global component and a local/object-based component to reflect motion relative to the background and other objects in the scene. This decomposition can be easily completed with a simple matrix transformation on the affine parameters. Without this distinction, the object motion would instead represent motion relative to the image frame. This decomposition provides a more meaningful and effective description for retrieval.

We summarize the object descriptors for video indexing in Figure 2-9. There have been many works about proposed object-based representation schemes [14, 19, 29, 69]. MPEG-7 standard adopted a similar object description scheme. In addition to object-based motion descriptors and a general description of motions or activities in shots, all the features discussed in the previous sub-section are accounted for in this representation. The shot-based motion feature set complements the object-based motion features, especially in shots where the key-objects cannot be reliably detected.

Shot
     Shot-ID: S_m (a unique index key of the shot m)
     Primitives: $\{f_s(1), f_s(2), ..., f_s(N)\}$

Key-frames: $\{KF_m(1), KF_m(2), ..., KF_m(J)\}$
Key-objects: $\{\mathbf{KO_m(1), Ko_m(2), ..., KO_m(K)}\}$ }
Object
    Object-ID: o_m (a unique index key of the object)
    Shape: α (alpha map of the object)
    Life: $\{F_b, F_d\}$
    Primitives: $\{f_o(1), f_o(2), ..., f_o(O)\}$
    Motion: motion model parameters and/or trajectory

Figure 2-9 Adding object-based primitives the shot representation scheme shown in Figure 2-5 and 2-6

2.4 Video Abstraction and Summarization

Video content abstraction refers to the process of extracting representative visual information about the landscape or structure of a video, which is shorter than the original video. In this section, we discuss approaches to visual abstraction of video data: key-frames and video highlights.

KEY-FRAME EXTRACTION
Key-frames are still images extracted from the original video data that best represent the content of shots in an abstract manner. When text documents are indexed, key words or summaries are used as index entries to sentences, paragraphs, chapters or entire documents. Similarly, in video indexing, we can extract key-frames and key-sequences as entries for shots, scenes or stories. In addition to content representation as discussed in the last section, key-frames are critically important for video browsing.

Key-frames have been frequently used to supplement the text of a video log [39], but in the past identifying them was done manually. The effectiveness of this method depends on how well the frames in the sequence were selected. The image frames within a sequence are not all equally descriptive. Certain frames may provide more information about the objects and actions within the shot than other frames. In some prototype systems and commercial products, the first frame in each shot is used as the only key-frame to represent the shot content. This representation reduces the data volume, however its representation power is very limited because it does not provide sufficient information about the actions presented in a shot, with the exception of shots with no change or motion.

Key-frame-based representation interprets video abstraction as a problem of mapping an entire segment (both static and motion content) to some small number of representative images. The challenge is that the extraction of key-frames needs to be automatic and content-based so that they maintain the important content of the video while removing all redundant information. In theory, semantic primitives of video, such as interesting objects, actions and events, should be used. However, such general semantic analysis is not feasible, especially when information from soundtracks and/or close caption is not available. In practice, we have to rely on low-level image features and other readily available information.

An effective approach to key-frame extraction, based on the temporal variation of low-level image features - such as color histograms and motion information - has been proposed by Zhang et al. [64, 65, 67, 68]. The number of key-frames needed to represent a segment is based on temporal variation of the video content in the segment: if there is a large temporal variation in content, there should be more key-frames, and vice versa. That is, after shot segmentation, key-frames in a shot will be selected based on the amount of temporal variation in the color histograms, and motion in reference to the first frame or the last selected key-frame of the shot. This approach achieves real-time processing speed, especially when an MPEG compressed video is used, with reasonable accuracy according to the user studies. [64, 68].

In this approach, frames in a shot will be compared sequentially - in term of color histogram changes against the last key-frame or the first frame of the shot - as they are processed, based on their similarities defined by the color histogram. If a significant content change occurs, the current frame will be selected as a new key-frame. Such a process will be iterated until the last frame of the shot is reached. In this way, any significant action in a shot will be captured by a key-frame, while static shots will result in only one key-frame. In addition, information about dominant or global motion, resulting from camera operations and large moving objects, is added into the selection process according to a set of rules. For a zooming-like sequence (zooming, dollying and perpendicular motion of large objects), the first and last frames will be usually selected as key-frame. One presents a global view, and the other a more focused view. For a panning-like sequence (panning, tilting and tracking), the number of frames to be selected will depend on the scale of panning: ideally, the spatial context covered by each frame should have little overlap, or each frame should capture a different, but sequential part of object activities.

Another way of key-frame selection is through the clustering of video frames [25], which employ a partitional clustering algorithm with cluster-validity analysis, to select the optimal numbers of clusters for shots. The resulting clusters are optimal in term of inter- and intra-cluster distance measures. The frame which is the closest to a cluster centroid is selected as the key-frame. While clustering is efficient, the curse of dimensionality needs to be handled carefully. In principle, the number of frames to be clustered has to be several times larger than the dimensions of the feature space. To be effective, feature dimensionality reduction can be done prior to clustering. For instance, Chang et al. [11], encoded the shot information in a matrix A and decomposed the matrix with a singular value decomposition (SVD), $A=U\Sigma V^T$. Each column in matrix A is the feature vector of a frame. The left singular matrix U consists of eigen images of a shot, while the right singular matrix V quantitatively measures the content richness of each frame. Since frames in a shot are highly correlated, normally only few principle components are selected. This not only reduces the feature space dimension but also suppresses undesired noise. As a result, clustering can be performed more effectively and efficiently in the projected feature space. Despite these advantages, SVD is computationally intensive.

Recent advances in key-frame selection utilize graph theory [12] and curve splitting [18, 70]. In general, these approaches represent a frame as a point in the high dimensional feature space. The idea is to select a subset of points that can either cover the remaining points within a feature distance or to capture the significant changes of content in a shot. Chang et al. [12] viewed a shot as a proximity graph and each frame is a vertex in the graph. The problem of key-frame selection can be equivalent to the vertex cover problem. Specifically, we want to find a minimal cover of vertices that minimizes the total feature distance between the vertices and their neighboring points. This problem is NP-complete. Chang et al., proposed sub-optimal solutions based on the greedy approach and rate-distortion performance. Unlike Chang, DeMenthon et al. [18] and Zhao et al. [70] turned the key-frame selection into a curve splitting problem. A shot is viewed as the feature trajectory or curve of high dimensional points. The task is to detect the junctions or break points of a curve as key-frames. Based on this idea, DeMenthon further proposed a hierarchical view of key-frames by recursively identifying curve junctions while simplifying a curve from fine to coarse levels.

In these approaches, the density of key-frames or the abstraction ratio can be controlled according to the user's need by adjusting the threshold for determining "significant" changes [64, 68]. However, the exact number of resultant key-frames will be determined a posteriori by the actual content of the input video, which is arguably a disadvantage of this type of key-frame extraction approach [24]. On the other hand, predefining the absolute number of key-frames without knowing the content of video may not be desirable: assigning two key-frames for a talking head sequence of 30 minutes should still be considered as having too much redundancy! In addition, assigning the same number of key-frames to, for instance, two video sequences of same length does not guarantee the same level of visual abstraction since the contents of the two sequences may have different levels of abstraction, and/or totally different levels of activities. Therefore, controlling the abstraction ratio or key-frame density is a more robust and useful approach.

A compromise to meet the need for a pre-defined number of key-frames, while maintaining the content-based selection criteria and a constant abstraction ratio amongst a given set of video sequences, is to set up a maximum number of key-frames. An initial set of key-frames can be selected at a given abstraction ratio using the approach discussed above. If the number of key-frames exceeds the maximum, frames with a high similarity to their immediate neighboring two frames can be filtered out.

This sub-section has focused on the extraction of key-frames, and criteria to ensure that key-frames are extracted for each shot. The key-frames represent the content of shots and should be referred a key-frame of shots. To extract an even smaller set of key-frames that can function as the visual abstract of scenes or stories, is an even more challenging problem, similar to that of detecting scene or story boundaries.

VIDEO HIHGLIGHTING

Video summarization highlights or summarizes sequences of the video contents. This is a relatively new research area and requires a high of level content analysis. A successful approach uses information from multiple sources, including sound, speech, transcripts and image analysis of video. Researchers, [47] working on documents with textual transcriptions have suggested producing video abstracts by first abstracting the text using classical text skimming techniques and then looking for the corresponding parts in the video.

A successful application of this type of approach is the *Informedia* project, in which text and visual content information were fused to identify video sequences that highlighted the important contents of video [26]. More specifically, low-level and mid-level visual features, including shot boundary, human face, camera and object motion and subtitles of video shots are integrated with keywords, spotted from text obtained from close caption and speech recognition, using the following procedures: (1) Keyword selection using the well-known TF-IDF technique to skim audio; (2) Sequence characterization by low-level and mid-level visual features; (3) Selecting a number of keywords according to the required skimming factor; (4) Prioritizing image sequences located in close proximity to each selected keyword, for instance, frames with faces or text; static frames following camera motion; frames with camera motion human faces or text; frame at the beginning of the scene; Finally, (5) composite a skimmed sequence with selected frames. Experiments using this skimming approach have shown impressive results in generating video summaries of limited types of documentary video which have very clean speech or text (close caption) contents, such as educational videos, news or parliamentary debates [26]. However, this approach is not appropriate for videos with a soundtrack containing more than just speech, such as home videos. Therefore, extracting a video summary from general video materials remains a challenging research topic.

Other work in video highlight extraction are more application tailored where domain knowledge are used. Two specific domains that have captured researchers' attention are sport videos [41] and scene surveillance [38]. In [41], a method was proposed to detect the highlight of baseball videos by audio features. The audio information being utilized are announcers' excited speech and the special sound effect like baseball hits. In surveillance applications, functionalities like detecting abnormal or dangerous objects for alarm generation require video content analysis. Stringa and Regazzoni [45] presented a surveillance system to detect abandoned objects and to highlight the people who left them in an indoor environment. In this system, foreground objects can be easily segmented since the background scene information is both known and static and the camera is mounted at fixed location. A multilayer perceptron is trained off-line to classify the foreground objects as abandoned objects, person, lighting or structural changes. By motion analysis, the person who left an abandoned object will be highlighted and the shape and color information of the object will be indexed for future reference.

2.5 Content-based Video Retrieval and Browsing

Once a collection of video sequences have been parsed into shots and scenes, and the content features of the sequences have been extracted and represented as described in Section 3.3, then, tools for content-based retrieval and browsing of video can be built to utilize these content meta-data. Also, video abstracts can be used in video browsing. Similar to the case of image retrieval, the retrieval process and, especially, browsing process, should be interactive and iterative, with the system accepting feedback. This section discusses similarity measures for video content comparison and schemes for content-based retrieval and browsing of video.

SIMILARITY MEASURES

Compared with feature-based image retrieval, combing multiple features to define the content similarity between two video sequences of shots for retrieval is even more challenging since more features, often with differing levels of importance are involved. Also content comparison can be performed based on key-frame-based features, shot-based temporal and motion features, object-based features, or a combination of the three.

When content comparison is performed using a key-frame-based content representation scheme, we define shot similarity based on the similarities between the two key-frame sets. If two shots are denoted as S_i and S_j, and their key-frame sets as $K_i = \{f_{i,m}, m = 1, ..., M\}$ and $K_j = \{f_{j,n}, n = 1, ... , N\}$, then the similarity between the two shots can be defined as:

$$\mathbf{s}_k(\mathbf{S}_i, \mathbf{S}_j) = \max[s_k(f_{i,1}, f_{j,1}), s_k(f_{i,1}, f_{j,2}), ..., s_k(f_{i,1}, f_{j,N}), ...,$$
$$s_k(f_{i,1}, f_{j,1}), s_k(f_{i,m}, f_{j,2}), ..., s_k(f_{i,m}, f_{j,N})] \tag{2-16}$$

where s_k is a similarity metric between two key-frames defined by any one or a combination of the image features; and there are a total of $M \times N$ similarity values, from which the maximum is selected. This definition assumes that the similarity between two shots can be determined by the pair of key-frames which are most similar, and it will guarantee that if there is a pair of similar key-frames in two shots, they are considered similar.

Another definition of key-frame based shot similarity is:

$$\mathbf{s}_k(\mathbf{S}_i, \mathbf{S}_j) = \frac{1}{M} \sum_{m=1}^{M} \max[s_k(f_{i,m}, f_{j,1}), s_k(f_{i,m}, f_{j,2}), ..., s_k(f_{i,m}, f_{j,N})] \tag{2-17}$$

This definition states that the similarity between two shots is the sum of the most similar pairs of key-frames. If only one pair of frames match, this definition is equivalent to Equation 2-16.

Similarly, when object-based content representation scheme, as illustrated in Figure 2-9, is used, the match between the query and candidate shots is based on the visual attributes of individual objects and/or their compositions in shots. That is, for queries, searching for the presence of one or more objects in a shot, the similarity between the query and the candidate is defined as the similarity between the key-objects. Formally, if two shots have key-object sets as $K_i\{o_{i,m}, m = 1, ..., M\}$ and $K_j\{o_{j,n}, n = 1, ... , N\}$, then the similarity between the two shots can be defined as:

$$\mathbf{s}_k(\mathbf{S}_i, \mathbf{S}_j) = \frac{1}{M} \sum_{m=1}^{M} \max[s_k(o_{i,m}, o_{j,1}), s_k(o_{i,m}, o_{j,2}), ..., s_k(o_{i,m}, o_{j,N})] \tag{2-18}$$

where s_k is a similarity metric between two objects. Similar to Equation 2-17, this definition states that the similarity between two shots is the sum of the similarities of the most similar key-object pairs.

In addition to key-frame or key-object based similarity, the shot-based feature set, $\{f_s(1), f_s(2), ..., f_s(N)\}$, as in the representation illustrated in Figure 2-5, is also used to define similarity between video sequences, similar to that in the feature-based image retrieval. For instance, when the pseudo-color-object representation is used, the similarity score between two shots is calculated by performing the histogram intersection between two dominant color histograms of the two shots [30].

Matching can also be done in a brutal force way, i.e., aligning and matching frames (or key-frames) across time. For instance Tan et al, employed dynamic programming to align two video sequences of different temporal length [48]. Another sophisticated ways of similarity measure include spatio-temporal matching [13, 15] and nearest feature line matching [70]. The spatio-temporal matching was proposed by Chang et al. to measure the similarity of video objects which are represented as trajectories and trails in the spatial and temporal domains [13]. Recently, Dagtas et al. further presented various trajectory and trail based models for motion-based video retrieval. Their proposed models emphasize both the spatial and temporal scale invariant properties for object motion retrieval. In addition, Zhao et al. [70] described the shot similarity measure by employing the concept of nearest feature line. Initially all frames in a shot are viewed as a curve. Frames that are located at the corners are extracted as key-frames. The key-frames are connected by lines and form a complete graph. Given a frame as a query, the distance from the frame to the graph is the nearest perpendicular projected distance among the frames to lines.

The overall similarity between two video sequences is often defined as a combination of weighted multi-feature-based similarities. The weight of each similarity, based on a particular feature, especially temporal features *vs.* image features, for a given query is currently left to the user [14, 64]. This is difficult for users, especially for inexperienced users. As a result, retrieval performance is often limited. Research on effective visual feature-based similarity for video retrieval is at an even earlier stage than that for image retrieval, and more progress is required before we can applied feature-based retrieval schemes to large scale, heterogeneous video databases.

SHOT CLUSTERING

Clustering can be used to abbreviate and organize the content of videos, and provide an efficient indexing scheme for video retrieval since similar shots are grouped under the same cluster. Clustering is also a necessary step in the detection of scenes in the video parsing process as discussed in Section 3.1. Partitioning-clustering methods are suited to the clustering of a large number of shots to build an index with different levels of abstractions, since they are capable of finding the optimal cluster at each level are more suited to abstracting data items [20]. A similar approach has been proposed for video shot grouping [63, 71]. This approach is flexible such that different feature sets, similarity metrics and iterative clustering algorithms can be applied at different levels.

One implementation of this approach is to use an enhanced *K*-means clustering algorithm incorporating fuzzy classification, which allows assignment of data items at the boundary of two classes according to the membership function of the data item of all the classes. This is useful especially at higher levels of hierarchical browsing, where users expect all similar data items to be under a smaller number of nodes [71]. The clustering can be based on key-frames and motion features of video shots. Implementation and evaluation of this clustering approach using Self-Organization method can be found in [63, 71]. The advantage of SOM is its good classification performance and learning ability without prior knowledge, which have been shown by many researchers. Another benefit of using SOM is that the similarities among the extracted classes can be seen directly from the two-dimensional map. This will allow horizontal exploring as well as vertical browsing of the video data, which is very useful when we have a large number of classes at lower levels.

Other proposed approaches that employ a clustering structure for retrieval include [12, 38]. For instance, Ngo [38] proposed a two-level hierarchical clustering structure to organize the content of sport videos. The top level is clustered by color features while the bottom level is clustered by motion features. The top level contains various clusters including wide-angle, medium-angle and close-up shots of players from different teams. The shots inside each cluster are partitioned to form sub-clusters in the bottom level according to their motion similarity. In this way, for example, the sub-cluster of a close-up shot can correspond either to "players running across the soccer field", or "players standing on the field". Such organization facilitates not only video retrieval and browsing, but also some high-level video processing tasks. For instance, to perform player recognition, only those shots in the cluster that correspond to close-up shots of players are picked up for processing. Through empirical results, Ngo showed that the cluster-based retrieval, in addition to speed up retrieval time, will generally give better results especially when a query is located at the boundary of two clusters.

BROWSING SCHEMES

Interactive browsing of full video contents is probably the most essential feature of new forms of interactive access to digital video. Content-based video browsing tools should support two different nonlinear approaches to accessing video source data: sequential and random access. In addition, these tools should accommodate two levels of granularity, overview and detail, along with an effective bridge between the two levels. Such browsing tools can only be built by using the structure and content featured obtained in the video parsing, feature extraction and content-based retrieval processes as discussed in the previous sections.

Four main types of browsing tools have been built based on different structural and content features of video: (1) time-line display of frames; (2) light-table of video icons; (3) hierarchical browser; (4) graph based storyboard. Time-line based browsers have been favored by users in video production and editing systems, for which time-line interfaces are classical. Some browsers rest on a single shot-based image component line [3, 65]; but the multidimensional character of video, calling for the multi-line representation of the contents, has been stressed by researchers working in the frame of the Muse toolkit [32, 63]. This has been systematized in the strata model proposed by Aguierre-Smith and Davenpoprt [1]. Time-line browsers are limited because it is difficult to zoom out while keeping a good image visibility, as a result the time-scope of what is actually displayed at a given moment on screen is relatively limited.

The light-table kind of video browser is often called a clip window, because the video sequence is spread in space and represented by video icons which function like a light table of slides [62, 65]. In other words, the display space is traded for time to provide a rapid overview of the content of a long video. A window may contain sequentially listed shots or scenes from a video program, a sequence of shots from a scene, or a group of similar or related shots from a stock archival. A clip window browser can also be constructed hierarchically so that it is just like the Windows file system used in PC operating systems. That is, each icon in a clip window can be zoomed in to open another clip window, and each icon represents the next level and a finer segment of video sequences [62].

One of the first attempts at building hierarchical browsers called the Video Magnifier [34] used successive horizontal lines, each of which offered greater time detail and a narrower time scope for selecting images from the video program. To improve the content accessibility of such browsers, the structural content of the video obtained in video parsing is utilized [64, 67]. As shown in Figure 2-10, videos are segmented and accessed as a tree. At the top of the hierarchy, an entire sequence or program is represented by five key-frames, each corresponding to one of five groups, each of which contain an equal number of consecutive shots. Any one of these segments may then be subdivided to create the next level (shots in the case shown in Figure 2-10) of the hierarchy. As we descend through the hierarchy, our attention focuses on smaller groups of shots, then single shots, and finally the key-frames of a specific shot, and all frames of a shot. We can also view sequentially any particular segment of a video selected from this browser at any level of the hierarchy by launching a video player.

The hierarchical browser shown in Figure 2-10 is a very powerful tool for the fast assessment of content of a video sequence or programs, since it uses the structure information of video. The shots at the higher levels are grouped only according to their sequential relations, not their content similarity. Thus, browsing through a large collection of video clips is inconvenient because similar shots are not grouped together. Therefore the browser is further extended to use similarity information between shots or sequences which have been pre-defined or clustered. That is, when a list of video programs or clips are accessed for browsing, the system clusters shots into classes, consisting of shots of similar visual content, using either key-frame and/or shot features. After clustering, each class of shots is represented by a key-frame determined by the centroid of the class, which is then displayed at the higher levels of the hierarchical browser. Figure 2-10 shows the data structure and browser layout for such a hierarchical browser. With this type of browser, the viewer can get a rough sense of the content of the shots in a class even without moving down to lower level of the hierarchy [68, 71].

An alternative approach to hierarchical browsers is the class based transition graph, proposed by Yeung et al [58, 59]. By clustering visually similar shots into scenes, a directed graph (whose nodes are scenes) is constructed, as shown in Figure 2-11. The resulting graph is displayed for browsing, and each node is represented by a key-frame extracted from one of the shots in the node. This graph can be edited for simplification by a human operator. The drawbacks of this approach are poor screen space use due to the graph layout problem, and in the fact that the linear structure of the document is no longer perceptible.

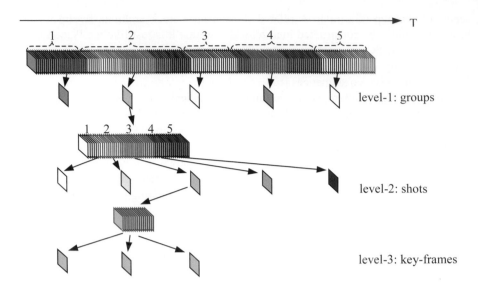

Figure 2-**10** Data structure and browser layout for key-frame-based hierarchical browser

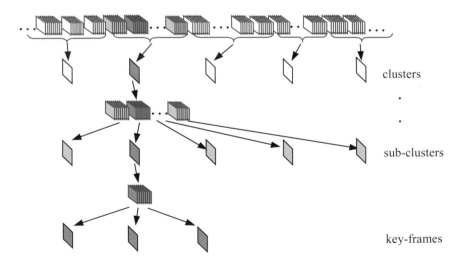

Figure 2-**11** Data structure and browser layout for similarity-based hierarchical browser

2.6 Conclusion

In this chapter, we have discussed a variety of existing techniques for video structure parsing, motion analysis, video content representation using visual features and content-based video retrieval and browsing. As for image retrieval, the current automated solutions video browsing and retrieval still depend heavily, if not totally, on use of low-level visual features. On the other hand, current structure parsing, content representation and browsing, and algorithms and tools can increase our productivity in video data browsing, management and applications significantly more than that for image databases. This is because conventional video management and browsing tools do not support non-linear and content-based access.

Information fusion from different sources, such as speech, sound and text, is as important as the visual data in understanding and indexing visual data. Keywords and conceptual retrieval techniques are and will always be an important part of visual information systems.

Application-oriented approaches are the most successful in visual data representation and retrieval researches. One of the most important lessons from the past research is that image and video content analysis, retrieval and management should not be thought of a process that can be fully automated. We should focus in developing video analysis tools to facilitate human users manage visual data more intelligently and efficiently.

Acknowledgement

The author would like to thank Yufei Ma of Microsoft Research Asia for his contribution to this chapter.

References

[1]   T.G. Aguierre-Smith and G. Davenport, "The Stratification system: A Design Environment for Random Access Video", Proc. 3rd Int. Workshop on Network and Operating System Support for Digital Audio and Video, La Jolla, CA, USA, November 1992, pp.250--261.

[2]   P. Aigrain and P. Joly, "The Automatic Real-Time Analysis of Film Editing and Transition Effects and its Applications," Computers & Graphics, January-February, 1994, Vol.18, No.1, pp.93-103.

[3]   P. Aigrain, H.J. Zhang and D. Petkovic, "Content-based Representation and Retrieval of Visual Media: A State-of-the-Art Review", International Journal of Multimedia Tools and Applications, Kluwer Academic Publishers, Vol.3, No.3, 1996.

[4]   P. Aigrain, P. Joly and V. Longueville, "Medium-Knowledge-Based Macro-Segmentation of Video into Sequences," Chapter 8, in Intelligent Multimedia Information Retrieval, M. T. Maybury (editor), AAAI/MIT Press, Cambridge, 1997, pp.159-173.

[5]   A. Akutsu and Y. Tonomura, Video Tomography: An Efficient Method for Camerawork Extraction and Motion Analysis, Proc. ACM Multimedia Conference, San Francisco, October, 1993.

[6]   F. Arman, A. Hsu and M. Y. Chiu, Feature Management for Large Video Databases, Proc. SPIE Conf. Storage and Retrieval for Image and Video Databases I, SPIE, Vol.1908, February, 1993, pp.2-12.

[7]   D. Bordwell and K. Thompson, Film Art: An Introduction, McGraw-Hill, New York, 1993.
[8]   J.S. Boreczky and L.A. Rowe, "Comparison fo video shot boundary detection techniques," Proc. SPIE Conf. Storage and Retrieval for Video Databases IV, San Jose, CA, USA, February, 1995.
[9]   P. Bouthemy and M. Gelgon and F. Ganansia, "A Unified Approach to Shot Change Detection and Camera Motion Characterization," IEEE Trans. on Circuits and Systems for Video Technology, Vol.9, No.7, pp.1030-1044, 1999.
[10]   M.G. Brown, et al, "Automatic content-based retrieval of broadcast news," Proc. of ACM Multimedia'95, San Francisco, November 1995, pp.35-43.
[11]   C.Y. Chang and A. A. Maciejewski and V. Balakrishnan, " Eigendecomposition-based Analysis of Video Images," Proc. of SPIE Conference on Storage and Retrieval for Image and Video Database VII", pp.186-191,1999.
[12]   H.S. Chang and S. S. Sull and S. U. Lee, "Efficient Video Indexing Scheme for Content-based Retrieval," IEEE Trans. on Circuits and Systems for Video Technology",Vol.9,No.8,Dec.1999.
[13]   S.F. Chang and W. Chen and H. J. Meng and H. Sundaram and D. Zhong, "A Fully Automatic Content-based Video Search Engine Supporting Multi-object Spatio-temporal queries," IEEE Trans. on Circuits and Systems for Video Technology,Vol.8, No.5, pp.602-615, 1998.
[14]   S.F. Chang, et al, "VideoQ: A automated content based video search system using visual cues," Proc. of ACM Multimedia'97, Seattle, November 1997, pp.313-324.
[15]   S. Dagtas and W. A. khatib and A. Ghafoor and R. L. Kashyap, "Models for Motion-based Video Indexing and Retreival," IEEE Trans. on Image Processing, Vol.9, No.1, pp. 88-101, Jan. 2000.
[16]   A. Dailianas, R. Allen and P. England, "Comparison of Automatic Video Segmentation Algorithms," Proceedings of SPIE Photonics East, Philadelphia, October 1995.
[17]   M. Davis, Media Streams: "An Iconic Visual Language for Video Annotation," Proc. of Symposium on Visual Languages, Bergen, 1993.
[18]   D. DeMenthon and V. Kobla and D. Doermann, "Video Summarization by Curve Simplification," Proc. of ACM Multimedia 1998,pp.211-218, 1998.
[19]   N. Dimitrova and F. Golshani, "Rχ for semantic video database retrieval," Proc. of ACM Multimedia'94, San Francisco, October 1994, pp.219-226.
[20]   R. Duda and P. Hart, Pattern recognition and scene analysis, Wiley, New York, 1973.
[21]   B. Furht, S. W. Smoliar and H. J. Zhang, Image and Video Processing in Multimedia Systems, Kluwer Academic Publishers, 1995.
[22]   Y. Gong, et al, "Automatic Parsing of TV Soccer Programs," Proc. Second IEEE International Conference on Multimedia Computing and Systems, Washington DC, 15-18 May 1995, pp167-174.
[23]   A. Hampapur, R. Jain and T. E. Weymouth, "Production Model Based Digital Video Segmentation," Multimedia Tools and Applications, 1995, Vol.1, No.1, pp.9-46.
[24]   A. Hanjalic and R. L. Langendijk, "A new key-frame allocation method for representing stored video streams," Proc. of 1st Int. Workshop on Image Databases and Multimedia Search, 1996
[25]   A. Hanjalic and H.J. Zhang, "An Integrated Scheme for Automated Video Abstraction Based on Unsupervised Cluster-Validity Analysis", IEEE Trans. on Circuits and Systems for Video Technology", Vol.9, No.8,pp.1280-1289,Dec.1999
[26]   A.G. Hauptmann and M. Smith, "Text,Speech and Vision for Video Segmentation: The Informedia Project," Working Notes of IJCAI Workshop on Intelligent Multimedia Information Retrieval, Montreal, August 1995, pp.17-22.
[27]   M.E. Hodges, R. M. Sassnett and M. S. Ackerman, "A Construction Set for Multimedia Applications," IEEE Software, January 1989, pp.37-43.
[28]   S.B. Jun and K. Yoon and H. Y. Lee,"Dissolve Transition Detection Algorithm Using Spatio-temporal Distribution of MPEG Macro-Block Types," Proc. of ACM Multimedia, 2000.
[29]   S.Y. Lee and H.M. Kao, "Video indexing – An approach based on moving object and track," Proc. SPIE Conf. Storage and Retrieval for Image and Video Databases, San Jose, February 1993, pp. 25-36.
[30]   T. Lin, H.J. Zhang, "Automatic Video Scene Extraction by Shot Grouping", Proceeding of 2000 IEEE International Conference on Pattern Recognition, 2000.
[31]   Y.F. Ma, H.J. Zhang, "A New perceived motion based shot representation", Proceeding of 2001 IEEE International Conference on Image Processing, 2001.
[32]   W.E. Mackay and G. Davenport, "Virtual Video Editing in Interactive Multimedia Applications," Communications of the A.C.M, Vol.32, No. 9, July 1989.
[33]   A. Merlino, D. Morey and M. Maybury, "Broadcast navigation using story segmentation," Proc. of ACM Multimedia'97, Seattle, November 1997, pp.381-388.

[34]   M. Mills, J. Cohen and Y. Y. Wong, "A Magnifier Tool for Video Data," Proc. INTERCHI'92, ACM, May 1992, pp.93-98.

[35]   A. Nagasaka and Y. Tanaka, "Automatic Video Indexing and Full-Search for Video Appearances," in E. Knuth and I.M. Wegener editors, Visual database Systems, Elsevier Science Publishers, Vol.II, Amsterdam, 1992, pp.113-127.

[36]   C.W. Ngo and T.C. Pong and R. T. Chin, "Detection of Gradual Transitions through Temporal Slice Analysis," Proc. of IEEE International Conference on Computer Vision and Pattern Recognition",Vol.1, pp.36-41, 1999.

[37]   C.W. Ngo, et al., "Motion Characterization by Temporal Slices Analysis," Proceeding of 2000 IEEE International Conference on Computer Vision and Pattern Recognition, pp.768-773, 2000.

[38]   C.W. Ngo, "Analysis of Spatio-Temporal Slice for Video Content Representation." Ph.D thesis, Hong Kong University of Science and Technology,2000.

[39]   C. O'Connor, "Selecting Key Frames of Moving Image Documents: A Digital Environment for Analysis and Navigation", Microcomputers for Information Management, 8(2), pp. 119-133, 1991.

[40]   S. Peleg, J. Herman, "Panoramic Mosaics by Manifold Projection," Computer Vision and Pattern Recognition, pp. 338-343, 1997.

[41]   Y. Rui and A, Gupta and A. Acero, "Automatically Extracting Highlights for TV Baseball Programs," Proc. of ACM Multimedia, 2000.

[42]   S. Sethi and N. Patel, "A Statistical Approach to Scene Change," Proc. SPIE Conf. Storage and Retrieval for Video Databases III, San Jose, February 1995.

[43]   B. Shahraray and D. Gibbon, "Automatic authoring of hypermedia documents of video programs," Proc. of ACM Multimedia'95, San Francisco, November 1995, pp.401-409.

[44]   B. Shahraray, Scene Change Detection and Content-Based Sampling of Video Sequences, IS\&T/SPIE'95 Digital Video Compression: Algorithm and Technologies, San Jose, 1995, February, Vol.2419, pp.2--13.

[45]   E. Stringa and C. S. Regazzoni, "Real Time Video-Shot Detection for Scene Surveillance Applications," IEEE Trans. on Image Processing, V9, No.1,Jan.2000.

[46]   D. Swanberg, C. F. Shu and R. Jain, "Knowledge guided parsing in video databases," Proc. of SPIE Conf. on Storage and Retrieval for Image and Video Databases, San Jose, CA, USA, February, 1993.

[47]   A. Takeshita, T. Inoue, and K. Tanaka, "Extracting text skim structures for multimedia browsing," Working Notes of IJCAI Workshop on Intelligent Multimedia Information Retrieval, Montreal, August 1995, pp.46-58.

[48]   Y.P. Tan and D. D. Saur and S. R. Kulkarni and P. J. Ramadge, "A Framework for Measuring Video Similarity and its Application to Video Query by Example," IEEE Int. Conf. on Image Processing, 1999.

[49]   Y. Taniguchi, A. Akutsu, and Y. Tonomura, "PanoramaExcepts: Extracting and packing panoramas for video browsing," Proc. ACM Multimedia Conference, Seattle, December 1997, pp.427-436.

[50]   L. Teodosio and W. Bender, "Salient Video Stills: Content and Context Preserved," Proc. ACM Multimedia'93, Anaheim, CA, USA, August, 1993.

[51]   Y. Tonomura, et al, "VideoMAP and VideoSpaceIcon: Tools for Anatomizing Video Content," Proc. InterChi'93, ACM, 1994, pp.131--136.

[52]   Y.T. Tse and R.L. Baker, "Global Zoom/Pan Estimation and Compensation for Video Compression," Proc. ICASSP'91, 1991, May, Vol.4.

[53]   H. Ueda, T. Miyatake and S. Yoshizawa, IMPACT: An interactive natural-motion-picture dedicated multimedia authoring system, Proc. CHI'91, ACM, 1991, pp.343--350.

[54]   N. Vasconcelos and A. Lippman, "Statistical Models of Video Structure for Content Analysis and Characterization," IEEE Trans. on Image Processing",Vol.9, No.1, pp.3-19, Jan. 2000.

[55]   J.Y.A. Wang and E. H. Adelson, "Representing Moving Images with Layers," IEEE Tran. on Image Processing, 3(5):625-638, September 1994.

[56]   B.L. Yeo and B. Liu, "A Unified Approach to Temporal Segmentation of Motion JPEG and MPEG Compressed Video," Proc. IEEE Inter'l Conf. on Multimedia Computing and Networking, Washington DC, May, 1995, pp.81-88.

[57]   M. Yeung and B.-L. Yeo, "Video content characterization and compaction for digital library applications," Proc. of SPIE Conf. on Storage and Retrieval for Image and Video Databases V, San Jose, CA, USA, February, 1997.

[58] M.M. Yeung, et al, Video Browsing using Clustering and Scene Transitions on Compressed Sequences, Proc. of IS\&T/SPIE'95 Multimedia Computing and Networking, San Jose, Vol.2417, February 1995, pp.399-413.

[59] M.M. Yueng, B.L. Yeo and B. Liu, "Extracting story units from long programs for video browsing and navigation," Proc. of IEEE International Conference on Multimedia Computing and Systems, June 1996.

[60] R. Zabih, K. Mai and J. Miller, A Robust Method for Detecting Cuts and Dissolves in Video Sequences," Proc. ACM Multimedia'95, San Francisco, November, 1995.

[61] H.J. Zhang A. Kankanhalli and S. W. Smoliar, "Automatic Partitioning of Full-Motion Video," Multimedia Systems, ACM-Sringer, Vol.1, No.1, 1993, pp.10-28.

[62] H.J. Zhang and S. W. Smoliar, "Developing Power Tools for Video Indexing and Retrieval," Proc. SPIE'94 Storage and Retrieval for Video Databases, San Jose, CA, USA, February, 1994.

[63] H.J. Zhang and D. Zhong, "A Scheme for visual feature-based image indexing," Proc. of SPIE conf. on Storage and Retrieval for Image and Video Databases III, pp. 36-46, San Jose, Feb. 1995.

[64] H.J. Zhang, et al, "Video parsing, retrieval and browsing: an integrated and content-based solution," Proc. ACM Multimedia'95, San Francisco, Nov. 5-9, 1995, pp.15-24.

[65] H.J. Zhang, et al, Video Parsing Using Compressed Data, Proc. SPIE'94 Image and Video Processing II, San Jose, CA, USA, February, 1994, pp.142-149.

[66] H.J. Zhang, et al, "Automatic Parsing and Indexing of News Video," Multimedia Systems, ACM, Vol.2, No. 6, 1995, pp.256-265

[67] H.J. Zhang, S. W Smoliar and J. H Wu. "Content-Based Video Browsing Tools", Proc. IS&T/SPIE Conf. on Multimedia Computing and Networking 95, San Jose, CA, 1995.

[68] H.J. Zhang, et al, "An Integrated System for Content-Based Video Retrieval and Browsing", Pattern Recognition, Pergomon Press/Pattern Recognition Society, May 1997.

[69] H.J. Zhang, J. Y. A. Wang and Y Altunbasak, "Content-based video retrieval and compression: A unified solution," Proc. of IEEE Int. Con.e on Image Processing, Santa Barbara, October 1997.

[70] L. Zhao and W. Qi and S. Z. Li and S. Q. Yang and H. J. Zhang, "Key-frame Extraction and Shot Retrieval using Nearest Feature Line (NFL)," *Proc. of ACM International Workshop on Multimedia Information Retrieval, 2000.*

[71] D. Zhong, H.J. Zhang and S.-F. Chang, "Clustering Methods for Video Browsing and Annotation," Proc. of SPIE Conf. on Storage and Retrieval for Image and Video Databases IV, San Jose, February, 1996.

[72] D. Zhong, "Visual Feature Based Image and Video Indexing and Retrieval," MS Thesis, Institute of Systems Science, National University of Singapore, August 1995.

[73] "Description of MPEG-4," ISO/IEC JTC1/SC29 /WG11 N1410, Oct. 1996.

3 RELEVANCE FEEDBACK IN CONTENT-BASED IMAGE RETRIEVAL

Dr. Hongjiang Zhang

In this chapter, we discuss relevance feedback technologies in content-based image retrieval systems. We firstly introduce the need and concept of relevance feedback technologies in content-based image retrieval systems. Then, key issues in relevance feedback as a learning process as well as a set of commonly used relevance feedback algorithms are reviewed in Section 3.2. After that, a framework for integrated relevance feedback and semantic learning in content-based retrieval is described in Section 3.3. Section 3.4 discusses some remaining research problems in relevance feedback for content-based image retrieval.

3.1 Introduction

As introduced in Chapter 1, content-based image retrieval (CBIR) attempts to automate the process of indexing an image database. CBIR approaches work with descriptions based on the inherent properties of images, such as color, texture and shape. However, the retrieval accuracy of today's CBIR algorithms is still limited. The main problem is the gap between low-level image features and semantic contents of images. This problem arises because visual similarity measures, such as color histograms, do not necessarily match the *semantics* of images and human *subjectivity*. Human perception of image similarity is subjective and task-dependent, that is, people often have different semantic interpretations of the same image. Furthermore, the same person may perceive the same image differently at different times. In addition, each type of visual feature tends to capture only one aspect of the image property and it is usually hard for a user to specify clearly how different aspects are combined to form an optimal query.

To address this problem, interactive relevance feedback techniques have been proposed. The key idea is that human perception subjectivity is incorporated into the retrieval process, providing users with the opportunity to evaluate the retrieval results. Queries or similarity measures are automatically refined on the basis of these evaluations. In the last few years, this research topic has become the focus in the CBIR research community.

Relevance feedback, originally developed for textual document retrieval [19], is a supervised active learning technique used to improve the effectiveness of information systems. It uses positive and negative examples from the user to improve system performance. For a given query, the system first retrieves a list of ranked images according to a predefined similarity metrics. Then, the user marks the retrieved images as relevant to the query (positive examples) or irrelevant (negative examples). The system will refine the query based on the feedback, retrieve a new list of images, and present them to the user. Hence, the key issue in relevance feedback is how to use positive and negative examples to refine the query and/or to adjust the similarity measure.

Studies on relevance feedback techniques in document retrieval are limited. However, this area of research has become active during the past three years after being introduced to the CBIR community. As the retrieval accuracy of the majority of CBIR algorithms is extremely low, directly applying the relevance feedback framework developed for textual document retrieval significantly improves the image retrieval accuracy.

3.2   Relevance Feedback Algorithms

There are many issues in relevance feedback approaches that are relevant to CBIR, such as learning schemes, feature selection, index structure and scalability. We focus our discussion on the consideration of relevance feedback in CBIR as a small sample machine learning problem. This is followed by a detailed description of the learning and search natures of each algorithm. We begin the discussion with an overview of the classic relevance feedback approaches in CBIR.

CLASSICAL ALGORITHMS

The early relevant feedback schemes for ICBR were adopted from feedback schemes developed for classical textual document retrieval. These schemes can be classified into two approaches: query point movement (query refinement) and re-weighting (similarity measure refinement) [1]. Both of these approaches were based on the vector model, the most popular model used in information retrieval [27].

The query point movement method tries to improve the estimate of the "ideal query point" by moving it towards positive example points and away from bad example points in the query space. There are various ways to update the query. A frequently used technique to iteratively improve this estimate is Rocchio's formula (see Equation 3-1). That is, for a set of relevant documents D'_R and non-relevant documents D'_N given by the user [24], the optimal query is defined as:

$$Q' = \alpha Q + \beta(\frac{1}{N_{R'}} \sum_{i \in D'_R} D_i) - \gamma(\frac{1}{N_{N'}} \sum_{i \in D'_N} D_i) \qquad (3\text{-}1)$$

where α, β, and γ are suitable constants; and $N_{R'}$ and $N_{N'}$ are the number of documents in D'_R and D'_N, respectively. This technique is also referred to as a learning query vector. It was used in the MARS system [19] to replace the document vector with visual feature vectors. Experiments show that retrieval performance can be improved by using these relevance feedback approaches.

The re-weighting method enhances the importance of a feature's dimensions that help retrieve relevant images and reduce the importance of the dimensions that hinder this process. This is achieved by updating the weights of the feature vectors in the distance metric. Consider a weighted metric defined as:

$$D = \sum_{j \in [N]} \omega_j \cdot \left| X_j^{(1)} - X_j^{(2)} \right| \qquad (3\text{-}2)$$

When an image of the query result is labeled as a positive example, the feature components that contribute more similarity to the match are considered more important, while the feature components with smaller contribution are considered to be less important. Therefore, the weight for a feature component, ω_i, is updated as:

$$\omega_i = \omega_i \cdot (1 + \overline{\delta} - \delta_i), \quad \delta = \left| f(Q) - f(A_j^+) \right| \qquad (3\text{-}3)$$

where $\overline{\delta}$ is the mean of δ. On the other hand, if an image is labeled as a negative example, the feature components that contribute more to the match should be depressed. That is, the weights are updated as:

$$\omega_i = \omega_i \cdot (1 - \overline{\delta} + \delta_i) \qquad (3\text{-}4)$$

This technique also referred as *learning the metric*, was proposed by Huang et al., [12]. The MARS system refinement of the re-weighting method is called the standard deviation method [25].

Instead of updating the individual components of a distance metric, we can also begin with a set of pre-defined distance metrics and use relevance feedback to automatically select the best one for the retrieval process. For instance, in the ImageRover system [28], appropriate L_p Minkowski distance metrics are automatically selected to minimize the mean distance between the relevant images specified by the user.

Another relevance feedback approach, proposed by Minka and Picard, updates the query space by selecting feature models. It is assumed that each feature model represents one aspect of the image content more accurately than others. Thus, the best way for effective content-based retrieval is to use "a society of models." This approach uses a learning scheme to dynamically determine which feature model or combination of models is best for subsequent retrieval.

Recently, more computationally robust methods that perform global feature optimization have been proposed. The MindReader retrieval system designed by Ishikawa et al. [13] formulates a minimization problem on the parameter estimating process. In traditional retrieval systems the distance function can be represented by ellipses aligned with the coordinate axis. In contrast, in the MindReader system a distance function is not necessarily aligned with the coordinate axis. Therefore, it allows correlations between attributes, in addition for different weights on each component.

A further improvement over the MindReader approach is given in [26]. In this approach, optimal query estimation and weighting functions are derived by a unified framework. Based on the minimization of the total distances of the positive examples from the revised query, the weighted average and a whitening transform in the feature space were found to be the optimal solutions. Assume that a query vector component q_i corresponds to the ith feature, an N element vector $r=[r_1,...r_N]$ represents the degree of relevance for each of the N input training samples, and there is a set of

N training vectors x_{ni} for each feature I then ideal query vector q_i^* for feature i, is the weighted average of the training samples for feature i given by:

$$q_i^{T*} = \frac{R^T X_i}{\sum_{n=1}^{N} r_n} \tag{3-5}$$

where X_i is the $N \times K_i$ training sample matrix for feature i, obtained by stacking the N training vectors x_{ni} into a matrix. It is interesting to note that the original query vector q_i does not appear in (3-5). This shows that the ideal query vector with respect to the feedbacks is not influenced by the initial query.

The optimal weight matrix W_i^* is given by:

$$W_i^* = (\det(C_i))^{\frac{1}{K_i}} C_i^{-1} \tag{3-6}$$

where C_i is the weighted covariance matrix of X_i. That is:

$$C_{i_{rs}} = \frac{\sum_{n=1}^{N} \pi_n (x_{nir} - q_{ir})(x_{nis} - q_{is})}{\sum_{n=1}^{N} \pi_n} \qquad r, s = 1, \ldots K_i \tag{3-7}$$

The critical inputs into the system are training vectors x_{ni} and the relevance matrix **R**. In this algorithm, initially, the user needs to input these data into the system. Another issue with this algorithm is that negative examples are not used in the updating of the query and similarity.

RELEVANCE FEEDBACK AS A LEARNING PRBELEM

Relevance feedback can be considered as a leaning problem — a user provides feedback about examples retrieved as a result of a query, and the system learns from such examples how to refine the retrieval results. The original query-movement method represented by the Rocchio's formula and re-weighting method [24] are both simple learning methods. According to Mitchell's definition [31], machine learning is about constructing computer programs that automatically improve with experience. Using this definition, any task that can be improved as a result of experience can be considered as a machine-learning task. In CBIR, relevance feedback improves the retrieval performance, and the *experience* is the feedback examples provided by the users. Hence, classic machine-learning methods, such as decision tree learning [19], artificial neural networks [16], Bayesian learning [36], and kernel based learning [35] can be and have been applied to relevance feedback in CBIR. However, as users are usually reluctant to provide a large number of feedback examples, the number of training samples is very small, typically less than ten in each round of a feedback session. Feature dimensions in CBIR systems are usually high. Hence, the crucial issue in performing relevance feedback in CBIR systems is how to learn from *small training samples in a very high dimension feature space*. This fact makes many learning methods, such as decision tree learning and artificial neural networks, unsuitable for CBIR.

The key issues in addressing relevance feedback in CBIR as a small sample learning problem include: first, quickly learn from small sets of feedback samples to improve retrieval accuracy effectively; second, how to accumulate the knowledge learned from the feedback; and third, how to integrate low-level visual and high-level semantic features in the query. Most of the published work has focused on the first issue. Compared with other learning methods, Bayesian learning is advantageous in addressing the first issue above, and almost all aspects of Bayesian learning have been explored in research about effective learning algorithms.

Vasconcelos and Lippman [36] treated feature distribution as a Gaussian mixture and used Bayesian inference for learning during feedback iterations in a query session. The Richer information captured by the mixture model also made image regional matching possible. The potential problems with their methods are computing efficiency and a complex data model that requires too many parameters to be estimated from very limited samples.

Active learning methods have been used to actively select samples which maximize the information gain, or minimize entropy/uncertainty in decision-making. These methods increase the speed of the learning process and enable faster convergence of the retrieval result which in turn increases user satisfaction. The approach proposed in [3] used Monte Carlo sampling to search for the set of samples that will minimize the *expected* number of future iterations. Tong and Chang [35] proposed using the SVM active learning algorithm to select the sample which maximizes the reduction in the size of the version space in which the class boundary lies. Without knowing *a priori* the class of a candidate, the best strategy is to halve the search space each time. They argued that selecting the points near the SVM boundary can almost achieves this goal, and it is more efficient than other more sophisticated schemes which require exhaustive trials on all the test items. Therefore, in their work, the points near the SVM boundary are used to approximate the most-informative points; and the most-positive images are chosen as the ones farthest from the boundary on the positive side in the feature space.

The following states some other issues in applying learning in relevance feedbacks in CBIR.

Relevance Feedback as Pattern Recognition?
Some researchers interpret the relevance feedback process in CBIR as a pattern recognition or classification problemin which, the positive and negative examples provided by the user can be treated as training examples. Therefore a classifier can be trained to separate all data into relevant and irrelevant groups. In this scenario many existing pattern recognition tools can be adopted for the task and experiments have been conducted with many kinds of classifiers , such as linear classifier [37], nearest-neighbor classifier [38], Bayesian classifier [36], support vector machines (SVM) [35], and so on. In this category, the most popular algorithm is presented in [35], where the SVM classifier is trained to separate the positive and negative examples, and for each query, classify all images in the database into two groups: relevant and irrelevant.

However, learning from relevance feedback in CBIR is different from the classical pattern recognition problem. In a typical pattern recognition case, there usually is a clear class structure. In other words, each item in a data set belongs to one or a number of clearly defined classes. The classification algorithm usually has a very clear goal: to separate different classes in the whole data set as much as possible. However, in most cases in CBIR, there is no pre-defined class structure. Even if there is a pre-defined class structure, it is still often difficult to determine which category an image item belongs to. This is due to human subjectivities and inconsistence in human perception. Hence, the classification often lacks clear goal. From the application point of view, such classification-based methods may improve the retrieval performance in some constrained contexts; but they will be limited when applied to general purpose image databases.

Category Search vs. Target Search
This topic is not directly related to feedback. The issue here is what a user needs when using a CBIR system. The user may look for similar images or a particular image. Before Cox, *et al.* [5] presented the idea of *target search, category search* was the only search style in CBIR system. In category search, the user provides the sample image and wants to find images with similar contents to the sample. Under a relevance feedback framework, the user can make the relevance judgment to iteratively refine the retrieval performance. The user can provide binary judgment of positive and negative examples [33, 36]. Single judgments have also been widely used where only positive examples are considered [13, 32]. Some systems provide more complex interactive methods where the degree of relevance and irrelevance is also considered [25]. Different levels are used to represent differing degrees of relevance and irrelevance of the image comparing with the query image.

Target search [3] presents another kind of scenario in which a user looks for a particular target image. The positive examples (s)he provides in one retrieval session are those close to the target image sought. However, there are a number of practical problems associated with this search style. First, identifying the image that is closer to the target is difficult for the user to determine. Second, similarities in image contents may be so far away from each other in a feature space that it is impossible for relevance feedback to solve the target search problem.

Feature vs. Semantics in Relevance Feedback
All the approaches described above perform relevance feedback at the low-level feature vector level by basically replacing keywords with features and adopting the vector model developed for document retrieval. While these approaches do improve the performance of ICBR, there are severe limitations. The inherent problem is that the low-level features are not as powerful at representing the complete semantic content of images, as keywords in representing text documents. Furthermore, users often pay more attention to the semantic content (or a certain object/region) of an image than to the background; furthermore, the feedback images may be partially similar in semantic content, but vary greatly in low-level features. Hence, using low-level features alone may not be effective in representing users' feedback and in describing their intentions.

In addition, there are typically two different modes of user interactions involved in image retrieval systems. In one case, the user types in a list of keywords representing the semantic contents of the desired images. In the other case, the user provides a set of example images as the input and the retrieval system will retrieve similar images. In most image retrieval systems, these two modes of interaction are mutually exclusive. However, combining these two approaches and allowing them to benefit from each other will yield a great advantage in terms of both retrieval accuracy and ease of use of the system.

There have been efforts to incorporate semantics in relevance feedback to image retrieval. The framework proposed in [17] (to be discussed in more detail later in this section) attempted to embed semantic information into a low-level feature based image retrieval process using a correlation matrix. The FourEye system by Minka and Picard [21] and the PicHunter system by Cox et al., made use of hidden annotations through the learning process [5]. However, they excluded the possibility of benefiting from good annotations, which may lead to a very slow convergence.

In terms of feature selection, unlike most CBIR systems that use image features such as a color histogram or moments, texture, shape, and structure features, Tieu and Viola [34] used a boosting technique totrain a classification function in a feature space of more than 45,000 features. The features were demonstrated to be sparse with high kurtosis, and were argued to be expressive for high-level semantic concepts. Weak 2-class classifiers were independently formulated along each feature space based on Gaussian assumption for both the positive and negative (randomly chosen) examples. The strong classifier is then a weighted sum of the weak classifiers as in AdaBoost.

The framework to be discussed in Section 3.3 integrates both semantics and low-level features into the relevance feedback process in a new way. Only when the semantic information is not available, the method is reduced to one of the previously described low-level feedback approaches as a special case.

RELEVANCE FEEDBACK WITH MEMORY

A disadvantage of the classic relevance feedback, as well as many learning-based approaches discussed above, is that the captured knowledge in the relevance feedback processes in one query session or one learning step is not memorized to continuously improve the retrieval accuracy. That is, even with the same query, a user will have to go through the same, often tedious, feedback process to obtain the same result, despite the fact the user has given the same query and feedbacks before. Strictly speaking, there is no learning or only limited learning in such systems as there is no knowledge accumulation across different query sessions. To overcome these limitations, another school of thought emerged, which is used learning approaches to memorize users' subjectivities in relevance feedback process. The challenge in this approach is how to memorize knowledge learned and how to handle inconsistent content subjectivities from different users and/or from different query sessions of the same user.

The approach proposed in [17] was the first attempt to explicitly memorize learned semantic information to improve CBIR performance. The basic idea of this approach is to accumulate semantic relevance between image clusters learnt from the user's feedback in correlation network. In other words, a correlation network is used to memorize the information. Figure 3-1 illustrates the correlation network. Mathematically, the correlation network is represented by a correlation matrix, *M*, defined as below:

$$M = \begin{bmatrix} w_{11} & w_{12} & \cdot & \cdot & \cdot & w_{1N} \\ w_{21} & w_{22} & \cdot & \cdot & \cdot & w_{2N} \\ \cdot & & & & & \cdot \\ \cdot & & & & & \cdot \\ \cdot & & & & & \cdot \\ w_{N1} & w_{N2} & \cdot & \cdot & \cdot & w_{NN} \end{bmatrix} \tag{3-8}$$

where the weight or coefficient, w_{ij}, represents the semantic correlation between images in cluster i and j.

The system works as follows. First, all images in a database are clustered into N clusters based on a visual feature similarity using, for instance, k-means algorithm. Obviously, the images in each cluster initially are only similar in term of the selected visual features, like in a typical CBIR system. Also, initially, all correlation coefficients for each pair of clusters are set to zero, meaning only images within the same cluster are correlated and images across clusters are uncorrelated. That is, the initial matrix is a unit one,

$$M_0 = I_{N \times N} \tag{3-9}$$

Then, for a given query, the initial retrieval is based on visual features. Assume that after a given iteration, $n+m$ images are displayed, and n images are marked relevant and m irrelevant. The relevant as well as irrelevant images may or may not be from difference clusters. This approach memorizes such feedbacks by updating the correlation matrix as below:

$$M_t = M_{t-1} + \sum_{i=1}^{m} F(q) F(p_i)^T - \sum_{i=1}^{n} F(q) F(n_i)^T \tag{3-10}$$

where q is the feature vector of the query, p_i and n_i are feature vectors of positive and negative feedback samples, and $F(x)$ is a transform function used to determine the update magnitude based on the feedback samples. In this way, the correlation between the cluster in which the query original falls and the cluster in which positive samples fall are increased, progressively embedding the information on semantic correlations between images. This correlation is used in subsequent retrievals, in which not only the visual features, but also the semantic correlations are used to determine the similarity of an image to the query. Experiments have shown that such a progressive learning approach effectively utilizes the knowledge learnt from previous queries to reduce the number of iterations to achieve high retrieval accuracy [17].

Also, if there are two distinct groups in one initial cluster which are semantically dissimilar, meaning that they are negative examples to each other, a split is performed to separate the initial cluster into two clusters. On the other hand, based on feedback, when two clusters that are close together in the feature space and have a high correlation between them according to M, the two initial clusters could be merged into one. That is, the correlation network dynamically updates its structure in addition to updating the correlation matrix through learning from the user's feedback.

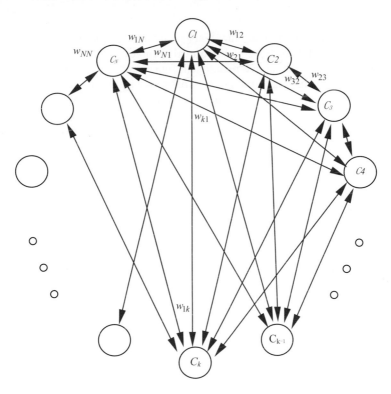

Figure 3-1 Correlation network to memorize semantic correlations between image groups

An important contribution of   this work is the proposed function, $F(x)$, used to incorporate feedback and determine the update magnitude. It was proposed that using the *Radial Basis Function*, a nonlinear transformation, for this purpose. That is,

$$[F(x)]_k = \phi(\| x - c_k \|) = exp\left\{\frac{-1}{2\sigma^2} \| x - c_k \|^2\right\} \qquad (3\text{-}11)$$

where c_k is centroid of the k'th cluster, $1 \leq k \leq N$; and x is the feature vector for feedback samples or the query. This function represents the membership of each sample in its cluster or group, which is inversely proportional to the distance between a sample and the centroid of its group k. In this way, a sample close to the centriod of its group will contribute more to the update as it is more likely to represent the correct direction of the increment.   Thus, the feature-based image similarity is utilized in the semantic upgrading process.

3.3 An Integrated Relevance Feedback Framework

As discussed in Section 3.2, an effective relevance feedback system should provide solutions about how to learn from small sets of feedback samples; accumulate learned knowledge; and integrate low-level visual and high-level semantic features in a query and feedback to achieve high retrieval accuracy.

To address all of the above-mentioned four issues, a CBIR framework with integrated relevance feedback and query expansion was proposed [18, 32, 39]. In this framework, the semantic-based index and relevance feedback are seamlessly integrated with those based on low-level feature vectors. Figure 3-2 illustrates the proposed CBIR framework. It consists of a semantic network which links images to semantic annotations in a database, a similarity measure that integrates both semantic features and image features, and a machine learning algorithm to iteratively update the semantic network and to improve the system's performance over time. The system supports both query-by-keyword and query-by-image-example through semantic network and low-level feature indexing. More importantly, the learning process propagates the keyword annotations from the labeled images to un-labeled ones during the feedback. In this way, more and more images are implicitly labeled with keywords through the semantic propagation process. This annotation propagation process also helps the system to accumulate knowledge learnt and improve the performance of future retrieval requests.

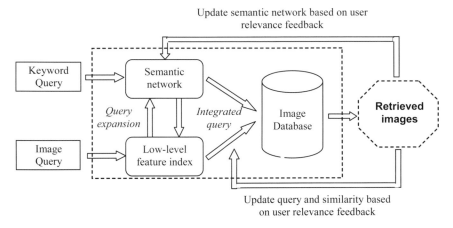

Figure 3-2 The proposed framework of integrated relevance feedback and query expansion

SEMANTIC NETWORK

The semantic network is a two-layered structure. The top layer is represented by a set of keywords with links to the images in the database. It can be considered an extension of the initial information embedding idea in the system shown in Figure 3-1. The degree of the relevance of the keywords to the semantic content of the associated images is represented as the weight on each link as shown pictorially in Figure 3-3. This layer is what we need in keyword relevance feedback and will be updated during the semantic propagation. The bottom layer is a keyword thesaurus to construct the connection between different keywords.

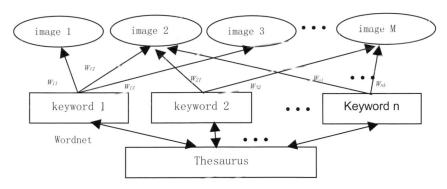

Figure 3-3 Semantic Network

The initial weights can be obtained by manual labeling. They can also be assigned according to its URL, filename, ALT text, hyperlinks, and surrounding text if the images are downloaded from websites. In the latter case, the initial value of the weight *wij* is calculated using the TF*IDF method [3, 4]. Of course, if no keyword information for the image is available, the corresponding feature vector is set to null.

With the semantic network, semantic based relevance feedback can be performed relatively easily compared with its low-level feature counterpart. This is performed by updating the weights w_{ij} associated with each link shown in Figure 3-3 without any user intervention. The weight updating process is described below.

1. The user submits a query, and the system retrieves similar images using cross-modality query extension, which will be explained later, in the next subsection.
2. The system collects the positive and negative feedback examples corresponding to the query;
3. For each keyword in the input query, check to see if any are absent from the keyword database. If so, add them into the database without creating any links;
4. For each positive example, check to see if any query keyword is not linked to it. If so, create a link with an initial weight from each missing keyword to this image. If so, create a link with an initial weight, from each missing keyword to this image OR If so, create a link with an initial weight from each missing keyword, to this image. (these sentences differ in meaning due to the punctuation, pls clarify –ed) For all other keywords that are already linked to this image, increase the weight by a predefined value or using the method defined by (3-10) and (3-11);
5. Similarly, for each negative example, check to see if any query keyword is linked with it. If so, decrease its weight, until it is zero.

Through this updating process, the keywords that represent the actual semantic content of each image will receive a larger weight. Also, it can be easily seen that as more queries are input into the system, the system is able to expand its vocabulary. Furthermore, a semantic propagation method is used to populate keywords to unlabeled images during the user's feedback iteration, which will be described later in this section.

INTEGRATED AND CROSS MODALITY QUERY AND RETRIEVAL

The proposed framework has an integrated relevance feedback scheme on which low-level features are based and high-level semantic feedbacks are performed. We define a unified metric function G to measure the relevance between query Q' and any image j within an image database in terms of both semantic and low-level feature content, where Q' includes the original query and users' feedback information.

$$G(j,Q') = \alpha \cdot sim_k(j,Q_k') + (1-\alpha) \cdot sim_f(j,Q_f') \qquad (3\text{-}12)$$

where $\alpha \in [0,1]$ is the weight of the semantic relevance in the overall similarity measure, which can be specified by users. The larger α is, the more important the semantic relevance will play in the overall similarity measurement. $Sim_f(j, Q_f')$ and $Sim_k(j, Q_k')$ are the semantic similarity and low-level feature similarity between image j and revised query Q', respectively.

The revised query Q' consists of two parts: the feature-based part, Q_f' and the semantic (keyword)-based part, Q_k'. Q_f' is defined by (3-5) based on feature vectors of the feedback images. With the semantic network, $Sim_k(j, Q_k')$ can be directly computedusing the updated weights.

To further improve the retrieval performance of the proposed framework, a cross-modality query expansion method is supported. That is, once a query is submitted in the form of keywords, the retrieved images based on the keyword search are considered as the positive examples, then the query is expanded by the features of these images. This is achieved by first searching the semantic network, shown in Figure 3-3, for the keywords. Then, the visual features of these images that contains these keywords (referred to as training images) are incorporated into the expanded query, q_i^*, defined by (3-5).

Since the expanded query q_i^* is defined by the feature vectors of all images associated with the query keywords, we need to determine the relevance vector $\boldsymbol{R}$ in (3-5) so the that proper relevance factor r is assigned to each image feature vector. Therefore, we introduce a relevance factor r_{ij} of the ith keyword association to the jth image, defined as:

$$r_{ij} = w_{ij} \left/ \sum_{l=1}^{N_j} w_{lj} \right. \qquad (3\text{-}13)$$

which is the relative weighting of the matched keyword i over all keyword weights of image j. This relevance factor is defined in accordance with information retrieval theory. That is, the importance of a keyword to an image that has links spreading over a large number of images in the database should be penalized. We use this relevance factor in (3-5) to compute the expanded query in the feature space.

One can consider this expanded query as a result of relevance feedback, except that the feedback images are obtained by semantic network search. Using this approach may seem dangerous at first since some images may have keyword association which the user does not intend to search for. However, the goal is to generate a set of queries that is guaranteed to contain the user's intended search results. The query can then be narrowed down by including more feedback images through the relevance feedback cycle.

For query-by-image-example, a similar procedure takes effect to extend the retrieval from feature space to semantic space through the semantic network. In this way, the user input information is utilized as much as possible to improve the retrieval performance.

Using the methods described above, we can perform the combined semantic and feature-based relevance feedback as follows.

1. Collect the user query and expand the query;
2. Compute the combined similarity according to (3-12) to retrieve the initial set of images using the expanded query;
3. Collect positive and negative feedbacks from the user;
4. Compute the feature vectors x_{ni} and the degree of relevance vector R for the retrieved images to form the revised feature-based query Q_f' defined by (3-5).
5. Update weights in the semantic network. The new weights implicitly define the revised keyword-based query Q_k' by defining $Sim_k(j, Q_k')$ in (3-12).
6. Use the revised query to compute the ranking score for each image based on (3-12) and sort the results;
7. Show new retrieval results and go to step 3.

From these query and combined feedback process, we can see that the system learns from the user's feedback both semantically and in a feature based manner. In addition, it can be easily seen that our method is simplified into the method of [26] when no semantic information is available.

PROBABLISTIC PROPAGATION SCHEME

As illustrated in Figure 3-3, the greater the number of correctly annotated images, the better the system retrieval performance. However, the reality is that human labeling of images is tedious and expensive, and not a feasible solution. This was what motivated CBIR research fifteen years ago. To address this issue, a probabilistic progressive keyword propagation scheme is used in our framework to automatically annotate images in the databases during the relevance feedback process utilizing a small percentage of annotated images.

We assume that initially only a few images in a database have been manually labeled with keywords and the retrieval is performed mainly based on low-level features. As stated before, the initial keyword annotation can be obtained from the Web through a crawler if the images are from the Web, or can be labeled by humans. While the user is providing feedback in a query session, a progressive learning process is activated to propagate the keyword annotation from the labeled images to un-labeled images so that an increasing number of images are implicitly labeled by keywords. Where there is user consensus, the semantic network is updated such that keywords become the dominant representation of the semantic content of their associated images. As more queries are input into the system, the system is able to expand its vocabulary. Also, through the propagation process, the keywords that represent the actual semantic content of each image will receive a large weight.

There are two major issues in keyword propagation: which images and which keyword(s) should be propagated during a query session. To answer the first question, a probability model based on Bayesian learning is proposed. We assume that (1) all positive examples in one retrieval session belong to the same semantic class with common semantic object(s) or meaning(s), and (2) the features from the same semantic class follow the Gaussian or Mixture Gaussian distributions. Therefore, all positive examples in a query session are used to calculate and update the parameters of the corresponding semantic Gaussian classes. Then the probability of each image in the database belonging to such semantic class is calculated. The common keywords in positive examples are propagated to the images with a very high probability of belonging to this class.

As we can see, the propagation framework uses the same procedure as the feedback algorithm in low-level features [32]. The only difference is that for low-level feature feedback, the calculated probability is used for the ranking of an image in the retrieval candidate list, whereas here, it is used to determine if an image should be in the propagation candidate list.

The propagation candidate set S is obtained as follows:

$$S = \{c_1, ..., c_k\}, \text{ where } p(c_j) > \psi \tag{3-14}$$

where $p(c_j)$ is the probability of image j in the database belonging to such semantic class, and ψ is a constant threshold which can be estimated by the training process.

The weight associated with the propagated keyword i and the image j is $w_{ij} = p(c_j)$. More complex distribution model, for example Mixture Gaussian, may be used in this propagation framework. However, because the user's feedback examples in practice are often very few, complex models will lead to more parameter estimation errors as there are more parameters to be estimated.

Also, to determine which keyword(s) should be propagated when an image is associated with multiple keywords, there are two approaches: using the relevance factor defined by (3-13) or using the region-based approach [14]. In the former approach, the relevance factor r_{ij} can be directly used to modify the weight with the propagated keyword. Obviously, the lower the relevance of a keyword to an image is, the less weight will be assigned to the keyword in the prorogation, and vice versa. When the region-based approach is used, unlabeled images to be propagated are firstly segmented into regions. By analyzing the feature distribution of the segmented regions, a probability association between each segmented region and the annotated keywords is set up for labeled images in region-based relevance feedback approach. Then, each keyword of the labeled image is assigned to one or several regions of the image with certain probabilities. The detail of the region-based feedback framework is in [14].

EXPERIMENTAL RESULTS

The image set used to evaluate the proposed framework described in this section is the Corel Image Gallery of 10,000 images, which are manually labeled into 79 semantic categories. Two hundred randomly selected images make up the test query set. Whether a retrieved image is correct or incorrect is judged according to the ground truth. Three types of color features and three types of texture features are used in our system. Feedback process is run as follows: Given a query from the test set, a different test image from the same category as the query is used in each round of feedback iteration as the positive example for updating the Gaussian parameters and revising the query. To incorporate negative feedback, the first two irrelevant images are assigned as negative examples. The accuracy is defined as:

$$Accuracy = \frac{\text{relevant images retrieved in top } N \text{ returns}}{N} \quad (3\text{-}15)$$

Several experiments have been performed as follows [39]. First, three feature-based feedback algorithms are compared. They are: a Bayesian feedback scheme by Su et al., in [32]; the scheme by Nuno [36]; and the scheme by Rui and Huang [26], as defined by (3-5) ~ ((3-7). This comparison is done in the same feature space. Figure 3-4 shows that the accuracy of Bayesian feedback scheme (referred as "our feedback approach) becomes higher than the other two methods after two feedback iterations. This demonstrates that the incorporated Bayesian estimation with the Gaussian parameter-updating scheme is able to improve retrieval effectively.

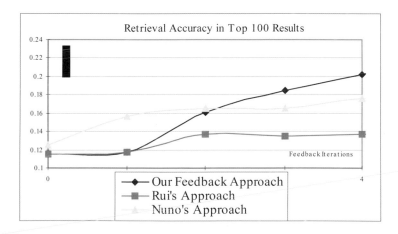

Figure 3-4 Retrieval accuracy for the top 100 results in the original feature space.

To demonstrate the performance of the semantic propagation, the following experiment was designed. Two hundred images in the query set were annotated by their category names. Only one keyword was associated with each query image, and the other images in database have no keyword annotations. During the test, each query image was used twice. The retrieval performance of the experiment is shown in Figure 3-5 with comparison to that without the propagation. The figure shows that for the feedback with propagation, the retrieval accuracy is much higher than the original one without it. This is because when a system has propagation ability, later queries can use the knowledge accumulated from previous feedback iterations. In other words, the system's learning ability will increase with users interaction.

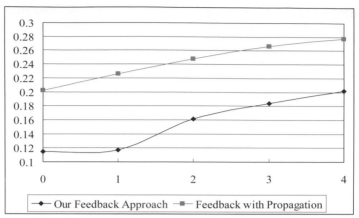

Figure 3-5 Retrieval accuracy for the top 100 results—the comparative performances of feedback without propagation and feedback with propagation scheme

3.4 Conclusion

In this chapter, we have discussed, in detail, relevance feedback technologies in content-based image retrieval systems. The key issues in relevance feedback in CBIR are extensively reviewed. We also presented a framework of integrated relevance feedback and semantic learning in content-based retrieval. This new framework makes the image retrieval system superior to both the classical CBIR and text-based systems.

Even though relevance feedback has been the most active research topic in content-based image retrieval, there remain many challenging research issues in this area, many of which have been discussed in this chapter. Among them are the key issues in addressing relevance feedback in CBIR as a small sample learning problem including: how to learn fast from small sets of feedback samples to improve retrieval accuracy effectively, how to accumulate knowledge learned from feedback, and how to integrate low-level visual and high-level semantic features in query. However, the majority of published work has focused only on the first issue.

Acknowledgement

The author would like to thank Dr. Zhong Su for his contribution presented in this chapter while he was a research intern in Microsoft Asia.

References

[1]   C. Buckley, and G. Salton, "Optimization of Relevance Feedback Weights," *Proc of SIGIR '95*.
[2]   S. Chandrasekaran, *et al.* "An Eigenspace Update Algorithm for Image Analysis". *CVGIP: Graphical models and Image Processing* , 1997.

[3]   Zheng Chen, Liu Wenyin, Feng Zhang, Minjing Li, Hongjiang Zhang, "Web Mining for Web Image Retrieval", Journal of the American Society for Information Science and Technology, Vol. 52, No. 10, pp. 831-839, August, 2001.

[4]   Zheng Chen, Liu Wenyin, Chunhui Hu, Mingjing Li, Hongjiang Zhang, "iFind. A Web Image Search Engine", Proc. SIGIR2001.

[5]   Cox, I. J., Minka, T. P., Papathomas, T. V. and Yianilos, P. N. "The Bayesian Image Retrieval System, PicHunter: Theory, Implementation, and Psychophysical Experiments," *IEEE Transactions on Image Processing -- Special Issue on Digital Libraries*, 2000.

[6]   Y. Deng, B. S. Manjunath and H. Shin, "Color Image Segmentation", *Proc. IEEE Computer Society Conference on Computer Vision and Pattern Recognition,* Vol.2, pp.446-51, 1999.

[7]   Diamantaras, I., and Kung, S.Y. *Principal Component Neural Networks, Theory and Applications*, John Wiley & Sons. Inc. 1996

[8]   Duda, R. O., and Hart, P. E. *Pattern Classification and Scene Analysis*, New York: John Wiley & Sons, 1973.

[9]   Faloutsos, C. and Lin, K., "Fastmap: A Fast Algorithm for Indexing, Data-mining and Visualization of Traditional and Multimedia " *Proc. of SIGMOD*, pp. 163-174, 1995.

[10]   Flickner, M. et al. "Query by Image and Video Content: The QBIC system." *IEEE Computer*. Vol 28, pp23-32, 1995

[11]   Fukunaga, K., *Introduction to Statistical Pattern Recognition*, 2nd Ed., Academic Press 1990

[12]   J. Huang, S. R. Kumar, and M. Metra, "Combining supervised learning with color correlograms for content-based image retrieval," *Proc. of ACM Multimedia '95*, pp. 325-334, Nov. 1997.

[13]   Ishikawa, Y., Subramanya R., and Faloutsos, C., "Mindreader: Query Databases Through Multiple Examples," *Proc. of the 24th VLDB Conference*, 1998.

[14]   Jing, F., Zhang, B., Lin, F., Ma, W. and Zhang, H. "A Novel Region-Based Image Retrieval Method Using Relevance Feedback", *Proc. of 3rd Intl Workshop on Multimedia Information Retrieval (MIR 2001)*, Ottawa, Canada, October 5, 2001.

[15]   Kirby, M. and Sirovich, L. "Application of the Karhunen-Loeve procedure for the characterization of human faces", *IEEE Trans. on Pattern Analysis and Machine Intelligence*, 12(1), pp. 103-108, 1990.

[16]   Laaksonen, J. Koskela, M. and Oja, E, "PicSOM: Self-Organizing Maps for Content-Based Image Retrieval", *Proc of International Joint Conference on Neural Networks*, 1999.

[17]   C Lee, W. Y. Ma and , H. J Zhang. "Information Embedding Based on user's relevance Feedback for Image Retrieval," *Proc. of SPIE International Conference on Multimedia Storage and Archiving Systems IV*, Boston, 19-22 Sept. 1999.

[18]   Lu, Y., Hu, C., Zhu, X., Zhang, H. and Yang, Q. "A Unified Framework for Semantics and Feature Based Relevance Feedback in Image Retrieval Systems", *Proc of 8th ACM Conference on Multimedia*, November 2000, Los Angeles, CA.

[19]   MacArthur, S.D.; Brodley, C.E.; Shyu, C.-R. "Relevance feedback decision trees in content-based image retrieval," *Proc of IEEE Workshop on Content-based Access of Image and Video Libraries*, pp.68 -72, 2000.

[20]   Meilhac, C. and Nastar, C. "Relevance Feedback and Category Search in Image Databases", *Proc. of IEEE International Conference on Multimedia Computing and Systems*, Italy, 1999.

[21]   Minka, T. and Picard, R. "Interactive Learning using a 'Society of Models'", *Pattern Recognition*, 30(4), 1997.

[22]   Mitchell. T. "Machine Learning". McCraw Hill, 1997

[23]   Ng, R. and Sedighian, A.. "Evaluating Multi-dimensional Indexing Structures for Images Transformed by Principal Component Analysis". *Proc. SPIE Storage and Retrieval for Image and Video Databases*, 1996.

[24]   Rocchio, Jr., J. J.(1971). Relevance Feedback in Information Retrieval. In *The SMART Retrieval System: Experiments in Automatic Document Processing* (Salton, G. eds) pp313-323. Prentice-Hall.

[25]   Rui, Y., Huang, T. S., and Mehrotra, S. "Content-Based Image Retrieval with Relevance Feedback in MARS," *Proc. of IEEE Int. Conf. on Image Processing*, 1997.

[26]   Rui, Y., and Huang, T. S. "A Novel Relevance Feedback Technique in Image Retrieval," *Proc of 7th ACM Conference on Multimedia*, 1999.

[27]   Salton, G., and McGill, M. J. *Introduction to Modern Information Retrieval,* McGraw-Hill Book Company, 1983.

[28]   S. Sclaroff, L. Taycher, and M. L. Cascia, "ImageRover: a content-based image browser for the World Wide Web," Boston University CS Dept. *Technical Report* 97-005, 1997.

[29]   Shaw, W. M. "Term-Relevance Computation and Perfect Retrieval Performance," *Information Processing and Management*.

[30]   Sheikholeslami, G., Chang, W. and Zhang, A. "Semantic Clustering and Querying on Heterogeneous Features for Visual Data." *Proc of 6th ACM Conference on Multimedia*, Bristol, UK, 1998.

[31]   Stone, H. S. and Li, C. S. "Image Matching by Means of Intensity and Texture matching in the Fourier domain," *Proc. of IEEE Int. Conf. Image Processing*, 1997.

[32]   Su, Z., Li, S., Zhang, H. "Extraction of Feature Subspaces for Content-Based Retrieval Using Relevance Feedback," *Proc of 9th ACM Conference on Multimedia*, Ottawa Canada, 2001.

[33]   Su, Z., Zhang, H and Ma, S. "Relevant Feedback using a Bayesian Classifier in Content-Based Image Retrieval," *Proc of SPIE Electronic Imaging 2001*, San Jose, CA, 2001.

[34]   Tieu K and Viola P (2000). Boosting image retrieval. IEEE Conf. Computer Vision and Pattern Recognition, 2000.

[35]   Tong S. and Chang E. "Support Vector Machine Active Leaning for Image Retrieval," *Proc of 9th ACM Conference on Multimedia*, Ottawa Canada, 2001.

[36]   Vasconcelos, N., and Lippman, A. "Learning from User Feedback in Image Retrieval Systems" *Prof of NIPS'99*, Denver, Colorado, 1999.

[37]   Wu, Y. Tian, Q. and Huang, T.S, "Discriminant EM algorithm with Application to Image Retrieval", *Proc of IEEE Int. Conf. on Computer Vision and Pattern Recognition*, South Carolina, 2000.

[38]   Wu, P. and Manjunath, B. S., "Adaptive Nearest Neighbour Search for Relevance Feedback in Large Image Database", *Proc of 9th ACM Conference on Multimedia*, Ottawa Canada, 2001.

[39]   H.J. Zhang and Z. Su, "Relevance Feedback in CBIR," *Proc of International Workshop on Visual Databases,* 2002.

[40]   Zhu, L. and Zhang, A. "Supporting Multi-example Image Queries in Image Databases," *Proc of IEEE International Conference on Multimedia and Expo*, July, New York City, NY.

4 VIDEO ANALYSIS AND SUMMARIZATION AT STRUCTURAL AND SEMANTIC LEVELS

Dr. Hari Sundaram and Prof. Shih-Fu Chang

In this chapter, we present an overview of some of the research issues related to three areas of audio-visual analysis — (a) segmentation, (b) event detection and (c) summarization.

We begin by presenting the idea of a computable scene, that comprises elementary homogeneous segments of audio visual data. We also discuss work on scene transition graphs and more recent work on segmentation using a computational model of memory. We define an event to be a change of state in an entity. We discuss recent work on event detection in soccer videos that uses sophisticated statistical models. We also discuss related work on detecting specific events in baseball using machine learning techniques. In our discussion on video summarization, we shall focus on recent work on an utility based framework for skim generation. A skim is an temporally condensed audio-visual clip that attempts to summarize the original video. In this chapter we shall discuss four specific sub-problems related to skim generation — (a) visual complexity and comprehension, (b) syntax preservation (c) audio-visual integration and (d) an utility framework for skim generation. We shall conclude the chapter with a discussion of the open issues.

4.1 Introduction

In this chapter, we discuss research issues and promising techniques related to three important aspects of audio-visual content analysis — (a) segmentation, (b) event analysis and (c) summarization. Each component plays an important role in the greater semantic understanding of the audio-visual data.

Segmenting audio-visual data into homogeneous segments in a manner suited for further processing (e.g. visual summaries) is an important first step. The segmentation of video data into manageable chunks is complicated by the presence of complex interactions between audio and video data (e.g. films). Domain dependent syntactical elements (e.g. dialogs) further complicate the segmentation task.

There has been prior work on video scene segmentation using image data alone [24, 59]. In [59], the authors derive scene transition graphs to determine scene boundaries. However, cluster thresholds are difficult to set and often have to be manually tuned.

In [24], the authors use an infinite, non-causal memory model to segment the video. We refine this idea of memory in our recent work [48], but in a finite, causal setting. Prior work [39, 41, 46] concerning the problem of audio segmentation dealt with very short-term (100 ms) changes in a few features (e.g. energy, cepstra). This was done to classify the audio data into several predefined classes such as speech, music ambient sounds etc. However, they do not explore the possibility of using the long-term consistency found in the audio data for segmentation.

Event analysis in video is important not only because we are often interested in detecting a specific phenomena, but also because it complements the segmentation procedure. There has been much work done in event analysis [1, 3, 11, 15, 21, 23, 29, 30, 46, 54, 57, 58, 62], and in this chapter we shall only be able to summarize some of the various methods and applications in event analysis. In particular we focus on the use of Hidden Markov Models (HMM's) [57] in soccer and an application for detecting important events in baseball [23].

Summarization is concerned with the problem of generating a drastically reduced representation of video data. This problem is important in many contexts, such as: (a) browsing digital libraries, (b) on demand summaries of the data stored in set-top boxes (interactive TV), (c) personalized summaries for mobile devices and (d) for news channels (e.g. CNN) that receive a tremendous amount of raw footage.

There has been much research on generating image-based storyboards [42, 52, 55, 59, 60, 61] and video skims [8, 18, 25, 27, 34]. Image based storyboards typically use time constrained clustering on the key-frames of the video shots to determine semantically representative images. However, since they are laid out in a static manner on an html page, they do not convey the underlying dynamism of the audio visual data. In the Informedia skimming project [8], important regions of the video were identified via a TF/IDF analysis of the transcript. They also used face detectors and performed motion analysis for additional cues. The MoCA project [25, 34] worked on automatic generation of film trailers. They used heuristics on the trailers, along with a set of rules to detect certain objects (e.g. faces) or events (e.g. explosions). Work at Microsoft Research [18] dealt with informational videos; there, they looked at slide changes, user statistics and pitch activity to detect important segments. Recent work [27] has dealt with the problem of preview generation by generating "interesting" regions based on viewer activity in conjunction with topical phrase detecting. However, in order to generate the preview, some viewers need to have seen the video. Now, we begin by discussing scene analysis.

4.2 Scene analysis

In this section, we begin by giving a computation definition of a *scene* that is based on low-level features alone, than on semantics. Then, we shall discuss two methods for detecting scenes — scene transition graphs [59, 60] and a memory model based approach [48, 51]. Finally, we shall conclude the section by discussing applications in films, sports video and for news.

THE COMPUTATIONAL SCENE DEFINITION

There are constraints on what we see and hear in films, due to *rules* governing camera placement, *continuity* in lighting as well as due to the *psychology* of audition.

In this chapter, we develop notions of a video and audio computable scenes by making use of these constraints. We adopt the following definition of audio and video scenes. A video scene is a continuous segment of visual data that shows *long-term[1]* consistency with respect to two properties: (a) chromaticity and (b) lighting conditions, while an audio scene exhibits a long terms consistency with respect to ambient sound. We denote them to be *computable* since these properties can be reliably and automatically determined using low-level features present in the audio-visual data. Note that these are *not* semantic scenes. We believe that these c-scenes are the first step in greater semantic understanding of a scene [51].

Type	Abbr.	Figure
Pure, no audio or visual change present.	P	
Audio changes consistent visual.	Ac-V	
Video changes but consistent audio.	A-Vc	
Mixed mode: contains unsynchronized audio and visual scene boundaries.	MM	

Table 4-1 The four types of c-scenes that exist between consecutive, synchronized audio-visual changes. solid circles: indicate audio scene boundaries, triangles indicate video scene boundaries

The a-scene and the v-scenes represent elementary, homogeneous chunks of information. We define a computable scene (abbreviated as c-scene) in terms of the relationships between a-scene and v-scene boundaries. It is defined to be a segment between two consecutive, synchronized[2] audio-visual scenes. This results in four cases of interest[3] (ref. Table 4-1)). We validated the computable scene definition, which appeared out of intuitive considerations, with actual film data. The data were

---

[1] Analysis of experimental data (one hour each, from five different films) indicates that for both the audio and the video scene, a minimum of 8 seconds is required to establish context. These scenes are usually in the same location (e.g. in a room, in the marketplace etc.) are and are typically 40~50 seconds long.
[2] In films, audio and visual scene changes will *not* exactly occur at the same time, since this is disconcerting to the audience. They make the audio flow "over the cut" by a few seconds [37], [40].
[3] Note that the figures for Ac_v, A-Vc and MM, in Table 4-1 show only one audio/visual change. Clearly, multiple changes are possible. We show only one change for the sake of clarity.

from three one-hour segments from three English language films[4]. The definition of a scene works very well in many film segments. In most cases, the c-scenes are usually a collection of shots that are filmed in the same location and time and under similar lighting conditions (these are the P and the Ac-V scenes).

The A-Vc (consistent audio, visuals change) scenes seem to occur under two circumstances. In the first case, the camera placement rules are violated. These are montage[5] sequences and are characterized by widely different visuals (differences in location, time of creation as well as lighting conditions) which create a unity of theme by the manner in which they have been juxtaposed. MTV videos are good examples of such scenes. The second case consists of a sequence of v-scenes that individually obey the camera placement rules (and hence each have consistent chromaticity and lighting). We refer to the second class as *transient* scenes. Typically, transient scenes occur when the director wants to show the passage of time e.g. a scene showing a journey, characterized by consistent audio track.

C-scene breakup	Count	Fraction
Pure	33	65%
Ac-V	11	21%
A-Vc	5	10%
MM	2	4%
Total	51	100%

Table 4-2 c-scene breakup from the film sense and sensibility

Mixed mode (MM) scenes are far less frequent, and can for example occur, when the director continues an audio theme well into the next v-scene, in order to establish a particular semantic feeling (joy/sadness etc.). Table 4-2 shows the c-scene type break-up from the first hour of the film *Sense and Sensibility.* There were 642 shots detected in the video segment. The statistics from the other films are similar. Clearly, c-scenes provide a high degree of abstraction, that will be extremely useful in generating video summaries. Note that while this paper focuses on computability, there are some implicit semantics in our model: the P and the Ac-V scenes, that represent c-scenes with consistent chromaticity and lighting are almost certainly scenes shot in the same location.

METHODS

In this section, we review two different approaches to determining computable scenes — scene transition graphs that only use visual features and a memory model based framework that performs multi-modal segmentation.

Scene Transition Graphs

---

[4] The English films: *Sense and Sensibility, Pulp Fiction, Four Weddings and a Funeral.*
[5] In classic Russian montage, the sequence of shots are constructed from placing shots together that have no immediate similarity in meaning. For example, a shot of a couple may be followed by shots of two parrots kissing each other etc. The meaning is derived from the way the sequence is arranged.

A scene transition graph [59, 61] is a compact representation of a video that is a

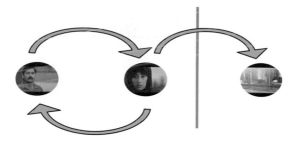

Figure 4-1 A scene transition graph is segmented into scenes by detecting cut edges. These cut edges divide the graph into disconnected sub-graphs.

directed graph. Each node of the graph represents a cluster of similar shots (under a suitable similarity metric). Two nodes i, j are connected via an edge if there exists a shot in cluster i that precedes a shot in cluster j. In [61], the authors perform time-constrained clustering on the shots using a time-window T to create the scene transition graph. Given two parameters δ and T, the maximum cluster diameter and the duration of the time window respectively, two shots belong to a cluster if temporally, they are within T sec. of each other and are within δ of each other with respect to the similarity metric [61].

In Figure 4-1, we have three nodes, and each node represents a time-constrained cluster. The presence of the cycle between the first two clusters indicates that these two nodes belong to the same scene. A scene transition occurs at a *cut-edge* i.e. when there is forward transition from a sub-graph to another sub-graph with no backwards transition. Note however, that the authors [59, 60, 61], attempt to segment the video at a *semantic* level. They do not have a computational model of a scene. An important concern in this work is the setting of the cluster threshold parameter δ and the time-window size T, both of which critically affect the segmentation result. Unfortunately, neither of these parameters can be set without taking the specific character of the data being analyzed into consideration.

Memory model

In order to segment data into computable scenes, we use a causal, first-in-first-out (FIFO) model of memory (Figure 4-2). This model is derived in part from the idea of coherence [24]. In our model of a listener, two parameters are of interest: (a) memory: this is the net amount of information (T_m) with the viewer and (b) attention span: it is the most recent data (T_{as}) in the memory of the listener (typical values for the parameters are T_m=32 sec. and T_{as}=16 sec.). This data is compared with the contents of the memory by the listener in order to decide if a scene change has occurred.

The work in [24] dealt with a non-causal, infinite memory model based on psychophysical principles, for video scene change detection. We use the same psychophysical principles to come up with a causal and finite memory model. Intuitively, causality and a finite memory will more faithfully mimic the human

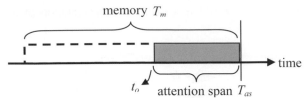

Figure 4-2 The attention span Tas is the most recent data in the memory. The memory (Tm) is the size of the entire buffer.

memory-model than an infinite model. We shall use this model for *both* audio and video scene change detection.

In order to segment the data into audio scenes, we compute correlations amongst the audio features in the attention-span with the data in the rest of the memory. The video data comprises shot key-frames. The key-frames in the attention span are compared to the rest of the data in the memory to determine a coherence value. This value is derived from a color-histogram dissimilarity. The comparison takes also into account the relative shot length and the time separation between the two shots. We locate maxima and minima respectively, to determine scene change points.

We introduce a topological framework that examines the local metric relationships

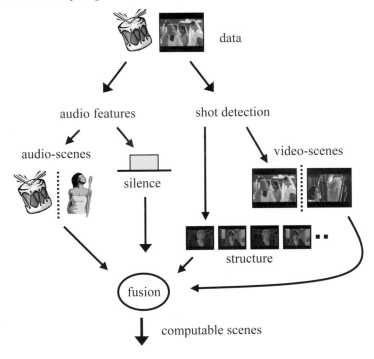

Figure 4-3 computable scene detection overview.

between images for structure detection. Since structures (e.g. dialogs) are independent of the duration of the shots, we can detect them independent of the v-scene detection framework. We exploit specific local structure to compute a function that we term the periodic analysis transform. We test for significant dialogs using the standard Students t-test. The silence is detected via a threshold on the average energy; we also impose minimum duration constraints on the detector.

A key feature of our work is the idea of imposing semantic constraints on our computable scene model. This involves fusing (see Figure 4-3) results from silence and structure detection algorithms. The computational model cannot distinguish between two cases involving two long and widely differing shots: (a) in a dialog sequence and (b) adjoining video scenes. However, human beings recognize the structure in the sequence and thus group the dialog shots together. Silence is useful in two contexts: (a) detecting the start of conversation by determining significant pauses [15]; and (b) in English films, the transitions between computable scenes may involve silence. Our experiments [51] show that the c-scene change detector and the structure detection algorithm work well.

4.3 Event analysis

In this section we briefly review work done on event analysis. We begin by defining an event, then we discuss three methods for event detection — Hidden Markov Models, graphical models and Bayesian inference. Then, we discuss two specific applications in baseball detection and echocardiogram videos.

In this work, we define an event to be a change of state or property of an entity. MPEG-7 has a rich description schemes (DS's) to describe entities, entity attributes and relationships between entities [28]. While these descriptions may just be textual, the framework is powerful since it supports reasoning and inference. The Event DS describes an event, which is a semantic activity that takes place at a particular time or in a particular location. The Event DS can describe either a perceivable or an abstract event in a narrative world. A perceivable event is a dynamic relation involving one or more objects taking place in time and space in a narrative world (e.g., "Alex shaking hands with Ana"). An abstract event results from the abstraction of a perceivable event (e.g., "A man shaking hands with a woman"). The Event DS includes elements that describe the composition of the event from sub-events, in addition to the location and time of the event.

METHODS
Hidden Markov Models

In this section we shall discuss how Hidden Markov Models (HMMs) a widely used statistical technique [36] can be used for event detection in soccer videos [57]. The technique is useful in automatic content filtering of soccer fans and professionals, and it is more interesting in the broader background of video structure analysis and content understanding. By structure, we are primarily concerned with the recurrent temporal sequence of high-level game states, namely *play* and *break*. The game is *in play* when the ball is in the field and the game is going on; *break*, or *out of play*, is the compliment set, i.e. whenever "the ball has completely crossed the goal line or touch line, whether on the ground or in the air" or "the game has been halted by the referee".

The states *play* (*P*) and *break* (*B*) consist of different sub-structures such as the switching of shots and the variation of motion. This is analogous to isolated word recognition [36] where models for each word are built and evaluated with the data likelihood. But as these domain-specific classes *P/B* in soccer are very diverse in themselves (typically ranging from 6 seconds up to 2 minutes in length), we use a set of models for each class to capture the structure variations.

We take a fixed-length sliding window (width 3 seconds, sliding by 1 second) and classify the feature vector into either one of the *P/B* classes. The feature stream is first smoothed by a temporal low-pass filter, normalized with regard to its mean and variance of the entire clip, then the segment of size 2xN in each time slice (2 is feature dimension, N is window length) is fed into the HMM-dynamic programming modules for classification. In our system, 6 HMM topologies are trained for *play* and for *break*, respectively. These include 1/2/3-state fully connected models, 2/3 state

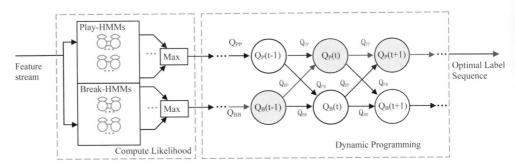

Figure 4-4 HMM-dynamic programming diagram. Only 2 out of 6 HMM topologies are shown; the Qs are HMM model likelihood or transition likelihood

left-right models and a 2-state fully connected model with an entering and an exiting state. The observations are modeled as mixture of Gaussians, and we have 2 mixtures per feature dimension per state in the experiments (Figure 4-4).

The HMM model parameters are trained using the EM algorithm. Training data are manually chopped into homogeneous *play/break* chunks; EM for the *play*-models are conducted over every complete *play* chunks, and vice versa for *break*-models. HMM training is not conducted over 3-second windows because we anticipate that the HMM structures will take a longer time correlation into account, and thus "tolerate" some less frequent events in a state, such as short *close-ups* within a *play*. Experiments show that the overall accuracy will be consistently 2~3% lower if models are trained on short segments, and the video tends to be severely over-segmented as some of the short close-ups and cutaways during a *play* will be misclassified as a *break*.

Extensive statistical analyses show that classification accuracy is about 83.5% over diverse data sets, and most of the boundaries are detected within a 3-second ambiguity window. It is encouraging that high-level domain-dependent video structures can be computed with high accuracy using compressed-domain features and generic statistical tools. We believe that the performance is because the features used are tuned to the domain syntax, as well as due to the power of the statistical tools in capturing the temporal dynamics of the video.

Learning object/scene detectors

In Visual Apprentice [23], we addressed a different research issue – how to efficiently develop classifiers or detectors for detecting video objects or scenes that correspond to specific events in video (such as the pitching scene in baseball). A user defines visual object/scene models via a multiple-level definition hierarchy: a scene consists of objects, which consist of object-parts, which consist of perceptual-areas, which consist of regions. The user trains the system by providing example images/videos and labeling components according to the hierarchy she defines (e.g., image of two people shaking hands contains two faces and a handshake). As the user trains the system, visual features (e.g., color, texture, motion, etc.) are extracted from each example provided, for each node of the hierarchy (defined by the user). Various machine learning algorithms are then applied to the training data, at each node, to learn classifiers. The best classifiers and features are then automatically selected for each node (using cross-validation on the training data). The process yields a visual object/scene detector (e.g., for a handshake), which consists of an hierarchy of classifiers as it was defined by the user. The visual detector classifies new images/videos by first automatically segmenting them, and applying the classifiers according to the hierarchy: regions are classified first, followed by the classification of perceptual-areas, object-parts, and objects. In [19] the concept of recurrent visual semantics is discussed, specifically on how it can be used to identify domains in which learning techniques such as the one presented in that work can be applied.

The technique proposed is specifically used to detect batting events in baseball video. This is done by building a hierarchy that represents the frontal view of the typical batting scene. The scene is represented by a hierarchy with the following elements: the scene contains a pitcher, a batter, and a field. The field contains two nodes, one for the sand and one for the grass. Each of the leaf nodes (pitcher, batter, sand-field, and grass-field) is represented by regions obtained from automatic segmentation. The approach in [23] can be used in video event detection by constructing Visual Object/Scene Detectors for specific domains. One possibility would be to construct several detectors and combine them to define higher-level semantic events. In baseball, a model for a batting scene from the VA, could be used with a model for the typical camera motion that follows a homerun (extending the VA or using another approach). The detection of the homerun event, then, would occur when the batting scene and corresponding motion are found. The VA was also use to detect handshake "events" in news images.

4.4 Video Summarization

In this section, we review work done in summarizing video data. A video summary is a drastically reduced temporal representation that attempts to capture the semantics of the underlying audio-visual data. There are two forms of summaries that we shall discuss here — image storyboards and visual skims.

IMAGE STORYBOARDS

Image based storyboards offer a key-frame based non-linear navigation of the video data [42, 52, 59]. All image based storyboard algorithms broadly share the following approach: (a) determine an appropriate feature space to represent the images (b) a clustering technique to cluster the image sequence in the feature space and (c) computing a measure of importance to determine the appropriate key-frame to represent the cluster.

Scene transition graphs

The scene transition graph (STG) offers a compact representation of the video. We can browse and navigate the video in a hierarchical, non-linear fashion. A STG shares similar characteristics to other image based storyboards in that it clusters frames in a feature space. However, by analyzing the shot transitions amongst the clusters, it provides for some of the temporal dynamics to be visualized in the storyboard. The analysis of the label transitions also allows the STG to detect elements of visual syntax, such as the dialog.

Enhancing image storyboards

Conventional image based storyboards do not capture the dynamism of the underlying audio-visual data. Hence there has been some effort to improve the interactivity of these schemes [53]. There, the image summary was enhanced with text (either from manual transcripts or OCR) and presented in a *manga*[6] like fashion. We outline four possible ways of enhancing current image summarization schemes.

1. **Text balloons:** If we have text aligned transcripts, then it may be possible to extract the important sentences corresponding to the cluster and then displaying them when the user moves the mouse over the relevant image.

2. **Audio segments:** If we perform an acoustic analysis of prosody [15], then we can identify important boundaries in the discourse (both for spontaneous and structured speech). Then, we could associate each key-frame with this audio segment. In the usage scenario, the user would click on the key-frame to hear the corresponding audio segment. In [50], we have automatically detected discourse boundaries for use in video skims.

3. **Animations:** At present image based storyboards are static i.e. the image representing each cluster does not change over time. An interesting variation would be to represent each image by an animated GIF, which cycles through other images in the cluster when the user moves the cursor over the storyboard key-frame. Another attempt at infusing dynamism is the *dynamic STG* [60] in which the shots comprising each cluster are rendered slowly over time.

4. **Structure and syntactical highlights:** Many domains possess very specific rules of syntax and characteristic structural elements that are meaningful in that domain. Examples of structure in films include the dialog and the regular anchor [51]. Specialized domains such as baseball, echocardiogram videos have a very specific syntactical description. These domains would

---

[6] Manga is the Japanese word for a comic book.

benefit from higher-order domain grouping rules that arranges the key-frames of the cluster in a manner highlighting the rules of syntax and the domain specific structures.

In the next section, we shall discuss the generation of audio visual skims.

VISUAL SKIMS

A video skim is a short audio-visual clip that summarizes the original video data. The problem is important because unlike the static, image-based video summaries [53], video skims preserve the dynamism of the original audio-visual data. Applications of audio-visual skims include: (a) on demand summaries of the data stored in set-top boxes (interactive TV); (b) personalized summaries for mobile devices; and (c) news channels (e.g. CNN) that receive a tremendous amount of raw footage.

There has been prior research on generating video skims. In the Informedia skimming project [8], important regions of the video were identified via a TF/IDF analysis of the transcript. They also used face detectors and performed motion analysis for additional cues. The MoCA project [34] worked on automatic generation of film trailers. They used heuristics on the trailers, along with a set of rules to detect certain objects (e.g. faces) or events (e.g. explosions). Work at Microsoft Research [18] dealt with informational videos; there, they looked at slide changes, user statistics and pitch activity to detect important segments. Recent work [27] has dealt with the problem of preview generation by generating "interesting" regions based on viewer activity in conjunction with topical phrase detecting. However, in order to generate the preview, some viewers need to have seen the video.

The goal of this work is the automatic generation of audio-visual skims for *passive*[7] tasks, that summarize the video. We make the following assumptions:

1. We do *not* know the semantics of the original.
2. The data is not a raw stream (e.g. home videos), but is the result of an editing process (e.g. films, news).

Since we work on passive tasks, the information needs of the user are a priori unknown. A decision to detect certain set of predefined events will induce a bias in the skim, thereby conflicting with the assumption that the user needs are unknown.

We first begin by discussing syntax preserving visual skims, then we discuss techniques for auditory analysis so as to integrate audio in the skim, and then we conclude by presenting our optimization framework for skim generation.

Syntax preserving

There are four important challenges to be overcome in skim generation:

1. What is the relationship between the visual complexity of a shot and its comprehension time?
2. How does the syntactical structure of the video data affect its meaning?
3. How to select audio segments optimally? And how to ensure that the resulting skim is coherent?

---

[7] A task is defined to be active when the user requires certain information to be present in the final summary (e.g. "find me all videos that contain Colin Powell."). In a passive task, the user does not have anything specific in mind, and is more interested in consuming the information e.g. set-top box previews.

4. Can we solve the skim generation problem in a general constrained utility
    maximization framework, so as to be able to easily add additional
    constraints easily?

First, we discuss the visual analysis that comprises two parts — visual complexity
and analysis of visual syntax.

Visual complexity

The intuition for relating image complexity and comprehension time comes from two
sources: (a) empirical observations from film theory and (b) experimental evidence
from psychology. Directors have long made use of the fact that the audience takes
less time to comprehend close-ups than long shots (as long shots usually have a lot of
detail) [44] to modulate the duration of each shot. Recent results in experimental
psychology [13] indicate the existence of an empirical law: the subjective difficulty
in learning a concept is directly proportional to the logical incompressibility of the
Boolean concept.

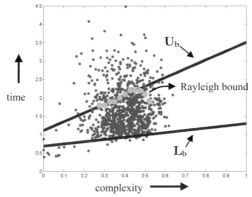

Figure 4-5 Avg. comprehension time (sec.) vs. normalized
complexity (x-axis) showing comprehension (upper/lower)
bounds. It also shows the Rayleigh (95th percentile) bounds.

We define the visual complexity of an shot to be its Kolmogorov complexity [10].
In [49], we showed that length of the Lempel-Ziv [8] codeword asymptotically
converges to the Kolgomorov complexity of the shot. The complexity is estimated
using a single key-frame. We conducted our experiments [49] over a corpus of over
3600 shots from six films. There, a shot was chosen at random (with replacement)
and then its key-frame presented to the subject (the first author). Then, we measured
the time to answer the following four questions (in randomized order), in an
interactive session: (a) who, (b) when, (c) what, and (d) where. The subject was
expected to answer the questions in minimum time *and* get all four answers right.

From an analysis of the complexity–average comprehension time density (see
Figure 4-5), we obtain lower and upper bounds on the density. The equations for the

---

[8] Lempel-Ziv encoding is a form of universal data coding that doesn't depend on the probability
distribution of the source [10].

lines are as follows:

$$U_b(c) = 2.40c + 1.11,$$
$$L_b(c) = 0.61c + 0.68,$$
(4-1)

where c is the normalized complexity and U_b and L_b are the upper and lower bounds respectively, in sec. The upper bound [49] implies that the for 95% of the shots, the average time for comprehension lies below this line; the second line just lower bounds the entire density. The lines were estimated for $c \in [0.25, 0.55]$ (since most of the data lies in this range) and then extrapolated. Hence, given a shot of duration t_o and normalized complexity c_S, we can condense it to at most $U_b(c_S)$ sec by removing the last $t_o - U_b(c_S)$ sec.

Analysis of film syntax

The phrase film syntax refers to the specific arrangement of shots so as to bring out their mutual relationship [49]. In practice, this takes on many forms [44] : (a) minimum number of shots in a sequence; (b) varying the shot duration, to direct attention; (c) changing the scale of the shot (there are "golden ratios" concerning the distribution of scale); (d) the specific ordering of the shots (this influences the meaning). These syntactical rules lack a formal basis, and have been arrived at by trial and error by film-makers. Hence, even though shots in a scene only show a small portion of the entire setting at any one time, the syntax allows the viewers to understand that these shots belong to the same scene. Visual syntax is important because film-makers do not think in terms of individual shots, but in phrases of shots. A shot can have a multitude of meanings, that gets clarified by its relationship to other shots. In [49], we used two elements of cinematic syntax for scene-level compression, and showed how one can exploit film-making rules in order to come up with a shot-dropping strategy.

Natural language integration

Audio skim generation aims at dramatic time reduction (up to 90%) while preserving perceptual coherence. There are some clear drawbacks to simple approaches to determining useful segments in the audio stream. Let us assume that we wish to compress an audio track that is 100 sec. long, by 90%. Then: (a) downsampling the audio by 90% will leave the audio severely degraded since the pitch of the speech segments will increase dramatically. (b) PR-SOLA [18] is a non-linear time compression technique that eliminates long pauses, and attempts to preserve the original pitch in the output. User studies indicate that speech sped up beyond 1.6x (i.e. ~40% compression) is disliked by users. (c) selecting only those segments that are synchronous with the pre-selected video shots makes the audio stream is choppy and difficult to comprehend [8].

Audio segment classification

We build a tree-structured classifier to classify each frame (100ms) into four generic classes: silence, clean speech, noisy speech and music / environmental sounds. We use 16 features in our approach. Silent frames are first separated from the rest of the audio stream using an adaptive threshold on the energy. Two SVM classifiers are then used in cascade: the remaining frames are separated into speech vs. non-speech (music or environmental sounds); and the speech class is further classified as clean and noisy speech. We then apply a modified Viterbi decoding algorithm [50] to smooth the sequence of frame labels. The decoder makes use of the class transition probabilities, classifier error likelihood and a duration utility (a function of the prior duration distribution of each class) to find the maximum likelihood class path.

Detecting significant phrases
In this work we focus on detecting segment beginnings (SBEG's) in speech[19, 20]. These are important as they serve as the introduction of new topic in the discourse. There has been much work in the computational linguistics community [19, 20, 43] to determine the acoustic correlates of the prosody in speech. Typically, SBEG's have a preceding pause that is significantly longer than for other phrases, higher initial pitch values (mean, variance), and smaller pauses that end the phrase than for other phrases [19][20]. In our algorithm, we extract the following features per phrase: pitch and energy values (min, max, mean, variance) for the (initial, last and complete) portions of the phrase, pause durations preceding and following the phrase. Once we've extracted the acoustic features per candidate phrase, the phrase is then classified using a SVM classifier [50].

Highlight generation

We use a constrained utility framework to create audio-visual skims [49]. There are three key components to our framework: (a) a utility model for video shots and audio segments (b) constraints stemming from audio-visual synchronization considerations and from minimum duration and (c) a constraint relaxation strategy to ensure a feasible solution. We shall only summarize our work here, the details can be found in [50].

Utility functions
In order to determine the skim duration, we need to measure the comprehensibility of a video shot and a audio segment as a function of its duration. The shot utility function, models the comprehensibility of a shot as a continuous function of its duration and its visual complexity. Note that the results dealing with visual complexity do not tell us how the comprehensibility of a shot *changes* when we decrease its duration. Hence the need for a shot utility function. While we do not have any experimental results indicating a similar complexity-time relationship for audio, it seems fairly reasonable to hypothecate its existence. Hence, the form of our audio utility function will be similar to the utility function for video shots [50]. We model the utility of a video shot (audio segment) independently of other shots (segments).

Constraints
There are three principal constraints in our algorithm: (a) audio-visual synchronization requirements, (b) minimum and maximum duration bounds on the video shots and the audio segments, and (c) the visual syntactical constraints.
    Audio-visual synchronization is achieved using the idea of tied multimedia

segments. A multimedia segment is said to be fully *tied* if the corresponding audio and video segments begin and end synchronously, and in addition are *uncompressed.* We only associate those speech segments that contain significant phrases with tied multimedia segments.

We focus on the generation of passive information centric summaries that have maximum coherence. Since we deem the speech segments to contain the maximum information, we shall seek to achieve this by biasing the audio utility functions in favor of the clean speech class. In order to ensure that the skim appears coherent, we do two things: (a) ensure that the principles of visual syntax are not violated and (b) have maximal number of tie constraints. These constraints ensure synchrony between the audio and the video segments. The target skim duration is met by successively relaxing the constraints. Relaxing the synchronization constraints has two effects: (a) the corresponding audio and video segments are no longer synchronized, and (b) they can be compressed and if necessary dropped. The details of the search strategy and the mathematical framework can be found in [50].

DOES AN "OPTIMAL" SUMMARY EXIST?

In this section, we attempt to answer a fairly intuitive question — are audio-visual skims optimal in some sense? Clearly, they preserve the dynamism of the original video, while attempting to preserve the semantics. The answer lies in looking at the relationship of the summary to the device, and the user's information needs.

The device on which the summary is to be rendered affects the skim in at least two ways: the nature of the user interface and the device constraints. The user interface can be complex (e.g. the PC), medium (e.g. a palm pilot) and simple (e.g. a cell phone). The user interface affects the resolution of the visual skim as well as the size of the thumbnails of the image storyboard. This is an important consideration, because on very small screens (e.g. cell phone) it would be very difficult for the user to comprehend the tiny thumbnails shown on the screen. Note that the user interface also influences the kinds of tasks that the user has in mind (e.g. it is difficult to input a query on a cell phone).

The computational resources available on the device — cpu speed, memory, bandwidth, availability of an audio rendering device, all effect the form of the summary. For example, a palm-pilot or a cell phone may not have the computational resources to render a video skim. The specific resources present will affect the resolution of the skim, and the decision to include video (as opposed to still images) in the skim. Hence, only when we know the nature of the user interface and the device resource capabilities can we come to a conclusion on the form of the summary.

4.5 Conclusion

In this chapter we have discussed three important aspects of video analysis — (a) scene analysis, (b) analysis of events and (c) schemes to summarize video.

In scene analysis we reviewed work done with scene transition graphs and the memory model based segmentation framework. We described a computational scene model for films. We showed the existence of four different types of computable scenes, that arise due to different synchronizations between audio and video scene boundaries. The computational framework for audio and video scenes was derived from camera placement rules in film-making and from experimental observations on the psychology of audition. We believe that the computable scene formulation is the first step towards deciphering the semantics of a scene.

Scene transition graphs are constructed using time-constrained clustering of video frames along with cluster label transition analysis. This results in a directed graph, that is then analyzed via analysis of cut edges for scene changes. We then showed how a causal, finite memory model formed the basis of our audio and video scene segmentation algorithm. In order to determine audio scene segments we determine correlations of the feature data in the attention span, with the rest of the memory. An important aspect of this work is the incorporation of high-level semantic constraints for merging information from different modalities (audio, video, silence and structure) to ensure that the resulting segmentation is consistent with human perception.

In event analysis, we first defined an event to be a change in state or property of an entity. Then we briefly discussed MPEG-7 event description schemes. We discussed in some detail the use of Hidden Markov Models for detecting states in soccer. We also gave an overview of an interactive framework, the Visual Apprentice, for learning video object/scene detectors and their applications for event detection in sports.

We discussed two summarization schemes — image based storyboards and video skims. Image based storyboards are typically constructed by clustering video shots in a feature space with an appropriate metric. The video shot key-frame closest to the cluster centroid is picked as the representative key-frame.

In this paper, we've presented a novel framework for condensing computable scenes. The solution has three parts: (a) analysis of visual complexity and film syntax, (b) robust audio segmentation and significant phrase detection via SVM's and (c) determining the duration of the video and audio segments via an constrained utility maximization. We defined a measure for visual complexity and then showed how we can map visual complexity of a shot to its comprehension time. After noting that the syntax of the shots influences the semantics of a scene, we devised algorithms based on simple rules governing the length of the progressive phrase and the dialog. We devised a robust audio segmentation algorithm using SVM classifiers in a tree structure, and imposed duration constraints on the segments using the a modified Viterbi algorithm. We also showed how we could analyze the prosody and detect significant phrases in the speech segments.

We have focused on generating information centric skims with maximum coherence. First, we developed utility functions for both audio and video segments, and we minimized an objective function that was based on the sequence utility. We introduced the idea of tied multimedia segments that imposes synchronization constraints on the skim. Additionally, the objective function is subject to video and audio penalty functions and minimum duration requirements on the audio and video segments.

OPEN ISSUES

We now discuss some of the interesting research issues connected to the ideas discussed here.

For video scene segmentation, there are several clear improvements possible to the work on memory model based scene analysis. The computational model for the detecting the video scene boundaries is limited, and needs to tightened in view of the model breakdowns discussed in [51]. One possible improvement is to do motion analysis on the video and prevent video scene breaks under smooth camera motion. Since shot misses can cause errors, we are also looking into using entropy-based irregular sampling of the video data in addition to the key-frames extracted from our shot-segmentation algorithm.

At a more conceptual level, regardless of the domain of analysis, we believe that general-purpose segmentation algorithms should be modified, keeping the following aspects in mind: (a) How does the specific domain affect the data that we are trying to analyze? Are there explicit (or implicit) constraints? (b) what are the specific syntactical structures in the domain? How do the viewers in that domain group these elements and what meanings do they assign to these structures. (c) What is the sensitivity of the final task to segmentation? In many cases, it may be likely that the final task (e.g. classification) is not very sensitive to the error and performance improvements can be found by fine tuning other aspects of the overall algorithm.

In summarization algorithms, there are several interesting avenues of research. Work on enhancing image based storyboards with animations, sound and text captions is an exciting area of research. The work on audio visual skims can be improved in many ways: (a) the creation of summaries for active tasks, in a constrained environment (b) computing time complexity curves with both audio and video (c) how to construct skims for raw video streams by modifying the skim generation mechanism discussed here? (d) given a domain, a systematic methodology for the assignment of utilities to the various entities of interest in that domain, and how these utilities are modified via their inter-relationships (e) summarization by changing the modality — for example, it may be possible to summarize a video in shorter period of time, by overlaying text captions that summarize the audio visual data (for example, the text captions could come from the transcript).

Acknowledgements

We would like to sincerely appreciate the help and valuable input provided by Ana Benitez, Shahram Ebadollahi, Alejandro Jaimes, and Lexing Xie.

References

[1]   B. Arons *Pitch-Based Emphasis Detection For Segmenting Speech Recordings,* Proc. ICSLP 1994, Sep. 1994, vol. 4, pp. 1931-1934, Yokohama, Japan, 1994.

[2]   B. Adams et. al. *Automated Film Rhythm Extraction for Scene Analysis,* Proc. ICME 2001, Aug. 2001, Japan.

[3]   A.B. Benitez, S.F. Chang, J.R. Smith *IMKA: A Multimedia Organization System Combining Perceptual and Semantic Knowledge,* Proc. ACM MM 2001, Nov. 2001, Ottawa Canada.

[4]   A.S. Bregman Auditory Scene Analysis: The Perceptual Organization of Sound, MIT Press, 1990.

[5]   B. Burke and F. Shook, "Sports photography and reporting", Chapter 12, in *Television field production and reporting*, 2nd Ed, Longman Publisher USA, 1996

[6]   M. Burrows, D.J. Wheeler *A Block-sorting Lossless Data Compression Algorithm,* Digital Systems Research Center Research Report #124, 1994.

[7]   C.-C. Chang, C.-J.Lin, *LIBSVM: a library for support vector machines,* http://www.csie.ntu.edu.tw/~cjlin/libsvm

[8]   M.G. Christel et. al *Evolving Video Skims into Useful Multimedia Abstractions,* ACM CHI '98, pp. 171-78, Los Angeles, CA, Apr. 1998.

[9]   N. Christianini, J. Shawe-Taylor *Support Vector Machines and other kernel-based learning methods,* 2000, Cambridge University Press, New York.

[10]   T.M. Cover, J.A. Thomas *Elements of Information Theory,* 1991, John Wiley and Sons.

[11]   S. Ebadollahi, S.F. Chang, H. Wu, *Echocardiogram Videos: Summarization, Temporal Segmentation And Browsing,* to appear in ICIP 2002, Sep. 2002, Rochester NY.

[12]   D.P.W. Ellis *Prediction-Driven Computational Auditory Scene Analysis,* Ph.D. thesis, Dept. of EECS, MIT, 1996.

[13]   J. Feldman *Minimization of Boolean complexity in human concept learning,* Nature, pp. 630-633, vol. 407, Oct. 2000.

[14]   Bob Foss *Filmmaking: Narrative and Structural techniques* Silman James Press LA, 1992.

[15]   Y. Gong; L.T. Sin; C. Chuan; H. Zhang; and M. Sakauchi, *Automatic parsing of TV soccer programs, Proc. ICMCS'95,* Washington D.C, May, 1995

[16]   B. Grosz J. Hirshberg *Some Intonational Characteristics of Discourse Structure,* Proc. Int. Conf. on Spoken Lang. Processing, pp. 429-432, 1992.

[17]   A. Hanjalic, R.L. Lagendijk, J. Biemond *Automated high-level movie segmentation for advanced video-retrieval systems,* IEEE Trans. on CSVT, Vol. 9 No. 4, pp. 580-88, Jun. 1999.

[18]   L. He et. al. *Auto-Summarization of Audio-Video Presentations,* ACM MM '99, Orlando FL, Nov. 1999.

[19]   J. Hirschberg, B. Groz *Some Intonational Characteristics of Discourse Structure,* Proc. ICSLP 1992.

[20]   J. Hirschberg D. Litman *Empirical Studies on the Disambiguation of Cue Phrases,* Computational Linguistics, 1992.

[21]   J. Huang; Z. Liu; Y. Wang, *Joint video scene segmentation and classification based on hidden Markov model, Proc. ICME 2000,* P 1551 -1554 vol.3, New York, NY, July 30-Aug3, 2000

[22]   J. Huang; Z. Liu; Y. Wang, *Integration of Audio and Visual Information for Content-Based Video Segmentation,* Proc. ICIP 98. pp. 526-30, Chicago IL. Oct. 1998.

[23]   A. Jaimes and S.F. Chang, *Concepts and Techniques for Indexing Visual Semantics,* book chapter in Image Databases, Search and Retrieval of Digital Imagery, edited by V. Castelli and L. Bergman. Wiley & Sons, New York, 2002

[24]   J.R. Kender E.L. Yeo, *Video Scene Segmentation Via Continuous Video Coherence,* CVPR '98, Santa Barbara CA, Jun. 1998.

[25]   R. Lienhart et. al. *Automatic Movie Abstracting,* Technical Report TR-97-003, Praktische Informatik IV, University of Mannheim, Jul. 1997.

[26]   L. Lu et. al. *A robust audio classification and segmentation method,* ACM Multimedia 2001, pp. 203-211, Ottawa, Canada, Oct. 2001.

[27]   T.S-Mahmood, D. Ponceleon, *Learning video browsing behavior and its application in the generation of video previews,* Proc. ACM Multimedia 2001, pp. 119 - 128, Ottawa, Canada, Oct. 2001.

[28]   MPEG MDS Group, *Text of ISO/IEC 15938-5 FDIS Information Technology — Multimedia Content Decsription Interface — Part 5 Multimedia Description Schemes,* ISO/IEC JTC1/SC29/WG11 MPEG01/N4242, Sydney, July 2001.

[29]   J. Nam, A.H. Tewfik *Combined audio and visual streams analysis for video sequence segmentation,* Proc. ICASSP 97, pp. 2665 –2668, Munich, Germany, Apr. 1997.

[30]   M. Naphade et. al. *Probabilistic Multimedia Objects Multijects: A novel Approach to Indexing and Retrieval in Multimedia Systems,* Proc. I.E.E.E. International Conference on Image Processing, Volume 3, pages 536-540, Chicago, IL, Oct. 1998.

[31]   M. Naphade et. al *A Factor Graph Framework for Semantic Indexing and Retrieval in Video, Content-Based Access of Image and Video Library* 2000 June 12, 2000 held in conjunction with the IEEE Computer Vision and Pattern Recognition 2000.

[32] R. Patterson et. al. *Complex Sounds and Auditory Images, in Auditory Physiology and Perception* eds. Y Cazals et. al, pp. 429-46, Oxford, 1992.

[33] S. Paek and S.-F. Chang, *A Knowledge Engineering Approach for Image Classification Based on Probabilistic Reasoning Systems* , IEEE International Conference on Multimedia and Expo. (ICME-2000), New York City, NY, USA, Jul 30-Aug 2, 2000.

[34] S. Pfeiffer et. al. *Abstracting Digital Movies Automatically,* J. of Visual Communication and Image Representation, pp. 345-53, vol. 7, No. 4, Dec. 1996.

[35] W.H. Press et. al *Numerical recipes in C,* 2nd ed. Cambridge University Press, 1992.

[36] L. R. Rabiner B.H. Huang *Fundamentals of Speech Recognition,* Prentice-Hall 1993.

[37] K. Reisz, G. Millar, *The Technique of Film Editing,* 2nd ed. 1968, Focal Press.

[38] C. Saraceno, R. Leonardi *Identification of story units in audio-visual sequences by joint audio and video processing,* Proc. ICIP 98. pp. 363-67, Chicago IL. Oct. 1998.

[39] E. Scheirer M.Slaney *Construction and Evaluation of a Robust Multifeature Speech/Music Discriminator* Proc. ICASSP '97, Munich, Germany Apr. 1997.

[40] S. Pfeiffer et. al. *Automatic Audio Content Analysis,* Proc. ACM Multimedia '96, pp. 21-30. Boston, MA, Nov. 1996,

[41] J.Saunders *Real Time Discrimination of Broadcast Speech/Music,* Proc. ICASSP '96, pp. 993-6, Atlanata GA May 1996.

[42] B. Shahraray, D.C. Gibbon *Automated Authoring of Hypermedia Documents of Video Programs,* in Proc. ACM MM 95, pp. 401-409, 1995.

[43] D. O'Shaughnessy *Recognition of Hesitations in Spontaneous Speech,* Proc. ICASSP, 1992.

[44] S. Sharff *The Elements of Cinema: Towards a Theory of Cinesthetic Impact,* 1982, Columbia University Press.

[45] L.J. Stifelman *The Audio Notebook: Pen and Paper Interaction with Structured Speech,* PhD Thesis, Program in Media Arts and Sciences, School of Architecture and Planning, MIT, Sep. 1997.

[46] S. Subramaniam et. al. *Towards Robust Features for Classifying Audio in the CueVideo System,* Proc. ACM Multimedia '99, pp. 393-400, Orlando FL, Nov. 1999.

[47] H. Sundaram S.F. Chang *Audio Scene Segmentation Using Multiple Features, Models And Time Scales,* ICASSP 2000, International Conference in Acoustics, Speech and Signal Processing, Istanbul Turkey, Jun. 2000.

[48] H. Sundaram, S.F. Chang *Determining Computable Scenes in Films and their Structures using Audio-Visual Memory Models,* Proc. Of ACM Multimedia 2000, pp. 95-104, Nov. 2000, Los Angeles, CA.

[49] H. Sundaram, S.F. Chang, *Condensing Computable Scenes using Visual Complexity and Film Syntax Analysis,* IEEE Proc. ICME 2001, Tokyo, Japan, Aug 22-25, 2001.

[50] H. Sundaram L. Xie Shih-Fu Chang *A framework work audio-visual skim generation.* Tech. Rep. # 2002-14, Columbia University, April 2002.

[51] H. Sundaram, S.F. Chang *Computable Scenes and structures in Films,* IEEE Trans. on Multimedia, Vol. 4, No. 2, June 2002.

[52] Y. Taniguchi et. al. *PanoramiaExcerpts: Extracting and Packing Panoramas for Video Browsing,* in Proc. ACM MM 97, pp. 427-436, Seattle WA, Nov. 1997.

[53] R. Tansley. *The Multimedia Thesaurus: Adding A Semantic Layer to Multimedia Information.* Ph.D. Thesis, Computer Science, University of Southampton, Southampton UK, August 2000.

[54] V. Tovinkere , R. J. Qian, *Detecting Semantic Events in Soccer Games: Towards A Complete Solution, Proc. ICME 2001,* Tokyo, Japan, Aug 22-25, 2001

[55] S. Uchihashi et. al. *Video Manga: Generating Semantically Meaningful Video Summaries* Proc. ACM Multimedia '99, pp. 383-92, Orlando FL, Nov. 1999.

[56] T. Verma *A Perceptually Based Audio Signal Model with application to Scalable Audio Compression,* PhD thesis, Dept. Of Electrical Eng. Stanford University, Oct. 1999.

[57] L. Xie et. al *Structure Analysis Of Soccer Video With Hidden Markov Models,* to appear in ICASSP 2002, Orlando, Fl, May 2002.

[58] P. Xu, L. Xie, S.F. Chang, A. Divakaran, A, Vetro, and H. Sun, *Algorithms and system for segmentation and structure analysis in soccer video, Proc. ICME 2001,* Tokyo, Japan, Aug 2001

[59] M. Yeung B.L. Yeo *Time-Constrained Clustering for Segmentation of Video into Story Units,* Proc. Int. Conf. on Pattern Recognition, ICPR '96, Vol. C pp. 375-380, Vienna Austria, Aug. 1996.

[60]   B.L. Yeo, M. Yeung *Classification, Simplification and Dynamic Visualization of Scene Transition Graphs for Video Browsing,* Proc. SPIE '98, Storage and Retrieval of Image and Video Databases VI, San Jose CA, Feb. 1998.

[61]   M. Yeung, B.L. Yeo and B. Liu, *Segmentation of Video by Clustering and Graph Analysis*, Computer Vision and Image Understanding, V. 71, No. 1, July 1998.

[62]   D. Yow, B.L.Yeo, M. Yeung, and G. Liu, "Analysis and Presentation of Soccer Highlights from Digital Video" *Proc. ACCV, 1995,* Singapore, Dec. 1995

[63]   T. Zhang C.C Jay Kuo *Heuristic Approach for Generic Audio Segmentation and Annotation,* Proc. ACM Multimedia '99, pp. 67-76, Orlando FL, Nov. 1999.

[64]   D. Zhong and S.F. Chang, "Structure Analysis of Sports Video Using Domain Models", *Proc. ICME 2001*, Tokyo, Japan, Aug. 2001

[65]   D. Zhong *Segmentation, Indexing and Summarization of Digital Video Content* PhD Thesis, Dept. Of Electrical Eng. Columbia University, NY, Jan. 2001.

5 CONTENT-BASED RETRIEVAL FOR DIGITAL AUDIO AND MUSIC

Dr. Changsheng Xu, Prof. David Dagan Feng, and Dr. Qi Tian

In this chapter, we summarize the research achievements in the area of content-based audio and music retrieval. This chapter covers the research aspects of audio feature extraction, generic audio classification and retrieval, music content analysis, and content-based music retrieval, providing an overview of current research in the area. In addition, two typical systems for content-based audio and music retrieval are discussed in detail. Finally, based on the current technology used in content-based audio/ music retrieval and the demand from real-world applications, future promising directions are identified.

5.1 Introduction

The rapid development of computer networks and the technologies related to the Internet have resulted in a rapid increase in the size of digital multimedia data collections. These data include information media such as audio, video, image and document. The effective organization of such information to enable efficient browsing, searching and retrieval has been an active research area during the past decades. In the content-based image and video retrieval area, various methods and several commercial systems have been developed since the early 1990's. Many special issues of leading journals have been dedicated to this topic [19, 38, 43, 45, 51, 57, 60].

Compared with the content-based image and video retrieval, content-based audio retrieval, especially content-based music retrieval, provides a special challenge because raw digital audio data is a featureless collection of bytes with the most rudimentary fields attached such as name, file format, sampling rate, which does not readily allow content-based retrieval. Current content-based audio retrieval methods are based on content-based image retrieval methods. Firstly, a feature vector is constructed by extracting acoustic and subjective features from the audio in the database. Secondly, the same features are extracted from the queries. Finally, the relevant audio in the database is ranked according to the feature match between the query and the database. Accuracy and speed are two important indexes to evaluate a retrieval method. An effective representation should be able to capture the most significant properties of the audio and such a representation must be robust under various circumstances and general enough to describe various audio classes to ensure a high level of accuracy in audio retrieval. A computerized method which allows efficient and automated content-based classification and retrieval for audio database is required, due to the rapid increase of audio/music database and Internet technology, which allows audio distribution on-line [13, 34, 58].

This chapter will cover all research areas related to content-based audio and music retrieval. In Section 5.2, audio feature extraction is described in detail. Section 5.3 introduces generic audio classification and retrieval methods. Music content analysis is presented in Section 5.4. Section 5.5 surveys content-based music retrieval methods. Section 5.6 introduces two case studies of content-based audio and music retrieval. Section 5.7 gives the concluding remarks and outlines promising future research direction.

5.2 Audio Feature Extraction

There are many features that can be used to characterize the audio signal. Generally, they can be divided into two categories: acoustic features and subjective/semantic features.

ACOUSTIC FEATURES
Acoustic features describe an audio in terms of commonly understood acoustical characteristics and can be computed directly from the audio file. These features include loudness, spectrum power, brightness, bandwidth, pitch, and cepstrum. Their descriptions and definitions are noted below.

Loudness
Loudness is approximated by the square root of the energy of the signal computed from the Short-Time Fourier Transform (STFT), in decibels. The loudness time series is highpass filtered to remove the long-term average volume level. A more accurate loudness estimate would account for the frequency response of the human ear; if desired, the necessary equalization can be added by applying the Fletch-Munson equal-loudness contours and even more detailed models of loudness summation. A 120-decibel range is audible to the human ear.

Spectrum Powers
The spectrum powers include total spectrum power and sub-band powers. They are all represented with logarithmic forms.
The total spectrum power can be calculated from:

$$P = \log(\int_0^{\omega_0} |F(\omega)|^2 \, d\omega) \tag{5-1}$$

where $|F(\omega)|^2$ is the power at the frequency ω and ω_0 is the half sampling frequency.
The sub-band power is represented as:

$$P_j = \log(\int_{L_j}^{H_j} |F(\omega)|^2 \, d\omega) \tag{5-2}$$

where L_j and H_j are lower and upper bound of sub-band j.

Brightness

Brightness is computered as the centroid of the STFT and is stored as a log frequency. It is a measure of the higher-frequency content of the signal. For example, putting your hand over your mouth as you speak reduces the brightness of the speech sound as well as the loudness. This feature varies over the same range as the pitch, although it can not be less than the pitch estimate at any given instant.

$$\omega_C = \frac{\int_0^{\omega_0} \omega \,|F(\omega)|^2 \, d\omega}{\int_0^{\omega_0} |F(\omega)|^2 \, d\omega} \qquad (5\text{-}3)$$

Bandwidth

Bandwidth is computed as the power-weighted average of the difference between the spectral components and the centroid. For example, a single sine wave has a bandwidth of 0 and ideal white noise has an infinite bandwidth.

$$B = \sqrt{\frac{\int_0^{\omega_0} (\omega - \omega_C)\,|F(\omega)|^2 \, d\omega}{\int_0^{\omega_0} |F(\omega)|^2 \, d\omega}} \qquad (5\text{-}4)$$

Pitch

Pitch is the fundamental period of a human speech waveform, and is an important parameter in the analysis and synthesis of speech signals. In an audio signal, which generally consists of pure speech as well as many other sounds, the physical meaning of pitch is lost. However we can still use pitch as a low-level feature to characterize changes in the periodicity of waveforms in different audio signals.

The monophonic pitch is estimated from the STFT. For each frame, the frequencies and amplitudes of the spectral peaks are computed using a parabolic interpolation. An approximate greatest common divisor algorithm is used to estimate the fundamental frequency. The pitch time-series is cleaned up by heuristics which remove harmonic jump and noise errors. The pitch algorithm also returns a pitch confidence value that can be used to weight the pitch in later calculations. There are many available pitch detection algorithms [21].

Cepstrum

The mel-frequency cepstra has proven to be highly effective in automatic speech recognition and in modeling the subjective pitch and frequency content of audio signals. Psychophysical studies have found the phenomena of the mel pitch scale and the critical band, and the frequency scale-warping to the mel scale has led to the cepstrum domain representation.

The cepstrum can be illustrated by use of the Mel-Frequency Cepstral Coefficients (MFCCs). These are computed from the FFT power coefficients [47]. The power coefficients are filtered by a triangular bandpass filter bank. The filter bank consists of $K=19$ triangular filters. They have a constant mel-frequency interval, and cover the frequency range of 0Hz – 4000Hz. Denoting the output of the filter bank by S_k $(k = 1,2,...,K)$, the MFCCs are calculated as:

$$c_n = \sqrt{\frac{2}{K}} \sum_{k=1}^{K} (\log S_k) \cos[n(k - 0.5)\pi / K] \quad n = 1,2,...,L \tag{5-5}$$

where L is the order of the cepstrum.

SUBJECTIVE/SEMANTIC FEATURES

Subjective features describe sounds using personal descriptive language. The system must be trained to understand the meaning of these descriptive terms. For example, a user might be looking for a "shimmering" sound.

Semantic features are high-level features that are summarized from the low-level features. Compared with low-level features, they are more accurately to reflect the characteristics of audio content. For example, a music score that includes events and instruments is a semantic feature of music signal.

Timbre

Timbre is determined by the harmonic profile of the sound source. It is also called tone color. It is generally defined as the quality which allows one to differentiate between sounds of the same level and loudness made by different musical instruments or voices. Every sound source has an individual quality that is determined by its harmonic profile. Music timbre contributes greatly to the effect of mood in music. Timbre depends upon the waveform, sound pressure, frequency location of the spectrum, and the temporal characteristics of the stimuli [3]. Investigations on building physical models for timbre perception in music analysis [4,20,23] have been conducted for a long time.

Rhythm

Rhythm is another important feature used to characterize music. It is the quality of happening at regular periods of time. It also can represent changes in patterns of timbre and energy in a sound clip. To set up the rhythm analysis, beat detection needs to be conducted. Various beat detection and rhythm analysis algorithms have been reported in the literature [18, 53].

Events

Score-level descriptions of music are typically in the form of events. These events are typically characterized by a starting time, a duration, and a series of parameters such as pitch, loudness, articulation, vibrato etc. All of these parameters can vary in time during the event. MIDI data is widely used as a score-level representation of music in computer environments. It is very limited in terms of the musical concepts it contains, but can be coerced into representing a wide range of music well enough.

It is possible to convert continuous frame-level data for pitch, loudness etc., into the MIDI values for the key number, pitch bend, key velocity and so on. The major difficulty is determining the event boundaries.

Instruments

It is useful to identify the type of the audio source. For musical examples, this would be typically be one of the standard musical instruments. Instrument identification can be accomplished using the histogram classification system. This requires that the system has been trained on all possible instruments.

5.3 Generic Audio Classification and Retrieval

Generally, audio can be classified into three categories: speech, music and other sounds. In this section, generic audio mainly refers to speech and other sounds. Music is treated as a subset of generic audio, because the structures specific to music have yet to be fully explored.

CONTENT-BASED AUDIO SEGMENTATION

It is useful to segment an audio stream into different semantic parts, such as speech, music, silence, and environment sounds. Segmentation is important not only for audio content analysis, but also for video structure parsing. Extracting the features from each segment of the audio stream and applying classification methods to obtain the audio scene achieves segmentation.

A lot of studies have been conducted using different features and different classification methods. The most successful achievement in this area is speech/music discrimination, because speech and music are quite different in spectral distribution and temporal change pattern. Saunders[50] used the average zero-crossing rate and the short time energy as features and applied a simple thresholding method to discriminate speech and music from the radio broadcast. Scheirer[52] used thirteen features in time, frequency and cepstrum domains and different classification methods (GMM, BP-ANN, KNN, MAP, etc.) to achieve a robust performance. Both approaches reported an accuracy rate for real-time classification greater than 95% when a window size of 2.4s was used. However, the performance decreases when a small window size is used or other audio scenes such as environment sounds are taken into consideration.

Further research work has been done to segment audio data into more categories. EI-Maleh[9] proposed a method to classify audio signals into speech, music and others for the purpose of parsing of news story. Music was first detected in terms of the average length of time that peaks exist in a narrow frequency region, then speech was separated out by pitch tracking. Kimber[28] proposed an acoustic segmentation approach that mainly applied to the segmentation of discussion recordings in meetings. Audio recordings were segmented into speech, silence, laughter and non-speech sounds by using cepstral coefficients as features and Hidden Markov Model (HMM) as the classifier. The accuracy rate varied with recording type. Zhang[61] proposed an approach to divide the generic audio data segmentation and classification task into two stages. In the first stage which was called the coarse-level audio segmentation and indexing, audio signals were segmented and classified into speech, music, song, speech with music background, environmental sound with music background, six types of environmental sound, and silence. For this level, physical audio features including the energy function, the average zero-crossing rate, the fundamental frequency, and the spectral peak tracks were used to ensure the feasibility of real-time processing. A rule-based heuristic procedure was built to classify audio signals based on these features. In the second stage which was called fine-level audio classification, further classification was conducted within each basic type. Speech was differentiated into the voices of men, women and children. Music was classified into classics, blues, jazz, rock and roll, music with singing and the plain song, according to the instruments or types. Environmental sounds were classified into semantic classes such as applause, bell ring, footstep, wind-storm, laughter, bird's cry, and so on. The accuracy rate reported was over 90%. Lu[37] proposed a robust two-stage audio classification and segmentation method to segment and classify an audio stream into speech, music, environment sound and silence. The first stage of classification was to separate speech from non-speech based on the KNN and LSP VQ classification scheme and simple features such as high zero-crossing rate ratio, low short time energy ratio, spectrum flux and LSP distance. The second stage further segmented non-speech class into music, environment sounds and silence with a rule-based classification scheme and two new features: noise frame ratio and band periodicity. The total accuracy rate was reported over 96%.

CONTENT-BASED AUDIO RETRIEVAL

Essentially, content-based audio retrieval is a pattern recognition problem in which two basic issues: future extraction, and classification based on extracted futures [15]. Considering the former issue, an effective representation should be able to capture the most significant properties of audio content, robust under various circumstances and general enough to describe various sound classes. For the latter issue, the formulation of a distance measure and the rule of classification are crucial [33].

A commercial system developed by Muscle Fish [58] has resulted in a compelling audio retrieval –by-similarity demonstration. In this system, various perceptual features, such as loudness, brightness, pitch, and timbre were used to represent an audio. Given enough training samples, a Gaussian classifier was constructed, or for retrieval, a covariance-weighted Euclidean (Mahalonbis) distance was used as a measure of similarity. For retrieval, the distance was computed between a given sound example and all other sound examples. Sounds were ranked by distance, with the closer sounds being more similar. A case study for this system will be given in detail in Section 5.6.

Foote[13] used 12 mel-frequency cepstral coefficients (MFCCs) plus energy as the audio features. A tree-structured vector quantizer was used to partition the feature vector space into a discrete number of regions or "bins". Euclidean or Cosine distances between histograms of sounds were compared and the classification is done by using the nearest neighbor (NN) rule. The best result was obtained with a supervised quantization tree with 500 bins, and a Cosine distance measure. This approach has been used for speaker identification as well as music and audio retrieval.

Li[33] proposed a method for content-based audio classification and retrieval in which feature selection, perceptual features, mel-cepstral features and their combinations were considered. While perceptual features such as brightness, bandwidth and sub-band energies capture the spectral characteristics of the sounds, some characteristic features were lost. The cepstral coefficients capture the shape of the frequency spectrum of a sound, from which most of the original sound signal can be reconstructed, and hence provide a complement to the perceptual features. Regarding classification, a new pattern classification method, called the Nearest

Feature Line (NFL), was used to explore information held in the audio database. A basic assumption for the NFL was that a prototype (training) set of sounds was available and there existed more than one prototype (feature point) for each sound class, which was generally a valid assumption. The NFL used information provided by multiple prototypes per class, in contrast to the commonly used NN in which classification was performed by comparing the query to each prototype individually.

There are other methods based on the low-level features of the audio content to classify and retrieval audio in a database. Liu[34] used similar features plus sub-band energy ratios. The separability of different classes was evaluated in terms of the intra- and inter-class scatters to identify highly correlated features, and a classification was performed by using a neural network. Feiten[11] used a self-organizing map (SOM) on perceptually-derived spectral features. The net effect was to organize a set of 100 sample-synthesizer sounds into a 2-D matrix such that similar sounds were closer and more disparate sounds were found further away on the grid. Pfeiffer[46] used a filter bank consisting of 256 phase-compensated gammaphone filters [6] to exact audio features. The audio signal was transformed into response probabilities. Such probability coefficients were used as audio features to classify audio content for application such as audio segmentation, music analysis and violent sound detection. A quick audio retrieval method using active search was presented in [54]. It was to search quickly through broadcast audio data to detect and locate sounds using reference templates, based on the active search algorithm and histogram modeling of zero-crossing features. The exact audio signal to be searched should be known *a priori* in this algorithm.

5.4 Music Content Analysis

The recent advances in computing power and Internet technology have made large amounts of raw digital music data available in the form of unstructured monolithic sound files, yet our ability to make automatic analyses of its content is rudimentary. Music content analysis will explore specific structures associated with music, such as harmonics, rhythm, musical instrument, etc., which may be used to automatically segment, classify, summarize, and transcript music content.

MUSIC SEGMENTATION
Music segmentation refers to the process of detecting when there is a change of "texture" in a music stream such as *intro, verse, chorus, ending,* etc. For example, the chorus of a song, the entrance of a guitar solo, or the change from music to singing voice are all examples of segmentation boundaries. To conduct segmentation, various features in different domains are selected and extracted to represent the characteristics of different segments of music. Some training algorithms including the Hidden Markov Model, neural network, and support vector machines may be able to help segmentation.

Bobrek[2] proposed tree-structured filter banks for music segmentation. An 11-stage tree-structured filter bank which led to wavelet packet analysis was designed and implemented. The tree-structured filter bank had sufficient frequency resolution over a wide range of frequencies, while the time resolution at high frequencies satisfied the minimum time resolution for the human ear. To improve selectivity, a new family of quadrature mirror filters (QMFs) with a narrow transition region and a low reconstruction error was implemented.

Tzanetakis[55] proposed a general methodology for music segmentation, which tracked multiple features including spectral centroid, spectral rolloff, spectral flux, zero-crossing rate, and root mean square (RMS). Intuitively the signal was viewed as a trajectory of points (feature vectors) in a high-dimensional space. Abrupt changes in this trajectory indicated segmentation boundaries. The segmentation algorithm worked in four stages:

- A time series of feature vectors V_t was calculated by iterating over the sound file. Each feature vector could be thought of as a short description of the corresponding frame of sound.

- A distance metric $\Delta_t = \|V_t - V_{t-1}\|$ was calculated between successive frames of sound. Mahalonobis distance was used:

$$D(x, y) = (x - y)^T \Sigma^{-1} (x - y) \qquad (5\text{-}6)$$

where Σ is the feature covariance matrix calculated from the whole sound file. This distance rotated and scaled the feature space so the contribution of each feature is equal. Other distance metrics, possibly using relative feature weighting, could also be used.

- The derivative $\dfrac{d\Delta_t}{dt}$ of the distance was taken. Thresholding was then used for finding the peaks of the result. The derivative of the distance would be low for slowly changing textures and high during sudden transitions. Therefore the peaks roughly corresponded to texture changes.

- Peaks were picked using simple huristics and used to create the segmentation of the signal into time regions. As a heuristic example, a minimum duration between successive peaks can be set to avoid small regions.

MUSIC GENRE CLASSIFICATION

The advent of audio compression algorithms is changing the world of music distribution. We are now moving toward a future in which all the world's music will be ubiquitously available. Musical genres are categorical descriptions that are used to characterize music in music stores, radio stations and now on the Internet. They are commonly used to organize the increasing amounts of music available in digital form on the World Wide Web and are important for music information retrieval. Musical genre categorization has traditionally been performed manually. Therefore, techniques for automatically classifying music into its various categories such as *song, instrumental, pop, jazz, classical, rock,* etc, based on different music features, would be a valuable addition to the development of music information retrieval systems. In order to get a good classification performance, robust classification and learning algorithms need to be investigated. It also requires a good knowledge of the music theory.

Logan[36] proposed a method to compare songs based solely on their audio content. A signature for each song was generated based on K-means clustering of spectral features. The signatures could then be compared using the Earth Mover's

Distance [Rubner *et al.*, 1998] which allows comparison of histograms with disparate bins. The distance measure was evaluated using a database of more than 8000 songs in different genres. It reported on average 2.5 out of the top 5 songs returned were perceptually similar for 20 songs judged by two users.

Tzanetakis[56] proposed an algorithm for automatic genre categorization of music signals. A feature set for representing music surface (centroid, rolloff, flux, zero-crossings, and low energy) and rhythm (full wave rectification, low pass filtering, downsampling, normalization, and autocorrelation) information was used to build automatic genre classification algorithms. The performance of proposed feature sets was evaluated by training statistical pattern recognition classifiers using real world music collections. The genre classifier used three classes/genres to describe data: classical, modern (rock, pop) and jazz. It accurately classified 75% of the music signals.

MUSIC SUMMARIZATION
The rapid development of affordable technologies for multimedia content capturing, data storage, high bandwidth/speed transmission and multimedia compression standards such as JPEG and MPEG, have resulted in a rapid increase of the size of digital multimedia data collections and greatly increased the availability of multimedia contents to the general user. The creation of a concise and informative extraction that accurately summarizes original digital content is extremely important in large-scale information organization and processing. Nowadays, the majority of summaries used commercially were produced from the original content manually. For example, a movie clip may provide a good preview of the movie. A book review describes a book in a short and concise fashion. An abstract of a paper provides the main results of the paper without providing detail. A biography tells the life story of a person without recording every single events of his/her life. However, as a large volume of digital contents has become publicly available on mediums such as the Internet during the recent years, efficient ways of automatic summarization have become increasingly important and necessary.

So far a number of techniques have been proposed and developed to automatically generate text [39,48], speech [22,31] and video [17,59] summaries. Compared with text, speech and video summarization, music summarization provides a special challenge because raw digital music data is a featureless collection of bytes, which is only available in the form of highly unstructured monolithic sound files. Music summarization refers to determining the most common and salient themes of a given music piece that may be used to represent the music and is readily recognizable by a listener. Automatic music summarization can be applied to music indexing, content-based music retrieval and web-based music distribution.

A summarization system [32] for MIDI data format used the repetition nature of MIDI compositions to automatically recognize the main melody theme segment for a given piece of music. A detection engine used algorithms that model melody recognition and music summarization problems as various string processing problems and efficiently processes the problems. The system recognized maximal length segments that had non-trivial repetitions in each track of the MIDI format of the musical piece. These segments were basic units of a music composition, and were the candidates for the melody in a music piece. However, MIDI format data are not sampled audio data (i.e., actual audio sounds), but instead contain synthesizer instructions, or MIDI notes, to reproduce the audio data. That is, a synthesizer generates actual sounds from the instructions in a MIDI format data. Compared with actual audio sounds, MIDI data can not provide a common playback experience and an unlimited sound palette for both instruments and sound effects. On the other hand, MIDI data is a structured format, so it is easy to create a summary according to its structure. Therefore, MIDI summarization is not practical in real applications. A music summarization system [35] parameterized each song using "Mel-cepstral" features that have found great success in speech recognition applications. Given these features for a song, various clustering techniques were used to discover the song structure. Then heuristics were used to extract the key phrase given this structure. This summarization method was suitable for certain genres of music such as rock or folk music, but it was less applicable to classical music. On the other hand, "Mel-cepstral" features can not uniquely reflect the characteristics of music content. To make summarization quality more accurate, other features should be taken into consideration.

MUSIC INSTRUMENT IDENTIFICATION

The ability of a normal human listener to recognize objects in the environment from only the sounds they produce is extraordinarily robust with regard to characteristics of the acoustic environment and other competing sound sources. In contrast, computer systems designed to recognize sound sources function precariously, breaking down whenever the target sound is degraded by reverberation, noise, or competing sounds. Automatic musical instrument recognition is a fascinating and essential subproblem in music indexing, retrieval and automatic transcription. It is closely related to computational auditory scene analysis. The basic ides is to develop a computational model that is capable of "listening" to a recording of musical instruments or human voices and identifying the instruments and voices. This model should be based on current models of signal processing in the human auditory system. It explicitly extracts salient acoustic features and uses a novel improvisational taxonomic architecture (based on statistical pattern recognition techniques) to identify the instrument and voice. This work has implications for research in music timbre, automatic media annotation, human speaker/singer identification and computational auditory scene analysis.

De Poli[7] constructed a series of Kohonen self-organizing-map (SOM) neural networks using inputs based on isolated tones. Various features calculated from tones (most often Mel frequency cepstrum coefficients) were used as inputs to the SOM, in some cases after the dimensionality of the feature space was reduce with principle components analysis. Kaminkyj[24] compared the classification abilities of a feed-forward neural network and a K-nearest neighbor classifier, both trained with features of the amplitude envelopes of isolated instrument tones. Both classifiers achieved nearly 98% correct classification of tones produced by four instruments (guitar, piano, marimba, and accordion) over a one-octave pitch range. Although this performance appears to be excellent, both the training and test data were recorded from the same instruments, performed by the same players in the same acoustic environment. Also, the four instruments chosen have very distinctive acoustic properties, so it is unlikely that the demonstrated performance would carry over to additional instruments or even to independent test data. Traditional pattern recognition techniques have been also applied to the isolated-tone classification problem. Fujinaga[14] trained a K-nearest neighbor classifier with features extracted from 1338 spectral slices representing 23 instruments playing a range of pitches. Using leave-one-out cross-validation with a genetic algorithm to identify good feature combinations, the system reached a recognition rate of approximately 50%. Kashino[25] used polyphonic pitch tracking and multiple features to identify the sources of notes produced by clarinet, flute, piano, trumpet, and violin in "random" chords. Dubnov[8] used a vector-quantizer based on MFCC features as a front-end to a statistical clustering algorithm. The system was trained with 18 short excerpts from many instruments. Marques[40] constructed a set of 9-way classifiers (bagpipes, clarinet, flute, harpsichord, organ, piano, trombone, violin, and "other") using several different feature sets and classifier architectures. Martin[41] proposed a theory of sound-source recognition, casting recognition as a process of gathering information to enable the listener to make inferences about objects in the environment or to predict their behaviour. Classifiers were built that were able to operate on test data that include samples played by several different instruments which are noisy and reverberant. Eronen[10] proposed a system for musical instrument recognition using a wide set of features to model the temporal and spectral characteristics of sounds. Signal processing algorithms were designed to measure these features in acoustic signals. The usefulness of the features was validated using test data that consisted of 1498 samples covering the full pitch ranges of 30 orchestral instruments from the string, brass and woodwind families. The correct instrument family was recognized with 94% accuracy and individual instruments in 80% of cases.

The main problem in evaluating musical instrument recognition system is that very few systems have been extensively evaluated with independent test data. We can not say that certain system has demonstrated any meaningful generality of performance until such testing is done on it.

MUSIC TRANSCRIPTION

Music transcription is defined to the process of listening to a piece of music and reconstructing musical notation for the notes that constitute the piece. The goal of automatic music transcription is to achieve a score-like (notes/pitches through time) representation from musical signals. Music melody extraction from polyphonic music signals is a challenge, although many methods have been reported for automatic monophonic speech and music pitch analysis. One of the remarkable characteristics of polyphonic musical signals is the huge range of the fundamental frequency and most frequency components that originate in different musical notes overlap. This results in an intrinsic ambiguity in the interpretation of musical signals. In addition, the identification of musical instruments becomes an essential issue in the multiple-instrument case.

The early work [44] for transcription of a single-pitched melody was based on frequency analysis. Later, recognition systems for multiple-pitched music performed by a single musical instrument (e.g. piano solos) were proposed [27]. Kashino[26] proposed a processing method for ensemble music recognition. The method consisted of two stages: adaptive template matching and musical context integration. The basic idea of the template matching involved a matched filter. As widely accepted, a match filter in the time-domain is a powerful tool to identify the signal of a specific sound in a mixture. Each template corresponded to a musical note played by a specific kind of instrument with a specific musical pitch (e.g. Piano-C4, Piano-D4). Then, by calculating correlation between the output from each filter and input signal, a specific sound source was detected. The output from adaptive template matching was treated as hypotheses rather than final results, and a hypothesis verification method was employed to integrate musical context. The evaluations using recordings of real ensemble performances reported that the integration of musical context improved the precision of source identification from 67.8% to 88.5% on average. Fernandez[12] proposed a computationally efficient method for automatic note recognition from a polyphonic musical signal by looking for correctly shaped (magnitude and phase wise) peaks in a time and frequency oversampled multiscale decomposition of the signal. Peaks (partial candidates) got accepted/discarded by their match to the window spectrum shape and continuity-across-scale constraints. The final partial list built a re-shaped and equalized spectrum. Note candidates were found by searching for harmonic patterns. Perceptual and source based rejection criteria helped discard false notes, frame-by-frame. Slightly non-causal post-processing used continuity to kill too short notes, fill in the gaps, and correct (sub)octave jumps. Klapuri[29] proposed two new methods for automatic music transcription. The first method suppressed the non-harmonic signal components caused by drums and percussive instruments. The second method estimated the number of concurrent voices by calculating certain acoustic features in the course of an iterative multipitch estimation system.

5.5 Content-based Music Retrieval

Compared with speech/sound retrieval, music retrieval is much more difficult because a music song consists of many types of sounds and instrument effects. Melody is the key content of music information, and is also a very important cue for music retrieval. Generally, the melodies of music pieces are stored in the database in the form of music score or music notes. A user produces a melody query by keying in a sequence of music notes, playing a few notes on an instrument, or singing/humming through a microphone. The query melody is usually incomplete, inexact and corresponding to anywhere of the targeting melody (not just at the beginning). However, in current stage, music contents in database side are MIDI files or music scores, which are easy to be represented by a symbolic sequence. To make content-based music retrieval more applicable, sound files of digital audio recordings such as WAV and MP3 files should be used to extract melody.

Humming is the most natural way to formulate music queries for people who are not trained or educated with music theory. The humming tune can be converted into a note-like representation by use of signal processing techniques. On the other hand, since music (MIDI or score) can be represented by a sequence of notes and this sequence can be converted into a string of letters, string matching can be applied to music retrieval.

It is difficult for lay people to sing or hum a song exactly, especially when singing from memory, therefore content based music retrieval techniques should be robust to various humming errors.

Ghias[16] reported surprisingly effective retrieval using query melodies that have been quantized to three levels, depending on whether each note was higher, lower, or similar pitch as the previous one. Besides simplifying the pitch extraction, this allowed for less-than-expert singing ability on the part of the user. McNab[42] used flexible string-matching algorithms to locate similar melodies located anywhere in a piece and provides a detailed design information and a prototype system encompassing all the aspects of a music retrieval facility. The system transcribed acoustic input, typically sung or hummed by the user, and retrieved music, ranked by how closely it matched the input. It operated on a substantial database of retrieval size, and retrieved information in at most a few seconds. It allowed for inaccurate singing or imperfect memory on the part of the user, for variation in the way music was performed, and for differences between music as it is performed. In order to take into account human inaccuracies of recall and of performance, the errors that people make in remembering melodies and in singing them were modeled. The flexible retrieval mechanisms that were tailored to the errors actually encountered in practice were devised. The dynamic programming matching (DP) was used to conduct string matching. DP matching was useful in music retrieval by humming because it allowed for a vague search key and could provide suitable calculation of the similarity. However, it was not applicable to a very large database, since the complexity of the DP matching was super-linear to the database size. In [5], music melody was represented by 4 types of segments according to the shape of the melody contour, the associated segment duration and segment pitch. Song retrieval was conducted by matching segments of melody contour. Kosugi[30] developed a retrieval system that enabled a user to hum part of a melody as a query to obtain the name of a desired song from an audio database. The combined of two musical features, tone transition and tone distribution, enhanced the accuracy of the retrieval system. These features were represented as high dimensional vectors to facilitate the retrieval process. Tone transition was one of the most important information for music recognition. However, this information by itself was not enough to search for a song using a hummed tune as a query, because there were variations in the tone and tempo of tune hummed by a casual singer. Thus, tone distribution information was used to decrease the effect of the variation. Multiple tone distribution information, such as the tone itself or tone differences between successive notes, was used. The information was recorded in the form of a histogram based on either the duration or the number of times that the tone sounds or tone differences appear. The tone distribution histograms based on the number of times decreased the effect of variations in the tempo of a hummed tune. In addition, fuzzy histograms were generated. When recording the tone in a histogram, extra accumulations were added to both sides of the true tone with n semitones. This decreased the effect of variation in tone in a hummed tune.

The basic idea for above method is similar. Pitch contour of the hummed query is detected and pitch changes are converted into strings according to the direction and/or magnitude of the pitch change. Similarly, the melody contour of MIDI is also converted into strings. String matching algorithms are employed to do similarity retrieval. The string matching approach requires precise detection of individual notes (onset and offset) out of the hummed query. However, it is not uncommon for people to substitute a long note with several short notes with same pitch value while humming a tune. When there are tied notes in the melody, it is likely wrong or incomplete notes will be detected. The string matching result would suffer drastically when the error in note detection is not minor.

To deal with above-mentioned issue, Zhu[63] proposed a new slope-based query-by-humming approach, in which the retrieval is robust to the inaccuracy in query and the use of the metronome is eliminated. A pitch tracking method was used to construct the melody curve from a user's humming. Curve features like melody slope pitch range, time duration, and note changes in the slopes were extracted from melody curves. A melody slope matching algorithm was applied to conduct retrieval. A case study for this system will be given in detail in Section 5.6.

5.6 Case Study

There are a lot of content-based audio/music retrieval systems that have been built. Here, we will select two representative systems for general audio and music retrieval respectively and highlight their distinct characteristics.

CONTENT-BASED AUDIO RETRIEVAL (MUSCLEFISH)

Many audio and multimedia applications would benefit from the ability to classify and search for audio based on its characteristics. MuscleFish [58] developed an audio analysis, search, and classification engine to reduce sounds to perceptual and acoustical features. This lets users search or retrieve sounds by any one feature or a combination of them, by specifying previously learned classes based on these features, or by selecting or entering reference sounds and asking the engine to retrieve similar or dissimilar sounds.

Acoustic Features

Loudness, pitch, brightness, bandwidth, and harmony are used to characterize audio signals. All of these aspects of sound vary over time. The trajectory in time is computed during the analysis but not stored as such in the database. However, for each of these trajectories, several features are computed and stored. These include the average value, the variance of the value over the trajectory, and the autocorrelation of the trajectory at a small lag. Autocorrelation is a measure of the smoothness of the trajectory. For example, it can distinguish between a pitch glissando and a wildly varying pitch , which the simple variance measure cannot. The average, variance, and autocorrelation computations are weighted by the amplitude trajectory to emphasize the perceptually important sections of the sound. In addition to the above features, the duration of the sound is stored. The feature vector thus consists of the duration plus the parameters of average, variance, and autocorrelation for each of the aspects of sound given above. Figure 5-1 shows a plot of the raw trajectories of loudness, brightness, bandwidth, and pitch for a recording of male laughter. After the statistical analyses, the resulting analysis record (shown in Table 5-1) contains the computed values. These numbers are the only information used in the content-based classification and retrieval of these sounds. It is possible to see some of the essential characteristics of the sound.

Table 5-1 Male laughter. Duration: 2.12571

Property	Mean	Variance	Autocorrelation
Loudness	-54. 4112	221. 451	0. 938929
Pitch	4. 21221	0. 151228	0. 524042
Brightness	5. 78007	0. 0817046	0. 690073
Bandwidth	0. 272099	0. 0169697	0. 519198

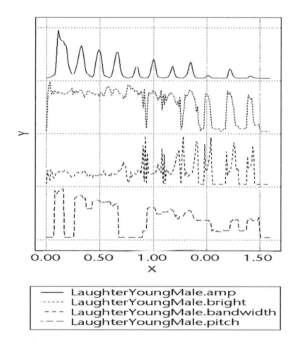

— LaughterYoungMale.amp
······ LaughterYoungMale.bright
- - - LaughterYoungMale.bandwidth
- · - LaughterYoungMale.pitch

Figure 5-1 Male laughter.

Training the System

It is possible to specify a sound directly by submitting constraints on the values of the *N*-vector described above directly to the system. For example, the user can ask for sounds in a certain range of pitch or brightness. However, it is also possible to train the system by example. In this case, the user selects examples of sounds that demonstrate the property the user wishes to train, such as "scratchiness." For each sound entered into the database, the *N*-vector, which is represented as a, is computed.

When the user supplies a set of example sounds for training, the mean vector μ and the covariance matrix R for the a vector in each class are calculated. The mean and covariance are given by:

$$\mu = (1/M)\sum_{j} a[j] \tag{5-7}$$

$$R = (1/M)\sum_{j} (d[j] - \mu)(a[j] - \mu)^T \tag{5-8}$$

where M is the number of sounds in the summation. In practice, one can ignore the off-diagonal elements of R if the feature vector elements are reasonably independent of each other. This simplification can yield significant savings in computation time. The mean and covariance together become the system's model of the perceptual property being trained by the user.

Classify Sounds
When a new sound needs to be classified, a distance measure is calculated from the new sound's a vector and the model above. A weighted L_2 or Euclidean distance can be used.

$$D = ((a - \mu)^T R^{-1}(a - \mu))^{1/2} \tag{5-9}$$

Again, the off-diagonal elements of R can be ignored for faster computation. Also, simpler measures such as an L_1 or Manhattan distance can be used. The distance is compared to a threshold to determine whether the sound is "in" or "out" of the class. If there are several mutually exclusive classes, the sound is placed in the class to which it is closest, that is, for which it has the smallest value of D. If it is known a priori that some acoustic features are unimportant for the class, these can be ignored or given a lower weight in the computation of D. For example, if the class models some timbre aspect of the sounds, the duration and average pitch of the sounds can usually be ignored.

Retrieving Sounds
Some example queries are:
- Retrieve the "scratchy" sounds. That is, retrieve all the sounds that have a high likelihood of being in the "scratchy" class.
- Retrieve the top 20 "scratchy" sounds.
- Retrieve all the sounds that are less "scratchy" than a given sound.
- Sort the given set of sounds by how "scratchy" they are.
- Classify a given set of sounds into the following set of classes.

For small databases, it is easier to compute the distance measure(s) for all the sounds in the database and then to choose the sounds that match the desired result. For large databases, this can be too expensive. To speed up the search, the sounds in the database are indexed by all the acoustic features. This allows people to quickly retrieve any desired hyper-rectangle of sounds in the database by requesting all sounds whose feature values fall in a set of desired ranges. Requesting such hyper-rectangles allows a much more efficient search.

As an example, consider a query to retrieve the top M sounds in a class. If the database has M_0 sounds total, one can first ask for all the sounds in a hyper-rectangle centered around the mean with volume V such that

$$V/V_0 = M/M_0 \tag{5-10}$$

where V_0 is the volume of the hyper-rectangle surrounding the entire database. The extent of the hyper-rectangle in each dimension is proportional to the standard deviation of the class in that dimension.

Then the distance measure is computed for all the sounds and the closest M sounds are returned. If enough sounds that matched the query from this first attempt are not retrieved, the hyper-rectangle volume is increased by the ratio of the number requested to the number found and try again.

Experiments

Above algorithms have been used at Muscle Fish on a test sound database that contains about 400 sound files. These sound files were culled from various sound effects and musical instrument sample libraries. A wide variety of sounds are represented from animals, machines, musical instruments, speech, and nature. The sounds vary in duration from less than a second to about 15 seconds.

A number of classes were made by running the classification algorithm on some perceptually similar sets of sounds. These classes were then used to reorder the sounds in the database by their likelihood of membership in the class. The following examples show the results of this process for several sound sets. These examples illustrate the character of the process and the fuzzy nature of the retrieval.

Example 1

Female speech. The test database contains a number of very short recordings of a group of female and male speakers. In this example, the female-spoken phrase "tear gas" was used. Figure 5-2 shows a plot of the similarity (likelihood) of each of the sound files in the test database to this sound using a default value for the covariance matrix R. The highest likelihood is for the other female speech recordings, with the male speech recordings following close behind.

Example 2

Touchtones. A set of telephone touchtones was used to generate the class in Figure 5-3. Again, the touchtone likelihood is clearly separated from those of other categories. One of the touchtone recordings that was left out of the training set also has a high likelihood, but the other one, as well as one of those included in the training set, returned very low likelihood. Upon investigation, two low-likelihood touchtone recordings were of entire seven-digit phone numbers, whereas all the high-likelihood touchtone recordings were of single-digit tones. In this case, the automatic classification detected an aural difference that was not represented in the user-supplied categorization.

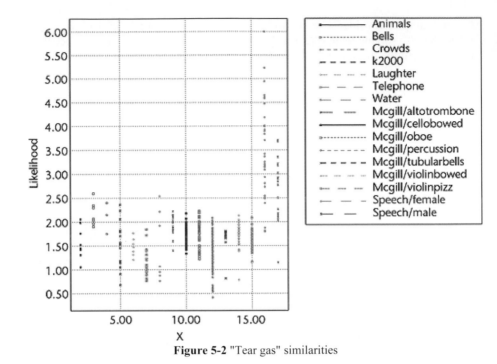

Figure 5-2 "Tear gas" similarities

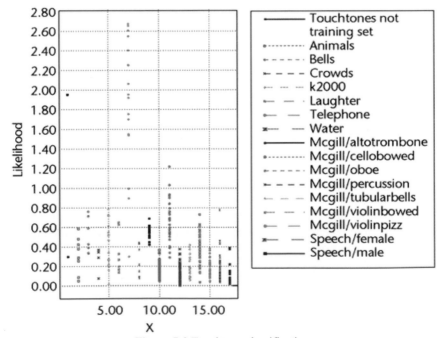

Figure 5-3 Touchtone classification

CONTENT-BASED MUSIC RETRIEVAL BY HUMMING (LIT)

Zhu[62] developed a content-based music retrieval humming system using melody contour matching. The system architecture is illustrated in Figure 5-4. The music files are in MIDI formats. To insert a MIDI file into the database, the melody track is firstly identified and the music notes are extracted from the melody track. Based on the music notes, a continuous melody contour is constructed, and geometrical features are extracted and stored with indexes into the database. To query the database, a user hums a tune through the microphone. The query tune can be any part of the target melody, i.e. not just the beginning. Pitch tracking is conducted on the humming query. A continuous contour is then constructed based on the detected pitches. Geometric contour features are then extracted from the continuous melody contour. A melody search engine is responsible for finding the similar melodies from the database.

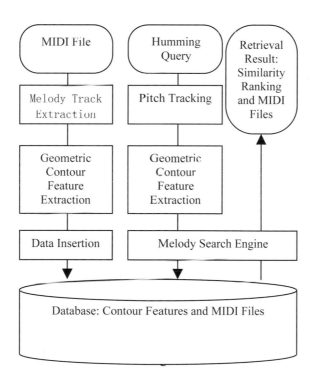

Figure 5-4 System architecture

The melody search engine operates in two levels. In level 1, contour feature matching and an index search finds the likely candidates in the database based on melody contour features. In level 2, alignment of contours and detailed contour comparison are conducted, where the similarity ranking is computed based on our proposed similarity metric. The retrieval result contains a list of the melodies, with a song name and the time position where there is a match in the target tune.

A Geometrical Contour Feature

A geometrical contour feature, called Melody Contour Token (MCT), is defined as a segment of continuous melody contour starting at a line segment with local extreme pitch value and ending at the next line segment with local extreme pitch value. The middle point of a local extreme line segment is both the starting point of a MCT and the ending point of the previous MCT. Pitch Span and Duration are two numerical feature values of a MCT. The pitch span is the difference of the pitch values of the two ends of the MCT. The pitch span can have positive values and negative values. It is positive for pitch value going up in the MCT, and it is negative for the other case. The duration of a MCT is the time duration from the starting point to the ending point of the MCT.

Contour Matching Using MCT

Generally, two similar melody contours will also have similar MCT feature values (pitch spans and durations). Because searching for similar MCTs is relatively inexpensive compared with direct contour similarity measurement, it is the first step in melody contour matching.

Suppose a sequence of m MCTs from the query is matched with m MCTs of a target melody. Denote the pitch span values and duration values of the query MCTs as $(Qs_1, Qs_2, ..., Qs_m)$ and $(Qd_1, Qd_2, ..., Qd_m)$. And denote pitch span values and duration values of the target MCTs as $(Ts_1, Ts_2, ..., Ts_m)$ and $(Td_1, Td_2, ..., Td_m)$. The two MCT sequences will have a match, if

$$\left| Qs_k - Ts_k \right| < Th_1, k = 1,...,m \qquad (5\text{-}11)$$

and

$$\left| \frac{\sum_{k=1}^{m}(Qd_k \times Td_k)}{\sqrt{\sum_{k=1}^{m} Qd_k^2 \sum_{k=1}^{m} Td_k^2}} - 1 \right| < Th_2 \qquad (5\text{-}12)$$

where Th_1 and Th_2 are thresholds.

The MCT sequence search process basically filters out the unlikely candidates from the whole melody database. The later detailed melody contour similarity measure is only conducted upon those likely candidates.

Contour Alignment and Similarity Metric

Proper alignment of two melody contours is important for the computation of similarity. Contour alignment contains two aspects: pitch value transposition and time scale. This can ensure the melody retrieval is independent of key and tempo. The pitch value transposition is done by making the mean pitch values lines of the two contours coincide. For time scale alignment, the corresponding MCTs are linearly scaled such that the starting times coincide and the ending times coincide. The MCTs of a query melody is scaled according to the MCTs of a target melody. The scaling factors for one alignment should not have big variation, since the humming is assumed to be in a constant tempo. This constraint is met in the MCT sequence searching process.

After contour alignment, for each line segment in a query MCT, a distance value can be computed. Any line segment in the MCT of the target melody contour that overlaps in time with this line segment, has a distance value with it. Among these values, the smallest distance value is kept for this line segment in query MCT.

For each MCT in the query melody contour a similarity score is computed as:

$$S_T(k) = \frac{|P(k)| - \sum_i |D_S(i)|}{|P(k)|} \qquad (5\text{-}13)$$

where $P(k)$ is the pitch span of the kth MCT in the query melody contour, and $D_S(i)$ is the distance value of the ith line segment in the query MCT. The summation of $|D_S(i)|$ corresponds to the sum of the distance values of all the line segments within the query MCT. $S_T(k)$ is the similarity score for the kth MCT in the query melody contour. The maximum value of $S_T(k)$ is 1.

The final similarity value for the whole query melody contour is computed as follows:

$$S = \sum_{k=1}^{L} S_T(k) W_T(k) \qquad (5\text{-}14)$$

where L is the number of MCTs in the query melody contour, and $W_L(k)$ is the weight of the kth MCT in the query melody contour. $W_L(k)$ is computed as follows:

$$W_L(k) = \frac{N_L(k)}{N_T} \qquad (5\text{-}15)$$

where $N_L(k)$ is the number of line segments in the kth slope and N_T is the total number of line segment in the contour.

Experiments

The melody database contained 12,000 MIDI files randomly downloaded from the Internet. Most of the files were western and eastern popular songs. There were a total of 2,841,000 music notes in the melodies and they generated about 1,136,000 melody contour tokens.

Robustness of MCT features in humming queries

Five human subjects participated in the experiment. Not one of them was a skilled singer. Ten melodies in the database are selected as target melodies. The subjects are familiar with the target melodies. Each subject hummed each melody twice, once in moderate tempo, and once in a fast tempo. The length of the hummed melody is in average four bars. The total time duration was about 16 seconds at a moderate tempo and about 10 seconds at a fast tempo. There were in total 100 hummed query melodies. MCT identification is performed on the hummed queries. For the ten target melodies, there were a total of 82 MCTs. Each identity of MCTs is manually verified in the query melodies. For the moderate tempo, on average 94% of the MCTs were correctly represented in the humming. For the fast tempo, the success rate was 87%.

MCT sequence search is conducted to identify the correct MCT sequence candidate in the target melody. Four MCT sequences in the queries are used to do the searching. The overall correct identification rate is 78%. But since each humming query usually contains four or five MCT sequence candidate, 94 of the 100 queries have at least one MCT sequence candidate correctly identified. Six of the 100 queries failed to match the desired MCT sequence candidates. These six queries are of relatively low quality, such as significantly out of tune.

In the robustness testing, the correct identification rate can be increased by loosening the threshold values. However, this may lead to a large number of candidates and degrading of the retrieval efficiency.

Efficiency of MCT sequence matching
MCT sequence matching is used to locate a likely candidate out of all possible candidates in the database. The aim of the experiment was to find how many unlikely candidates were successfully rejected. A hundred humming queries were used in the experiment. Searching was conducted for all the candidates in the database. Figure 5-5 shows the average number of candidates returned from the database versus the number of MCTs used in a MCT sequence matching: the more MCTs are used, the smaller the number of candidates returned.

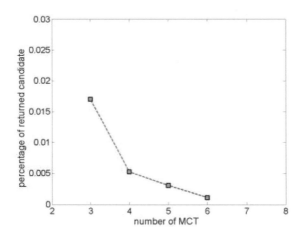

Figure 5-5 Percentage of returned candidate vs. number of MCT

Retrieval performance of similarity metric
Two experiments were conducted to examine the performance of the similarity metric. In the first experiment, the proposed method is compared with the method in [1] which was also based on continuous melody contour. In [1], 20 queries were used to search 77 tunes: the retrieval result was 85% for the top 1, and 90% for the top 3. In our comparison experiment, a hundred queries were used to search 100 tunes, including the 10 target melodies. The other 90 melodies were selected randomly from the database. Eighty-eight percent were correctly returned for the top 1, 89% for the top 2 and 92% for the top 3: which is a better retrieval result than that achieved by [1].

In the second experiment, the retrieval performance was examined using the

12,000 melodies in the database. Figure 5-6 shows the rates of correct retrieval for the tunes present in the top 10 rank list. An eighty percent correct retrieval rate was achieved for the top 10 rank list.

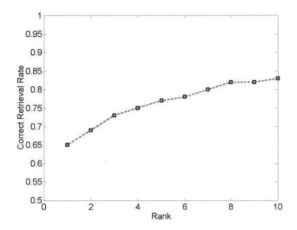

Figure 5-6 Retrieval rate for the top 10 list

5.7 Conclusion

This chapter reviewed current technical achievements in audio feature extraction, generic audio classification and retrieval, music content analysis and content-based music retrieval. Two typical systems for content-based audio and music retrieval were described in detail. The open research issues and future research directions were also identified.

Although many advances have been made in various research areas of content-based audio/music retrieval, there are still many research issues that need to be explored to make content-based audio/music retrieval system more applicable in practice.

MELODY EXTRACTION FROM SOUND RECORDINGS
In current content-based music retrieval systems, melody extraction in database side is conducted from MIDI files or music scores, which have well-defined structures and are easy to be represented as a symbolic string. However, both MIDI files and music scores are not real digital sound recordings, which are desired and accepted by users when they perform retrieval tasks. To make the content-based music retrieval truly applicable in practice, efficient techniques for melody extraction direct from sound recordings need to be explored. At the current stage, how to extract melodic information from polyphonic music, which consists of unstructured raw music data, is a big challenge. It primarily depends upon the achievements of music content analysis.

RETRIEVAL ACCURACY

Retrieval accuracy is the most important factor to evaluate the performance of a retrieval method. However, it is a very difficult task for content-based multimedia retrieval, especially for audio/music retrieval. Feature extraction processing either on the database side or on the query side will introduce noise which decreases the accuracy of the retrieval. Moreover, for content-based audio/music retrieval, it is difficult for a novice to sing a song accurately. Therefore, increasing the retrieval accuracy will continue to be a research direction of researchers.

WEB-BASED SEARCH ENGINE

The expansion of the World Wide Web is astonishing. Each day thousands of music products are added to the web. Web-based search engines are highly desirable for improving the organization and retrieval of the almost unlimited information on the Web. There are a lot of successful text-based engines such as Yahoo, Infoseek, Alta Vista, etc. Some technical breakthroughs are required before content-based search engines become comparable in standard to their text based counterparts.

Acknowledgement

The authors would like to thank Mr. Yongwei Zhu for the contribution and valuable comments on the drafts of the chapter.

References

[1] W.P. Birmingham and R.B. Dannenberg (2001), MUSART: music retrieval via aural queries, In *Proc. Second International Symposium on Music Information Retrieval*.

[2] M. Bobrek and D.B. Koch (1998), Music signal segmentation using tree-structured filter banks, *Journal of Audio Engineering Society*, Vol.46, No.5, pp.412-427.

[3] A. Bregman (1990), *Auditory scene analysis*, Cambridge: MIT Press.

[4] G.R. Charbonneau (1981), Timbre and the perceptual effects of three types of data reduction, *Computer Music Journal*, Vol.5, No.2, pp.10-19.

[5] A. Chen, M. Chang, J. Chen, J.L. Hsu, C.H. Hsu and S. Hua (2000), Query by music segments: an efficient approach for song retrieval, In *Proc. ICME2000*, pp.889-892.

[6] M.P. Cook (1993), *Modelling Auditory Processing and Organization,* Cambridge University Press, Cambridge, UK.

[7] G. De Poli and P. Prandoni (1997), Sonological models for timbre characterization, *Journal of New Music Research*, Vol.26, pp.170-197.

[8] S. Dubnov and X. Rodet (1998), Timbre recognition with combined stationary and temporal features, In *Proc. International Computer Music Conference*, pp.102-108.

[9] K. El-Maleh, M. Klein, G. Petrucci and P. Kabal (2000), Speech/music discrimination for multimedia application, In *Proc. ICASSP00*.

[10] A. Eronen and A. Klapuri (2000), Musical instrument recognition using cepstral coefficients and temporal features, In *Proc. IEEE International Conference on Acoustics, Speech and Signal Processing*.

[11] B. Feiten and S. Gunzel (1994), Automatic indexing of a sound database using self-organizing neural nets. *Computer Music Journal* **18**, pp. 53–65.

[12] C. P. Fernandez and Q.F.J. Casajus (1999), Multi-pitch estimation for polyphonic musical signal, In *Proc. ICASSP-99*.

[13] J. Foote (1997), Content-based retrieval of music and audio. In *Multimedia Storage and Archiving Systems, Proc. of SPIE*. **3229**, pp. 138–147.

[14] I. Fujinaga (1998), Machine recognition of timbre using steady-state tone of acoustic musical instruments. In *Proc. International Computer Music Conference*, pp.207-210.

[15] K. Fukunage (1990), *Introduction to statistical pattern recognition*, Academic Press, Boston.

[16] A. Ghias (1995), Query by humming, In *Proc. ACM Multimedia 95*, San Francisco, USA.

[17] Y. Gong and X. Liu (2001), Summarizing video by minimizing visual content redundancies, In *Proc. IEEE International Conference on Multimedia and Expo*, Tokyo, Japan, pp. 788-791.

[18] M. Goto and Y. Muraoka (1994), A beat tracking system for acoustic signals of music, in *Proc. ACM Multimedia 1994*, San Francisco, ACM.

[19] A. Gupta and R. Jain (1997), Visual information retrieval. *Communications of ACM* **40**, pp.35–42.

[20] S. Handel (1995), Timbre perception and auditory object identification, In *Hearing*, Moore B. C. J., ed., New York: Academic Press.

[21] W. Hess (1983), *Pitch determination of speech signals*, Springer-Verlag.

[22] C. Hori and S. Furui (1998), Improvements in automatic speech summarization and evaluation methods, In *Proc. International Conference on Spoken Language Processing*, Sydney, Australia.

[23] A.J.M. Houstsma (1997), Pitch and timbre: definition, meaning and use, *Journal of New Music Research*, Vol.26, pp.104-115.

[24] I. Kaminkyj and A. Materka (1995), Automatic source identification of monophonic musical instrument sounds, In *Proc. IEEE International Conference on Neural Network*, pp.189-194.

[25] K. Kashino and A. Makerka (1997), Sound source identification for ensemble music based on the music stream extraction, In *Proc. International Joint Conference on Artificial Intelligence*.

[26] K. Kashino and H. Murase (1999), Music recognition using note transition context, In *Proc. ICASSP-99*.

[27] H. Katayose and S. Inokuchi (1989), An intelligent transcription system, In *Proc. Int'l Conf. Music Perception and Cognition*, pp.95-98.

[28] D. Kimber and L. Wilcox (1996), Acoustic segmentation for audio browsers, In *Proc. Interface Conference*, Sydney, Australia.

[29] A. Klapuri (2001), Eronen A., Seppanen J. and Virtanen T., Automatic transcription of music, In *Proc. Symposium on Stochastic Modeling of Music*, 22th of October, Ghent, Belgium.

[30] N. Kosugi, Y. Nishihara, S. Kon'ya, M. Yamamuro and K. Kushima (1999), Music retrieval by humming, In *Proc. of IEEE PACRIM'99*.

[31] K. Koumpis and S. Renals (1998), Transcription and summarization of voicemail speech, In *Proc. International Conference on Spoken Language Processing*, Sydney, Australia.

[32] R. Kraft, Q. Lu and S. Teng (2001), Method and apparatus for music summarization and creation of audio summaries, US Patent 6,225,546.

[33] S.Z. Li (2000), Content-based classification and retrieval of audio using the nearest feature line method, *IEEE Transactions on Speech and Audio Processing*, September.

[34] Z. Liu, J. Huang, Y. Wang and T. Chen (1997), Audio feature extraction and analysis for scene classification. In *IEEE Signal Processing Society 1997 Workshop on Multimedia Signal Processing*, pp. 523–528.

[35] B. Logan and S. Chu (2000), Music summarization using key phrases, In *Proc. IEEE International Conference on Audio, Speech and Signal Processing*, Orlando, USA.

[36] B. Logan and A. Salomon (2001), A music similarity function based on signal analysis, In *Proc. ICME2001*, Japan, pp.952-955.

[37] L. Lu, H. Jiang and H.J. Zhang (2001), A robust audio classification and segmentation method, In *Proc. ACM Multimedia 2001*, Ottawa, Canada.

[38] W.Y. Ma and H.J. Zhang (1999), Content-based image indexing and retrieval. In *Handbook of Multimedia Computing*, ed. by Furht B. CRC Press, Florida, pp. 227–244.

[39] I. Mani and M.T. Maybury (eds.) (1999), *Advances in Automatic Text Summarization*, Cambridge, Massachusetts: MIT Press.

[40] J. Marques (1999), *An Automatic Annotation System for Audio Data Containing Music*, Master's Thesis, Massachusetts Institute of Technology, Cambridge, MA, USA.

[41] K.D. Martin (1999), *Sound-source Recognition: A Theory and Computational Model*, Ph.D Thesis, Massachusetts Institute of Technology, Cambridge, MA, USA.

[42] R. McNab, L. Smith, I. Witten, C. Henderson and S. Cunningham (1996), Towards digital music library: Tune retrieval from acoustic input, In *Proc. Digital Library '96*, pp. 11–18.

[43] A.D. Narasimhalu (1995), Special section on content-based retrieval, *ACM Multimedia Sys.* **3**, pp. 1–41.

[44] T. Niihara and S. Inokuchi (1986), Transcription of sung song, In *Proc. ICASSP-86*, pp.1277-1280.

[45] A. Pentland and R. Picard (1996), Special issue on digital library, *IEEE Trans. Patt. Recog. And Intell.* **18**, pp. 673–733.

[46] S. Pfeiffer, S. Fischer and W.E. Elsberg (1996), Automatic audio content analysis, Tech. Rep. No. 96-008, University of Mannheim, Mannheim, Germany.

[47] L. Rabiner and B.H. Juang (1993), *Fundamentals of speech recognition.* Prentice Hall, Englewood Cliffs, N.J., pp. 189.

[48] F. Ren and Y. Sadanaga (1998), An automatic extraction of important sentences using statistical information and structure feature, In *Proc. NL98-125*, pp. 71-78.

[49] Y. Rubner, C. Tomasi and L. Guibas (1998), The Earth Mover's Distance as a metric for image retrieval, Tech. Rep., Stanford University.

[50] J. Saunders (1996), Real-time discrimination of broadcast speech/music, In *Proc. ICASSP96*, Vol.2, pp. 993-996.

[51] B. Schatz and H. Chen (1996), Building large-scale digital libraries. *IEEE Comput. Mag.* **29**, pp. 22–77.

[52] E. Scheirer and M. Slaney (1997), Construction and evaluation of a robust multifeature music/speech discriminator, In *Proc. ICASSP97*, Vol.2, pp.1331-1334.

[53] E. Scheirer (1998), Tempo and beat analysis of acoustic musical signals, in *J. Acoust. Soc. Am.* 103(1), pp 588-601.

[54] G. Smith, H. Murase and K. Kashino (1999), Quick audio retrieval using active search, In *Proc. ICASSP99*, Turkey.

[55] G. Tzanetakis and P. Cook (1999), Multifeature audio segmentation for browsing and annotation, In *Proc. IEEE Workshop on Applications of Signal Processing to Audio and Acoustics*, New Paltz, New York.

[56] G. Tzanetakis, G. Essl and P. Cook (2001), Automatic musical genre classification of audio signals, In. *Proc. Int. Symposium on Music Information Retrieval (ISMIR)*, Bloomington, Indiana, USA.

[57] N.G. Venkat and V.R. Jijay (1995), Special issues on content-based image retrieval systems. *IEEE Comput. Mag.* **28**, pp. 18–62.

[58] E. Wold, T. Blum, D. Keislar and J. Wheaton (1996), Content-based classification, search and retrieval of audio, *IEEE Multimedia Mag.* **3**, pp. 27–36.

[59] I. Yahiaoui, B. Merialdo and B. Huet (2001), Generating summaries of multi-episode video, In *Proc. IEEE International Conference on Multimedia and Expo*, Tokyo, Japan, pp. 792-795.

[60] H.J. Zhang (1999), Content-based video browsing and retrieval, In *Handbook of Multimedia Computing,* ed. by Furht B. CRC Press, Florida, pp. 255–280.

[61] T. Zhang and C.-C. Kuo (1999), Video content parsing based on combined audio and visual information, In *Proc. SPIE 1999*, San Jose, USA, Vol.4, pp.78-89.

[62] Y. Zhu, C. Xu and M. Kankanhalli (2001), Melody curve processing for music retrieval, In *Proc. ICME2001*, Japan, pp.401-404.

[63] Y. Zhu, M. Kankanhalli and C. Xu (2001), Pitch tracking and melody slope matching for song retrieval, In *Proc. PCM2001*, Beijing, China, pp.530-537.

6 MPEG-7 MULTIMEDIA CONTENT DESCRIPTION STANDARD

Dr. John Smith

In this chapter, we describe the recently developed MPEG-7 Multimedia Description Interface Standard. MPEG-7 provides a standardized metadata system for describing multimedia content using XML. MPEG-7 allows interoperable indexing, searching, and retrieval of video, images, audio, and other forms of multimedia data. In this chapter, we introduce the description tools standardized by MPEG-7 and examine their use in multimedia applications.

6.1 Introduction

MPEG-7 is a standard developed by International Standards Organization (ISO) and International Electrotechnical Commission (IEC), which specifies a "Multimedia Content Description Interface." MPEG-7 provides a standardized representation of multimedia metadata in XML. MPEG-7 describes multimedia content at a number of levels, including features, structure, semantics, models, collections, and other immutable metadata related to multimedia description. The objective of MPEG-7 is to provide an interoperable metadata system that is also designed to allow fast and efficient indexing, searching and filtering of multimedia based on content.

Several key-points about MPEG-7 include the following: MPEG-7 is not a video coding standard. MPEG has a history in developing video coding standards, including MPEG-1, -2, and -4. However, MPEG-7 addresses only metadata aspects of multimedia. The MPEG-7 standard specifies an industry standard schema using XML Schema Language. The schema is composed of Description Schemes (DS) and Descriptors. Overall, the MPEG-7 schema defines over 450 simple and complex types. MPEG-7 produces XML descriptions but also provides a binary compression system for MPEG-7 descriptions. The binary compression system allows MPEG-7 descriptions to be more efficiently stored and transmitted. The MPEG-7 descriptions can be stored as files or within databases independent of the multimedia data, or can embedded within the multimedia streams, or broadcast along with multimedia data.

MPEG-7 STANDARD SCOPE
The scope of the MPEG-7 standard is shown in Figure 6-1. The normative scope of MPEG-7 includes Description Schemes (DSs), Descriptors (Ds), the Description Definition Language (DDL), and Coding Schemes (CS). MPEG-7 standardizes the syntax and semantics of each DS and D to allow interoperability. The DDL is based

on XML Schema Language. The DDL is used to define the syntax of the MPEG-7 DSs and Ds. The DLL allows the standard MPEG-7 schema to be extended for customized applications.

The MPEG-7 standard is "open" on two sides of the standard in that the methods for extraction and use of MPEG-7 descriptions are not defined by the standard. As a result, methods, algorithms, and systems for content analysis, feature extraction, annotation, and authoring of MPEG-7 descriptions are open for industry competition and future innovation. Likewise, methods, algorithms, and systems for searching and filtering, classification, complex querying, indexing, and personalization are also open for industry competition and future innovation.

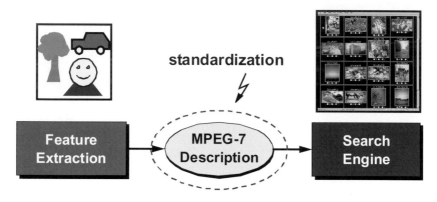

Extraction	MPEG-7 Scope	Use
Content analysis (D, DS)	Description Schemes (DSs)	Searching & filtering
Feature extraction (D, DS)	Descriptors (Ds)	Classification
Annotation tools (DS)	Language (DDL)	Complex querying
Authoring (DS)	Coding Schemes (CS)	Indexing

Figure 6-1 Overview of the normative scope of MPEG-7 standard. The methods for extraction and use of MPEG-7 descriptions are not standardized.

6.2 Context and Applications

MPEG CONTEXT
The MPEG-7 standard represents a major step in progression of standards developed by MPEG. As shown in Figure 6-2, it functionality is complementary to that provided by MPEG-1, MPEG-2, and MPEG-4 standards. These standards address compression, coding, transmission, retrieval, and streaming. MPEG-7 addresses indexing, searching, filtering, and content management, browsing, navigation, and metadata related to acquisition, authoring, editing, and other events in content life cycle.

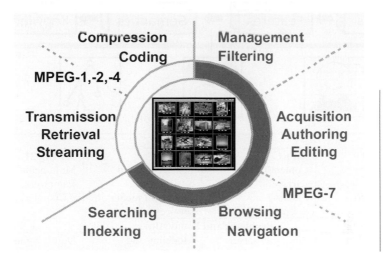

Figure 6-2 Overview of normative scope of MPEG-7 standard. The methods for feature extraction and search are not standardized.

The focus of the MPEG standards has changed over time to address the requirements of industry. As shown in Figure 6-2, the traditional problems dealt with audio-visual data at the signal level and included focus on compression, coding, and communications. MPEG developed several audio-visual coding standards, namely, MPEG-1, MPEG-2, and MPEG-4, to address the need for interoperability around coding formats for applications such as video on-demand and storage, broadband and streaming video delivery. Requirements of some applications, however, created new challenges for dealing with audio-visual data at the feature level. For example, applications such as content-based retrieval and content adaptation created requirements for similarity searching based on audio-visual features and object- and feature-based coding of audio and video. MPEG addressed many of the object- and feature-based coding requirements in the MPEG-4 standard. MPEG-7 is the first standard to address description of audio-visual features of multimedia content. MPEG-7 descriptors allow indexing and searching based on features such as color, texture, shape, edges, motion, and so forth.

While content-based retrieval is useful for many applications such as multimedia databases, intelligent media services, and personalization, many applications require an interface at the semantic level. This involves, for example, a description of scenes, objects, events, people, places, and so forth. MPEG-7 provides rich metadata for describing semantics of real-world scenes related to the content. Beyond MPEG-7, MPEG is developing MPEG-21 to address the requirements of a multimedia framework that allows transactions of digital items, which contain, multimedia content, associated metadata, and rights descriptions.

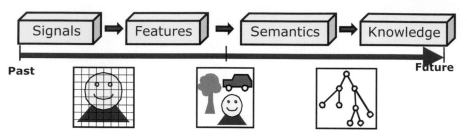

Applications			
MPEG-1,-2,-4	**MPEG-4,-7**	**MPEG-7**	**MPEG-21**
Video storage	Content-based	Semantic-based	Multimedia
Broadband	retrieval	retrieval	framework
Streaming video	Filtering	Intelligent media	e-Commerce
delivery	Adaptation	services (iTV)	
Problems and Innovations			
Compression	Similarity search	Modeling and	Digital rights
Coding	Object- and feature-	classification	management
Communications	based coding	Personalization and	Digital item
		summarization	adaptation

Figure 6-3 Progression of MPEG standards and related applications, problems, and innovations.

MPEG-7 APPLICATIONS

MPEG-7 addresses a wide diversity of application types and application domains as shown in Figure 6-4. The standard is not geared towards any specific industry, rather, it provides a fairly generic and extensible metadata system that can be further adapted and customized to particular application domains. MPEG-7 applications can be broadly categorized as pull applications (searching and browsing), push applications (filtering), and content adaptation (universal multimedia access and perceptual QoS), as follows:

Pull Applications

Pull applications involve searching or browsing in which the user is actively seeking multimedia content or information. Example pull applications include Internet search engines and multimedia databases. The benefits of MPEG-7 for pull applications result from interoperability, which allows queries to be based on standardized descriptions.

Push Applications

Push applications involve filtering, summarization or personalization in which a system or agent selects or summarizes multimedia content or information. Example push applications include digital television and digital video recording. The benefits of MPEG-7 for push applications result from standardized description of content, which allows intelligent software agents to filter content or channels based on standardized descriptions.

Universal Multimedia Access

Universal Multimedia Access applications involve adapting multimedia content according to usage context, which includes user preferences, device capabilities, network conditions, user environment, and spatial, temporal and operational context. Example Universal Multimedia Access applications include adaptation and delivery of multimedia content for wireless cell phone users in which, for example, video content is adapted for user preferences, device capabilities, and time-varying bandwidth. The benefits of MPEG-7 for Universal Multimedia Access applications result from the standardized representation of the content description, transcoding hints, and user preferences.

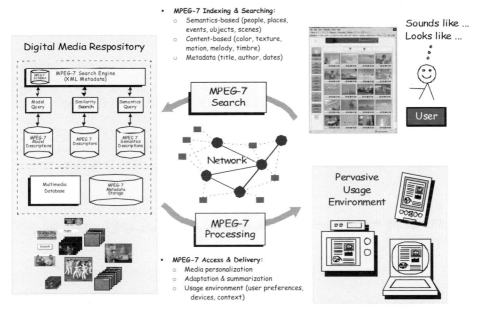

Figure 6-4 MPEG-7 applications include pull-type applications such as multimedia database searching, push-type applications such as multimedia content filtering, and universal multimedia access

Other Application Domains

The creation of the MPEG-7 standard was inspired from a large number of application domains. The process involved identifying requirements from these application domains and forming a generic requirement set for the standard. MPEG-7 can be applied to a large number of application domains, such as the following:

- Education (e.g., distance learning)
- Journalism (e.g. searching for speeches by voice or face)
- Cultural services (history museums, art galleries, etc.)
- Entertainment (e.g. searching a game, karaoke)
- Investigation services (human characteristics recognition, forensics)

- Geographical information systems (GIS)
- Remote sensing (cartography, ecology, natural resources management, etc.)
- Surveillance (traffic control, surface transportation)
- Bio-medical applications
- E-commerce and shopping (e.g. searching for clothes/patterns)
- Architecture, real estate, and interior design
- Film, video and radio archives

6.3 MPEG-7 Constructs

The MPEG-7 standard consists of several parts. The parts specify the basic constructs of the multimedia content description standard and provide information about implementation and extraction and use of MPEG-7 descriptions.

MPEG-7 PARTS
The MPEG-7 standard is comprised of a number of different parts, each one specifying a different aspect of the standard. The naming and role of each of the parts is given as follows:

Part 1 – Systems: specifies the tools for preparing descriptions for efficient transport and storage, compressing descriptions, and allowing synchronization between content and descriptions.

Part 2 – Description Definition Language: specifies the language for defining the standard set of description tools (DSs, Ds, and datatypes) and for defining new description tools.

Part 3 – Visual: specifies the description tools pertaining to visual content.

Part 4 – Audio: specifies the description tools pertaining to audio content.

Part 5 – Multimedia Description Schemes: specifies the generic description tools pertaining to multimedia including audio and visual content.

Part 6 – Reference Software: provides a software implementation of the standard.

Part 7 – Conformance: specifies the guidelines and procedures for testing conformance of implementations of the standard.

Part 8 – Extraction and Use: provides guidelines and examples of the extraction and use of descriptions.

MPEG-7 BASIC CONSTRUCTS
The basic constructs of MPEG-7 include the Description Definition Language (DDL), Description Schemes (DS), and Descriptors (D). The relationships among these constructs are shown in Figure 6-5.

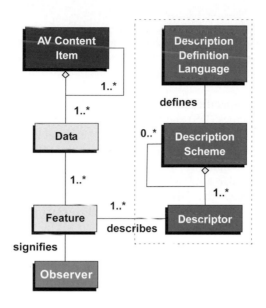

Figure 6-5 Basic constructs of MPEG-7 and their relationships

The constructs are defined as follows:

- The Description Definition Language (DDL) is the language specified in MPEG-7 for defining the syntax of Description Schemes and Descriptors. The DDL is based on the XML Schema Language.

- Description Schemes (DS) are description tools defined using DDL that describe entities or relationships pertaining to multimedia content. Description Schemes specify the structure and semantics of their components, which may be Description Schemes, Descriptors, or datatypes. Examples of Description Schemes include: MovingRegion DS, CreationInformation DS, and Object DS.

- Descriptors (D) are description tools defined using DDL that describe features, attributes, or groups of attributes of multimedia content. Example Descriptors include: ScalableColor D, SpatioTemporalLocator D, AudioSpectrumFlatness D.

- Features are defined as a distinctive characteristic of multimedia content that signifies something to a human observer, such as the "color" or "texture" of an image. This distinguishes Descriptions from Features as follows: consider color to be a feature of an image, then the ScalableColor D can be used to describe the color feature.

- Data (Essence, Multimedia Data) is defined as a representation of multimedia in a formalized manner suitable for communication, interpretation, or processing by automatic means. For example, the data can correspond to an image or video.

The MPEG-7 standard specifies the Description Definition Language (DDL) and the set of Description Schemes (DS) and Descriptors that comprise the MPEG-7 schema, as shown in Figure 6-6. However, MPEG-7 is also extensible in that the DDL can be used to define new DSs and Descriptors and extend the MPEG-7 standard DSs and Descriptors. For example, if a given medical imaging application requires description of a particular kind of feature of an imaging artifact, such as texture patterns in MRI images, the DDL can be used to define a new MRI texture descriptor. The MPEG-7 schema is defined in such a way that would then allow this descriptor to be used together with the standardized MPEG-7 DSs and Descriptors, for example, to include the MRI descriptor within an MPEG-7 image description.

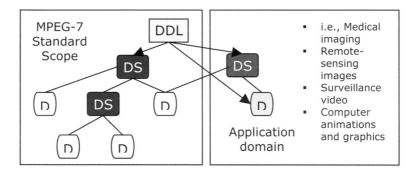

Figure 6-6 The normative scope of MPEG-7 includes the DDL and a standard set of Description Schemes and Descriptors. MPEG-7 can be extended by using the DDL to specify the syntax of specialized description tools.

MPEG-7 EXTENSIBILITY

The normative scope of MPEG-7 includes the DDL and a standard set of Description Schemes and Descriptors, as shown in Figure 6-6. However, MPEG-7 is designed to be extensible in that the DDL can be used to define syntax of new description tools that are outside of the standard. For example, consider that in medical imaging or remote-sensing imaging applications a specific kind of shape or texture descriptor may be needed to effectively describe the corresponding image features. The DDL can be used to define these new Descriptors. It is then possible to include instances of those Descriptors within MPEG-7 descriptions.

The possible extension mechanisms are the following: (1) XML Schema "extension," in which new attributes and/or elements are added to a type definition originally specified in the MPEG-7 standard, and (2) XML Schema "restriction," in which values in a type definition originally specified in the MPEG-7 standard are constrained to a specified set. The following example shows the specification of a MedicalImageRegion DS, which extends the StillRegion DS defined in the MPEG-7 standard.

```
<complexType name="MedicalImageRegionType">
   <complexContent>
      <extension base="mpeg7:StillRegionType">
         <attribute name="param" type="positiveInteger"
            use="optional"/>
      </extension>
   </complexContent>
</complexType>
```

6.4 MPEG-7 Conceptual Model

The MPEG-7 conceptual model was developed during the process of making the MPEG-7 standard. The role of the conceptual model was to provide a top-down design of the entities and relationships in the multimedia domain that was used to inform the work on developing the specific description tools that comprise the MPEG-7 standard, as shown in Figure 6-7. The MPEG-7 conceptual model consists firstly of a list of principal concepts that were defined using descriptions of MPEG-7 applications. The principal concept list includes a definition of each concept.

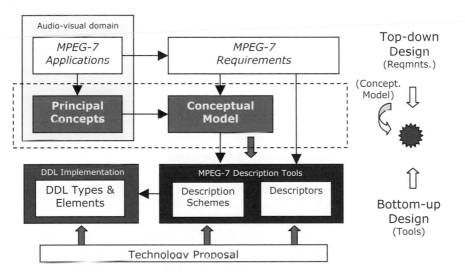

Figure 6-7 The MPEG-7 conceptual model provides a high-level of the principal concepts of the MPEG-7 standard

The conceptual model is defined by mapping the principal concepts to entities, attributes and relationships in an entity-relationship model. An example is shown in Figure 6-8, which shows five entities: video, shot, frame, key-frame, and image, and the relationships among them. Three types of relationships are shown: aggregation, which expresses that a video contains shots; association, which expresses that a key-frame is associated with a shot; and generalization, which expresses that a frame is a generalization of a key-frame.

The MPEG-7 conceptual model was used to identify gaps and overlaps in the MPEG-7 description tools. Although there is not a direct one-to-one mapping of principal concepts to Description Schemes and Descriptors, the MPEG-7 conceptual model was used to determine the designation of whether a description tool would be a DS or Descriptor. Basically, any description tool mapping to an entity or relationship in the conceptual model was designated as a DS. Whereas, any description tool mapping to an attribute in the conceptual model was designated a Descriptor. As a result, the syntax definition of Description Schemes and Descriptors differs in that Description Schemes contain an id attribute, which allows them to participate in relationships, whereas Descriptors do not.

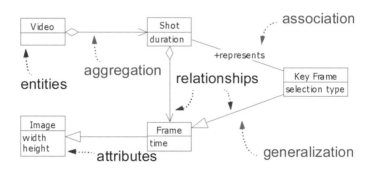

Figure 6-8 The example entity-relationship model shows five entities: video, shot, frame, key-frame, and image, and their relationships.

6.5 MPEG-7 Description Definition Language

The MPEG-7 DDL is used to define the syntax of the MPEG-7 description tools. The DDL is based on the XML Schema Language. The following shows an example definition of a description tool in MPEG-7 using the DDL. This example defines the `StructuredAnnotation` datatype. The `StructuredAnnotation` datatype consists of seven elements, all of type `TermUseType`. Basically, `TermUseType` allows a string value to be entered where the term may come from a classification or controlled term list or be a free term. Each of the seven elements is optional in that the definition allows a minimum of zero occurrences of each element. Furthermore, there may be more than one instance of each element within a description since the definition allows an unbounded maximum occurrence of each element. The

StructuredAnnotation datatype also contains an attribute xml:lang which specifies the language used in the description.

```
<complexType name="StructuredAnnotationType">
   <sequence>
      <element name="Who" type="mpeg7:TermUseType"
         minOccurs="0" maxOccurs="unbounded"/>
      <element name="WhatObject" type="mpeg7:TermUseType"
         minOccurs="0" maxOccurs="unbounded"/>
      <element name="WhatAction" type="mpeg7:TermUseType"
         minOccurs="0" maxOccurs="unbounded"/>
      <element name="Where" type="mpeg7:TermUseType"
         minOccurs="0" maxOccurs="unbounded"/>
      <element name="When" type="mpeg7:TermUseType"
         minOccurs="0" maxOccurs="unbounded"/>
      <element name="Why" type="mpeg7:TermUseType"
         minOccurs="0" maxOccurs="unbounded"/>
      <element name="How" type="mpeg7:TermUseType"
         minOccurs="0" maxOccurs="unbounded"/>
   </sequence>
   <attribute ref="xml:lang" use="optional"/>
</complexType>
```

The following shows the use of StructuredAnnotation datatype in describing a scene in a video. The StructuredAnnotation datatype is used to describe information about "who", "what object", "what action", "where", and "when" as it relates to the description of the video scene. The description shows the use of a controlled term list for "who".

```
<Mpeg7>
   <Description xsi:type="ContentEntityType">
      <MultimediaContent xsi:type="VideoType">
         <Video>
            <MediaLocator>
               <MediaUri>video.mpg</MediaUri>
            </MediaLocator>
            <StructuredAnnotation xml:lang="en">
               <Who href="urn:people:NYMets:2000">
                  <Name>Piazza</Name>
               </Who>
               <WhatObject>
                  <Name>Splintered bat</Name>
               </WhatObject>
               <WhatAction>
                  <Name>Clemens throws bat</Name>
               </WhatAction>
               <WhatAction>
                  <Name>Piazza ducks</Name>
```

```
                    </WhatAction>
                    <Where>
                        <Name>NY</Name>
                    </Where>
                    <When>
                        <Name>Subway series</Name>
                    </When>
                </StructuredAnnotation>
            </TextAnnotation>
        </Video>
      </MultimediaContent>
   </Description>
</Mpeg7>
```

6.6   MPEG-7 Multimedia Description Schemes

The MPEG-7 Multimedia Description Schemes (MDS) specify the generic description tools pertaining to multimedia including audio and visual content. The MDS description tools are categorized as (1) basic elements, (2) tools for describing content and related metadata, (3) tools for describing content organization, navigation and access, and user interaction, and (4) classification schemes.

Basic Elements
The basic elements form the building blocks for the higher-description tools. The following basic elements are defined:

- Schema tools. Specifies the base type hierarchy of the description tools, the root element and top-level tools, the multimedia content entity tools, and the package and description metadata tools.

- Basic datatypes. Specifies the basic datatypes such as integers, reals, vectors and matrices, which are used by description tools.

- Linking and media localization tools. Specifies the basic datatypes that are used for referencing within descriptions and linking of descriptions to multimedia content, such as spatial and temporal localization.

- Basic description tools. Specifies basic tools that are used as components for building other description tools such as language, text, and classification schemes.

Content Description Tools
The content description tools describe the features of the multimedia content and the immutable metadata related to the multimedia content.

The following description tools for content metadata are defined:

- Media description. Describes the storage of the multimedia data. The media features include the format, encoding, and storage media. The tools allow multiple media description instances for the same multimedia content.

- Creation & production. Describes the creation and production of the multimedia content. The creation and production features include title, creator, classification, purpose of the creation, and so forth. The creation and production information is typically not extracted from the content but corresponds to metadata related to the content.

- Usage. Describes the usage of the multimedia content. The usage features include access rights, publication, and financial information. The usage information may change during the lifetime of the multimedia content.

The following description tools for content description are defined:

- Structure description tools. Describes the structure of the multimedia content. The structural features include spatial, temporal or spatio-temporal segments of the multimedia content.

- Semantic description tools. Describes the "real-world" semantics related to or captured by the multimedia content. The semantic features include objects, events, concepts, and so forth.

The content description and metadata tools are related in the sense that the content description tools use the content metadata tools. For example, a description of creation and production or media information can be attached to an individual video or video segment in order to describe the structure and creation and production of the multimedia content.

Content organization, navigation and access, user interaction

The tools for organization, navigation and access, and user interaction are defined as follows:

- Content organization. Describes the organization and modeling of multimedia content. The content organization tools include collections, probability models, analytic models, cluster models, and classification models.

- Navigation and Access. Describes the navigation and access of multimedia such as multimedia summaries and abstracts; partitions, views and decompositions of image, video, and audio signals in space, time and frequency; and relationships between different variations of multimedia content.

- User Interaction. Describes user preferences pertaining to multimedia content and usage history of users of multimedia content.

Classification Schemes
A classification scheme is a list of defined terms and their meanings. The MPEG-7 classification schemes organize terms that are used by the description tools. Applications need not use the classification schemes defined in the MPEG-7 standard. They can use proprietary or third party ones. However, if they choose to use the MPEG-7 standard classification schemes defined, no modifications or extensions are allowed. Furthermore, MPEG-7 has defined requirements for a registration authority for MPEG-7 classification schemes, which allows third parties to define and register classification schemes for use by others. All of the MPEG-7 classification schemes are specified using the `ClassificationScheme DS`, that is, they are themselves MPEG-7 descriptions.

Examples
The following examples illustrate the use of different MPEG-7 Multimedia Description Schemes in describing multimedia content.

Creation Information

The following example gives an MPEG-7 description of the creation information for a sports video.

```
<Mpeg7>
    <Description xsi:type="CreationDescriptionType">
        <CreationInformation>
            <Creation>
                <Title type="popular">Subway series</Title>
                <Abstract>
                    <FreeTextAnnotation> Game among city rivals
                    </FreeTextAnnotation>
                </Abstract>
                <Creator>
                    <Role
                    href="urn:mpeg:mpeg7:cs:RoleCS:2001:PUBLISHER"/>
                    <Agent xsi:type="OrganizationType">
                        <Name>Sports Channel</Name>
                    </Agent>
                </Creator>
            </Creation>
        </CreationInformation>
    </Description>
</Mpeg7>
```

Free Text Annotation

The following example gives an MPEG-7 description of a car that is depicted in an image.

```
<Mpeg7>
    <Description xsi:type="SemanticDescriptionType">
        <Semantics>
            <Label>
                <Name> Car </Name>
            </Label>
            <Definition>
                <FreeTextAnnotation>
                    Four wheel motorized vehicle
                </FreeTextAnnotation>
            </Definition>
            <MediaOccurrence>
                <MediaLocator>
                    <MediaUri> image.jpg </MediaUri>
                </MediaLocator>
            </MediaOccurrence>
        </Semantics>
    </Description>
</Mpeg7>
```

Collection Model

The following example gives an MPEG-7 description of a collection model of "sunsets" that contains two images depicting sunset scenes.

```
<Mpeg7>
    <Description xsi:type="ModelDescriptionType">
        <Model xsi:type="CollectionModelType" confidence="0.75"
            reliability="0.5" function="described">
            <Label>
                <Name>Sunsets</Name>
            </Label>
            <Collection xsi:type="ContentCollectionType">
                <Content xsi:type="ImageType">
                    <Image>
                        <MediaLocator xsi:type="ImageLocatorType">
                            <MediaUri>sunset1.jpg</MediaUri>
                        </MediaLocator>
                    </Image>
                </Content>
                <Content xsi:type="ImageType">
                    <Image>
                        <MediaLocator xsi:type="ImageLocatorType">
                            <MediaUri>sunset2.jpg</MediaUri>
                        </MediaLocator>
                    </Image>
                </Content>
            </Collection>
```

```
        </Model>
    </Description>
</Mpeg7>
```

Video Segment

The following example gives an MPEG-7 description of the decomposition of a video segment. The video segment is first decomposed temporally into two video segments. The first video segment is decomposed into a single moving region. The second video segment is decomposed into two moving regions.

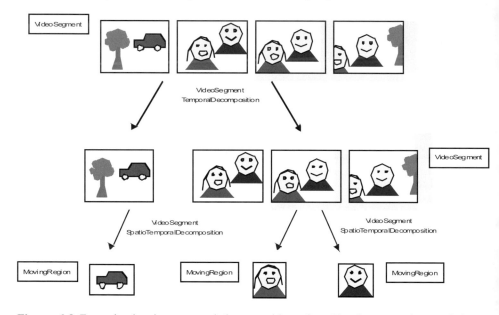

Figure 6-8 Example showing temporal decomposition of a video into two shots and the spatio-temporal of each shot into moving regions.

```
<Mpeg7>
    <Description xsi:type="ContentEntityType">
        <MultimediaContent xsi:type="VideoType">
            <Video>
                <TemporalDecomposition gap="false"
                    overlap="false">
                    <VideoSegment id="shot1">
                        <SpatioTemporalDecomposition>
                            <MovingRegion id="car">
                                <!-- more elements -->
                            </MovingRegion>
                        </SpatioTemporalDecomposition>
                    </VideoSegment>
                    <VideoSegment id="shot2">
                        <SpatioTemporalDecomposition>
```

```
                        <MovingRegion id="person1">
                            <!-- more elements -->
                        </MovingRegion>
                    </SpatioTemporalDecomposition>
                    <SpatioTemporalDecomposition>
                        <MovingRegion id="person2">
                            <!-- more elements -->
                        </MovingRegion>
                    </SpatioTemporalDecomposition>
                </VideoSegment>
            </TemporalDecomposition>
        </Video>
    </MultimediaContent>
  </Description>
</Mpeg7>
```

Semantic Event

The following example gives an MPEG-7 description of the event of a handshake between people. Person A is described as the agent or initiator of the handshake event and Person B is described as the accompanier or joint agent of the handshake.

Figure 6-9 Example image showing two people shaking hands

```
<Mpeg7>
    <Description xsi:type="SemanticDescriptionType">
        <Semantics>
            <Label>
                <Name> Shake hands </Name>
            </Label>
            <SemanticBase xsi:type="AgentObjectType" id="A">
                <Label href="urn:example:acs">
                    <Name> Person A </Name>
                </Label>
            </SemanticBase>
            <SemanticBase xsi:type="AgentObjectType" id="B">
                <Label href="urn:example:acs">
                    <Name> Person B </Name>
                </Label>
            </SemanticBase>
            <SemanticBase xsi:type="EventType">
                <Label>
```

```
                    <Name> Handshake </Name>
                </Label>
                <Definition>
                    <FreeTextAnnotation>
                        Clasping of right hands by two people
                    </FreeTextAnnotation>
                </Definition>
                <Relation
        type="urn:mpeg:mpeg7:cs:SemanticRelationCS:2001:agent"
                    target="#A"/>
                <Relation
type="urn:mpeg:mpeg7:cs:SemanticRelationCS:2001:accompanier"
                    target="#B"/>
            </SemanticBase>
        </Semantics>
    </Description>
</Mpeg7>
```

6.7   MPEG-7 Audio Description Tools

The MPEG-7 Audio description tools describe audio data. The audio description tools are categorized as low-level and high-level. The low-level tools describe features of audio segments. The high-level tools describe the structure of audio content or provide application-level descriptions of audio.

Low-level audio tools
The following low-level audio tools are defined in MPEG-7:

- Audio Waveform: Describes the audio waveform envelope for display purposes.

- Audio Power: Describes temporally-smoothed instantaneous power, which is equivalent to the square of waveform values.

- Audio Spectrum: Describes features such as the audio spectrum envelop (spectrum of the audio according to a logarithmic frequency scale), spectrum centroid (center of gravity of the log-frequency power spectrum), spectrum spread (second moment of the log-frequency power spectrum), spectrum flatness (flatness properties of the spectrum of an audio signal within a given number of frequency bands), and spectrum basis (basis functions that are used to project high-dimensional spectrum descriptions into a low-dimensional representation).

- Harmonicity: Describes the degree of harmonicity of an audio signal.

- Silence: Describes a perceptual feature of a sound track capturing the fact that no significant sound is occurring.

High-level audio tools

The following high-level audio tools are defined in MPEG-7:

- Audio Signature: Describes a signature extracted from the audio signal that is designed to provide a unique content identifier for purposes of robust identification of the audio signal.

- Timbre: Describes the perceptual feature of an instrument that makes two sounds having the same pitch and loudness sound different. The Timbre Descriptors relate to notions such as "attack", "brightness" or "richness" of a sound.

- Sound Recognition and Indexing: Supports applications that involve audio classification and indexing. The tools include the Description Schemes for Sound Model and the Sound Classification Model, and allow description of finite state models using the Description Schemes for Sound Model State Path and the Sound Model State Histogram.

- Spoken Content: Describes the output of an Automatic Speech Recognition (ASR) engine including the lattice and speaker information.

- Melody: describes monophonic melodic information that facilitates efficient, robust, and expressive similarity matching of melodies.

Examples

The following example describes a melody contour of a song:

```
<Mpeg7>
    <Description xsi:type="ContentEntityType">
        <MultimediaContent xsi:type="AudioType">
            <Audio>
                <AudioDescriptionScheme xsi:type="MelodyType">
                    <Meter>
                        <Numerator>3</Numerator>
                        <Denominator>4</Denominator>
                    </Meter>
                    <MelodyContour>
                        <Contour>2 -1 -2 1 -1 1 -1</Contour>
                        <Beat>1 4 5 7 8 9 9 10</Beat>
                    </MelodyContour>
                </AudioDescriptionScheme>
            </Audio>
        </MultimediaContent>
    </Description>
</Mpeg7>
```

The following example describes a continuous Hidden Markov Model of audio sound effects. Each continuous Hidden Markov Model has five states and represents a sound effect class. The parameters of the continuous density state model can be estimated via training, for example, using the Baum-Welch algorithm. After training, the continuous HMM model consists of a 3x3 state transition matrix, a 3x1 initial state density matrix, and 3 multi-dimensional Gaussian distributions defined in terms of the mean and variance parameters. Each multi-dimensional Gaussian distribution has six dimensions corresponding to audio features described by the AudioSpectrumFlatness D.

```
<Mpeg7>
    <Description xsi:type="ModelDescriptionType">
        <Model xsi:type="ContinuousHiddenMarkovModelType"
            numOfStates="3">
            <Initial mpeg7:dim="5"> 0.1 0.2 0.1 </Initial>
            <Transitions mpeg7:dim="3 3">
                0.2 0.2 0.6
                0.1 0.2 0.1
                0.4 0.2 0.1
            </Transitions>
            <State>
                <Label>
                    <Name> State 1 </Name>
                </Label>
            </State>
            <State>
                <Label>
                    <Name> State 2 </Name>
                </Label>
            </State>
            <State>
                <Label>
                    <Name> State 3 </Name>
                </Label>
            </State>
            <DescriptorModel>
                <Descriptor xsi:type="AudioSpectrumFlatnessType"
                    loEdge="250" highEdge="1600">
                    <Vector> 1 2 3 4 5 6 </Vector>
                </Descriptor>
                <Field>Vector</Field>
            </DescriptorModel>
            <ObservationDistribution
                xsi:type="GaussianDistributionType" dim="6">
                <Mean mpeg7:dim="6"> 0.5 0.5 0.25 0.3 0.5 0.3
                </Mean>
                <Variance mpeg7:dim="6">
                    0.25 0.75 0.5 0.45 0.75 0.3
                </Variance>
            </ObservationDistribution>
            <ObservationDistribution
                xsi:type="GaussianDistributionType" dim="6">
                <Mean mpeg7:dim="6">
```

```
              0.25 0.4 0.25 0.3 0.2 0.1
            </Mean>
            <Variance mpeg7:dim="6">
              0.5 0.25 0.5 0.45 0.5 0.2
            </Variance>
          </ObservationDistribution>
          <ObservationDistribution
            xsi:type="GaussianDistributionType" dim="6">
            <Mean mpeg7:dim="6">
              0.2 0.5 0.35 0.3 0.5 0.5
            </Mean>
            <Variance mpeg7:dim="6">
              0.5 0.5 0.5 0.5 0.75 0.5
            </Variance>
          </ObservationDistribution>
        </Model>
      </Description>
</Mpeg7>
```

6.8   MPEG-7 Visual Description Tools

The MPEG-7 Visual description tools describe visual data such as images and video. The tools describe features such as color, texture, shape, motion, localization, and faces.

Color
The color description tools describe color information including color spaces and quantization of color spaces. Different color descriptors are provided to describe different features of visual data. The DominantColor D describes a set of dominant colors of an arbitrarily shaped region of an image. The ScalableColor D describes the histogram of colors of an image in HSV color space. The ColorLayout D describes the spatial distribution of colors in an image. The ColorStructure D describes local color structure in an image by means of a structuring element. The GoFGoPColor D describes the color histogram aggregated over multiple images or frames of video.

Texture
The HomogeneousTexture D describes texture features of images or regions based on the energy of spatial-frequency channels computed using Gabor filters. The TextureBrowsing D describes texture features in terms of regularity, coarseness and directionality. The EdgeHistogram D describes the spatial distribution of five types of edges in image regions.

Shape
The RegionShape D describes the region-based shape of an object using Angular Radial Transform (ART). The ContourShape D describes a closed contour of a 2D object or region in an image or video based on Curvature Scale Space (CSS)

representation. The `3DShape` D describes an intrinsic shape description for 3D mesh models based on a shape index value.

Motion

The `CameraMotion` D describes 3-D camera motion parameters, which includes camera track, boom, and dolly motion modes; and camera pan, tilt and roll motion modes. The `MotionTrajectory` D describes motion trajectory of a moving object based on spatio-temporal localization of representative trajectory points. The `ParametricMotion` D describes motion in video sequences including global motion and object motion by describing the evolution of arbitrarily shaped regions over time in terms of a 2-D geometric transform. The `MotionActivity` D describes the intensity of motion in a video segment.

Localization

The Localization Descriptors describe provides the location of regions of interest in the space and jointly in space and time. The `RegionLocator` describes the localization of regions using a box or polygon. The `SpatioTemporalLocator` describes the localization of spatio-temporal regions in a video sequence using a set of reference regions and their motions.

Face

The `FaceRecognition` D describes the projection of a face vector onto a set of 48 basis vectors that span the space of possible face vectors.

Examples

The following example uses the `ScalableColor` D to describe a photographic image depicting a sunset.

```
<Mpeg7>
    <Description xsi:type="ContentEntityType">
        <MultimediaContent xsi:type="ImageType">
            <Image>
                <MediaLocator>
                    <MediaUri>image.jpg</MediaUri>
                </MediaLocator>
                <TextAnnotation>
                    <FreeTextAnnotation> Sunset scene
                    </FreeTextAnnotation>
                </TextAnnotation>
                <VisualDescriptor xsi:type="ScalableColorType"
                    numOfCoeff="16"
                    numOfBitplanesDiscarded="0">
                    <Coeff> 1 2 3 4 5 6 7 8 9 0 1 2 3 4 5 6
                    </Coeff>
                </VisualDescriptor>
            </Image>
        </MultimediaContent>
    </Description>
```

```
</Mpeg7>
```

The following example uses the GoFGoPColor D to describe a video segment.

```
<VideoSegment>
    <VisualDescriptor xsi:type="GoFGoPColorType"
        aggregation="Average">
        <ScalableColor numOfCoeff="16"
            numOfBitplanesDiscarded="0">
            <Coeff> 1 2 3 4 5 6 7 8 9 0 1 2 3 4 5 6 </Coeff>
        </ScalableColor>
    </VisualDescriptor>
</VideoSegment>
```

The following example uses the RegionShape D to describe a probability model that characterizes oval shapes.

```
<Mpeg7>
    <Description xsi:type="ModelDescriptionType">
        <Model xsi:type="ProbabilityModelClassType"
            confidence="0.75" reliability="0.5">
            <Label relevance="0.75">
                <Name> Ovals </Name>
            </Label>
            <DescriptorModel>
                <Descriptor xsi:type="RegionShapeType">
                    <MagnitudeOfART> 3 5 2 5 . . . 6
                    </MagnitudeOfART>
                </Descriptor>
                <Field>MagnitudeOfART</Field>
            </DescriptorModel>
            <ProbabilityModel
                xsi:type="ProbabilityDistributionType"
                confidence="1.0" dim="35">
                <Mean dim="35"> 4 8 6 9 . . . 5 </Mean>
                <Variance dim="35"> 1.3 2.5 5.0 4.5 . . . 3.2
                </Variance>
            </ProbabilityModel>
        </Model>
    </Description>
</Mpeg7>
```

6.9 Beyond MPEG-7

MPEG is continuing to work on MPEG-7 to define new Descriptors and Description Schemes and to complete the specification of MPEG-7 conformance. MPEG is also working to amend the MPEG-2 standard to allow carriage of MPEG-7 descriptions within MPEG-2 streams. An MPEG-7 industry focus group has also been formed to bring together organizations interested in development and deployment of MPEG-7 systems and solutions.

References

[1]   G. Akrivas, at. Al, "An Intelligent System for Retrieval and Mining of Audiovisual Material Based on the MPEG-7 Description Schemes," in *Proc. of the European Symposium on Intelligent Technologies, Hybrid Systems and their implementation on Smart Adaptive Systems (EUNITE),* Tenerife, Spain, 12-14 December 2001.

[2]   O. Avaro, P. Salembier, MPEG-7 Systems: overview, *IEEE Transactions on Circuits and Systems for Video Technology,* 11(6):760-764, June 2001.

[3]   A. B. Benitez, et al., "Object-based multimedia content description schemes and applications for MPEG-7," *Journal of Signal Processing: Image Communication,* 1-2, pp.235-269, Sept. 2000.

[4]   A. B. Benitez, D. Zhong, S.-F. Chang, J. R. Smith, "MPEG-7 MDS Content Description Tools and Applications," W. Skarbek (Ed.): *Computer Analysis of Images and Patterns, CAIP 2001,* Sept. 5-7, 2001, Proceedings. Lecture Notes in Computer Science 2124 Springer 2001.

[5]   A. B. Benitez and J. R. Smith, "MediaNet: A Multimedia Information Network for Knowledge Representation," *Proc. SPIE Photonics East, Internet Multimedia Management Systems,* November, 2000.

[6]   M. Bober, "MPEG-7: Evolution or Revolution?," W. Skarbek (Ed.): *Computer Analysis of Images and Patterns, CAIP 2001* Warsaw, Poland, Sept. 5-7, 2001, Proceedings. Lecture Notes in Computer Science 2124 Springer 2001.

[7]   M. Bober, "MPEG-7 visual shape descriptors," *IEEE Trans. on Circuits & Syst. for Video Tech. (CSVT),* 6, pp.716-719, July.. 2001.

[8]   M. Bober, W. Price, J. Atkinson, "The Contour Shape Descriptor for MPEG-7 and Its Application," *Proc. IEEE Intl. Conf on Consumer Electronics (ICCE),,* 13-8, June 2000.

[9]   L. Cieplinski, "MPEG-7 Color Descriptors and Their Applications," W. Skarbek (Ed.): *Computer Analysis of Images and Patterns, CAIP 2001,* Sept. 5-7, 2001, Proceedings. Lecture Notes in Computer Science 2124 Springer 2001.

[10] Y.-C. Chang, M.-L. Lo, and J. R. Smith, "Issues and Solutions for Storage, Retrieval, and Search of MPEG-7 Documents," *Proc. SPIE Photonics East, Internet Multimedia Management Systems,* November, 2000.

[11] S.-F. Chang, T. Sikora, A. Puri Overview of the MPEG-7 standard, *IEEE Trans. on Circuits & Syst. for Video Tech. (CSVT),* 6, pp.688-695, July.. 2001.

[12] N. Dimitrova, L. Agnihotri, C. Dorai, R. M. Bolle, "MPEG-7 Videotext Description Scheme for Superimposed Text in Images and Video," *Journal of Signal Processing: Image Communication,* 1-2, pp.137-155, Sept. 2000.

[13] A. Divakaran, "An Overview of MPEG-7 Motion Descriptors and Their Applications," W. Skarbek (Ed.): *Computer Analysis of Images and Patterns, CAIP 2001,* Sept. 5-7, 2001, Proceedings. Lecture Notes in Computer Science 2124 Springer 2001.

[14] T. Ebrahimi; Y. Abdeljaoued; R. Figueras i Ventura; O. Divorra Escoda "MPEG-7 Camera," *Proc. on IEEE Intl. Conf. Image Processing (ICIP),* vol. III, pp. 600-603, Oct.. 2001.

[15] T. Echigo, K. Masumitsu, M. Teraguchi, M. Etoh, and S. Sekiguchi, "Personalized Delivery of Digest Video Managed on MPEG-7," *IEEE ITCC-2001*, pp. 216-220, 2001.

[16] N. Fatemi, O.A. Khaled, "Indexing and Retrieval of TV News Programs Based on MPEG-7," *Proc. IEEE Intl. Conf on Consumer Electronics (ICCE),* 20-6, June 2001

[17] J .Hunter, "An overview of the MPEG-7 description definition language (DDL)," *IEEE Trans. on Circuits & Syst. for Video Tech. (CSVT),* 6, pp.765-772, July.. 2001.

[18] J. Hunter, F. Nack, "An overview of the MPEG-7 Description Definition Language (DDL) proposals, *Journal of Signal Processing: Image Communication,* 1-2, pp.271-293, Sept. 2000.

[19] S. Jeannin, A. Divakaran, "MPEG-7 visual motion descriptors," *IEEE Trans. on Circuits & Syst. for Video Tech. (CSVT),* 6, pp.720-724, July.. 2001.

[20] E. Kasutani, A. Yamada, "The MPEG-7 Color Layout Descriptor: A Compact Image Feature Description for High-Speed Image/Video Segment Retrieval," *Proc. on IEEE Intl. Conf. Image Processing (ICIP),* vol. l, pp. 674-677, Oct.. 2001.

[21] S. Krishnamachari, A. Yamada, M. Abdel-Mottaleb, E. Kasutani, "Multimedia content Filtering, Browsing and Matching using MPEG-7 Compact Color Descriptors," *Proc of VISUAL2000,* Nov. 2000.

[22] P. Kuhn, T. Suzuki, A. Vetro, "MPEG-7 Transcoding Hints for Reduced Complexity and Improved Quality," *Proc. of Packet Video 2001,* MO3-2, April 2001

[23] P. Kuhn, T. Suzuki, "MPEG-7 Metadata for Video Transcoding: Motion and Difficulty Hints," *Proc. of SPIE Storage and Retrieval for Media Databases 2001,* pp.4315-4338, Jan. 2001

[24] K. Lee ,. H.Chang, S. Chun; H. Choi; S. Sull, "Perception-based Image Transcoding for Universal Multimedia Access," *Proc. on IEEE Intl. Conf. Image Processing (ICIP),* vol. II, pp. 474-477, Oct.. 2001.

[25] A. T. Lindsay, S. Srinivasan, J. P. A. Charlesworth, P. N. Garner, W. Kriechbaum, "Representation and linking mechanisms for audio in MPEG-7," *Journal of Signal Processing: Image Communication,* 1-2, pp.193-209, Sept. 2000.

[26] B.S. Manjunath, J. R. Ohm, V. V. Vasudevan, A. Yamada., "Color and texture descriptors," *IEEE Trans. on Circuits & Syst. for Video Tech. (CSVT),* 6, pp.703-715, July.. 2001.

[27] D. Messing; P. van Beek; J. Erric, "The MPEG-7 Color Structure Descriptor: Image Description Using Color and Local Spatial Information," *Proc. on IEEE Intl. Conf. Image Processing (ICIP),* vol. l, pp. 670-673, Oct.. 2001.

[28] M. Naphade, C.-Y. Lin, J. R. Smith, B. Tseng, and S. Basu, "Learning to Annotate Video Databases," *IS&T/SPIE Symposium on Electronic Imaging: Science and Technology - Storage & Retrieval for Image and Video Databases X,* San Jose, CA, January, 2002.

[29] A. Natsev, J. R. Smith, Y.-C. Chang, C.-S. Li, J. S. Vitter, "Constrained Querying of Multimedia Databases: Issues and Approaches," *IS&T/SPIE Symposium on Electronic Imaging: Science and Technology - Storage & Retrieval for Image and Video Databases IX*, San Jose, CA, January, 2001.

[30] J.-R. Ohm, "The MPEG-7 Visual Description Framework - Concepts, Accuracy, and Applications," W. Skarbek (Ed.): *Computer Analysis of Images and Patterns , CAIP 2001*, Sept. 5-7, 2001, Proceedings. Lecture Notes in Computer Science 2124 Springer 2001.

[31] J-R. Ohm, "Flexible Solutions for Low-Level Visual Feature Descriptors in MPEG-7," *Proc. IEEE Intl. Conf on Consumer Electronics (ICCE)*, 13-6, June 2000.

[32] F. Pereira and R. Koenen, "MPEG-7: A Standard for Multimedia Content Description", *International Journal of Image and Graphics*, pp527-546, Vol. 1, No. 3, July 2001.

[33] P.Philippe , "Low-level musical descriptors for MPEG-7," *Journal of Signal Processing: Image Communication,* 1-2, pp.181-191, Sept. 2000.

[34] S. Quackenbush, A. Lindsay, "Overview of MPEG-7 audio," *IEEE Trans. on Circuits & Syst. for Video Tech. (CSVT)*, 6, pp.725-729, July.. 2001.

[35] Y. Rui, T. S. Huang, and S. Mehrotra, "Constructing Table-of-Content for Videos," *ACM Multimedia Systems Journal*, Special Issue Multimedia Systems on Video Libraries, Vol.7, No.5, Sept, 1999, pp 359-368.

[36] P. Salembier, R. Qian, N. O'Connor, P. Correia, I. Sezan, P. van Beek, Description Schemes for Video Programs, Users and Devices, *Signal Processing: Image communication*, Vol.16(1-2):211-234, September 2000.

[37] P. Salembier and J. R. Smith, "MPEG-7 Multimedia Description Schemes," *IEEE Trans. Circuits and Systems for Video Technology*, Vol. 11, No. 6, June, 2001.

[38] I. Sezan, P. van Beek, "MPEG-7 Standard and Its Expected Role in Development of New Information Appliances," *Proc. IEEE Intl. Conf on Consumer Electronics (ICCE)*,, 13-2, June 2000.

[39] T. Sikora, "The MPEG-7 visual standard for content description-an overview," *IEEE Trans. on Circuits & Syst. for Video Tech. (CSVT),* 6, pp.696-702, July.. 2001.

[40] J. R. Smith, B. S. Manjunath, N. Day, "MPEG-7 Multimedia Content Description Interface Standard," *IEEE Intl. Conf. On Consumer Electronics (ICCE),* June, 2001.

[41] J. R. Smith, "MPEG-7 Standard for Multimedia Databases," *ACM Intl. Conference on Data Management (ACM SIGMOD),* May, 2001.

[42] J. R. Smith, "Content-based Access of Image and Video Libraries," *Encyclopedia of Library and Information Science*, Marcel Dekker, Inc., Eds. A. Kent, 2001.

[43] J. R. Smith, "Content-based Query by Color in Image Databases," *Image Databases*, John Wiley & Sons, Inc., Eds. L. D. Bergman and V. Castelli, 2001.

[44] J. R. Smith, S. Srinivasan, A. Amir, S. Basu, G. Iyengar, C.-Y. Lin, M. Naphade, D. Ponceleon, B. Tseng, "Integrating Features, Models, and Semantics for TREC Video Retrieval," *Proc. NIST Text Retrieval Conference (TREC-10)*, November, 2001.

[45] J. R. Smith, S. Basu, C.-Y. Lin, M. Naphade, B. Tseng, "Integrating Features, Models, and Semantics for Content-based Retrieval," *Proc. NSF Workshop in Multimedia Content-Based Indexing and Retrieval*, September, 2001.

[46] J. R Smith, Y.-C. Chang, and C.-S. Li, "Multi-object, Multi-feature Search using MPEG-7," *Proc. IEEE Intl. Conf. On Image Processing (ICIP)*, Special session on Multimedia Indexing, Browsing, and Retrieval, October, 2001. Invited paper.

[47] J. R. Smith and V. Reddy, "An Application-based Perspective on Universal Multimedia Access using MPEG-7," *Proc. SPIE Multimedia Networking Systems IV*, August, 2001.

[48] J. R. Smith, "MPEG-7's Path for an Intelligent Multimedia Future," *Proc. IEEE Intl. Conf. On Information Technology for Communications and Coding. (ITCC)*, April, 2001.

[49] J. R. Smith and B. Lugeon, "A Visual Annotation Tool for Multimedia Content Description," *Proc. SPIE Photonics East, Internet Multimedia Management Systems*, November, 2000.

[50] J. R. Smith and A. B. Benitez, "Conceptual Modeling of Audio-Visual Content," *Proc. IEEE Intl. Conf. On Multimedia and Expo (ICME-2000)*, New York, NY, July, 2000.

[51] N. Takahashi, M. Iwasaki, T. Kunieda, Y. Wakita, N. Day, "Image retrieval using spatial intensity features," *Proc. on Signal Processing. Image Communication* 16 (2000) pp 45-57, June 2000.

[52] B. Tseng, C.-Y. Lin, and J. R. Smith, "Video Summarization and Personalization for Pervasive Mobile Devices," *IS&T/SPIE Symposium on Electronic Imaging: Science and Technology - Storage & Retrieval for Image and Video Databases X*, San Jose, CA, January, 2002.

[53] T. Walker, "Content-Based Navigation for Television Programs Using MPEG-7 Description Schemes," *Proc. IEEE Intl. Conf on Consumer Electronics (ICCE)*,, 13-1, June 2000

[54] P. Wu, Y. Choi, Y.M. Ro, and C.S. Won, "MPEG-7: Texture Descriptors", *International Journal of Image and Graphics*, pp547-563, Vol. 1, No. 3, July 2001.

[55] S. J. Yoon, D. K. Park, C. S. Won, S-J. Park, "Image Retrieval Using a Novel Relevance Feedback for Edge Histogram Descriptor of MPEG-7," *Proc. IEEE Intl. Conf on Consumer Electronics (ICCE)*, 20-3, June 2001.

7 MULTIMEDIA AUTHENTICATION AND WATERMARKING

Dr. Bin B. Zhu and Dr. Mitchell D. Swanson

Most multimedia signals today are in digital formats which are easy to reproduce and modify without any trace of manipulations. In this chapter, we present a comprehensive review of current technologies for multimedia data authentication and tamper detection. We first introduce in Section 7.2 the problem issues and general requirements for a multimedia authentication system. This is followed in Section 7.3 with detailed descriptions of the current algorithms for the three types of multimedia authentication: hard or complete authentication, soft authentication, and content-based authentication. The conclusion and future research directions are presented in Section 7.4.

7.1 Introduction

Nowadays the vast majority of multimedia signals are in digital formats which offer significant advantages over analog formats in terms of quality and efficiency in creation, storage and distribution. They can also be easily edited without obvious trace of manipulations with modern PC and software. In today's digital world seeing is no longer believing! People seek for information with integrity and authenticity. This makes detection of illegal manipulations an important issue for digital multimedia. Authentication technologies fulfill an increasing need for trustworthy digital data in commerce, industry, defense, etc.

General data authentication has been well studied in cryptography. The classical scenario for an authentication problem can be described as follows: A sender S sends a digital message M to a receiver R. An effective authentication system must ensure that the message M received by R: (1) is exactly the same as that sent by S (*integrity verification*); (2) is really sent by S (*origination verification*); and (3) S cannot deny having sent M (*non-repudiation*). Two types of authentication techniques have been developed in cryptography: digital signature and message authentication codes (MACs) [1, 2]. The digital signature approach applies a cryptographic hash function to the message M. The hash value is then encrypted with the sender S's private key. The result is the digital signature of the message M. The digital signature is sent to R together with the message M it authenticates. The receiver R uses the sender's public key to decrypt the digital signature to obtain the original hash value, which is compared with the hash value calculated from the received message M. If they are the same, then the message R received is authentic, otherwise it has been tampered. MACs, on the other hand, are constructed by taking as inputs the message M and a secret key shared by both creator S and verifier R. Cryptographic hash functions and block ciphers are normally used to construct MACs. MACs are sent together with the message M to R for integrity verification. If the MACs calculated from the received message agree with the received MACs, R concludes the message is authentic, otherwise manipulated. It is clear that both digital signature and MACs can satisfy the first requirement (integrity verification), but only the digital signature can meet the last two requirements (origination verification and non-repudiation), due to the public/private key encryption it uses.

Authentication methods developed for general data can be applied directly to multimedia authentication. Friedman [3] proposed in 1993 a "trustworthy digital camera" scheme which applies the camera's unique private key to generate digital signatures to authenticate its output images. Unlike other digital data, multimedia data contains high redundancy and irrelevancy. Some signal processing like compression is usually or even required to be applied to multimedia signals without affecting the authenticity of the data. Classical authentication algorithms will reject such a signal modification as the signal has been manipulated. In fact, classical authentication can only authenticate the binary representation of digital multimedia instead of its perceptual content. In multimedia authentication, it is desirable that the perceptual content is authenticated so an authenticator remains valid across different representations as long as the underlying perceptual content has not changed. In addition to perceptual content authentication, it is also desirable for a multimedia authentication system to indicate tamper places and severity if the media is manipulated. The classical authentication can generate only a binary output (tampered or not) for the whole data, irrespective of whether the manipulation is minor or severe.

All manipulations on multimedia can be classified into two categories, incidental and malicious manipulations:

- *Incidental manipulations*: Incidental manipulations do not change the authenticity of the perceptual content of multimedia, and should be accepted by an authentication system. Common ones include format conversions, lossless and high-quality lossy compression, A/D and D/A conversions, resampling, etc.
- *Malicious manipulations*: Manipulations in this category change the perceptual quality or semantic meaning to a user, and thus should be rejected. They include cropping, dropping, inserting, replacing, reordering perceptual objects or video frames, etc.

Different applications may have different criteria to classify manipulations. A manipulation that is considered as incidental in one application could be considered as malicious in another. For example, JPEG image compression is generally considered as an incidental operation in most applications but may be rejected for medical images since loss of details to lossy compression may render a medical image useless.

Multimedia authentication can be classified according to integrity criteria into three types: hard (or complete), soft, and content-based authentications. *Hard authentication* detects any modification to the content representation of digital multimedia. The only incidental manipulation accepted by the hard authentication is lossless compression or format conversions in which the visual pixel values or audio samples do not change. This is similar to the classical authentication except that those lossless operations are rejected by the classical authentication. *Soft authentication* detects any manipulations that lower the perceptual quality below an acceptable level. *Content-based authentication* detects any manipulations that change the semantic meaning of the content to a user. This is normally achieved by authenticating perceptual features extracted from the media. It is clear that the content-based authentication has the largest tolerance to distortion since perceptible distortion may not change the semantic meaning, and the hard authentication has the least tolerance. Different applications may require different types of authentication. A medical image database may need hard authentication, while audiovisual entertainment may require soft or content-based authentication. In entertainment, audiovisual data may be processed by various middle stages from the sender to a receiver to maximize the received quality with resources available. Common operations by middle stages include rate control and transcoding. Soft or content-based authentication that is robust to the middle stage processing is desirable in this application since otherwise each middle stage has to check the integrity of the data it receives and re-authenticate the processed data on behalf of the original signer. This increases the complexity for a middle stage and may lower the security of an authentication system since the authentication secret may have to be shared with middle stages in order for them to re-authenticate the processed data.

Multimedia authentication, especially image authentication, has been under active study in recent years. All the proposed authentication methods can be classified into two types of approaches: external signatures and watermarking. *External signature* approaches attach the authentication data generated from a digital multimedia signal to the media by concatenation or in the format's header field. Most approaches of this type do not change the media they authenticate, but some do, especially for soft authentication. *Watermarking*, on the other hand, embeds authentication data into the media to be authenticated, due to the redundancy and irrelevancy contained in multimedia signals. Three types of watermarking have been developed: *fragile watermarking* which is designed to be fragile to any modification to the content, *semi-fragile watermarking* that is robust to some perceptual quality preserving manipulations but fragile to others, and *robust watermarking* which is robust to both signal processing and intentional attacks. There exist several excellent reviews focused on robust watermarking [4-7], and also some brief reviews on tamper detection watermarks [8-10]. Most proposed algorithms for hard or soft authentication are based on fragile or semi-fragile watermarking, and those for content-based authentication are based on either external signatures or robust/semi-robust watermarking. The watermarking used for content-based authentication is called *content-fragile watermarking*. We note here that hard or content-based authentication can use either external signatures or watermarking approaches, but it is more convenient for soft authentication to use watermarking approaches. Both types of approaches offer certain advantages and disadvantages. The authenticator in an external signature approach is detached from the media it authenticates, and remains untouched when the signal undergoes distortion. It provides simpler verification with a better prospect for achieving robustness to incidental manipulations. It may also provide origination verification and non-repudiation capability. The drawback is that the external authenticator is easy to strip off, and may be lost after a format conversion. In some applications, it may require a complex system to manage authentication data. A watermarking approach, on the other hand, provides a persistent link between the authenticator and the content it authenticates. No or minimal management of additional data is needed for authentication. It is also much easier to locate tampered areas with watermarks. But a watermarking approach is generally more complex and harder to design, and the media to be authenticated has to be modified to insert a watermark. It is also much harder in a watermarking approach to provide origination verification and non-repudiation capability. Watermarking is normally much less robust to intentional attacks than encryption. An integrated approach that combines the strengths of both types of approaches seems to be the key to flexible and secure digital multimedia authentication systems.

7.2 The multimedia authentication paradigm

REQUIREMENTS

There are some general requirements or desirable features for multimedia authentication. We try to give a comprehensive list here. It should be noted that the actual requirements or relative importance of these requirements depends on the application.

- **Integrity verification**. This is the same requirement as in the classical authentication. Multimedia authentication should be the authentication of its perceptual content instead of its binary representation. Three levels of integrity verifications, hard, soft, and content-based authentications, are possible in a multimedia authentication application. They are discussed in detail in the previous section.
- **Tamper localization and estimation**. Multimedia signals generally consist of large amounts of data. When tampering occurs, it is desirable to locate tampered regions so that the non-tampered parts may still be used, and to determine if the semantic meaning of the content has been preserved. For a tampered region, it is desirable to know the severity of tampering, and sometimes to characterize or recover from the alteration. Localization capability implies that signal integrity is checked part by part and that synchronization between the authenticator generation and verification is an important issue, especially for audio authentication. It may also have severe consequences on security of an authentication algorithm (see below). Tamper localization should be carefully balanced with security when designing an authentication algorithm.

- **Fragment authentication**. In some cases, only a portion of the digital multimedia signal is received. It is desirable to know the integrity of the received part.
- **Security**. This is the same as the classical authentication. It should be hard for an unauthorized party who has full knowledge of the algorithm to fake a valid authentication, to derive the authentication secret from the public data and knowledge, or undertake a malicious manipulation without detection. This means that all authentication systems should depend on some secret key with a large key space. It is desirable in many cases that authentication creation and verification use different keys so a verifier does not need to know the creation secret for verification.
- **Origination verification**. Same as the classical case.
- **Non-repudiation.** Same as the classical case.

There are some specific requirements for watermarking-based authentication:
- **Perceptual transparency**. The embedded watermark should not be perceptible under normal observation or interfere with the functionality of the multimedia.
- **Statistically invisible**. This is a requirement only for the approach whose security depends on secret watermarking. Many watermarking schemes use public watermarking so that everybody can extract the embedded data.
- **Blind extraction.** For media authentication, the original signal is not available to the verifier. This means that the watermarking extraction should be *blind*, i.e., no access to the original signal. This is very different from some robust watermarking schemes where the original signal is used for watermark extraction. In many applications, there is an additional requirement that the secret passed to a verifier to check integrity should be independent of the signal to be authenticated. This means that any signal dependent features used for verification should be contained in the verification algorithm instead of passing as secret parameters. This requirement is called *strictly blind* verification requirement. A strictly blind authentication system makes secrets maintenance much simpler. This chapter will focus on strictly blind watermarking authentication algorithms.

There are also other issues to be considered when designing an authentication system, such as the complexity of authenticator generation and verification, robustness to incidental manipulations, size of the authentication data for external signature approaches, etc. All these aspects might be mutually competitive, and a reasonable compromise is always necessary in a practical authentication system design.

ATTACKS ON MULTIMEDIA AUTHENTICATION SYSTEMS

Security is one of the most important issues for an authentication system. It may be practically impossible to design a system robust to all forms of possible attacks, especially attacks drawing on advances in new technologies and hardware. However, knowledge of common attack modes is essential in designing a good authentication system. In addition to the attacks discussed in [1, 2] for classical authentication, some attacks are designed to exploit specific features in the multimedia authentication and the watermarking process for watermarking-based authentication. Some of the common attacks are listed below.

Undetected modification. An attack may try to make changes to the authenticated media without being detected by the authentication system. Tamper localization capability, acceptance of incidental manipulations, and the watermarking process for watermarking-based algorithms open the door for an attacker to mount an undetected malicious manipulation.

Mark transfer. This attack is designed for block-wise watermarking-based authentication. Attacks may use available watermarked signals to forge a valid mark for another, arbitrary media. One of the famous mark transfer attacks is the vector quantization attack proposed by Holliman and Memon [11] . Mark transfer attack can also be performed in the following way: the mark is first removed, then the signal is modified, and finally the mark is reinserted.

Information leakage. An attacker may try to deduce some secret information or the key. Recovered secret information may be used to reduce the key search space. Once the key is deduced, the attacker can then forge the authenticator to any arbitrary multimedia content.

7.3 Multimedia authentication schemes

As we mentioned in Section 7.1, multimedia integrity can have three different levels: hard, soft, and content-based authentication. All the algorithms proposed in the literature can be classified into two types: external signatures and watermarking. While there is no mandatory relationship between the authentication type and the approach type, and either type of approach can implement each authentication type, there are definitely some preferences in proposed algorithms. In particular, most proposed hard and soft authentication schemes are watermarking-based, and content-based authentication schemes are about equal in both types of approaches. We follow this pattern in the descriptions of the multimedia authentication schemes in this section. In the following subsections, hard and soft authentication will focus on fragile and semi-fragile watermarking approaches, while both types of approaches will be described for content-based authentication.

HARD MULTIMEDIA AUTHENTICATION WITH FRAGILE WATER-MARKING

Fragile watermarking used for hard multimedia authentication has been well studied. In fragile watermarking, the inserted watermark is fragile so that any manipulation to the multimedia content disturbs its integrity. One can readily detect tampered parts by checking for presence and integrity of this fragile watermark. One of first fragile watermarking techniques proposed for image tamper detection was based on inserting check-sums of gray levels determined from the seven most significant bits into the least significant bits (LSB) of pseudo-randomly selected pixels [12] . It provides very limited tamper detection: only odd number of bits changed for a pixel can be detected. It is also easy to fake a valid watermarking after malicious modification.

Yeung and Mintzer [13] proposed a simple and fast fragile image watermark scheme that provides integrity check for each image pixel. It can be described as follows. A secret key is used to generate a binary lookup table or function f that maps image pixel values to either 1 or 0. For color images, three such functions, f_R, f_G, and f_B, one for each color channel, are generated. These binary functions are used to encode a secret binary logo L. The logo can be generated from a secret key, or have some graphical meaning. For a gray image I, pixel values are perturbed to satisfy the following equation

$$L(i, j) = f_g(I(i, j)) \quad \text{for each pixel } (i, j). \tag{7-1}$$

For an RGB image, the three color channels I_R, I_G, and I_B are perturbed to obtain

$$L(i, j) = f_R(I_R(i, j)) \oplus f_G(I_G(i, j)) \oplus f_B(I_B(i, j)) \quad \text{for each pixel } (i, j), \tag{7-2}$$

where $\oplus$ denotes the exclusive OR. The pixels are updated sequentially in a row-by-row manner with an error fusion scheme (the original pixel value is adjusted by the diffused error caused by watermarking the previous pixel) to improve the watermarked image quality. Integrity of a test image is easily verified by checking the relationship expressed in Eq. 7-1 or Eq. 7-2 for each pixel (i, j). The same scheme has been extended to quantized DCT coefficients for image authentication in the JPEG compression domain [14] .

The logo in Yeung-Mintzer watermark must be kept secret. Otherwise the search space for the mapping functions f_R, f_G, and f_B is dramatically reduced [15] . Fridrich et al. [16] proposed an attack on the Yeung-Mintzer watermark without knowledge of the key and logo if they are reused to watermark multiple images. Given two 8 bit gray images, I_1 and I_2, of size $M \times N$, which are watermarked with the same key and logo L, we have

$$f_g(I_1(i, j)) = L(i, j) = f_g(I_2(i, j)) \quad \text{for all pixel } (i, j).$$

The last equation constitutes $M \times N$ equations for 256 unknown f_g. It was reported that only two images are needed on average to recover 90% of the binary f_g [16] . Once the binary function is estimated, the logo can be easily derived. Actually, if the logo is a real image rather than a randomized picture, we can use this additional information to recover the rest of the binary function f_g. A similar attack can be mounted to color images even though it is more complex and more images with reused key and logo are needed. Making the mapping functions dependent on pixel positions as proposed in [15] can dramatically increase the complexity of the above attack at the cost of much higher complexity. Fridrich et al. [17] proposed replacing the mapping function f_g by a symmetric encryption E_K with a private key K on a block of size $a \times a$ with the current pixel at the lower right corner, and the pixels are modified by rows. More precisely, Eq. 7-1 is replaced by the equation:

$$L(i, j) = Parity(E_K \{I(i-u, j-v)\} \mid 0 \le u, v < a) \quad \text{for each pixel } (i, j)$$

This neighborhood dependent scheme can effectively thwart the attack described above. However, both modified schemes are still vulnerable to the so-called "vector quantization" attack described later in this section. Making the mapping function dependent on a statistically unique image ID generated either randomly or from the image itself can completely thwart such an attack. More details can be found later in this subsection.

Wong [18-20] proposed two similar versions of a block-based fragile image watermarking scheme. The only difference between the two versions is that one uses cryptographic MACs and the other uses a cryptographic digital signature. The public/private key-based version is described here since it does not need to disclose the secret key for integrity checking. To embed a logo into a host image I of size $m \times n$, a binary watermark W of the same size as the host image is generated from the logo by periodical replication, zero-appending, or some other method. It is then partitioned into blocks $\{W_r\}$ of size $k \times l$ pixels. The host image is partitioned likewise. For each image block I_r, the least significant bit of every pixel in the block is set to zero to form a new image block $\hat{I}_r$. A cryptographic hash $H(m, n, \hat{I}_r)$ is calculated. The hash bits are either truncated or extended to the length that equals the number of pixels $k \times l$ in a block and then XORed bit by bit with the corresponding watermark block W_r. The output is encrypted with the private key. The result is inserted into the LSBs of pixels in the block I_r. At the receiver side, a hash value is calculated in the same way as the sender side, and then XORed with original hash value extracted from the LSBs of the block. If the key is right and the image is untouched, the original binary watermark map can be observed. If some blocks have been modified, the corresponding blocks of the output watermark map should be random due to the property of the hash function whose output varies wildly for any change to the input. By looking at the output binary watermark map, an image can be checked for authenticity, and tampered blocks can be identified. The block size should be large enough, say 12×12 pixels, to prevent the birthday attack [2] to the hash function which requires roughly $2^{p/2}$ trials to be successful, where p is the number of bits in the hash output (less than or equal to the number of pixels in a block for the approach).

Both Yeung-Mintzer and Wong watermarks embed a logo for authentication. The logo can carry some useful visual information about the image or its creator, and makes tamper detection and localization much simpler. The logo can be designed so that the extracted watermark has obvious patterns which can be used to find tampered regions without knowledge of the original logo. Both schemes belong to a type of watermarking called *block-wise independent watermarking* in which each watermarked block X'_i depends only on the original block X_i, the watermark W_i, and the embedding key K_i for the block [11] . A block-wise independent watermarking scheme may be vulnerable to the K-equivalent block-swap attack. Given a key K, two signal blocks X_i and X_j are called K-equivalent if both blocks use the same key and watermark for embedding. If two K-equivalent blocks are swapped in a watermarked signal, the extracted watermark will remain the same. This can be exploited to mount a K-equivalent block-swap attack: if two signals are watermarked with the same key and binary watermark in either the Yeung-Mintzer or the Wong scheme, a hacker can fake a new signal with the same watermark and key by picking up blocks with the right location randomly from either one of the images. Even for the same signal, if a hacker knows the watermark, he can find out K-equivalent blocks and swap those blocks without affecting the watermark.

Block-wise independent watermarking schemes are vulnerable to a more general attack called the *vector quantization* attack proposed first by Holliman and Memon [11] which requires knowledge of the embedded logo. Such a requirement was removed later by Fridrich et al. [16] but more images with reused logos and embedding keys are needed to mount a successful attack. In fact, the security when the logo is unknown is the same as when the logo is known but the block index is used to generate the hash value. The vector quantization attack basically uses the available watermarked blocks to vector-quantize an arbitrary signal block so it carries the same watermark. If the embedded watermark is available to a hacker such as in the case of Wong's public key encryption scheme, a hacker, Oscar, first partitions watermarked signal blocks $\{X'_i\}$ into K-equivalent classes. Even if the watermark is unknown, if Oscar can access multiple signals watermarked with the same watermark and key, he can use the block positions among the signals to form K-equivalent classes. To embed a watermark W_i into a block Y_i of the signal of his choosing, he finds out the K-equivalent class C_i associated with the required watermark W_i, and then replaces Y_i with a block in the class C_i that approximates Y_i the best. This is equivalent to vector-quantizing Y_i with the vectors in the class C_i. The resulting signal carries the same watermark even though Oscar does not know the watermark embedding key. The perceptual quality of the resulting signal depends on the vectors in each equivalent class. Smaller size blocks and a larger number of entries for each equivalent class yield better approximation, and thus better perceptual quality of the signal with forged watermark. The Yeung-Mintzer watermark is the most vulnerable watermarking scheme to a vector quantization attack due to its smallest block size (just 1×1) and only two equivalent classes, each associated with either watermark bit 0 or 1. The resulting forged image is expected to have a very good perceptual quality, esp. when error diffusion is used with the attack [16] . When the logo is unknown or the logo is known but the block index is used to generate hash values, the number of vectors in each equivalent class is determined by the number of watermarked images available to an attacker. When the logo is known and no block index is used for the hash function, this number is inversely proportional to the total number of equivalent classes which is determined by the number of distinct watermark blocks. For the latter case, the more complex the watermark block, the greater the number of distinct watermark blocks, and thus the harder for a successful vector quantization attack. However, a complex logo can make visual recognition of tampering much harder for Wong's scheme. Increasing the number of watermark bits for each block achieves the same effect at the cost of larger embedding distortion. Increasing block size reduces the number of image blocks that can be obtained from each image, but the tamper localization capability is reduced. It also makes it less likely that a good approximation of a block will be made from available watermarked blocks, and thus results in poorer perceptual quality of a forged image. Making the hash value in Wong's scheme dependent on a block index can dramatically increase the complexity of vector quantization attacks when the logo is known, but cannot thwart such attacks if an attacker has access to enough watermarked signals embedded with the same embedding key and the same logo. The complexity of the attack, in this case, is the same as when the logo is unknown.

A more efficient way to thwart the vector quantization attack is to make the watermarking of a block dependent on itself and also its neighborhood. Several remedies to Wong's scheme have been proposed in the literature. One can calculate a hash value based on the current block as well as some of its neighboring blocks [11] . Another method is to use a sufficiently larger surrounding neighborhood (say 16×16) to calculate the hash value and embed into a smaller (say 8×8) block [21] . It is clear that the greater the dependence a watermarked block has on the remaining blocks in a signal, the harder it is to counterfeit a watermark image. The cost for greater dependence on neighborhood is the reduced tamper localization capability. A third method is to introduce a unique signal ID, along with the block index, to the input of the hash function for each block [20] . This can effectively prevent a hacker from finding K-equivalent blocks at different locations or from different signals without sacrificing the resolution of tamper localization. The drawback is that the unique signal ID is needed for authentication of each signal. This brings the complexity of management of these IDs, which may be a big burden for many applications. The unique ID can be generated from the signal itself, say by hashing the whole image with LSBs zeroed-out for Wong's scheme, but any error in the signal may result in a wrong ID which completely impairs the tamper localization capability of Wong's watermarking scheme. A robust mechanism for extracting bits from a signal content which is robust to watermarking embedding and other manipulations is always desirable in most of watermarking schemes [22] . Visual hash bits extracted from an image from the algorithm proposed by Fridrich [23] might be employable as the image ID. The verifier will extract the visual bits from the test image so that no ID management is needed. The signal ID can also be embedded repeatedly into small blocks of a multimedia signal with some robust watermark algorithms. The robust watermark must be applied before the fragile watermark since robust watermark embedding usually leads to large distortion which may cause a false alarm for the fragile watermark if the embedding order is reversed. The drawback for the last two methods is that if the distortion to a signal is large enough, the extracted ID may be wrong, in which case the tamper localization capability of a fragile watermark algorithm may be greatly compromised. The fourth solution is the hierarchical block signature proposed by Celik et al. [24] which builds a tree structure for blocks where each parent block consists of four neighboring blocks as its child blocks, and the whole image is the root of the tree. Digital signatures are found for all the blocks in the same way as Wong's watermark. They are then inserted to the LSBs of the leaf blocks so that the watermark for a leaf block consists of the signature bits from itself, one fourth of the signature bits from its parent block, and one sixteenth from its great parent block, etc. The drawback is that the leaf block size must be large enough to embed the signature bits from the block itself and its ancestors, but the number of signature bits for a block cannot be too small since otherwise the hash function can be attacked by the birthday attack [2] . This indicates its resolution for tamper localization should be worse than that of the original Wong's scheme. The problem can be alleviated by using more LSBs per pixel for watermark embedding at the expense of worse perceptual quality of watermarked images.

Li and Feng [25] proposed an image authentication algorithm that uses the parity of quantized DCT coefficients of randomly selected blocks to embed MAC bits generated from lower order quantized DCT coefficients in the selected blocks and all the quantized DCT coefficients from other blocks. Lee and Won [26] applied Reed-Solomon error correction codes to rows and columns of the scrambled version of the input image with LSB zeroed-out. The resulting parity bits were then inserted into the LSBs of the original image for authentication and tamper restoration. Fridrich et al. [27] embedded a low resolution version of the image into itself for authentication and error recovery. Schneider and Chang [28] applied two layer hash values from a video to generate an external digital signature for hard video authentication.

Almost all watermarking techniques proposed in the literature introduce some distortion to the original signal, and the distortion is permanent and irreversible. It is often desirable to embed a watermark in a reversible way so the original signal can be recovered from the watermarked signal. The first reversible data embedding scheme was proposed by Honsinger et al. [29] which uses a spatial additive, image-independent robust watermark into an 8 bit image using addition modulo 256. Fridrich et al. [30-32] have proposed several invertible watermarking schemes of hard image authentication. The basic idea for these algorithms is to embed the authentication data of a signal into itself through some invertible embedding scheme. At the verification side, the embedded authentication data is first extracted from the test signal, and then the inverse operation is applied to the test signal to get the candidate signal. Authentication data is calculated from the candidate signal, and compared with the extracted authentication data. If they agree, the signal is authentic, and the candidate signal is the original signal. Otherwise the signal has been modified. One scheme [30] losslessly compresses each bit plane from the LSB in the spatial domain or a set of quantized DCT coefficients until there is enough room to append the authentication data to the compressed bit stream to replace the original data being compressed. Another scheme [30] uses an additive, non-adaptive robust invertible watermark like that proposed by Honsinger et al. [29] to embed the authentication data. A third scheme [31, 32] partitions an image into small blocks and defines a flipping function F on pixel values such that $F^2 = Identity$, and a discrimination function that captures the regularity of each block. All blocks are classified into three categories: R-type, S-type, or U-type, depending if the regularity increases, decreases, or is the same when the flip function is applied to the block. Each R-type block is assigned bit 1, and S-type block bit 0. A RS-vector is generated by scanning the bit values of R- and S- type blocks in a certain order. The RS-vector is losslessly compressed and appended with the authentication data. The resulting bit stream is used to adjust the block type, and the flipping function is applied if needed, which flips an R-type block to S- type or vice versa. In this case, the RS-vector of the watermarked image is the bit stream embedded. We note here that the above authentication schemes' output is a binary decision for the whole signal, similar to classical authentication schemes, and no tamper localization can be provided. The watermarked signal may contain perceptual distortion, even though the original signal can be recovered if the signal has not been modified. Finally, we note that not every signal can be authenticated by invertible watermarking schemes since some signals cannot be invertible watermarked.

SOFT MULTIMEDIA AUTHENTICATION WITH SEMI-FRAGILE WATERMARKING

Most proposed algorithms for soft authentication use semi-fragile watermarking. Semi-fragile watermarks are designed to be robust to incidental modifications but sensitive to malicious modifications. The most common incidental modification is high-quality compression. Semi-fragile watermarks calculate some local or global distance between the original and extracted watermarks and compare them with a preset threshold to measure distortion and detect local or global tampering. Van Schyndel et al. [33] modified the LSB of image pixels by adding extended m-sequences to rows of pixels. A simple cross-correlation was used to test for the presence of the watermark. Wolfgang and Delp [34, 35] extended van Schyndel's scheme by embedding a bipolar m-sequence to improve robustness and localization capability. Their scheme requires the unaltered watermarked image for authentication, which is not practical for real applications.

The team of Zhu, Swanson, and Tewfik [36-40] were the first to propose to use quantization of a vector space to embed a watermark into digital multimedia for both robust and fragile purposes. The vector space, either some signal domain itself, such as spatial or frequency domain, or a projection to some random vectors, is partitioned into disjoint segments, with each segment assigned a watermarking digit. Data is embedded by using a vector in a segment to represent a signal vector falling into the segment. This is very similar to vector quantization. The quantization step is based on masking values calculated from the host signal, and thus guarantees transparency of watermarked signals. If the position a vector is perturbed to in a segment is known, the same scheme can be used to estimate local distortion and integrity by using the distance between the received quantization position and the original embedding position for each segment. A simple way to do this is to perturb a signal with a key-based pseudo-random pattern [36, 40]. More specifically, if we use an image as an example, the image is first divided into blocks of $M \times N$ pixels, and then a suitable transform is applied to the block. For embedding in spatial domain where a spatial masking model is used, no transform is needed; for DCT domain embedding where a frequency masking model is used, the block is transformed into the DCT domain. Suppose we use a DCT domain approach, and let $F(i, j)$ be the signal DCT value and $M(i, j)$ the corresponding masking value for frequency bin (i, j), then the embedding procedure can be expressed as:

$$F(i, j) \to F_w(i, j) = M(i, j)\{\left\lfloor \frac{F(i, j)}{M(i, j)} \right\rfloor + r(i, j)\, sign(F(i, j))\}, \tag{7-3}$$

where $r(i, j)$ is a key-based random signal in the interval $(0, 1)$, $\lfloor \cdot \rfloor$ rounds towards 0, and $sign(x)$ is the sign of x defined as:

$$sign(x) = \begin{cases} 1, & if\ x \geq 0; \\ -1, & if\ x < 0. \end{cases}$$

For a received image block with DCT coefficients $F_r(i, j)$, the masking values $M_r(i, j)$ are calculated, the error e at (i, j) is estimated by the following equation

$$\hat{e} = F_r - M_r \cdot \{r \cdot sign(F_r) + \left\lfloor \frac{F_r}{M_r} - (r - 1/2) \cdot sign(F_r) \right\rfloor\},$$

where all the values are evaluated at the same frequency bin (i, j). It has been shown [36, 40] that if the true error e at (i, j) is smaller in absolute value than $M(i, j)/2$, and assume that for small distortions the masking values for the received signal are the same as for the original signal, $M_r(i, j) = M(i, j)$, then the estimated error $\hat{e}$ is the true error, i.e., $\hat{e} = e$. It has been further shown that the error estimates are fairly accurate for small distortions, such as high quality JPEG compression, and are good indicators for large distortions.

To increase security, a key-based random selection of signal values or transform coefficients can be applied to prevent an unauthorized person from knowing the embedding positions. An even better way is to use the projection method to project signal vectors to key-based random directions before embedding with Eq. 7-3 [38, 40] or a combination of the random selection of blocks with the projection to random directions. They are straightforward to extend to video authentication in the raw signal domain, or audio authentication in the MDCT or FFT (magnitude) domain where a synchronization mark is embedded with the robust audio watermark version with the same quantization scheme for synchronization purpose.

Kundur and Hatzinakos [41, 42] proposed a similar but less sophisticated quantization-based watermarking scheme in the wavelet domain of an image for authentication and error measurement with both spatial and frequency information and localization, due to the special time-frequency property of the wavelet transform. The image is first decomposed using the wavelet transform to the Lth level into the detail image components $f_{k,l}(m,n)$ and the lowest resolution approximation $f_{u,L}(m,n)$ where $l = 1, 2, ..., L$, $k = h, v, d$ for horizontal, vertical, and diagonal detail components, and (m, n) is the wavelet coefficient index of the image. A quantization function Q is used to assign binary values to the wavelet coefficients f at resolution l, which are selected pseudo-randomly for modifications by a secret key

$$Q_{\Delta,l}(f) = \begin{cases} 0, & if \left\lfloor \dfrac{f}{\Delta 2^l} \right\rfloor is \ even, \\ 1, & if \left\lfloor \dfrac{f}{\Delta 2^l} \right\rfloor is \ odd. \end{cases} \tag{7-4}$$

If the ith wavelet coefficient chosen to be modified by a secret-key-based random process is $f_{k,l}(m,n)$, then it is modified to be $f_{k,l}^w(m,n)$ so that

$$Q_{\Delta,l}(f_{k,l}^w(m,n)) = w(i) \ XOR \ qkey(m,n), \tag{7-5}$$

where $w(i)$ is the ith watermark bit and $qkey$ is a bit generated from the image and a secret key. If Eq. 7-5 holds for $f_{k,l}^w(m,n) = f_{k,l}(m,n)$, then no change is needed, otherwise $f_{k,l}^w(m,n) = f_{k,l}(m,n) - \Delta 2^l \cdot sign(f_{k,l}(m,n))$. Thus the maximum modification at resolution l is $\pm \Delta 2^l$. The watermark is extracted by evaluating the expression

$$w(i) = Q_{\Delta,l}(f_{k,l}'(m,n)) \ XOR \ qkey(m,n), \tag{7-6}$$

where f' is the wavelet coefficient of a test image. The extent of tampering is evaluated using the number of correctly recovered watermark bits $w(i)$. The wavelet transform is usually the Haar transform, in which the coefficients at each resolution level l are rational numbers of the form $r/2^l$ where $r \in Z$. For Haar transform, the embedding scheme described in Eq. 7-5 guarantees that the watermarked image is of integer values in the spatial domain, thus no rounding error occurs. That means that, if a watermarked image undergoes no distortion, the extracted watermark from Eq. 7-6 is identical to the embedded watermark. This is an advantage over Zhu et al.'s approach (Zhu-watermark in the following) described in Eq. 7-3 which may show slight error $\hat{e} \neq 0$ for a watermarked image without any distortion, due to the error introduced by rounding watermarked signal values to integers, and the small difference of masking values calculations at the embedding side (on the original signal) and receiver side (on the watermarked signal). Signal independent quantization steps like Eq. 7-4 can remove such a difference, at the expense of watermarking transparency or robustness (i.e., measuring range which is proportional to the quantization steps). Zhu-watermark can easily be extended to the wavelet domain of an image by using the masking values from [43] .

Other than the differences mentioned above, it is interesting to compare the embedding and error measurement methods of the two methods. It is easy to see that the Zhu-watermark always replaces the signal's quantization residue by a random but known residue (similar to dithering) to maximize the measure distance. In the Kundur-watermark, the distance between the embedding positions of bit 0 and 1 is not maximized. It is possible that the value after watermarking is very close to the boundary that separates 0 and 1, and a small distortion will flip the extracted watermark bit value, thus affect the accuracy of the distortion measurement. For example, suppose wavelet coefficient $f_{k,l}^{w}(m,n)$ after watermark embedding is $f_{k,l}^{w}(m,n) = 2r \cdot \Delta \cdot 2^{l} + \varepsilon$ where $r \in Z$ and ε is a small positive number, $0 < \varepsilon << \Delta 2^{l}$, then a small perturbation $\delta > \approx \varepsilon$, which is much less than the quantization step, will create a slightly modified wavelet coefficient $f_{k,l}^{w}(m,n) - \delta$ yet still flip the watermarking bit from 0 to 1. The distortion measurement is also different for the two approaches. In the Zhu-watermark, the distortion on a single signal coefficient embedded with watermark is measured by the distance of the extracted residue and the embedding residue, and thus can measure a slight distortion which does not flip the watermarking bit value. On the contrary, Kundur-watermark's distortion measure depends on the extracted watermark bit value, and thus can detect distortion only when the extracted watermark bit is different from what is embedded. That is why it cannot set the watermarked signal coefficient to fixed positions to maximize the distance between embedding positions of watermarking bit 0 and 1, otherwise a perturbation smaller than Δ will not be detectable. It is interesting to point out that the distortion measure in the Zhu-watermark is a measure of the actual distortion incurred locally, no matter what kind of noise it is, while the Kundur-watermark's decision is probabilistic since the distortion is a statistical measure based on the assumption of Gaussian noise distortion. In summation, Zhu-watermark's quantization and distortion measure should give superior authentication and distortion measurement to Kundur's approach.

In Kundur's approach, *qkey*, a bit sequence generated from the host image and a secret key, is XORed with watermark bits, and plays a critical role to thwart forgery. If the sequence *qkey* is generated from the received signal, then its sensitivity is critical to the extracted watermark bits given by Eq. 7-6. If it is too sensitive, then a small distortion on the part that *qkey* is based on will reflect to the extracted watermark bits, thus distorting the accuracy of the error measurement; if it is not very sensitive, its role to thwart forgery is greatly compromised. If *qkey* is passed to the verifier, then it is the same as a signal independent bit sequence which does not help to thwart forgery of the same image (e.g., modifying the image and then placing all detail wavelet coefficients to the same quantization residue values as those of the unaltered watermarked image, assuming the quantization parameter Δ is known to an attacker). Note that Δ is not too hard to get by a brute force attack since Δ is signal independent and has a small range of possible values. If it is too large, embedding distortion at smooth regions can be visible; if it is too small, it cannot survive incidental manipulations such as high-quality compression. In addition, passing *qkey* to the verifier makes the scheme not strictly blind, which is very undesirable for many applications. Another possible security problem in using the Kundur watermark for authentication is that, by exploiting the fact that a modification of a multiple of the quantization step cannot be detected by the receiver, a hacker with knowledge of Δ may carefully engineer a modification of a watermarked signal without raising any suspicion from the verifier, although such an attack may be useless in reality. Projecting wavelet coefficients to random directions before the quantization step [38, 40] mentioned earlier make such an attack much more difficult.

Many semi-watermarking algorithms based on quantization have been proposed in the literature. They are more or less variations of the above two schemes. Eggers et al. [44] used a scaling factor and a pseudo-random number for their quantization-based watermark. A random number randomizes the embedding residue, which was also used in Zhu et al.'s quantization-based robust and fragile watermarking schemes [38, 40]. Wu and Kuo [45] applied the Kundur-watermark to log values of the magnitude of the low frequency DFT coefficients for speech authentication, with the non-adaptive quantization step at each coefficient determined according to the SNR of a speech codec at that frequency. Salient points, where the audio signal energy is fast climbing to a peak, and a synchronization mark by robust audio watermarking are used for synchronization. Tefas and Pitas [46, 47] applied chaotic mixing procedures to generate a watermark from a small logo and then embedded it for image and video authentication. We conclude this subsection by pointing it out that semi-fragile watermarking, as a special case of watermarking, is vulnerable to the vector quantization attack if embedded block-wise independently.

CONTENT-BASED AUTHENTICATION

A set of perceptual characteristics or features of a multimedia signal is called *content*. Content determines how human beings interpret a multimedia signal, i.e., the semantic meaning of media. Content-based authentication is used to authenticate the content extracted from a multimedia signal instead of the signal itself. Content is represented by a vector called the *feature vector*. Content authentication is gauged by a distance of the feature vector of the original signal s_o from the feature vector of the test signal s_t whose authenticity is to be tested:

$$d = \left\| feature(s_o) - feature(s_t) \right\|$$

If the distance d is larger than a preset, application-dependent threshold T, i.e., $d > T$, then the content is modified, otherwise the content is authentic. By measuring localized feature distances, a content-based authentication scheme may be able to give local tamper measurement.

All content-based authentication approaches need to first extract content from a multimedia signal. This is called *feature extraction*. These features are then used as the hash value in the classical signature-based authentication approach or embedded directly into the signal by applying some robust or semi-robust watermarking schemes, or both. The goal of content-based authentication schemes is to accept all manipulations that preserve the content of a multimedia signal while rejecting all other manipulations that modify the content. Features used as a signal's content determine directly what manipulations it can accept and what manipulations it rejects. Thus feature extraction plays a critical role for the performance of a content-based authentication method. Content features should be robust enough to achieve insensitivity to small modifications that preserve perceptual quality or semantic meaning of a media signal while being able to detect large changes. Depending on the application, different sets of features have been proposed for content-based authentication, such as image edges, histogram, etc. Due to the complexity of the human auditory and visual perception systems, feature extraction is still in its infancy. Currently, no universally applicable features exist for all applications. One of the key issues in designing a content-based authentication scheme is to find a set of features which adequately describes the content of a multimedia signal, while meeting the desired property that the features are fragile to the set of content modifying manipulations but robust to content preserving manipulations for a certain application.

Content-based Multimedia Authentication with External Signatures

The general procedure for generating an external content-based digital signature is shown in Figure 7-1. To generate a content-based digital signature, features are first extracted from the signal. This is followed by an optional data reduction stage such as down-sampling to reduce the amount of feature data. This stage may be necessary since features may correspond to a large amount of data. Feature data can be further reduced by an optional lossless compression at the next stage. The result is then encrypted together with other parameters such as image size, threshold for integrity verification, etc. with the private key to form the content-based digital signature of the signal. The digital signature is sent out together with the signal to the receiver for integrity verification.

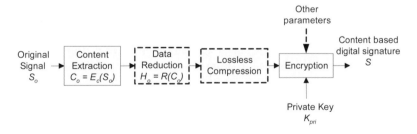

Figure 7-1 Generating an external content-based digital signature

The general procedure to verify a signal's authentication is shown in Figure 7-2. Features are first extracted from the test signal, and the optional data reduction stage may be applied, in exactly the same way as what was done in Figure 7-1 for digital signature generation. The original features are extracted from the separate digital signature by decrypting it first with the public key, and then decompressing if lossless compression was applied when generating the digital signature. The two feature vectors are then compared to find their distance d with some distance measure. If the distance d is larger than a preset threshold, it concludes that the content of the signal is modified, otherwise the content is authentic. The distance measure may be applied to local regions to verify the integrity of local regions.

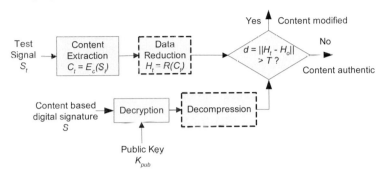

Figure 7-2 Verifying an external content-based digital signature

There are some variations to the above general procedures shown in Figure 7-1 and Figure 7-2. One possible variation is that instead of encrypting the feature data, it is possible to apply a cryptographic hash function to the data, and the resulting hash value is encrypted with the private key. In this case, the feature data together with the encrypted hash value form the digital signature for a multimedia signal. The advantage of this variation is that the data to be encrypted is much smaller and thus much faster and possibly a little securer for the encryption algorithm since the less data to be encrypted, the securer the encryption. It is possible to use the encrypted hash value alone as the digital signature to send out. In this case, the verifier cannot reverse the hash value to get the feature data from the digital signature. A verifier can only compare if the hash value from the test signal is equal to the original hash value from the digital signature. In this way, any changes to the content will be rejected, even the incidental manipulations for which the content is preserved. It is very similar to the classical digital signature. This is usually undesirable except in some applications like medical images where even a slight modification is not allowed. In most applications we need to tell the difference between content-preserving manipulations, such as high-quality lossy compression, and content modifying manipulations. For these applications, the threshold T must be larger than 0, and the feature data must be passed to the verifier.

Content-based authentication schemes differ from each other mainly on the implementation of "feature extraction" and the corresponding "comparison" shown in Figure 7-2. Features used in such a scheme should satisfy some general requirements, such as:

1. Univocally identifying the signal.
2. Invariant under or insensitive to mild manipulations such as compression that preserve the perceptual quality and content of the signal.
3. Highly sensitive to content-changing modifications.
4. Localized tamper detection capability is usually desirable.
5. Feature data size should be reasonably small.
6. Complexity is low enough for designed applications.

One of the early content-based image authentication approaches, proposed by Schneider and Chang [28] uses block histograms as features. An image is first partitioned into blocks, and the intensity histogram is calculated for each block. This can be further improved for spatial tamper detection by using variable block sizes: starting with small blocks, their histograms are combined to form histograms for larger blocks. This allows fine details to be protected by small blocks and large details by larger blocks. The Euclidean distance between intensity histograms is proposed as the content distortion measurement function. Some reduced distance function such as the difference of means of the histograms can also be used as distance measurement, which improves the robustness to JPEG compression at the cost of less sensitivity to subtle image manipulations. This approach requires storage of private key encrypted histograms, which may be considerably large. A histogram can be represented by moments. Lower order moments can be used as features for authentication [49] . They can capture significant characteristics of a histogram, but with much less data. The lowest order of meaningful moments is average, which is also proposed as an image feature for authentication [50] . An image is first divided into blocks, and the mean of each block is calculated and then quantized. The resulting block mean map is compressed before encryption with the private key. The data amount for the digital signature is greatly reduced as compared with the histogram approach. A little different block mean-based image authentication scheme is described in [51] . Each block mean is quantized to a binary value with either a uniform threshold or some variable local thresholds. The resulting binary image is then softly authenticated by applying a general message soft authenticator called approximate MAC [52] to each row and column. This is different from other content-based authentication schemes based on non-watermarking where the signal itself does not change when its authentication data is generated. This scheme has to modify an image whose authentication data is to be generated such that the pixels whose block mean is within a preset distance to the block's quantization threshold will be modified to make sure that the image transmitted does not contain any block whose mean is within the preset distance to its quantization threshold. In addition to this drawback, it is much easier to attack than using the block mean directly as proposed in [50] since the pixel value can be modified in a larger range without changing the authentication codes. In fact, as long as the block mean keeps the same polarity to its quantization threshold, the scheme will not be able to detect the modification. All these drawbacks trade for one gain in the approach: the authentication data is much less than the block mean-based approach proposed in [50] . The most significant drawback of all the approaches described in this paragraph, which are based on either block histograms or means, is that it is trivial to modify an image with possible content changes without altering its block histograms or means.

A better approach is to extract locations of "perceptually interesting" feature points for multimedia authentication. Bhattacharjee and Kutter [53] proposed a scale interaction model in the image wavelet transform domain to extract "salient" image feature points for image authentication. A continuous Mexican-Hat wavelet transform is used in their method. Their feature extraction procedure can be described as the following:

1. Define a feature detection function $P_{ij}(\cdot)$ as:

$$P_{ij}(\vec{x}) = \left| M_i(\vec{x}) - \gamma M_j(\vec{x}) \right|,$$

where $M_i(\vec{x})$ and $M_j(\vec{x})$ represent the Mexican-Hat wavelet transform coefficients at image location $\vec{x}$ for scales i and j, respectively, and $\gamma = 2^{(i-j)}$ is a normalization constant. The Mexican-Hat wavelet $\psi(\vec{x})$ is given by

$$\psi(\vec{x}) = (2 - |\vec{x}|^2) \exp(-\frac{\vec{x}^2}{2}).$$

For an image I, the wavelet transform coefficients at scale i and location $\vec{x}$ is given by:

$$M_i(\vec{x}) = \left\langle 2^{-i} \psi(2^{-i} \cdot \vec{x}; I) \right\rangle,$$

where $\langle \cdot; \cdot \rangle$ denotes the convolution of its operands. The following parameters $i = 2$ and $j = 4$ are used in [53].

2. Determine all maxima points of $P_{ij}(\cdot)$ in their corresponding local neighborhood.

A point of local maxima is accepted as a feature point if the variance of the image intensity in the neighborhood of the point is higher than a threshold. This is necessary to avoid spurious local maxima used as feature points. These feature points are then encrypted with a private key or lossless-compressed by Variable Length Code (VLC) or others to reduce data amount before encryption. The result is used as the content-based digital signature of the image. At the verification stage, locations of feature points are compared. A small deviation of location less than a threshold, say 1 pixel, is generally allowed to make the system less sensitive to content-preserving manipulations. If the number of unmatched feature points is larger than a threshold, it claims the image content is modified, otherwise the content is authentic.

Edges of an image can also be used as features for image/video content-based authentication [48, 49, 54], as well as zero-crossings [55], which can be considered as a special case of edges. Edges are first detected from an image with a Sobel operator or Canny edge detector, and then compared with a threshold to get the binary edge map. The resulting map can be optionally spatially subsampled to reduce the data amount. Lossless compression such as JBIG [56] or Variable Length Code (VLC) is then applied to further reduce the data amount. The result is encrypted with the private key to form the digital signature of the image. To check integrity, binary edge maps are generated from the test image and from the digital signature by decrypting with the public key followed by decompression, and then compared. Some content-preserving manipulations like JPEG may generate some distortion to the edge map. For example, if a block contains an edge, the quantization applied to the DCT coefficients of the block in JPEG will cause smoothing of the edge. Independent block DCT transform and quantization in JPEG can also generate some distortion to the extracted edge map. To make the feature vector comparison more robust to JPEG compression, Queluz [48, 49] proposed an edge map comparison scheme that is robust to JPEG compression: first the error pixels are found for the two binary edge maps to be compared. For each error pixel, a confidence measure is calculated based on how likely that point is an edge point, and on the spatial distance from that error position to a good matching position. The confidence measure is compared to a threshold to get a binary value with 1 for high confidence and 0 for low confidence of error for each error point. Then the following iterative search is applied to those error positions of value 0 until no change occurs: if a 0 error pixel has at least three neighbor error pixels of value 1, then its value is changed to 1. After the iterative procedure, all those error points with value 0 are removed from error pixel map. Integrity violation is decided if the maximum connected region in the resulting error image exceeds a pre-defined threshold.

The advantage of the "perceptually interesting" features by Bhattacharjee and Kutter [53] is its compact signature size. But the relevancy of the feature points selected by their method to the perceptually important feature points is not well established. These feature points may not be adequate for cropping and replacing manipulations on a small region. It may also be less accurate when trying to indicate which local region is tampered. Edge-based features tend to catch most of perceptually important features, but its signature may be long, and the edges detected may not be consistent for the original image and the image after acceptable manipulations. A major problem for both edge-based features and "perceptually interesting" features is that they do not capture some image attributes such as colors that do not depend on edges. A change to these attributes that modifies the content is not detectable by edge-based content-based authentication schemes.

Lin and Chang [57, 58] proposed a content-based image authentication scheme that exploits the fact that the same quantization table is applied to all image blocks in JPEG compression. The scheme accepts JPEG compression and rejects all other manipulations. It is based on the fact that if there is an inequality between two DCT coefficients at the same frequency from two arbitrary DCT blocks, then the same inequality still holds or they are equal after being quantized by the same quantization step. To extract the features from an image, the image is first divided into blocks and DCT transformed, exactly the same as in the JPEG compression. Then DCT blocks are randomly mapped into pairs based on some secret key. For each pair of DCT blocks, some frequency bins including dc are selected for comparison. For each pair of values at a selected frequency bin, if the comparison result is greater than or equal to 0, then feature bit 1 is generated; otherwise 0 is generated. The generated bits from all the block pairs together with the image size and mean values of the DCT coefficients at the selected frequency bins for all the DCT blocks are encrypted with the private key to form a digital signature. The mean values are included in the digital signature since otherwise it cannot detect if the image is uniformly changed by a constant value. In the image integrity verification stage, it follows the signature generation steps to compare the values at each frequency bin for each pair of blocks. If the original and the test images have a different sign in the equality beyond a tolerance bound $\tau \geq 0$, for example, if the signature bit is 1, i.e., the corresponding inequality in the original image is $\Delta_{i,j}^{orig}(f_k) \equiv DCT_i^{orig}(f_k) - DCT_j^{orig}(f_k) \geq 0$ at frequency f_k for the DCT blocks i and j; and the corresponding inequality for the test image is $\Delta_{i,j}^{test}(f_k) < \tau$, then it concludes that the test image has been manipulated by some operation other than JPEG. The small tolerance bound τ is used here to avoid small difference caused by rounding spatial image values to integers, different DCT implementations, etc. The tampered blocks can be identified by the mismatched inequalities between the test image and the signature, but if pairs of blocks do not overlap, there is no way to tell which block of the tampered pair has been modified. Overlapping pairs have to be used in this case to identify which block has been tampered. They have also proposed to compare the difference with some nonzero values in generating feature bits so the difference can be refined. Unfortunately the verification stage needs the quantization table for authentication, which is not practical for most applications since it requires that the test image is in JPEG format, and the received JPEG is the only JPEG applied. The second generation of JPEG compressed image is not allowed since the quantization table for the first generation JPEG is lost. In general, an integrity verifier has no knowledge of the test image's manipulation history. Rejecting two consecutive JPEG compressions while accepting one JPEG compression does not make much sense since one JPEG compression may have the same or lower perceptual quality than that compressed twice by JPEG.

They [59, 60] extended the same approach for video authentication which adds the time codes of each frame to the feature codes extracted from each frames with the above mentioned method. They have also designed an authentication scheme in MPEG compression domain to survive certain MPEG transcoding operations [59, 60]. The data to be encrypted with the private key which forms the digital signature for each GOP (group of pictures) includes the feature data obtained from compression pairs of quantized DCT coefficients in each macroblock, where three values 1, 0, or -1 are used to indicate the comparison result of greater than, equal to, or less than 0 (instead of two values for image case), the hash value of the combination of the GOP header and the hash values of the motion vectors and control codes (i.e., all the data except the codes of DCT coefficients), the quantization scale in the Slice or MB header (and their control codes) for each picture, ordered according to the position of each picture in the GOP. The resulting digital signature for a GOP is placed in the *user_data* area of the GOP header.

The scheme smartly exploits some invariance properties in JPEG compression. It also has some obvious drawbacks. The algorithm cannot reliably tell JPEG quality level, and thus will accept JPEG compression at any quality level. Even though counting the number of signature bits that change to "equal" in the test image can give a rough estimation of JPEG quality, the method itself is not very reliable. This problem can be addressed successfully by combining the digital signature with a semi-fragile watermarking scheme described below. In addition, the DCT blocks in signature generation have to be exactly the same as in JPEG compression. If there is a small mismatch of the DCT blocks, for example, by image shifting, the scheme will reject as malicious manipulations. Of course, by design, the scheme cannot accept incidental manipulations other than JPEG, even though by introducing an appropriate tolerance bound τ, this limitation can be alleviated to some extent.

In addition, the scheme may not be secure if the same key is used to authenticate multiple images. The security of the scheme lies on the key-based secret formation of pairs for comparison. Once the secret mapping function is known, a hacker can easily modify DCT coefficients while keeping unchanged the original inequalities of the DCT coefficients for each pair of DCT blocks and also keeping the mean values for each DCT coefficients over all the blocks. Radhakrishnan and Memon [61] has proved that a hacker Oscar only needs access to $O(\log N)$ images and their corresponding signatures generated by the same key (i.e., using the same pair of DCT blocks), where N is the number of DCT blocks in a given image. The attack is as follows: from the first image and its signature bits, he can use the first signature bits belonging to the first DCT block pairs to form a set of all possible pairs of blocks. Then he can use the corresponding signature bits from the second image to remove impossible pairs from the set of candidate pairs since all images use the same pairs of blocks for authentication. He can continue this way until the set has only one candidate. He then uses the same method and the rest of the signature bits to find out other pairs of blocks. A hacker needs roughly $2 \cdot \log N$ to make the attack successful with complexity about $N^3 \log N$ if only the dc value is used for comparison to generate signature bits. For 512×512 images which have 4096 blocks of 8 by 8 pixels, Oscar needs roughly 24 images to deduce the block pairs being compared to generate the signature. If ac values are also compared to generate signature, or DCT blocks are grouped into non-overlapping pairs, the number of images to mount a successful attack of this kind is less than $2 \cdot \log N$. To prevent such an attack, the key has to be changed frequently, or a randomly generated number is used together with the key to generate the pairs, and the random number is encrypted as part of the signature [61]. This can prevent images using the same pairs to generate signatures when the same key is used.

For speech signals, short-time energy and zero-crossings are candidate features for content-based authentication, but the size of these features tends to be large, and thus impractical. Wu and Kuo [45, 62] proposed a content-based speech authentication scheme which has much smaller overhead to survive speech coding and other incidental operations. Three types of features are extracted from a speech signal. They are pitch information, the changing shape of the vocal tract, and the energy envelope. They proposed to combine CELP coding, which includes ITU G.723.1, GSM-AMR, and

MPEG4-CELP, with the extraction of the first two feature types: the first three LSP coefficients except the silent portion will be used as pitch information, and one pitch coefficient, used as the changing shape of the vocal tract, is obtained from each frame as the average of the "lag of pitch predictors" of all subframes except the non-tonal part. The starting and ending points of silent periods and also non-tonal regions are included in the authentication data as parameters since LSP coefficients in silent periods do not model the shape of the speaker vocal tract, and the pitch coefficients extracted from the CELP codec have no physical meaning in non-tonal regions. For synchronization purposes, the locations of the first 15 salient points (the positions where the audio signal energy is quickly climbing to a peak) are also concatenated to the feature codes. At the verification phase, the feature difference between the decrypted and extracted for each type of feature is calculated independently. The difference for LSP coefficients is the difference of the weighted average of the three LSP coefficients. The feature difference is only calculated for the LSP and the pitch coefficients in non-silent periods and tonal regions, respectively. A low-pass filter is applied to the resulting difference sequences before being compared with a threshold to determine a signal's authenticity.

Content-fragile Watermarking

Content-fragile watermarking can also be used for content-based multimedia authentication. Features extracted from digital media can be embedded into the media itself with a robust or semi-robust watermarking scheme instead of generating an external digital signature. Although it uses different ways to place the feature data, content-fragile watermarking shares other steps with its external signature counterpart. In content-fragile watermarking, the whole space of signal coefficients is divided into the three subspaces: signature generating, watermarking, and ignorable subspaces. Signal coefficients in the signal generating subspace are used to generate content-features which will be converted to authentication bits to be embedded into the signal itself. The coefficients in the watermarking subspace will be used to embed the authentication bits into the signal. Those coefficients in the ignorable subspace will not be used in either way. The division of subspaces should be kept secret or determined pseudo-randomly based on some secret key. The signature generating subspace can be overlapped with the watermarking subspace. If they are not overlapped, then there is no interference between the two procedures. This is much simpler than the overlapping case. The drawback is that it may be less accurate in locating tamper regions. The overlapping approach is much more complex since the watermarking procedure modifies the signal slightly which may affect the accuracy of feature extraction from the watermarked signal. Care has to be taken to make sure that the watermarking procedure does not distort the feature extraction significantly enough to have a false alarm, esp. when the signal is under acceptable manipulations. A general approach to addressing this problem for the overlapping case is to carefully design the feature extraction algorithm so it is robust to the watermarking procedure, and/or to design the watermarking procedure which does not distort the feature extraction much. Another common approach is to use some iterative procedures to guarantee the extracted signatures, before and after watermarking, match each other. Rey and Dugelay [63] have proposed the following iterative procedures: to sign a multimedia signal, extract features from the newly watermarked signal, and then embed the newly obtained authentication data to the original signal (in order to avoid cumulating distortions) until the difference between the current extracted features and the previous extracted features is smaller than a threshold. Whether such an iterative method converges is yet to be proved, but they claim that in practice three iterations are enough. Some constraints are imposed to the watermarking scheme used for content-based authentication: it must allow blind extraction; it should be robust to allow lossless extraction; its embedding capacity must be large enough to embed a signature usually consisting of a large amount of data.

Lin and Chang proposed a semi-watermarking scheme to embed the feature bits extracted from an image [64] . The method accepts JPEG up to a certain quality level, and rejects other manipulations. An image is first divided in DCT blocks in exactly the same way as in JPEG. These DCT blocks are randomly divided into signature generation and watermarking subspaces, and there is no overlapping between the two subspaces. Feature bits are extracted from the signature generation subspace in the same way as their external signature approach described in Section 7.3.3.1. These feature bits are then embedded with a DCT-based quantization scheme into the selected frequency bins in the DCT blocks in the watermarking subspaces. The quantization step for each frequency bin is one larger than the corresponding JPEG quantization step at the coarsest JPEG quality level it wants to accept. This would guarantee lossless extraction of the watermark bits as long as the JPEG quality is not below the designated JPEG compression quality. For JPEG compression with quality level lower than the designated level, the extracted signature bits may be wrong since the JPEG quantization step is larger than the embedding quantization step, and thus rejected by the scheme. This is an advantage over their external signature counterpart described in Section 7.3.3.1: the watermarked scheme can reliably accept JPEG compression only up to a designated quality level. The gain comes with a cost that the perceptual quality of the watermarked image is close to that of the JPEG compression at the lowest quality level it accepts. The watermarking perceptual quality may not be acceptable if the admissible JPEG quality is too low.

Dittmann et al. [54] embedded, with a robust watermarking algorithm, a watermark pattern generated from extracted edge features, and checks if the watermark pattern generated from the test image exists in the test image for content-based authentication. This method cannot give localized tamper indication, and may not be robust to some incidental manipulations. Bassali et al. [65] embedded the encrypted mean of a macroblock into the mean of each of its constituent microblocks by a quantization-based watermarking scheme. Fridrich et al. [66] divided an image into medium size blocks and inserted a robust spread-spectrum watermark into each blocks. It compares the watermark detectability from each block to determine malicious manipulations from incidental ones.

7.4 Conclusion

We reviewed the current and emerging state of multimedia data authentication and tamper detection technologies. An overview of the requirements and challenges for authentication systems indicated that, while some general rules apply, they are often application dependent. We found that multimedia integrity could be best categorized into three types of multimedia authentication. Hard authentication, usually based on fragile watermarks to detect any modification to the underlying signal, has received significant coverage in the literature and is the most mature authentication approach to date. Soft authentication, frequently based on semi-fragile watermarks to measure signal modification within perceptual tolerance, is still in a relatively early research stage. Additional work is needed to improve robustness to incidental changes while remaining sensitive to malicious modifications. Content-based authentication, using external signatures or watermarks, detects any manipulations that change the semantic meaning of the content. Employing signal feature vectors and tolerance measures, content-based authentication is still emerging. More work on defining features which adequately describe the perceptual content of a multimedia signal is needed to meet the desired property of being fragile to content modifying manipulations yet robust to content preserving modifications. Content-based authentication may have the widest potential in commercial applications.

REFERENCES

[1] D. R. Stinson, *Cryptography, Theory and Practice*, CRC Press, 1995.

[2] J. Menezes, P. C. van Oorschot, and S. A. Vanstone, *Handbook of Applied Cryptography*, CRC Press, 1996.

[3] G. L. Friedman, "The Trustworthy Digital Camera: Restoring Credibility to the Photographic Image," *IEEE Trans. Consumer Electronics*, vol. 39, no. 4, pp. 905-910, 1993.

[4] M. D. Swanson, M. Kobayashi, and A. H. Tewfik, "Multimedia Data-Embedding and Watermarking Technologies," *Proc. IEEE*, vol. 86, no. 6, pp. 1064-1087, 1998.

[5] F. Hartung and M. Kutter, "Multimedia Watermarking Techniques," *Proc. IEEE*, vol. 87, no. 7, pp. 1079-1107, 1999.

[6] R. B. Wolfgang, C. I. Podilchuk, and E. J. Delp, "Perceptual Watermarks for Digital Image and Video," *Proc. IEEE*, vol. 87, no. 7, pp. 1108-1126, 1999.

[7] C. D. Vleeschouwer, J. Dekaugke, and B. Macq, "Invisibility and Application Functionalities in Perceptual Watermarking – An Overview," *Proc. IEEE*, vol. 90, no. 1, pp. 64-77, 2002.

[8] E. T. Lin, and E. J. Delp, "A Review of Fragile Image Watermarks," *Proc. Multimedia and Security Workshop at ACM Multimedia' 99,* Orlando, FL. Oct. 1999, pp. 25-29.

[9]   J. Fridrich, "Methods for Tamper Detection in Digital Images," *Proc. Multimedia and Security Workshop at ACM Multimedia' 99,* Orlando, FL, Oct. 1999, pp. 19-23.

[10] M. G. Albanesi, M. Ferretti, F. Guerrini, "A Taxonomy for Image Authentication Techniques and Its Application to the Current State of the Art," *Proc. 11th Int. Conf. Image Analysis and Processing,* 2001, pp. 535 -540.

[11] M. Holliman and N. Memon, "Counterfeiting Attacks on Oblivious Block-wise Independent Invisible Watermarking Schemes," *IEEE Trans. Image Processing,* vol. 9, no. 3, pp. 432-441, 2000.

[12] S. Walton, "Information authentication for a slippery new age," *Dr. Dobbs Journal,* vol. 20, no. 4, pp. 18-26, 1995.

[13] M. Yeung and C. Mintzer, "An invisible watermarking technique for image verification," *IEEE Int. Conf. Image Processing,* 1997, vol. 2, pp. 680-683.

[14] M. Wu and B Liu, "Watermarking for Image Authentication," *IEEE Int. Conf. Image Processing,* vol. 2, pp. 437-441, 1998.

[15] N. Memon, S. Shende, and P. W. Wong, "On the Security of the Yeung-Mintzer Authentication Watermark," *Proc. IS&T PICS Symp.* Savannah, Georgia, March, 1999.

[16] J. Fridrich, M. Goljan, and N. Memon, "Further Attacks on Yeung-Mintzer Fragile Watermarking Scheme," *Proc. SPIE Photonic West, Electronic Imaging 2000, Security and Watermarking of Multimedia Contents,* San Jose, California, January 24-26, 2000.

[17] J. Fridrich, M. Goljan, A. C. Baldoza, "New fragile authentication watermark for images," *IEEE Int. Conf. Image Processing,* vol. 1, pp. 446-449, 2000.

[18] P. W. Wong, "A watermark for image integrity and ownership verification," *Proc. IS&T PIC,* Portland, Oregon, 1998.

[19] P. W. Wong, "A public key watermark for image verification and authentication," *Proc. IEEE Int. Conf. Image Processing,* vol. 1, pp. 455-459, 1998.

[20] P. W. Wong and N. Memon, "Secret and Public Key Image Watermarking Schemes for Image Authentication and Ownership Verification," *IEEE. Trans. Image Processing,* vol. 10, no. 10, pp. 1593-1601, 2001.

[21] D. Coppersmith, F. Mintzer, C. Tresser, C. W. Wu, and M. M. Yeung, "Fragile Imperceptible Digital Watermark with Privacy Control," Proc. SPIE/IS&T Int. Symp. Electronic Imaging: Science and Technology, San Jose, CA, 1999, vol. 3657, pp. 79-84.

[22] M. Holliman, N. Memon, and M. M. Yeung, "On the Need for Image Dependent Keys for Watermarking," *Proc. Content Security and Data Hiding in Digital Media,* Newark, NJ, May 14, 1999.

[23] J. Fridrich, "Robust bit extraction from images," *Proc. IEEE Int. Conf. Multimedia Computing & Systems,* vol. 2, pp. 536-540, 1999.

[24] M. U. Celik, G. Sharma, E. Saber, and A. M. Tekalp, "A Hierarchical Image Authentication Watermark with Improved Localization and Security," *IEEE Int. Conf. Image Processing*, vol. 2, pp. 502-505, 2001.

[25] D. Li and D. Feng, "A DCT Robust Multimedia Authentication Scheme," *7th Australian and New Zealand Intelligent Info. System Conf.* Perth, Western Australia, Nov. 2001.

[26] J. Lee and C. S. Won, "A Watermarking Sequence Using Parities of Error Control Coding for Image Authentication and Correction," *IEEE Trans. Consumer Electronics*, vol. 46, no. 2, pp. 313-317, May 2000.

[27] J. Fridrich and M. Goljan, "Images with Self-Correcting Capabilities", *Proc. IEEE Int. Conf. Image Processing*, 1999, vol. 3, pp. 792-796.

[28] M. Schneider and S.-F. Chang, "A robust content based digital signature for image authentication," *IEEE Int. Conf. Image Processing*, vol. 3, 1996, pp. 227-230.

[29] C. W. Honsinger, P. W. Jones, M. Rabbani, and J. C. Stoffel, "Lossless recovery of an original image containing embedded data," *U. S. Patent No. 6,278,791*, Aug. 2001.

[30] J. Fridrich, M. Goljan, and R. Du, "Invertible Authentication," *Proc. SPIE, Security and Watermarking of Multimedia Contents*, San Jose, California January, 2001.

[31] M. Goljan, J. Fridrich, and R. Du, "Distortion-free Data Embedding," *4th Information Hiding Workshop*, Pittsburgh, Pennsylvania, April, 2001.

[32] J. Fridrich, M. Goljan, and R. Du, "Lossless Data Embedding? New Paradigm in Digital Watermarking," *Special Issue on Emerging Applications of Multimedia Data Hiding*, vol. 2002, no.2, February 2002, pp. 185-196.

[33] R. G. van Schyndel, A. Z. Tirkel, and C. F. Osborne, "A digital watermark," *Proc. IEEE Int. Conf. Image Processing*, vol. 2, pp. 86-90, 1994.

[34] R. B. Wolfgang and E. J. Delp, "A Watermarking for Digital Images," *Proc. IEEE Int. Conf. Image Processing*, vol. 3, pp. 219-222, 1996.

[35] R. B. Wolfgang and E. J. Delp, "Fragile Watermarking Using the VW2D Watermarking," *Proc. SPIE Security and Watermarking of Multimedia Contents*, San Jose, California, Jan. 25-27, 1999, pp. 204-213.

[36] B. Zhu, M. D. Swanson, A. H. Tewfik, "A Transparent Authentication and Distortion Measurement Technique for Images," *Proc. 7th IEEE Digital Signal Processing Workshop*, Loen, Norway, pp. 45-48, September, 1996.

[37] M. D. Swanson, B. Zhu, A. H. Tewfik, "Robust Data Hiding for Images," *Proc. of 7th IEEE Digital Signal Processing Workshop*, Loen, Norway, pp. 37-40, September, 1996.

[38] M. D. Swanson, B. Zhu, A. H. Tewfik, "Data Hiding for Video-in-Video," *Proc. IEEE Int. Conf. on Image Processing*, vol. 2, pp. 676-679, Santa Barbara, CA, October, 1997.

[39] M. D. Swanson, B. Zhu, A. H. Tewfik, "Video Data Hiding for Video-in-Video and Other Applications," *Proc. of SPIE Multimedia Storage and Archiving Systems*, vol. 3229, pp. 32-43, Dallas, TX, November, 1997.

[40] B. Zhu, *Coding and Data Hiding for Multimedia*, Ph. D. Thesis, Univ. of Minnesota, Dec. 1998.

[41] D. Kundur, and D. Hatzinakos, "Towards a Telltale Watermarking Technique for Tamper Proofing," *Proc. IEEE Int. Conf. Image Processing*, vol. 2, 1998, pp. 409-413.

[42] D. Kundur, and D. Hatzinakos, "Digital watermarking for telltale tamper proofing and authentication," *Proc. IEEE*, vol. 87, no. 7, 1999, pp. 1167-1180.

[43] A. B. Watson, G. Y. Yang, J. A. Solomon, and J. Villasenor, "Visibility of Wavelet Quantization Noise," *IEEE Trans. Image Processing*, vol. 6, no. 8, pp. 1164-1175.

[44] J. J. Eggers and B. Girod, "Blind Watermarking Applied to Image Authentication," *IEEE Int. Conf. Acoustics, Speech, and Signal Processing*, vol. 3, pp. 1977-1980, 2001.

[45] C.-P. Wu and C.-C. J. Kuo, "Comparison of Two Speech Content Authentication Approaches," *SPIE Int. Symp. Electronic Imaging*, San Jose, California, January 2002.

[46] A. Tefas and I. Pitas, "Image Authentication Using Chaotic Mixing Systems," *Proc. IEEE Int. Symp. Circuits and Systems (ISCAS 2000)*, vol. 1, pp. 216-219, 2000.

[47] F. Bartolini, A. Tefas, M. Barni, and I. Pitas, "Image authentication techniques for surveillance applications," *Proc. IEEE* vol. 89, no. 10, pp. 1403-1418, 2001.

[48] M. P. Queluz, "Towards Robust, Content Based Techniques for Image Authentication," *IEEE 2nd Workshop on Multimedia Signal Processing*, pp. 297 -302, 1998.

[49] M. P. Queluz, "Content-Based Integrity Protection of Digital Images," *SPIE Conf. Security and Watermarking of Multimedia Contents*, vol. 3657, pp. 85-93, Jan. 1999.

[50] D.-C. Lou and J.-L. Liu, "Fault Resilient and Compression Tolerant Digital Signature for Image Authentication," *IEEE Trans. Consumer Electronics*, vol. 46, no. 1, pp. 31-39, 2000.

[51] L. Xie, G. R. Arce, and R. F. Graveman, "Approximate Image Message Authentication Codes," *IEEE Trans. Multimedia*, vol. 3, no. 2, pp. 242-252, 2001.

[52] R. F. Graveman and K. Fu, "Approximate Message Authentication Codes," *Proc. 3rd Annual Fedlab Symp. Advanced Telcomm./Info. Distribution*, vol. 1, College Park, MD, Feb. 1999.

[53] S. Bhattacharjee and M. Kutter, "Compression Tolerant Image Authentication," *IEEE Int. Conf. Image Processing*, vol. 1, pp. 435-439, 1998.

[54] J. Dittmann, A. Steinmetz, and R. Steinmetz, "Content-based Digital Signature for Motion Pictures Authentication and Content-Fragile Watermarking," *IEEE Int. Conf. Multimedia Computing and Systems*, vol. 2, pp. 209-213, 1999.

[55] C.-T. Li, D.-C. Lou, and T -H. Chen, "Image Authentication and Integrity Verification via Content-Based Watermarks and a Public Key Cryptosystem," *IEEE Int. Conf. Image Processing*, vol. 3, pp. 694-697, 2000.

[56] ISO/IEC JTC1/SC2/WG9, "Progressive Bi-level Image Compression, Revision 4.1," *CD 11544*, September 16, 1991.

[57] C.-Y. Lin and S.-F. Chang, "A Robust Image Authentication Method Surviving JPEG Lossy Compression," *SPIE Storage and Retrieval of Image/Video Database*, San Jose, Jan 1998.

[58] C.-Y. Lin and S.-F. Chang, "A Robust Image Authentication Method Distinguishing JPEG Compression from Malicious Manipulation," *IEEE Trans Circuits and Systems of Video Tech.*, vol. 11, no. 2, pp. 153-168, 2001.

[59] C.-Y. Lin and S.-F. Chang, "Generating Robust Digital Signature for Image/Video Authentication," *Multimedia and Security Workshop at ACM Multimedia 98*, Bristol, UK, Sep. 1998.

[60] C.-Y. Lin and S.-F. Chang,, "Issues and Solutions for Authenticating MPEG Video," *SPIE Security and Watermarking of Multimedia Contents*, EI '99, San Jose, CA, Jan. 1999.

[61] R. Radhakrishnan and N. Memon, "On the Security of the SARI Image Authentication System," *IEEE Int. Conf. Image Processing*, vol. 3, pp. 971-974, 2001.

[62] C.-P. Wu and C.-C. J. Kuo, "Speech Content Authentication Integrated with CELP Speech Coders," *IEEE Int. Conf. on Multimedia and Expo (ICME2001)*, Aug. 2001.

[63] C. Rey and J.-L. Dugelay, "Blind Detection of Malicious Alterations on Still Images Using Robust Watermarking," *IEE Seminar on Secure Images and Image Authentication* (Ref. No. 2000/039), pp. 7/1-7/6, 2000.

[64] C.-Y. Lin and S.-F. Chang, "Semi-Fragile Watermarking for Authenticating JPEG Visual Content," *SPIE Security and Watermarking of Multimedia Contents II*, San Jose, CA, Jan. 2000.

[65] H. S. Bassali, J. Chhugani, S. Agarwal, A. Aggarwal, and P. Dubey, "Compression Tolerant Watermarking for Image Verification," *IEEE Int. Conf. Image Processing*, vol. 1, pp.430-433, 2000.

[66] J. Fridrich, "Image Watermarking for Tamper Detection," *Proc. IEEE Int. Conf. Image Processing*, vol. 2, pp. 404-408, 1998.

8 INDEXING AND RETRIEVING HIGH DIMENSIONAL VISUAL FEATURES

Dr. Jesse S. Jin

In this chapter, we review high-dimensional retrieval by examining the chronological evolution of various indexing techniques. We then discuss the R-tree which leads to the development of CSS+-tree. Two important aspects of high dimensional index and retrieval, namely varying distance metrics and dimension reduction are, also discussed and some creative solutions proposed.

8.1 Introduction

Multimedia databases contain non-textual objects which may be images, digitized sounds, or videos. At present, the primary search method for locating objects in the database is based on the textual descriptions accompanying the images. A purely text-based approach poses significant limitations. Some features in images, such as irregular shapes and jumbled textures, are extremely difficult, if not impossible, to describe in text. The description limits the scope of searching to that predetermined by the author of the system or the current application, and leaves no means for using the data beyond that scope. Description can be subjective, and different people may give quite different descriptions for the same image. There is also the common problem of understanding natural language description. Descriptive text is entered manually and hence, the indexing process is slow.

It would be much easier if the computer could compare these objects using the type of visual or auditory cues used by humans. A way to do this is to extract features (representing these cues) from the objects, based on what is known about human perception. These feature vectors are extracted from the objects using filters corresponding to the ones used by humans. For example, to characterize texture we can use a set of Gabor filters which are similar to the simple cells in the human visual system [53]. A collection of texture images can then be represented by a collection of feature vectors, resulting from the application of the Gabor filters to the images [32, 29, 2].

The vectors resulting from the feature extraction process are usually in high dimension. Due to the "curse of dimensionality", it is not possible to efficiently search the exact k-nearest-neighbors of a point in a high dimensional space. Although the size of a feature vector database might be large, if the dimensionality is

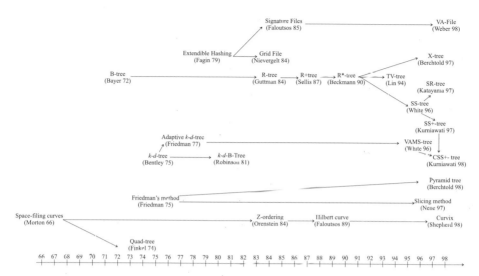

Figure 8-1 The developments of multidimensional access methods

high enough, most of the space will be empty. This condition will cause most of the feature vector space to be covered during the k-nearest-neighbor search. When this happens, the search degrades to a linear search instead of a *log(n)* search (*n* is the number of vectors in the database), as predicted by the asymptotic lower-bound for proximity queries. This is because the spherical area covered by the k-nearest-neighbors intersects the bounds of most of the subspace resulting from any partitioning of the data.

In a high-dimensional space, therefore, we will generally have to abandon the goal of finding the exact k-nearest-neighbors. A similarity search is a kind of k-nearest-neighbor search where we are interested in getting k vectors close to a given vector, but not necessarily the exact closest k. In interactive querying a multimedia database, having this kind of error is not totally disastrous since the user will have some degree of tolerance. They will have the impression that the system does not quite retrieve what they want, but it would be more acceptable than having to wait for the program to scan the whole or a large portion of the database before returning any results.

Access methods for multidimensional data have been investigated since 1972, starting with the "post-office" tree method [35]. This field has been growing rapidly and a large number of methods have been proposed since then. In this chapter, a detailed survey of the area will not be attempted, and the reader is referred to [8, 39, 23, 48]. Only the main developments in the field and methods that specifically address high-dimensional problems will be described in this chapter.

The main developments in the field can be seen in Figure 8-1 (To make the figure more readable, only the first author is cited in Figure 8-1. All the structures listed in the figure are mentioned in the text with full reference details). While the early structures (before 1990) were not actually designed to deal with high-dimensional spaces, they are included because of their importance in the development of more recent high-dimensional structures.

Based on how the method copes with multidimensional spaces, multidimensional access methods can be classified into three general categories: tree-based, hash-based, and projection-based methods.

Tree-based methods, including R-tree [24, 55, 56, 38], k-d-tree [7, 21, 1], quadtree [17, 48, 26] or clustering [22, 11], divide the data hierarchically. The whole collection of vectors is divided into a collection of leaf nodes. Several leaf nodes are then arranged into an internal node. The process is continued for internal nodes until we only have one node, the root node, in the end (see Figure 8-2).

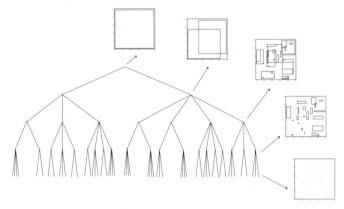

Figure 8-2 A multidimensional tree structure

Hash-based methods associate each vector with a disk block where the vector (and other data associated with it) is kept. This is, in effect, similar to dividing the space where the vectors are embedded. The association can be done via a directory residing in memory [43, 19] or by using some kind of signature calculated from the vector's value on each dimension [54]. As an example, a signature-based method, namely VA-File is shown in Figure 8-3.

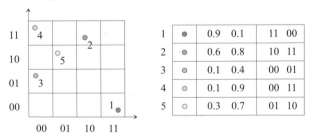

Figure 8-3 A signature-based method (VA-File)

Projection-based methods "flatten" the multidimensional space into the one-dimensional space and organize the space using one-dimensional access methods. They project all vectors to one-dimensional points by mapping each vector onto one or several lines [20, 42] or one or more fractal curves [44, 28, 50]. Some of these methods actually project a point into several one-dimensional points by projecting it onto several lines or fractal curves. Fractal curves attract much attention because the dimensionality of these curves is actually greater than one although the curves themselves are one-dimensional lines. Figure 8-4 shows one of the projections using the Hilbert curve.

Figure8-4 The Hilbert curve in 2-D and 3-D spaces

There are also approaches that combine two or more of these approaches, e.g. the Pyramid tree [9] which combines projections and tree structures.

Most of these structures are developed for low-dimensional spaces. They will have problems when the dimensionality is high enough with respect to the amount of data, since the k-nearest-neighbor disk will cover an area that often is as large as the total data space itself. Besides this, there are also problems inherent in some of the multidimensional access structures that prevent them from being used for similarity searches in high-dimensional spaces. This chapter will introduce a new index structure, CSS+-tree, and discuss issues of varying distance metrics and dimension reduction.

8.2 The R-Tree Family

Of all the multidimensional access structures, the R-tree and its variants have been regarded as the most successful in dealing with high dimensionality and have been used in several commercial systems, such as Postgres/Illustra [15] and IBM QBIC [18]. The R-tree family [24, 49, 5; 55, 56, 38] is a generalization of the B-tree [4] for multidimensional data. If M and m are the maximum and minimum number of entries in a node respectively, the properties of such trees are:

o   Every node contains between m and M index entries unless it is the root;

o   Each node is bounded by a spatially minimum bounding rectangle that contains all objects (feature vectors or child nodes) within the node;

o   The root node has at least two children unless it is a leaf;

o   All leaves appear at the same level [24].

The R-tree uses bounding rectangles. The main problem with this kind of bounding envelope is that there are large overlapping areas. If a query falls into an overlapping area, such as query a in Figure 8-5(a), search has to go through all rectangles containing a (branches c_1 and c_3 as shown in Figure 8-5(b)). Another problem is that the bounding rectangle is not consistent with the similarity measurement, the distance. For example, query *a* falls into clusters c_1 and c_3 in Figure 8-5(a). Due to the inconsistency, cluster c_{23} will not be the first candidate in the search.

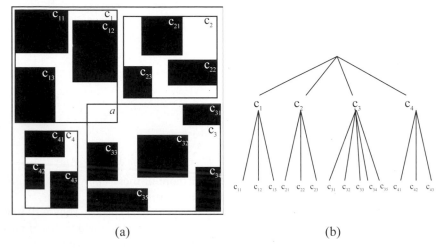

| (a) | (b) |

Figure 8-5 Overlapping and inconsistency problem of the bounding rectangle

The k-d-tree uses a hyper-plane to divide the space along one dimension at a time. It could produce elongated spaces, as shown in Figure 8-6(a). The k-d-B-tree solves the problem by allowing the hyper-plane in any direction, as shown in Figure 8-6(b). The k-d-tree and k-d-B-tree solve the overlapping problem but may worsen the inconsistent distance problem. SS-tree and X-tree use bounding spheres, as shown in Figure 8-6(c). The distance in query is consistent with the bounding envelope distance but the overlapping could be bigger than bounding rectangles especially at the top levels of the tree.

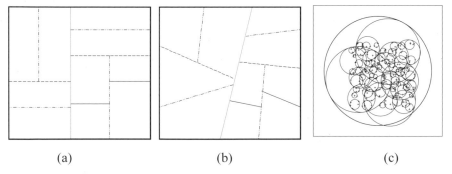

| (a) | (b) | (c) |

Figure 8-6 Node separation surfaces of k-d-tree (a), k-d-B-tree (b) and SS-tree (c)

The k-d-tree based structures (k-d-tree, BSP-tree, BBD-tree) are not suitable as external memory structures as their nodes do not utilize disk blocks efficiently, causing the trees to become too deep. A disk-based version of k-d-tree, k-d-B-tree [46], could not perform well because of the excessive amount of dead space. All other tree-structures are closely related to the R-tree [24], striving to maintain a balanced tree by doing the insertion from the bottom-up. When building an index of an existing database, a more efficient index structure can be obtained if the tree is built using one of the bulk-loading techniques. A statically built (or bulk-loaded) version of k-d-B-tree, VAMSplit-k-d-tree, could perform much better than other trees built incrementally. This is due to the global nature of bulk-loading techniques, the partitioning algorithm of bulk-loading techniques uses more information than just local (within a node) information available to incremental partitioning heuristics.

8.3 The CSS+-Tree

The issues of improving an R-tree based multi-dimensional access structure include: selecting the shape of bounding envelopes, splitting heuristics, the heuristics to reduce overlapping, the updating criteria, and efficient search strategy. These issues are not independent of each other. The splitting heuristic and sub-tree selection criteria will determine the shape of each node and hence the goodness of the bounding envelope. The goodness of the bounding envelope will also determine the extent of overlapping between nodes. The heuristics used to reduce overlapping will also affect the shape of the nodes. We present our CSS+-tree by addressing four issues: bounding envelope, splitting heuristics, updating criteria and searching strategy. Overlapping often happens during the updating process, and we address it as an updating criterion.

GENERAL STRUCTURE OF CSS+-TREE
The CSS+-tree is a tree-based structure which uses a collection of unbalanced trees produced by multi-resolution adaptive k-means clustering. The data organization within each cluster is similar to an R-tree, with similar insertion/deletion operations. For system utilizing random-access with a disk block as the minimum transfer size, hierarchical organization that preserves the spatial locality well will have a better similarity search performance. The CSS+-tree organizes the data by following inherent clustering that exists in the data. Within each cluster, it is desirable to have a balanced hierarchical structure in R-tree/B-tree fashion.

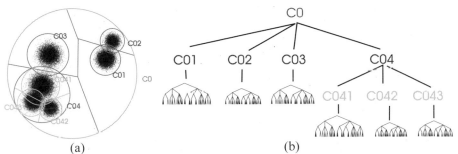

Figure 8-7 Clustering the data (a) and the hierarchical structure (b) in a CSS+-tree

For example, for the dataset in Figure 8-7(a), the first level of clustering detected is shown in Figure 8-7(b). Each cluster defines a Voronoi region delineated by the lines in the figure. These regions and the bounding circles together define the cluster areas. The next level of clustering can be seen in gray lines in Figure 8-7(a) and the resulting tree structure can be seen in Figure 8-7(b).

BULK LOADING OF THE CSS+-TREE

The building of the CSS+-tree consists of three main stages, sampling, clustering and tree building.

Sampling. In the sampling stage, a proportion of the population is randomly selected. Applying the clustering stage directly to the database will not be a viable option if the size is large. A multilevel (multi-resolution) sampling is used, as shown in Figure 8-8. This sampling method will also provide a way to find the clustering structure of the data in hierarchical stages.

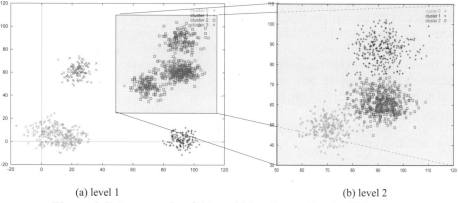

(a) level 1   (b) level 2

Figure 8-8 An example of the multi-level sampling in CSS+-tree

Clustering. In this step, an effort is made to find the natural grouping of the data. Within this stage, no effort is made to produce a balanced structure, since a balanced partition is unlikely to be the best partition of the data. An example of a case where a balanced split destroys much of the locality information can be seen in Figure 8-9.

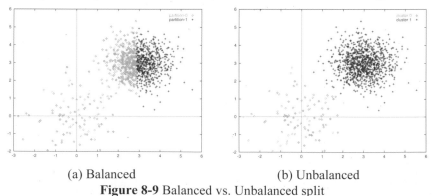

(a) Balanced   (b) Unbalanced

Figure 8-9 Balanced vs. Unbalanced split

The clustering and sampling stage is performed in an iterative manner. After each sampling and clustering process, a further sampling within each cluster is performed to find finer clusters. Then, clustering is applied to the samples within each cluster. We apply an adaptive k-means algorithm using the Duda and Hart [13] rule. It is designed to test whether there is any reason to believe that a set S containing n samples consists of more than one cluster. The null hypothesis is that all n samples belong to a normal population with mean μ and covariance matrix $\sigma^2 I$. For large n, μ can be estimated by $\mu = \frac{1}{n}\sum_{x \in S} x$ and σ^2 can be estimated by

$\sigma^2 = \frac{1}{nd}\sum_{x \in S}\|x - m\|^2$. If the null hypothesis is true, then the sum-square error for

the cluster $J_e(1) = \sum_{x \in S}\|x - m\|^2$ can be estimated by $nd\sigma^2$ and its variance is $2nd\sigma^4$. If this cluster is divided by a hyper-plane through the sample mean then the sum of squared error will be $J_2(2) = \sum_{i=1}^2 \sum_{x \in S_i}\|x - m\|^2$. Duda and Hart [13] proved that $J_e(2)$ is distributed normally with mean $n(d - 2/\pi)\sigma^2$ and variance $2n(d - 8/\pi)\sigma^4$. If the population consists of two clusters, then the resulting sum-squared error should be significantly less than $J_e(2)$. This can be stated by a one sided test with confidence p, $0 \le p \le 1$.

$$J_e(2)/J_e(1) < 1 - \frac{2}{\pi d} - \alpha\sqrt{\frac{2(1 - 8/\pi^2 d)}{nd}} \qquad (8\text{-}1)$$

The iteration will stop when no new clusters are recovered or if the cluster size is smaller than a prescribed minimum. In high dimensional space, preserving locality information is more important than keeping a balanced tree.

Tree building. Within each cluster, any method to structure the data can be used. For supporting similarity searches, a structure that will maintain data locality is needed. The SS+-tree [38] with minor modifications is chosen, since this structure will allow data to grow in a scalable way while maintaining their localities.

Algorithm 1 Test to determine whether to split a cluster

Input: A set clusters centers C, some or all of them might be marked as tried
Output: A cluster that should be split according to Equation 8-1.
  while There are unmarked centers in C **do**
  $C_{max} \leftarrow$ cluster with the maximum variance that has not been marked
  $\sigma \leftarrow$ current cluster variance
  $\sigma_{after} \leftarrow$ the variance if the cluster is split
  if $\sigma_{after} / \sigma < 1 - \frac{2}{\pi d}\,\alpha\sqrt{\frac{2(1 - 8/\pi^2 d)}{nd}}$ **then**
    {reject the null hypothesis that the cluster is only one cluster}
    Mark C_{max}
  else
    return C_{max}
  end if

end while

Algorithm 2 Adaptive k-means algorithm

Input: A set S of d-dimensional points and a minimum cluster size, $|S| = n$.

Output: The set of cluster centers and cluster assignments of each point

$C \leftarrow \varnothing$; {C is the set of centroids found}

$$m \leftarrow \frac{1}{n}\Sigma_{x \in S}\, x$$

$C \leftarrow C \cup \{m\}$

while There are clusters to split in C **do** {See *Algorithm 1*}

$v \leftarrow$ centroid of the cluster to be split.

Find the direction with the maximum variance $d_{\sigma_{max}}$ in cluster and centroid v

$v_1 \leftarrow v + \sqrt{\sigma_{max}}$ in dimension $d_{\sigma_{max}}$

$v_2 \leftarrow v - \sqrt{\sigma_{max}}$ in dimension $d_{\sigma_{max}}$

$C \leftarrow C - \{v\}$

$C \leftarrow C - \{v_1, v_2\}$

Perform k-means iterations to assign each data points to the centers and to move the new cluster centers to the middle of each formed clusters

end while

return C and the assignments of each data points

DYNAMIC BEHAVIOR OF THE CSS+-TREE

In this section, the CSS+-tree's dynamic behaviors, insertion, deletion, and reorganization will be described.

Insertion within the CSS+-tree is handled differently at different tree levels. The cluster configuration within the tree is not changed with insertions. After the initial bootstrap or reorganization, the cluster configuration will remain the same until it is regarded necessary reorganization it. Insertion within a cluster is handled in a similar way to an R-tree [24], but the insertion procedure and splitting heuristics are chosen to minimize the increase in the tree's variance. Since local minimization often results in sub-optimal configuration, a way to prevent the tree structure from diverging too far from optimality is required. The method can be seen in the reorganization discussed later.

Almost every tree-based multidimensional access structure, e.g. the R-tree and its variants [24, 5, 49], SS-tree [56], X-tree [10], uses a splitting plane that is perpendicular to one of the coordinate axes. With the exception of the SS-tree, SS+-tree, CSS+-tree, and VAMSplit-R-tree, the split criteria used are mainly topological ones, such as the minimization of the perimeter, area, the increase of node's area, or overlapping areas. As the tree structure aims to support similarity searches, it is advantageous if nearby vectors to be collected reside in the same or nearby nodes. This means that a division of the data that reflects the data clustering is preferred and that it is better with the less variance among nodes.

A variance minimizing splitting rule similar to that used in SS-tree [56] is used in the CSS+-tree. Experiments have demonstrated that using the k-means splitting rule will not provide a significant advantage if the data is not clustered. Splitting the data along the dimension with maximum variance will yield results similar to those of the k-means splitting rule and the partitions will be more suitable for bounding using hyper-boxes, because the split will use hyper-planes perpendicular to the coordinate axis.

The algorithm used to insert new data points into the CSS+-tree is described in Algorithms 3. It differs from the algorithms for the R*-tree [5] in the reorganization steps that might follow a node overflow (Algorithm 4). The algorithms for the SS-tree and splitting the data in an overflowed node are the same [56].

Algorithm 3 Insert

Input: A CSS+-tree t and a point x to be inserted, m is the minimum number
of entries in a node, M is the maximum number of entries in a leaf node.
Output: A modified CSS+-tree with x inserted
  $p \leftarrow$ ChooseInsertionLeaf(t, x)
  if numChildren(p) $< M$ **then** add(p, x)
  else
    $p_s \leftarrow$ HandleOverflow(p, x)
    if $p_s =$ **then return**
    end if {Try to insert p_s in the parent}
    $s \leftarrow$ parent(p)
    while $s \neq$ NULL **do**
      if p or p_s overlaps heavily with other sibling **then**
        Reorganization(s, p_s)
      end if
      if numChildren(s) $< M$ **then**
        add(s, p_s)
        return
      end if
      $x \leftarrow p$
      $p \leftarrow s$
      $p_s \leftarrow$ HandleOverflow(p, x)
      if $p_s =$ NULL **then return**
      end if
      $s \leftarrow$ parent(p)
    end while
    Create a new root node with the old root and p as its children.
  end if

Algorithm 4 Handle Node Overflow

Input: A node p, size(p) = M, M is the maximum node size. x is the new
child to be inserted in p. If the overflow can be handled using re-insertion,
the return value will be NULL.
Output: The sibling of the node if the node is split, otherwise return NULL.
{Check for forced reinsert}
if This is the first overflow for p **then**
    ForcedReinsert(p, x)
    return NULL
else
    Divide all $M+1$ data into p & p_s by splitting the maximum variance dimension
end if
return ps

When deletion does not cause a node to underflow, it will be done immediately. Otherwise, all of its children will be force-reinserted.

Reorganization of the CSS+-tree is composed of two parts: a gradual reorganization due to forced-reinsertions and a local reorganization where node's grandchildren are rearranged.

The CSS+-tree itself is always trying to reorganize itself gradually by forcing the reinsertion of over-flowed and under-flowed nodes. This "forced-reinsert" strategy was first proposed in Beckmann et al. [5]. For an over-flowed node, instead of splitting the entries into two nodes, a proportion of the children of the over-flowed node are reinserted at the same level. For an under-flowed node, the node is deleted and all its children are reinserted. The deletion of an under-flowed node can make its parent become under-flowed. The algorithm describing the handling of this "under-flow chain" can be seen in deletion.

Apart from the local reorganization due to the forced reinsertions, a local reorganization with a larger scope can be triggered when the children of a node overlap each other badly. This reorganization will involve all grandchildren of the reorganized node (two levels down the affected node).

COMPARISON TO OTHER HIERARCHICAL STRUCTURES

The properties of the CSS+-tree and other well-known R-tree variants are compared in Table 8-1.

Table 8-1 A comparison of the dynamic R-tree variants

Name	Bounding envelopes	Splitting heuristic	Insertion	Reorganization	Balanced
CSS+-tree	Sphere and plane at cluster level, rectangle below cluster level	Maximizing cluster separation and minimizing total variance	Into subtree with closest centroid	Forced reinsert and local reorganization	No

R tree	Rectangle	Minimizing total area of bounding boxes (linear, quadratic, and exponential split heuristic)	Minimizing bounding box area enlargement	None	Yes
R+-tree	Rectangle	Minimizing total area, perimeter, number of split propagated, distance between rectangle centers	The new feature vector will be inserted to each node whose bounding box overlaps it	None	Yes
R*-tree	Rectangle	Minimizing total perimeter of bounding rectangles	Minimizing bounding rectangles overlap enlargement	Forced reinsert	Yes
SS-tree	Sphere	Minimizing variance (split plane perpendicular to one of the coordinate axis)	Minimizing distance of the new feature vector to the node centroid	Forced reinsert	Yes
SR-tree	Sphere and rectangle	Minimizing variance (split plane perpendicular to one of the coordinate axis)	Minimizing distance of the new feature vector to the node centroid	Forced reinsert	Yes
SS+-tree	Sphere or rectangle	Minimizing variance (k-means heuristic)	Minimizing distance of the new feature vector to the node centroid	Forced reinsert, local reorganization based on k-means clustering	Yes
TV-tree	Sphere or rectangle	Minimizing distance between the new feature vector within the trees nodes	Minimizing increase in overlap, decrease in dimensionality, increase in radius, distance of the center of the node to the new vector	Forced reinsert	Yes
X-tree	Rectangle	Minimizing overlap, constructing super-nodes if an overlap	Minimizing bounding boxes overlap enlargement	Forced reinsert	Yes

		minimi-zing split cannot be found			

8.4 Handling Varying Distance Metrics

Similarity search using tree-based methods assumes that the distance metric used for building the structures, and queries is the same. However, many applications require varying distance metrics. For example, due to the subjective nature of similarity measurements, the distance metrics can vary between different users. Even with the same user, there may be a change of focus during retrieval. For example, in retrieval with feedback [31], there is a need to adjust weights on different features. The afore-mentioned tree structures cannot be used easily if the distance metric in query is inconsistent with the distance metric in the index tree. If we map the index tree into the new metric space, it is equivalent to rebuilding the tree, and therefore, computationally expensive. If we map the k-nn space into the index metric, it may have an arbitrary shape. For example, a k-nn hyper-sphere in the Euclidean distance with different weighting will produce a hyper-ellipse in the new metric space, as shown in a two-dimensional case in Figure 8-10. Calculating the intersection of the new k-nn bounding ellipse and d-dimensional bounding envelopes (rectangles in Figure 8-10) of index tree nodes involves solving a $(d-1)$-order polynomial. It is not trivial as there is no analytic solution beyond third-order polynomials. A numerical solution is normally employed.

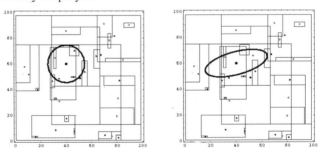

Figure 8-10 Varying distance metrics affects k-nn search

In this section, we present an efficient similarity search algorithm using weighted Euclidean distance with an existing spatial access tree. We start by describing a commonly used approach [15]. We identify the drawback of the approach and present a new approach with a two-dimensional implementation. An extension of the algorithm into high dimensional spaces is also given. The experiments were carried out on an R-tree variant, the CSS+-tree. However, the proposed approach is not restricted to this particular tree. They can be used in any hierarchical structures with a bounding envelope for each node.

METHODS FOR K-NN SEARCH IN WEIGHTED EUCLIDEAN DISTANCE
When the distance metric is inconsistent with the metric in the tree structure, a commonly used technique to find the k-nearest-neighbors [16, 15, 36] is noted

below. To assist comprehension, we illustrate it with an example, where the index tree is built using the unweighted Euclidean distance and the query is issued using the weighted Euclidean distance, and illustrate the process in Figure 8-11.

o First, we define a distance metric which is the lower bound to the weighted distance. It can be a simple transformation of both metrics used in the query and in building the index tree. The lower bound ensures that false dismissals do not occur. A proof can be found in Faloutsos [15]. Usually, we use the same metric as the one used to build the index tree.

o We then search for *k* nearest neighbors to obtain a bound (*LBD* in Figure 8-11).

o Transferring the distance metric to the query distance metric, a new bounding shape for the *k*-nn disk is obtained (*ED* in Figure 8-11) and the maximum distance *D* can be calculated.

o Next, we issue a range query to find all vectors within the distance D from the query point (all vectors in *RQD*, as shown in Figure 8-11). The number of vectors returned from the range query will be greater or equal to *k*.

o We sort the set from the last step and find the true *k* nearest neighbors.

From Figure 8-11, we can see that the circular range query area (*RQD*) is much larger than the actual *k-*nearest-neighbor disk. This is not desirable, since a larger query area will intersect many tree nodes, i.e., we have to examine more tree nodes than necessary. Another drawback of the approach is that we have to search the tree twice. The first scan finds the *k*-nearest-neighbors in the original metric and the second scan finds all the vectors located within the range *D* (a range query). Sorting is also needed to select the first *k* nearest neighbors in the varied metric within range *D*.

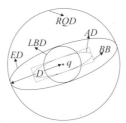

Figure 8-11 The evolution of search areas in Faloutsos' and our approach

We propose an efficient *k*-nn search in variable metrics. The approach is based on the observation that a transform between the query metric and the index metric exists, as shown in Figure 8-12, and transforming the *k*-nn disk is much simpler than transforming all tree nodes, the latter is equivalent to rebuilding the tree.

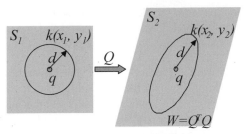

Figure 8-12 The transformation between a normal and a weighted Euclidean space

Using this transformation, we can find the length of the major axis of the k-nearest-neighbor disk. Assume W is the weight matrix, and I is the identity matrix, q is the query point, k is the kth nearest neighbor found, and $q, k \in S_2$ (see Figure 8-12), the bounding rectangle of the nearest-neighbor disk can be calculated as follows.

- Let $\vec{e}_1$ and $\vec{e}_2$ be the eigen-vectors of W and let λ_1 and λ_2 be the corresponding eigen-values sorted according to their magnitudes.

- Let Q be the result of Cholesky decomposition on W, i.e., $W = Q^T Q$.

- Then the bounding rectangle R is defined by the center q and directional vectors $\vec{v}_1$ and $\vec{v}_2$ where $r = d(q, k, W)$, and $\vec{v}_1 = \vec{e}_1 r / \sqrt{\lambda_1}$ and $\vec{v}_2 = \vec{e}_2 r / \sqrt{\lambda_2}$. The corners of the rectangles are:

$c_1 = q + \vec{v}_1 + \vec{v}_2$

$c_2 = q + \vec{v}_1 - \vec{v}_2$

$c_3 = q - \vec{v}_1 - \vec{v}_2$

$c_4 = q - \vec{v}_1 + \vec{v}_2$

EXTENSION TO HIGH DIMENSIONAL SPACES

In the high dimensional space, care must be taken in calculating the intersection between the bounding hyper-box and tree's bounding envelopes. A brute-force calculation is exponential to the dimension. We provide an algorithm of complexity $O(d^3)$, where d is the dimension.

Suppose the query's bounding hyper-box B is defined by its center point q and d orthogonal vectors $\vec{v}_i$, ie, $B(q, \vec{v}_1, ..., \vec{v}_d)$. The bounding hyper-box of a tree node is A and its closest corner to the origin is c_{min} and the farthest corner to the origin is c_{max}. We use $r(j)$ and $\vec{r}(j)$ to represent the projections of point r and vector $\vec{r}$ in dimension j respectively in the following description.

- For all dimensions $j = 1..d$, find the minimum coordinate of the query's bounding hyper-box (can be done in $O(d)$ by subtracting $\vec{v}_i(j)$ from $q(j)$). If any minimum is not inside the half-space defined by the equation $x \leq c_{max}(j)$, then there is no intersection. Otherwise, continue checking. This process takes $O(d^2)$ comparisons.

o For all dimensions $j = 1..d$, find the maximum coordinate of the query's bounding hyper-box (can be done in $O(d)$ by adding $\bar{v}_i(j)$ to $q(j)$). If any minimum is not inside the half-space defined by the equation $x \geq c_{min}(j)$, then there is no intersection. Otherwise, continue checking. This process also takes $O(d^2)$ comparisons.

o If above checking fails, it does not mean that there is a non-empty intersection area. It means that B is not intersected with the hyper-box $c_{min} \leq x \leq c_{max}$. The check should also be made from the B rectangle in the query space using all the half-spaces defined by $x \leq q + \bar{v}_i$ and $x \geq q - \bar{v}_i$. Transformation to the query space takes $O(d^3)$ multiplications and comparisons are of order $O(d^2)$.

The details of implementation can be found in Jin and Kurniawati [30]. The total complexity is $O(d^2) + O(d^2) + O(d^3) = O(d^3)$. The cost to calculate the eigen-matrix and Cholesky decomposition of a dxd matrix is $O(d^3)$ (A fast implementation of order $O(d^{2.367})$ has been reported [12]). These matrices are recalculated whenever the weight matrix changes, which can only happen between queries or after updating the tree.

EMPIRICAL RESULTS AND DISCUSSION

The algorithm has been compared with Faloutsos' method using the unweighted Euclidean distance as the lower bound of the weighted distance. The result is shown in Table 8-2. All experiments are texture feature vectors of dimension 30 extracted from 14,016 images in PhotoDisc library using Gabor filters [41]. The experiments were performed using five non-diagonal but symmetrical weight matrices [31] whose ratios of the largest and smallest eigen-values ranged from 2 to 32. The fan out of the tree's leaf nodes was 66 and the fan out of internal nodes was 22.

Table 8-2 A comparison over the texture vectors of PhotoDisc images

	Proposed method			Faloutsos' method		
ratio	ltouch	lused	lnode	ltouch	lused	inode
2	25.32	7.06	4.22	36.62	13.78	8.16
4	23.38	6.43	4.17	32.54	12.67	8.09
8	26.04	6.45	4.45	35.63	12.66	8.51
16	24.53	6.76	4.08	33.92	13.43	7.97
32	22.89	6.49	4.15	30.08	12.59	8.01

We measured the number of leaf nodes that had to be examined (*ltouched*), the number of internal nodes accessed (*inode*), and the number of leaf nodes that actually contained any of the nearest-neighbor set (*lused*). Note that the running time is not used in comparison as it depends on the machine architecture, memory structures, buffer size, etc. It is widely accepted that touched node number is an objective measurement for comparing performance [47, 55]. The ratio column is the ratio between square root of the largest and the smallest eigen-values. All experiments are 21-nearest-neighbor searches and the tabulated results are the average of 100 trials using a random vector chosen from the dataset.

The algorithm uses the bounding hyper-box of the k-nn hyper-sphere instead of calculating intersections directly. It has $O(d^3)$ complexity. Comparing with Faloutsos' algorithm, the bounding rectangle is more compact than Faloutsos' query disk (see *RQD* and *BB* in Figure 8-11). The saving is significant if the dimension is high because the peripheral volume increases exponentially with the dimension.

8.5 Dimension Reduction

Many of currently available multi-dimensional indexing techniques fail to handle high-dimensional data efficiently [40, 10, 55]. An obvious solution to the problem of exponential growth of multi-dimensional indexing structures with increasing dimension of data is to reduce the dimension.

A popular technique of dimension reduction is principal component analysis (PCA) [33] which is widely used because of its simplicity and the availability of efficient computational algorithms. It is a mapping technique that involves transforming data such that variables are de-correlated and a few variables account for most of the variances. Dimension reduction is achieved by dropping variables with insignificant variances. An improvement on linear PCA is provided by non-linear generalizations of PCA which extend the ability of PCA to incorporate non-linear relationships in the data [37].

Neural networks, in particular, associative neural networks, form another class of solutions to the problem of dimensionality reduction. There are two kinds of associative neural networks, auto-associative and hetero-associative networks. The former takes the training patterns, memorizes them and tries to reproduce them as the output. The latter takes input patterns distinct from the training patterns and the output gives the best classification of the input in terms of the training categories. Note that both are supervised networks.

Some of the auto-associative networks, the so-called "PCA-type" networks, produce results very similar but not superior to PCA [3] in terms of reconstruction error. Both linear and non-linear PCA transforms can be implemented by multi-layer feed-forward neural networks [37, 34]. We will use "multi-layer network" throughout the rest of the paper because our discussion is restricted to networks which have only feed-forward connections from input to output. However, readers should keep in mind that there are other more general multi-layer networks, namely recurrent networks, with connections allowed both ways between a pair of units, and even from a unit to itself. The auto-associative network performs the identity mapping through a compression layer and the goal is to keep as much information in this layer as possible in order to reconstruct the input data at the output layer. The use of one hidden layer in these networks allows dimensionality reduction akin to PCA while increasing the number of hidden layers to three allows non-linear PCA mapping. Auto-associative networks can a achieve high dimension reduction rate. However, the similarity preservation is poor because the network is only sensitive to

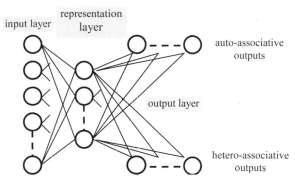

Figure 8-13 Hybrid-associative networks

the training patterns. When there is more than one hidden layer, the network is difficult and slow to train.

Unlike auto-associative networks, hetero-associative neural networks develop arbitrary internal representations in the hidden layers in order to associate inputs to class identifiers usually in the context of pattern classification. The hetero-associative network preserves similarity well. However, it has a complicated structure. The training can be easily trapped into local minima, especially when there are overlaps among classes and training starts from patterns in the overlapping area. It is difficult to avoid this by selecting the training patterns because of the high dimensionality. Besides, the dimension reduction rate is not as high as the auto-associative networks.

The adaptive property of auto-associative networks and the classifying property of hetero-associative networks remind us that a more robust scheme can be achieved by combining auto-associative and hetero-associative functions in one network, a hybrid network. In this way, more complete information is embedded in the compression layer, which is adjusted to jointly satisfy both the requirements of dimensionality reduction and the similarity in the application domain. This section demonstrates the usefulness of the hetero and the hybrid networks in image indexing and retrieval. We included results of one-hidden-layer auto-associators and the standard PCA to demonstrate how well the hetero and the hybrid networks fare against them. The use of better auto-associators (eg, three-hidden layer networks) has been left to future investigation.

REDUCING DIMENSION USING HYBRID-ASSOCIATIVE NEURAL NETWORKS

We propose a hybrid hetero-associative/auto-associative (hybrid-associative) neural network, which combines a hetero-associative network and an auto-associative network into one, as shown in Figure 8-13. Both sets of output nodes are fully connected to the same bottleneck hidden layer. The network is trained to simultaneously implement an identity mapping from the input layer to the auto-associative subset of the output layer, and a hetero-associative mapping to the rest of the output layer. After training, the hybrid network should have learned both the hetero-associative and the auto-associative functions. Consequently, the advantages to similarity retrieval due to hetero-associative reorganization are attained and the representation retains much of the input information since it also allows de-mapping to a set of auto-associative outputs. The hybrid network may thus provide a more robust approach to dimensionality reduction than a hetero-associative network.

Our hybrid-associative neural networks are trained using the back-propagation algorithm [25, 45], a gradient descent learning rule. Starting with an initial set of small random weights, the algorithm searches through the weight space moving in the direction of the steepest descent, after presentation of each training pattern, to minimize the cost function

$$E = \frac{1}{2} \sum_{k=1}^{n} (y_k - d_k)^2 \qquad (8\text{-}2)$$

where y_k is the response of output node k, d_k is the desired output response, and n is the total number of output nodes. The weights are adjusted according to the update rule

$$w_{ij}(t+1) = w_{ij}(t) + \eta \delta_j x_i \qquad (8\text{-}3)$$

where w_{ij} is the weight from node i in layer L-1 to node j in layer L, h is the learning parameter, d_j is the error at node j and x_i is response of node i. If L is the output layer,

$$\delta_j = (d_j - x_j)\, x_j\, (1 - x_j) \qquad (8\text{-}4)$$

or if L is a hidden layer,

$$\delta_j = x_j\, (1 - x_j) \sum_k \delta_k\, w_{jk}$$

$$(8\text{-}5)$$

where d_k is the error at node k in layer L+1. The response of a node j is given by

$$x_j = \frac{1}{1 + e^{-\sum_m w_{mj} x_m}} \qquad (8\text{-}6)$$

where m is a node in the layer preceding the node j layer. The training cycle continues until the cost function reaches a specified level.

DIMENSION REDUCTION RESULTS

We apply four methods, PCA, auto-associators, hetero-associators and hybrid-associators, to dimension reduction and compare their performances. We use the Karhunen-Loève transformation in the PCA method to map initial features x to uncorrelated components y. The first component exhibits the largest variance and the last component, the least variance. This allows insignificant components to be dropped off with minimal resultant residual error and data reduction is achieved. Thus, it is possible to map the original d-dimensional data onto an l-dimensional subspace spanned by l most significant covariance eigenvectors where $l < d$.

The auto-associator used in the experiment is one hidden layer associative multi-layer perceptron (MLP); the hetero-associator used is one hidden layer hetero-associative MLP; and the hybrid-associator used is one hidden layer hybrid hetero-associative/auto-associative MLP.

There are four databases used in the experiments. Databases A, B or C consists of 530 texture images (128 x 128 pixels), 10 homogeneous members from each of 53 distinct classes. Database D consists of 1110 images (128 x 128 pixels), 10 homogeneous members from each of 111 distinct classes. Database A is the training set for the neural network methods and the base set for PCA computations. Database B is basically the same as A except that the images are shifted 10 pixels to right and 10 pixels down. Database C has the images of A rotated 90 degrees clockwise. The feature data for each of the images are normalized to the range [0, 1] separately per dimension. Because similarity is subjective, we choose these test data to eliminate the side effect of the similarity ranking problem. Database D comes from Volume 3 (111 images) of PhotoDisc image database. Each image in Volume 3 is divided into 10 subimages. It serves as the major testing set.

Each type of neural network was trained with back-propagation using all the images in database A until the root-mean-square error is 0.01 or the number of epochs reaches 5000. Both thresholds are sufficient enough to distinguish between converging and non-converging situations, as we are interested in the comparison of the reduction rate. For the hetero-associative network, 1-of-53 encoding is used, that is, there are 53 outputs, each corresponding to a class. If the class corresponding to it is detected all of the outputs with the exception of one are low. To illustrate, if an input vector for an image belonging to class 1 is presented, output 1 is trained to go high (value 1) while all the other are trained to output low (value 0). On the other hand, if an input for an image belonging to class 2 is presented, output 2 is trained to go high, and so on. Training proceeds until all output nodes respond correctly (i.e. *rms* error reaches the desired level) for each input in the training set or until the maximum allowed number of epochs has elapsed.

The hybrid hetero-associative/auto-associative network has a total of 83 outputs of which 30 are auto-associative and 53 are hetero-associative. It simply combines the required outputs of the auto-associative network and the hetero-associative network into one larger network. It is trained so that both parts are jointly satisfied. Once trained, the hybrid network can function either as an auto-associator or hetero-associator, or both.

Once each network is trained, the feature data from each of the databases are used to activate the network and the transformed or reduced data is obtained from the hidden layer. The reduced data is used to form the texture index.

Similarity retrieval proceeds by the algorithm described in Section 8.3 over the CSS+-tree. Euclidean distances between the reduced texture data of the query image and the *k-n*earest-neighbor images are calculated and sorted in ascending order. Retrieval results are listed in an increasing order of distance which serves as a similarity measure. The retrieval performance is measured in terms of recall rate or percentage of correct retrieval (ie. belonging to the appropriate class) when each image in the query set is presented, and recall precision (recall rate vs the number of top matches considered). Ideally, if each member of a class is presented as a query, all the members of the class should be retrieved before members of the other classes are.

Table 8-3 Retrieval performance in terms of recall rate (% correct retrieval)

		No. of Hidden Units					
		5	10	15	20	25	30
GROUP I	Raw	65.64					
	Hetero-asso.	59.45	75.77	81.30	78.85	78.85	77.75
	Auto-asso.	60.30	62.89	58.30	62.42	60.23	63.23
	Hybrid-asso.	55.77	73.62	73.42	75.06	72.75	75.68
	PCA	63.25	65.43	65.83	65.62	65.60	65.64
Group II	Raw	62.91					
	Hetero-asso.	53.53	66.28	68.36	72.06	72.38	72.43
	Auto-asso.	55.72	60.45	56.00	56.66	56.38	60.68
	Hybrid-asso.	50.32	62.96	67.49	68.79	67.53	69.17
	PCA	62.42	64.26	64.58	64.68	64.72	64.70
Group III	Raw	48.98					
	Hetero-asso.	41.94	60.09	57.06	57.32	60.36	57.68
	Auto-asso.	43.75	46.94	41.25	41.15	40.87	43.58
	Hybrid-asso.	47.75	57.89	51.06	55.04	50.17	54.28
	PCA	51.38	52.85	52.02	52.25	51.57	51.53
Group IV	Raw	50.72					
	Hetero-asso.	42.25	47.36	50.12	54.19	56.88	59.42
	Auto-asso.	40.17	42.54	41.75	43.34	43.68	43.79
	Hybrid-asso.	44.70	48.11	54.47	55.16	54.69	55.79
	PCA	45.32	47.37	47.54	48.02	48.34	49.25

Table 8-3 shows a comparison of the retrieval scores for each of the dimensionality reduction methods used. In summary,

o   Due to the linearity, PCA and auto-associative networks retains much of the original information. They can achieve a very high reduction rate. For example, down to dimension 5, they perform better than hetero-associators and hybrid-associators.

o   PCA performance stays close to those obtained with the raw data but maintained it quite consistently down to 5 dimensions.

o   The auto-associative network registers the poorest results among the network-based methods but stays within 2% and 8% off the performance using the raw data at lower dimensions.

o   The hetero-associative network performs reasonably well. It is able to maintain a retrieval performance better than that with the normalized raw data using the training set as query set even down to 10 dimensions – a reduction of 33 percent. This performance, however, is not sustained when the query images do not belong to the training set (e.g., dataset D).

o   The hybrid network performs better than PCA overall but does not really match the performance of the purely hetero-associative network in the training set. However, the hybrid network produces higher inter-class distance than the hetero network (Table 8-4). It is a very important property in developing an index tree of image features because range search and nearest-neighbor search are based on distance.

Table 8-4 Average inter-class distances normalized by dimension

	Raw	Auto-asso.	Hetero-asso.	Hybrid-asso.
Average Normalized Inter-class distance	0.022	0.070	0.019	0.059

8.6 Conclusion

This chapter touched on three important issues in multidimensional indexing and retrieval. We reviewed multidimensional index and retrieval research and presented our CSS+-tree. It outperformed the P-tree (the late development of the X-tree), the VAMS-tree (an improved descendant of the R-tree, SS-tree and TV-tree), and the VA-file: the P-tree and VA-file are "state-of-the-art" methods that are recently proposed and reported to outperform other methods by a wide margin [9, 54]. Another landmark achievement is dimension reduction using a hybrid-associate neural network. Our algorithm can reduce the dimension by half without degrading or even improving the performance of the signature. We also raise the issue of varying similarity metrics in retrieval and introduce our solution to the problem.

References

[1]   Arya, S. (1995). *Nearest Neighbor Searching and Applications*. PhD thesis, Computer Vision Laboratory, University of Maryland, 20742-3275.

[2]   Bai, X., Xu, G., Jin, J. S., & Kurniawati, R. (1997). Content based retrieval and related techniques. *Robot* 19:231-240.

[3]   Baldi P & Hornik K (1989). Neural networks and principal component analysis: Learning from examples without local minima. *Neural Networks*; 2: 53-58.

[4]   Bayer, R. & McCreight, E. (1972). Organization and maintenance of large ordered indexes. *Acta Informatica*, 1(3):173-189.

[5]   Beckmann, N., Kriegel, H.-P., Schneider, R., & Seeger, B. (1990). The R-tree: an efficient and robust access method for points and rectangles. *Proceedings of the ACM SIGMOD Int. Conf. on Management of Data*, pp.322-331.

[6]   Bentley, J. L. (1975). Multidimensional binary search trees used for associative searching. *Communications of the ACM*, 18(9):509-517.

[7]   Bentley, J. L. & Stanat, D. F. (1975). Analysis of range searches in quad trees. *Information Processing Letters*, 3(6):170-173.

[8]   Berchtold, S. & Keim, D. A. (1998). Indexing high-dimensional space: database support for next decade's applications. In *ACM-SIGMOD'98 Tutorial*, Seattle, Washington. ACM.

[9]   Berchtold, S., Böhm, C., & Kriegel, H.-P. (1998). The pyramid-tree: Breaking the curse of dimensionality. *Proceedings of the ACM SIGMOD International Conference on Management of Data*, pp.78-86, Seattle, Washington.

[10] Berchtold, S., Keim, D. A., & Kriegel, H.-P. (1996). The X-tree: An index structure for high-dimensional data. *Proc. 22th Int. Conf. on Very Large Data Bases*, pages 28-39, Bombay, India.

[11] Brin, S. (1995). Near neighbor search in large metric spaces. *In VLDB 1995*.

[12] Cormen, T., Leiserson, C., & Rivest, R. (1990). *Introduction to Algorithms*. The MIT Press, MIT, Cambridge.

[13] Duda, R. O. & Hart, P. E. (1973). *Pattern Classification and Scene Analysis*. John Wiley & Sons, New York.

[14] Ellis, T., Roussopoulos, N., & Faloutsos, C. (1987). The R+tree: A dynamic index for multi-dimensional objects. *Proc. the 13th Conf. on VLDB*, pp.507-518.

[15] Faloutsos, C. (1996). Searching Multimedia Databases by Content. Advances in Database Systems. Kluwer Academic Publishers, Boston.

[16] Faloutsos, C., Equitz, W., Flickner, M., Niblack, W., Petkovic, D., & Barber, R. (1994). Efficient and effective querying by image content. *J. of Intelligent Information Systems*, 3:231-262.

[17] Finkel, R. A. & Bentley, J. L. (1974). Quad trees: A data structure for retrieval on composite keys. *Acta Informatica*, Springer Verlag (Heidelberg, FRG and NewYork NY, USA) Verlag, 4.

[18] Flickner, M., Sawhney, H., Niblack, W., Ashley, J., Huang, Q., Dom, B., Gorkani, M., Hafner, J., Lee, D., Petkovic, D., Steele, D., & Yanker, P. (1995). Query by image and video content: The QBIC system. *IEEE Computer*, pp.23-32.

[19] Freeston, M. (1987). The bang file: a new kind of grid file. *Proceedings of the ACM SIGMOD Int. Conf. on Management of Data*, pages 260-269.

[20] Friedman, J. H., Baskett, F., & Shustek, L. H. (1975). An algorithm for finding nearest neighbors. *IEEE Trans. on Computers*, C-24:1000-1006.

[21] Friedman, J. H., Bentley, J. L., & Finkel, R. (1977). An algorithm for finding best matches in logarithmic expected time. *ACM Trans. on Math. Software*, 3(3):209-226.

[22] Fukunaga, K. & Narendra, P. M. (1975). A branch and bound algorithm for computing k-nearest neighbors. *IEEE Trans. on Computers*, C-24(7):750-753.

[23] Gaede, V. & Günther, O. (1995). Multidimensional access methods. *Technical Report ISS-16*, Institute of Information Systems, Humboldt-Universit"at zu Berlin, Spandauer Str. 1 D 10178 Berlin.

[24] Guttman, A. (1984). R-trees: A dynamic index structure for spatial searching. *Proc. the ACM SIGMOD Int. Conf. on Management of Data*, pp.47-57, Boston.

[25] Hertz J, Krogh A & Palmer R. *Introduction to the theory of Neural Computation*. Addison-Wesley, Redwood City 1991.

[26] Hjaltason, G. R. & Samet, H. (1995). Ranking in spatial databases. In Egen hofer, M. J. & Herring, J. R., editors, *Advances in Spatial Databases*, 4th International Symposium, SSD'95, volume 951 of Lecture Notes in Computer Science, pp.83-95, Berlin. Springer-Verlag.

[27] Ievergelt, J., Hinterberger, H., and Sevcik, K. C. (1984). The grid file: an adaptable symmetric multikey file structure. *ACM Transactions on Database Systems*, 9(1).

[28] Jagadish, H. (1990). Linear clustering of objects with multiple attributes. *Proc. the ACM SIGMOD Int. Conf. on Management of Data*, pp.332-342, Atlantic City.

[29] Jin, JS; Greenfield, H & Kurniawati, R (1997). CBIR-VU: a new scheme for processing visual data in multimedia systems. *Lecture Notes in Computer Science: Visual Information Systems*, pp.40-65.

[30] Jin, J S & Kurniawati, R (2001). Varying similarity metrics in visual information retrieval. *Pattern Recognition Letters*, 22(5):583-592.

[31] Jin, J. S., Kurniawati, R., and Xu, G. (1996). A scheme for intelligent image retrieval in multimedia databases. *Journal of Visual Communication and Image Representation*, 7(4):369-377.

[32] Jin, J., Tiu, L. S., and Tam, S. W. S. (1995). Partial image retrieval in multimedia databases. *Proceedings of Image and Vision Computing New Zealand*, pages 179-184, Christchurch. Industrial Research Ltd.

[33] Joliffe I. *Principal Component Analysis*. Springer-Verlag, New York 1986.

[34] Kambhatla N & Leen T. Fast Non-linear dimension reduction. In: Cowan J, Tesauro G, Alspector J (eds). *Advances In Neural Information Processing Systems 6*. Morgan Kaufman, San Mateo, CA 1994.

[35] Knuth, D. (1998). *The Art of Computer Programming*, Vol. 3: Sorting and Searching. Addison-Wesley, Reading, Mass, 2nd edition.

[36] Korn, F., Sidiropoulos, N., Faloutsos, C., and Siegel, E. (1996). Fast nearest neighbor search in medical image databases. *International Conference on Very Large Data Bases*, Bombay, India.

[37] Kramer M. Nonlinear principal component analysis using autoassociative neural networks. *AIChe Journal* 1991; 37: 233-243.

[38] Kurniawati, R., Jin, J. S., and Shepherd, J. A. (1997). The SS+-tree: An improved index structure for similarity searches in a high dimensional feature space. *Proceedings of the SPIE: Storage and Retrieval for Image and Video Databases V*, volume 3022, pages 110-120, San Jose, CA.

[39] Kurniawati, R., Jin, J., and Shepherd, J. A. (1997). Techniques for supporting efficient content-based retrieval in multimedia databases. *The Australian Computer Journal*, 29(4):122-130.

[40] Lin, K.-I., Jagadish, H. V., and Faloutsos, C. (1994). The TV tree: An index structure for high-dimensional data. *VLDB Journal*, 3(4):517 549.

[41] Ma, N. L. & Jin, J. S. (1998). A generalized content-based image retrieval system. *Proc. of ACM Symposium on Applied Computing*, pp.460-461, Atlanta.

[42] Nene, S. A. & Nayar, S. K. (1997). A simple algorithm for nearest neighbor search in high dimensions. *IEEE Transactions on Pattern Analysis and Machine Intelligence* 19.

[43] Nievergelt, J; Hinterberger, H; & Sevcik, K C (1984). The grid file: an adaptable symmetric multikey file structure. *ACM Transactions on Database Systems*, 9(1).

[44] Orenstein, J. & Merrett, T. (1984). A class of data structures for associative searching. *Proc. of 3rd SIGACT-SIGMOD symposium on principles of database systems*, pages 181-190, Waterloo, Ontario, Canada.

[45] Pandya A & Macy R. Pattern Recognition With Neural Network In C++. CRC Press 1996.

[46] Robinson, J. T. (1981). The K-D-B-tree: A search structure for large multidimensional dynamic indices. *Proceedings of the ACM SIG MOD International Conference on Management of Data*, pp.10-18, Ann Arbor, MI.

[47] Roussopoulos, N., Kelley, S., and Vincent, F. (1995). Nearest neighbor queries. pages 71-79, San Jose, California.

[48] Samet, H. (1990). The Design and Analysis of Spatial Data Structures. Addison Wesley.

[49] Sellis, T; Roussopoulos, N & Faloutsos, C (1987). The R+-tree: a dynamic index for multi-dimensional objects. *Proceedings of the Thirteenth Conference on Very Large Databases*, Los Altos, CA, pp.507–518.

[50] Shepherd, J. A., Megiddo, N., and Zhu, X. (1998). Making QBIC faster. *Technical Report*, IBM Almaden Research Center.

[51] Smoliar, S. W. & Zhang, H. J. (1994). Content-based video indexing and retrieval. *IEEE Multimedia*, 1(2):62-72.

[52] Sproull, R. F. (1991). Refinements to nearest-neighbor searching in k dimensional trees. *Algorithmica*, 6:579-589.

[53] Watson, A. B. (1983). Detection and recognition of simple spatial forms. In Braddick, O. J. and Sleigh, A. A., editors, *Physical and biological processing of images*, pages 101-114. Springer-Verlag, New York.

[54] Weber, R; Schek, H J & Blott, S (1998). A quantitative analysis and performance study for similarity-search methods in high-dimensional spaces. *The Proceedings of the 24th Int. Conf. on VLDB*, pp.194-205.

[55] White, D. A. & Jain, R. (1996). Algorithms and strategies for similarity retrieval. *Technical Report VCL-96-01*, Visual Computing Laboratory, University of California, San Diego.

[56] White, D. A. & Jain, R. (1996). Similarity indexing with the SS-tree. *Proc. 12th IEEE International Conference on Data Engineering*, New Orleans, Louisiana.

9 A SEMANTIC DATA MODELING MECHANISM FOR A MULTIMEDIA DATABASE

Dr. Qing Li and Prof. Yueting Zhuang

Semantic data modeling of multimedia data is a fundamental problem in multimedia databases. The variety, context-dependency, and ambiguity of the semantics of multimedia data make it difficult to apply conventional data modeling techniques, which usually deal with explicit, self-contained alphanumeric data. Many semantic data modeling techniques are either designed for multimedia or can be used for modeling multimedia data. Specifically, we describe *MediaView* as a novel object-oriented view mechanism that can successfully capture the elusive nature of multimedia data.

In this chapter, we firstly describe the problem of semantic multimedia modeling and the dilemma of applying traditional data modeling techniques to multimedia data. We then present a brief review of related techniques for semantic and multimedia data modeling. The *MediaView* mechanism is described in detail and some of its applications are discussed.

9.1 Introduction

The explosive growth of multimedia objects (e.g., images, video clips, and audio segments) has created the need to manipulate them in an organized, efficient, and scalable manner, preferably, using multimedia databases. However, the problem of semantic data modeling, which is essential for a multimedia database, has not been fully understood and successfully solved. The difficulty comes from the following properties of multimedia objects in regard of their semantics:

- **Variety**. Many multimedia objects can be semantically interpreted in a variety of ways. That is, an object[1] may have multiple latent meanings.
- **Context-dependency**. The "appropriate" meaning for a multimedia object depends on the particular application or user who manipulates the object, the perspective from which it is interpreted, and the other objects that interact with it in the same environment. All these factors comprise the context indispensable for the interpretation of an object's semantics.

---

[1] If not indicated explicitly, "object" refers to "multimedia object" in this chapter.

- **Ambiguity**. When considered in isolation rather than in a specific context, a multimedia object does not have a "default" meaning — an explicit and dominant one among all of its latent meanings.

Santini et al. [24] have made the following observation about the semantics of images: "meaning is not an *intrinsic* property of the images that the database filters during the query, but an *emergent* property of the interaction between the user and the database". As a proof of this argument, consider the interpretation of van Gogh's famous painting "Sunflower", the far left image in both Figure 9-1 and Figure 9-2. When it is placed with the images in Figure 9-1, which are another four paintings of van Gogh, the meaning of "van Gogh's paintings" is implied. When the same image is interpreted in the context of Figure 9-2, however, the meaning of "flower" is more evident. In fact, this argument on the image semantics can be generalized to other types of media, i.e., a multimedia object must be interpreted in a *context* specific to the application and the database.

Figure 9-1 "Sunflower" in the context of "van Gogh's paintings"

Figure 9-2 "Sunflower" in the context of "flower"

The nature of multimedia data is fundamentally different from that of traditional data (e.g., numbers and text) whose semantics are explicit, objective, and self-contained. This large distinction accounts for the failure of applying conventional data modeling techniques on multimedia. For example, consider an object-oriented data model, which is more powerful than other traditional models in terms of semantic modeling capability. Conventionally, database designers rely on the conceptual schema—the middle level of the traditional ANSI/SPARC three-schema database architecture—to capture the semantics of multimedia data. A common way to do this is to define many classes, each of which represents an explicit semantic concept, and classify multimedia objects into appropriate classes. However, this approach conflicts with the nature of multimedia in several aspects: (1) No matter how many classes are defined, they are unlikely to cover all the eligible semantic concepts, which are unlimited. (2) It rules out the possibility that an object has various semantic interpretations (which imply that this object is eligible for more than one class), given that multiple class membership is not permitted, as is the case of a strongly typed object model. (3) As a consequence of (2), the data cannot be shared among applications that prefer different interpretations of the same data.

Alternatively, some data models choose to model only the primitive features of a multimedia object, such as the data size, format, and source, rather than its semantics. These models overcome the limitations mentioned in the last paragraph. Nevertheless, this approach is a sacrifice of modeling capability for generality and flexibility: Although in theory such data models are compatible with a broad range of applications, in practice no application is adequately supported due to the absence of data semantics. A typical multimedia application, say, authoring of an e-magazine, is unlikely to conduct queries like "find all JPEG images with size less than 200 KB", but is more reasonable to "find all the paintings drawn by van Gogh in his last year", which obviously goes beyond the capability of such data models.

The failure of conceptual data models to describe multimedia data can be ascribed to the absence of contexts in which their semantics can be properly interpreted. Therefore, it is natural and desirable to shift the task of semantic modeling from the conceptual level to the external (view) level, which is much closer to the applications. In this chapter, we propose *MediaView* as an object-oriented view mechanism to achieve this functional shift in the ANSI/SPARC three-level schema architecture. In this mechanism, the conceptual schema no longer characterizes the semantics of the multimedia data; instead, their semantics are described at the external level by a set of views, called *media views*, which provide the application-specific context for the interpretation of multimedia data. Specialized view derivation mechanisms, such as heuristic enumeration and object algebra, are proposed for the definition of media views. Media views can facilitate powerful multimedia-flavored queries and semantics-oriented navigation functionality, and suggest the semantic correlations between objects, according to which the physical storage and indexing (of data) can be optimized.

9.2 Semantic and Multimedia Data Modeling Techniques

This section presents a brief review on the data modeling techniques related to semantic modeling of multimedia, which are classified into three categories, (1) multimedia database techniques, (2) previous object-oriented view mechanisms, and (3) object-oriented data models with dynamic functions.

MULTIMEDIA DATABASE TECHNIQUES
The proliferation of multimedia data imposes a great challenge on conventional database technology, which is inadequate to model their characteristics, such as the huge data size, spatial and temporal nature, etc. To address these limitations, much work has been proposed towards different aspects regarding the management

ofmultimedia, including data models [2, 12, 18], presentation [13], indexing [9], and query processing.

Among all the data models, the object-oriented approach is generally regarded as the most suitable choice for modeling multimedia data, mainly because of its great modeling capacity and its extensibility (for new types of data) by means of inheritance. For example, OVID (Object Video Information Database) system [18] was proposed based on an object data model for video management. Similarly, the STORM [2] object-oriented database management system integrates structural and temporal aspects for managing different presentations of multimedia objects. The ORION project [12] proposed by Kim et al. also adopts object-oriented methodology for multimedia management.

In addition to data modeling, there is also research work addressing the retrieval and presentation issues in multimedia databases. For example, in [9] Faloutsos surveyed the work on multi-dimensional indexing and suggested a fast searching method for multimedia objects. The work of Klas et al. [13] focused on the presentation issue by proposing an efficient and generic playout service for multimedia application. Bertino et al. [] presented an overview on issues concerning query processing of multimedia and described MULTOS as a multimedia server with advanced retrieval capability.

Although the above-mentioned efforts are successful in their respective domains, they fall short of modeling the semantic aspect of multimedia data, which is of vital importance to multimedia information system (MMIS) applications. Motivated by this problem, *MediaView* provides an effective mechanism for semantic modeling of multimedia, which is orthogonal and complementary to all the previous works. In fact, it can be easily integrated with the existing techniques, provided that an object model is used.

OBJECT-ORIENTED VIEW MECHANISM

There have been many successful previous research projects on view mechanism in object-oriented data models [1, 11, 12, 23, 25,28]. Most of them utilize the query language defined for their respective object model to derive a view (or virtual class), e.g., Abiteboul et al. [1] proposed a general framework for view definition based on a clear semantics. These approaches differ from each other in the way that they treat the derived view in the global schema. For example, Heiler's approach [11] treats each view as a standalone object, rather than integrating into the schema. In Kim's work [12], the derived views are attached to the schema root as its direct subclasses. The approaches of Scholl et al. [25] and Tanaka et al. [28] address the issue of incorporating the derived views into the global schema. However, the consistency of the schema is not guaranteed in their approaches. Rundensteiner's proposal of MutliView [23] is a more systematic solution for object view mechanism, which not only provides an algorithm for integrating the derived virtual classes into the global schema, but also allows the generation of multiple view schemata, with the consistency and closure property enforced by automatic tools. The issue of view materialization is also addressed in MultiView [15].

The fundamental distinction between *MediaView* and past work is that all the existing view solutions are "information customization" mechanisms, i.e., hiding properties and instances of the base classes from a customized view, while *MediaView* is an "information augmentation" mechanism, in which objects are empowered with new properties when they are included into views.

"DYNAMIC" OBJECT MODELS
There has also been a significant research undertaken on the development of OODBMSs that can support advanced "dynamic functions", including *object evolution and migration, dynamic conceptual clustering*, and *multiple perspectives and representations*. Two radically different approaches aimed at supporting dynamic functions in OODBMSs can be distinguished: one is based on language-independent or weakly-typed object models; and the other is based on strongly-typed object models extended with *role* facilities. These dynamic functions enhance the semantic modeling capability of an object model and make it more suitable for multimedia data.

- **Language-Independent and Weakly-Typed Models**
This approach supports the various dynamic functions for OODBMSs by employing a weakly-typed programming language (such as Smalltalk), where an object can have multiple classes, or by adopting a language-independent object model where the OODBMS defines its own object model and appropriate mappings are provided from languages to this object model. An example of the former is GemStone [6], whereas OODAPLEX [7], O2 [8] and TIGUKAT [20] are examples of the latter. Consequently, these systems can support object migration as well as allow an object to be viewed from different perspectives. An important disadvantage of this approach is the inherent performance penalty due to the inefficiency of weakly typed languages.

- **Strongly Typed Models with Role Extensions**
Examples of work belonging to this group include conventional OODBMSs developed from strongly-typed programming languages such as C++. In particular, these systems use a static, classification-based approach in defining the structural and behavioral properties of data objects, where each object in the database is assumed to have exactly one class, namely, the one in which the object was created. Such an assumption, however, imposes some serious restrictions on modeling dynamic and/or multi-faceted real world objects, which is especially true for multimedia applications. To relax this modeling restriction, *role* facilities have been proposed and/or incorporated recently into the "statically-typed" OODBMSs [16, 21, 22, 26, 27], resulting in improved modeling power and flexibility for conventional object-oriented models.

   In the context of object-oriented databases, roles have been demonstrated to be effective in supporting *object evolution and migration* [21,27], as well as *conceptual clustering* modeling [16]. In particular, [21] describes a role model that facilitates the description of different behaviors throughout an object's evolution (and state-transitions). The work on *object migration patterns* [27] considered allowable patterns for object migration (i.e., change of class) represented as dynamic constraints, and used the valid "role set" histories of objects for this purpose. For the work on *conceptual clustering* models [16], roles have been used as "threads" through which objects are grouped together in forming ad hoc, dynamic collections (called *clusters*) as derived "meta-data". Roles have also been found useful in supporting multiple perspectives/representations of objects. Work in this direction includes *object specialization* [26], *aspects* [22] and *Fibonacci* [3], which all use different approaches. The work described in [26] took the combined approach of a prototype system and an object-oriented system, where classes are viewed (and realized) as individual objects (prototypes) auxiliary roles (perspectives), and objects define their own inheritance paths. By contrast, in [22], roles are attached to object classes directly as *extensions*, with each extension providing a specialized perspective (called an *aspect*) for the objects in the class. Independently, the Fibonacci system by Albano et al. [3] allows a role hierarchy to be associated to a base class so that its objects can play any role from that hierarchy.

9.3 *MediaView*: A Semantic Modeling Mechanism

MediaView is not the first piece of work on object-oriented view mechanism, a "old" topic addressed by many previous works [1, 11, 12, 23, 25, 28]. Nevertheless, *MediaView* is the first view mechanism (to the best of our knowledge) that has been designed especially for multimedia databases. Consequently, the definition, derivation, and usage of media views are characteristic of a multimedia flavor and are fundamentally different from all the previous work.

AN OVERALL PICTURE

As shown in Figure 9-3, the MediaView mechanism suggests a novel definition of the ANSI/SPARC three-level schema architecture for multimedia databases. This redefinition is based on the observation that a multimedia object can be characterized by features at two independent levels—primitive features at low level, as well as semantic features at high level. In MediaView, these two sets of features are respectively modeled at the conceptual level and at the external level. Note that this is a departure from the convention that the conceptual schema takes charge of all the semantic modeling tasks, while views are only used to define customized interface of the data.

The primitive features of multimedia are basic and context-independent ones that are unlikely to have diverse interpretations in different contexts. Examples of primitive features include generic features such as data size and format, as well as media-specific features such as color histogram for images, motion vector for videos, and pitch contour for audios. The nature of primitive features determines that they can be suitably modeled by the conceptual schema, which provides an objective foundation for a broad range of applications. Specifically, these features are characterized through a collection of *base classes* defined in the conceptual schema during the initial schema definition process. For example, "*Bitmap Image*" is a typical base class defined with attributes such as "*color histogram*", "*texture*", etc.

APPLICATIONS (USERS)

EXTERNAL LEVEL

CONCEPTUAL LEVEL

INTERNAL LEVEL

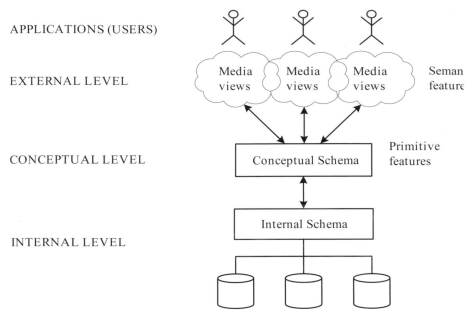

Figure 9-3 "Redefinition" of ANSI/SPARC three-schema database architecture

In addition to the primitive features, each multimedia object possesses semantic features that are usually subjective and context-dependent. Such features vary with the specific contexts in which an object is interpreted, e.g., van Gogh's "Sunflower" can be described either as "flower" or "van Gogh's painting" depending on the specific situations. Therefore, we deem that the external (view) schema is in a better position than the conceptual schema to model such semantic features. By defining its own views, each application (user) can specify a customized context in which the multimedia objects are interpreted for their specific purposes. Moreover, as stated by Rundensteiner [23], views in an object-oriented data model have additional advantages for semantic data modeling: (1) an object can maintain its unique identity even if it is included in more than one view, such that no new objects are created as the result of applying views; (2) type-specific attributes and operations can be assigned to objects when they are modeled by views, which are defined as virtual classes.

At the external level, the specific tool for modeling the semantic features of multimedia is a set of media views. On the surface, like many existing object-oriented view mechanisms, a media view is defined as a virtual class with specific attributes and operations. In essence, each media view corresponds to a semantic concept and is a cluster of the multimedia objects that can be interpreted by that semantic concept. Therefore, a media view can consist of heterogeneous but semantically related objects selected from different base classes, instead of homogeneous objects (from the same class) as in the traditional view mechanisms. In addition, while a multimedia object belongs to only one base class, it can be included into more than one media view simultaneously. This corresponds to the fact that a multimedia object may have multiple semantic interpretations, but its primitive features are uniform.

FUNDAMENTALS OF *MEDIAVIEW*

In the following, we firstly present a formal definition of media view and some related concepts. After that, we introduce a semantic graph model that describes the data reorganization as the outcome of applying media views.

Basic Concepts

Definition 1: Set C as the set of base classes. A **base class** $C_i \in C$ has a unique class name, a type description, and a set of objects associated with it. The type of C_i is referred to as **type(C_i)**, which defines a set of properties as the common interface of all the instances of C_i. The set of properties are referred to as **properties(C_i)**, and each property in it can be a value of a simple type, an instance of a certain class, or a method. The set of objects associated with C_i is defined as **extent(C_i)**= $\{o \mid o \in C_i\}$.

Definition 2: A **media view** MV_i is a virtual class that has a unique view name, a type description, and a set of objects associated with it. The type of MV_i is referred to as **type(MV_i)**, which defines a set of properties **properties** (MV_i) as the common interface of all its instances. Similarly, a property can be a value of a simple type, an instance of a media view, or a method. The set of objects associated with MV_i is defined as **extent(MV_i)**= $\{o \mid o \in MV_i\}$.

Definition 3: A base class C_i is defined as a **subclass** of another base class C_j if and only if the following two conditions hold: (1) **properties(C_j)** $\subseteq$ **properties(C_i)**, and (2) **extent(C_i)** $\subseteq$ **extent(C_j)** If C_i is the subclass of C_j, we also say that there is an **is-a** relationship from C_i to C_j. A **base schema** (BS) is a directed acyclic graph $G=(V, E)$, where V is a finite set of vertices and E is a finite set of edges as a binary relation defined on $V \times V$. Each element in V corresponds to a base class C_i. Each edge in the form of $e=<C_i, C_j> \in E$ represents an is-a relationship from C_i to C_j (or C_i is a subclass of C_j).

Definition 4: A media view MV_i is a **subview** of another media view MV_j (or there is an **is-a** relationship from MV_i to MV_j) if and only if **properties(MV_j)** $\subseteq$ **properties(MV_i)** and **extent(MV_i)** $\subseteq$ **extent(MV_j)**. A **view schema** (VS) is a directed acyclic graph $G=\{V, E\}$, where a vertex in V corresponds to a media view MV_i, and an edge $e=<MV_i, MV_j> \in E$ represents an is-a relationship from MV_i to MV_j (or MV_i is a subview of MV_j).

One can draw a parallel line between the definitions of media view and base class, as well as between base schema and view schema. Actually, a media view behaves like a base class except the following differences: (1) The instances of a media view are selected from the base classes; one cannot create a view instance from scratch. (2) A media view may consist of heterogeneous objects selected from different base classes. (3) An object must belong to a unique base class, while it can be included into zero, one, or multiple media views.

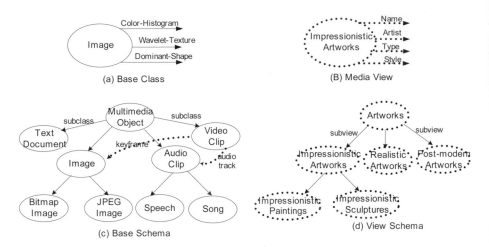

Figure 9-4 Examples of base class, base schema, media view and view schema

In addition to the **is-a** relationship, conventional object-oriented data models also support the **composition** relationship between two base classes, which defines an instance of a class as a property of the instances of another class. Similarly, there is also a **composition** relationship between two media views, through which we can create a complex view instance by referring to simple view instances as its properties.

Figure 9-4 exemplifies the basic concepts defined above, including: (a) a base class "Image"; (b) the media view "Impressionistic Artworks"; (c) a base schema; and (d) a view schema, from which one can appreciate the main distinction between media views and base classes in regard of modeling functionality. Suppose both the class "Image" and the media view "Impressionistic Artworks" are used to model van Gogh's painting "Sunflower". The base class describes the painting by its low-level features, such as color histogram and dominant shape, while the media view interprets it from a higher level semantic perspective, say, its style. Accordingly, the base schema is mainly constructed on the type or structural relationship, e.g., "JPEG Image" is a kind of "Image", and "Video Clip" refers to a set of "Image" and "Audio Clip" as its properties. In contrast, the view schema is based on semantic relationships, e.g., "Realistic Artworks" is one category of "Artworks".

The definition of media view differs from that proposed in conventional object-oriented view mechanism [23] in several fundamental aspects: Firstly, a media view can accommodate objects from different base classes (heterogeneous objects), while a conventional view can only contain objects from a single class (homogeneous objects). This distinction is desirable for multimedia databases, since multimedia objects of different types may suggest the same semantic concept. Secondly, new properties can be defined for a media view, such that an object is empowered with these new properties (in addition to the properties defined in its base class) when it is identified into that media view. This offers a significant improvement in modeling capacity over the traditional view mechanisms, which only allows view to inherit or

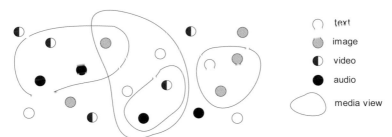

Figure 9-5 Semantic-based data reorganization using media views

derive properties from base classes. Last but not the least, the view schema and the base schema in the *MediaView* mechanism, which model different aspects of multimedia data, are completely independent of each other. In contrast, previous work such as [23] treated views and base classes as integral parts of a global schema.

Semantic Graph Model

Applying media views to a database corresponds to a semantics-based reorganization of multimedia objects (which are initially organized according to their types by base classes). As illustrated in Figure 9-5, a media view clusters multimedia objects of various types (typically, text, image, video, and audio) that have the same or closely related semantics as its instances. In this way, it defines an application-specific context in which the semantics of these objects can be interpreted. Since an object may have various semantic interpretations and thus belong to multiple media views, a media view may overlap or even subsume other views in terms of their instances.

As the objects of a media view are semantically correlated, we can create a link between any pair of them to represent their semantic correlation. For each media view, such links constitute a complete graph among all its instances. Therefore, all the objects of the database can be interconnected into a huge graph (not necessarily a connected graph) by a set of such complete sub-graphs corresponding to media views. This leads us to the following definition:

> **Definition 5**: *The **semantic graph** (SG) is an undirected graph G={V, E}, where V is a finite set of vertices and E is a finite set of edges. Each element $V_i \in V$ corresponds to a multimedia object O_i in the database. E is a ternary relation defined on $V \times V \times N$. Each $e=<V_i, V_j, n> \in E$ represents a **semantic link** of degree n between object O_i and O_j, where n is the number of media views to which both objects belong. We define n as the **correlation factor** between O_i and O_j.*

> **Definition 6**: *The **correlation matrix** $M=[m_{ij}]$ is an adjacency matrix of the semantic graph. Specifically, each element m_{ij} contains the correlation factor between O_i and O_j, with all the diagonal elements set to zero.*

O_1 Biography of van Gogh

O_2 An audio guide

O_3 "Potato Eaters" (by van Gogh)

O_4 "Sunflower" (by van Gogh)

O_5 Ohter impressionistic artwork

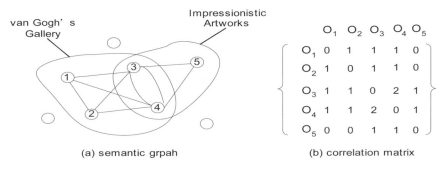

(a) semantic grpah   (b) correlation matrix

Figure 9-6 Semantic graph and correlation matrix in an e-publishing scenario

The semantic graph model is illustrated in the example in Figure 9-6. Consider the scenario of an e-publisher who plans to compose two special issues of an e-magazine - on the topics of van Gogh and impressionistic artworks. To collect related materials, the editor creates two media views, "van Gogh's Gallery" and "Impressionistic Artworks" (see Figure 9-6(a)), and inserts multimedia objects into them. For "van Gogh's Gallery", the materials include a biography of van Gogh, an audio guide, and two of his paintings "Potato Eaters" and "Sunflower". Both paintings are also eligible for the other media view, "Impressionistic Artworks", which also contains the works of other impressionistic artists. A fraction of the semantic graph corresponding to these two views is shown in Figure 9-6(a), along with the corresponding correlation matrix in Figure 9-6(b). The advantage of the semantic graph and the correlation matrix lies in their revelation of the semantic correlation between objects for a certain application. Later, we will demonstrate their use in view derivation and application in the similar scenarios in e-publishing.

VIEW OPERATORS

We define a set of operators that take media views and view instances as operands. Our intention is not to come up with a complete set of operators, but to focus on those that are indispensable in supporting query and navigation over multimedia objects. Table 9-1 summarizes these operators with a description of their syntax and semantics.

The view operators are classified into two categories according to the type of operand: *type-level* operators that manipulate media views (types) as operands, and *instance-level* operators with view instances (object) as operands. The type-level

type-level	v-overlap	syntax	$<boolean>:=$ **v-overlap** ($<media\ view1,\ media\ view2>$)
		semantics	true, if and only if ($\exists o \in O)(o \in$ **extent**($<media\ view1>$) **and** $o \in$ **extent**($<media\ view2>$))
	cross	syntax	$\{<object>\}:=$ **cross** ($<media\ view1,\ media\ view2>$)
		semantics	$\{<object>\} := \{o \in O \mid o \in$ **extent**($<media\ view1>$) **and** $o \in$ **extent**($<media\ view2>$)\}
	sum	syntax	$\{<object>\}:=$ **sum** ($<media\ view1,\ meida-view2>$)
		semantics	$\{<object>\} := \{o \in O \mid o \in$ **extent**($<media\ view1>$) **or** $o \in$ **extent**($<media\ view2>$)\}
	subtract	syntax	$\{<object>\}:=$ **subtract** ($<media\ view1,\ media\ view2>$)
		semantics	$\{<object>\}:= \{o \in O \mid o \in$ **extent**($<media\ view1>$) **and** $o \notin$ **extent**($<media\ view2>$)\}
instance-level	class	syntax	$<base\ class> :=$ **class**($<view\ instance>$)
		semantics	$<view\ instance>$ is a instance of $<base\ class>$
	components	syntax	$\{<object>\} :=$ **components** ($<view\ instance>$)
		semantics	$\{<object>\} := \{ o \in O \mid o$ is a component (direct or indirect) of $<$view instance$>$\}
	i-overlap	syntax	$<boolean> :=$ **i-overlap** ($<view\ instnace1>,\ <view\ instance2>$)
		semantics	**true**, if and only if $\exists o \in O)\ (o \in$ components ($<$view instance1$>$) **and** $o \in$ components($<$view instance2$>$))

Table 9-1 Media view operators

operators include: **v-overlap**, which examines whether two media views have common instances, and three set operators **cross**, **sum**, and **subtract** which locate objects that are instances of both of the media views, instances of either of the two views, or instances of the first view but not the second view. The instance-level operator **class** is used to retrieve the name of the base class of the specified view instance. As introduced in Section 9.3.2, complex view instances can be composed by taking in simple view instances (via composition relationships) as its components. The **components** operator is designed to find all the components (either direct or indirect) of a complex view instance, i.e., the objects that link to the specified view instance through one or more composition relationships. The **i-overlap** operator is a Boolean operator that returns true when two complex view instances have some common components.

As can be seen, these view operators mainly manipulate the instances of media views and investigate the relationship between media views based on shared instances. The common objects shared by two media views suggest an ad hoc, implicit relationship between these two views, which cannot be explicitly defined or even expressed. As an example, consider the relationship between media view "van Gogh's Gallery" and "Impressionistic Artworks" (see Figure 9-6) through sharing of van Gogh's paintings. Such implicit relationships offer an important supplement to the explicitly defined is-a and composition relationships. The set of view operators proposed here provides a tool for making use of these relationships.

VIEW DERIVATION MECHANISM

We propose two approaches for the derivation of media view: heuristic enumeration and object algebra. In the former approach, the user identifies the instances of a media view by enumeration with the recommendation based on the content (low-level features) or semantic links (embodied in the semantic graph) among the candidate objects. In the latter approach, new media views are derived from existing media views by applying object algebra operators on them.

Heuristic Enumeration

To define a media view using this approach, a user, usually the application designer, first specifies the properties of the media view, and then enumerates all its instances. The enumeration process can be conducted manually, or with the recommendation of promising candidates objects based on the content or semantic similarity between objects, or using a combination of the above strategies.

- **Blind enumeration**. Blind enumeration is a blunt approach, which requires the user to enumerate all the instances of a media view without any hints from the database. To create a media view, the user browses the database (in a haphazard manner), picks up the relevant objects, and labels them as the instances of the media view. Not surprisingly, this method is not scalable to large databases due to the enormous manual effort and processing time required.

- **Content-based enumeration**. This is a heuristic enumeration approach which recommends candidate objects from the system based on the content (expressed as primitive features) of the objects. In this approach, the user chooses a representative object (from the objects already identified as view instances or any other objects at hand) as a sample and searches for similar objects. In this case, the similarity of the sample object with every object in the database is calculated based on primitive features. The primitive features and similarity functions can be whatever is defined in the base classes and are dependent on the type of the sample object. For example, for images we choose color histogram, dominant shape as the primitive features, and Euclidean distance as the similarity metric. After the similarity calculation is finished, a list of candidate objects ranked by their similarity score is returned to the user as the recommendation, from which he select relevant ones as view instances. The whole process is essentially in the spirit of query-by-example (QBE) paradigm in content-based retrieval.

- **Semantics-based enumeration**. This approach is the counterpart of the content-based enumeration at the semantic level. Specifically, it takes the advantage of the semantic links among multimedia objects in the semantic graph (*SG* for short). A simple version of this approach works as follows: When the user identifies an object, say, O_i, as the instance of the media view being defined, we recommend the objects that are directly linked with O_i in SG (i.e., objects that appear together with O_i in some media views) as the candidate objects for further enumeration. The candidates are ranked by their correlation factor (viz., degree of the link) with the "seed" object O_i. To do this, we look into ith row of the correlation matrix M and sort the objects according to the value of each m_{ij}.

The justification of the above approach is that two objects in the same media view are semantically correlated and therefore stand a greater chance (than a random pair of objects) of being selected together into another media view. Again, consider the scenario of e-publishing illustrated in Figure 9-6: Suppose that the editor wants to create the third media view "van Gogh's paintings" based on the two defined views, "van Gogh's Gallery" and "Impressionistic Artworks". As the first step, he selects O_4, van Gogh's famous painting "Sunflower", into this new media view. Then, by examining the semantic graph, the system suggests the candidate objects O_3, O_1, O_2, and O_5 to him, with O_3 appearing in the top since its correlation factor is the largest among the four. In this case, the editor can quickly confirm O_3, which is another masterpiece of van Gogh, as the member of the new media view.

The above approach is straightforward and intuitive. However, it suffers from an obvious setback: the objects that are indirectly linked to the seed object are not considered for recommendation. These objects are likely to be promising candidates of the media view as well. To overcome this limitation, we introduce the following concepts:

Definition 7. *The **n-level correlation matrix** M(n) is derived from **correlation matrix** M by the following formula:*

$$M(n) = M + kM^2 + k^2 M^3 + \cdots + k^{n-1} M^n \quad (0<k<1)$$

*where n is a positive integer and k is a constant between 0 and 1. Each element $[M(n)]_{ij}$ is defined as the **n-level correlation factor** between objects O_i and O_j.*

The n-level correlation matrix is a more sophisticated model that defines the semantic correlations between objects, because it takes into account the relationship between indirectly linked objects (up to n intermediate objects). A closer examination of its definition reveals that it is defined by a formula similar to the one used for calculation of the transitive closure of a matrix, except that an additional parameter k is included. This difference can be explained by the fact that our model is mainly concerned with the semantic correlation between objects. Therefore, it is natural to consider the length of the path between two objects by introducing k to penalize the contribution of long paths to the correlation factor. In comparison, the transitive closure only deals with the connectivity between two objects, and therefore it treats long paths and short paths equally.

Similar to the simple version of semantics-based enumeration, we can make recommendations according to the n-level correlation factor between the candidate object and the seed object. By taking into account the indirect links, this advanced model is expected to return better quality candidates to the user. In practice, we can adjust parameters k and n to control the contribution of indirect links.

Object Algebra

In addition to enumeration-based view derivation strategies, we also define a set of object algebra through which a media view can be derived by reusing the definition of existing views. Object algebra is a mechanism commonly used to derive views in many previously suggested object-oriented view mechanisms. However, while in traditional approaches a virtual class can be derived from either base classes or existing virtual classes, in our mechanism a media view can only be derived from existing views, due to the gap in semantics between media views and base classes. Table 9-2 displays a set of object algebra operators, which are adapted from those defined in [23].

As can be seen from table 9-2, each operator specifies both the properties and the instances of a new media view on the top of existing ones. The **select** operator derives a new media view by filtering the properties and instances of the source view. The properties of the new view are specified by the <property list>, and its instances are those in the source view that satisfy the <predicate>. The **intersection** operator defines a new media view whose properties are "inherited" from both of the source

select	syntax	*<derived view>*:= **select** *<property list>* **from** (*<source view>*) **where** *<predicate>*
	semantics	**properties**(*<derived view>*):= { $p \in P$ | $p \in$ **properties**(*<source view>*) **and** $p \in$ *<property list>*} **extent**(*<derived view>*) = { $o \in O$ | $o \in$ **extent**(*<source view>*) **and** *<predicate>*(*o*)=true }
intersection	syntax	*<derived view>*:= **intersection** (*<source view1>*, *<source view2>*)
	semantics	**properties**(*<derived view>*) :={$p \in P$ | $p \in$ **properties**(*<source view1>*) **or** $p \in$ **properties**(*<source view2>*))} **extent**(*<derived view>*) := {$o \in O$ | $o \in$ **extent**(*<source view1>*) **and** $o \in$ **extent**(*<source view2>*))}
union	syntax	*<derived view>*:= **union** (*<source view1>*, *<source view2>*)
	semantics	**properties**(*<derived view>*):={$p \in P$ | $p \in$ **properties**(*<source view1>*) **and** $p \in$ **properties**(*<source view2>*))} **extent**(*<derived view>*) := {$o \in O$ | $o \in$ **extent**(*<source view1>*) **or** $o \in$ **extent**(*<source view2>*))}
difference	syntax	*<derived view>*:= **difference** (*<source view1>*, *<source view2>*)
	semantics	**properties**(*<derived view>*) := {$p \in P$ | $p \in$ **properties**(*<source view1>*))} **extent**(*<derived view>*) := {$o \in O$ | $o \in$ **extent**(*<source view1>*) **and** $o \notin$ **extent**(*<source view2>*))}

Table 9-2 Object algebra operators

views, and instances are those common members of both views. On the contrary, the properties of the media view defined by **union** are the intersection of the properties of two source views, and the instances of the new view are the objects that belong to either of the source views. The media view defined by **difference** has the same properties with the first source view, and its instances are those belonging to the first source view but not in the second view.

Despite the similarity in syntax, object algebra operators are fundamentally different from the view operators defined in Section 0. The two main differences are: (1) Object algebra operators create new media views while view operators only return simple values (e.g., Boolean) or set of objects. (2) Object algebra operators manipulate not only the instances but also the properties (type) of a media view, while view operators only deal with view instances. For example, the object algebra **intersection** is syntactically similar to the view operator **cross**, both of which identify the common member of two media views. However, the **intersection** also derives a set of properties from two source views, from which it defines a new media view, while the **cross** returns the common instances.

Object algebra operators provide useful tools for deriving media views on the top of existing views. For example, we can create a media view called "Impressionistic Paintings" through the query "select (name, artist, style) from 'Impressionistic Artworks' where type = 'painting'". As another example, a media view called "van Gogh's Artworks" can be derived by "intersection ('Impressionistic Artworks', 'van Gogh's Gallery')". This approach not only reduces manual efforts required for view definition, but also saves the storage cost for the definitions of media views.

Both of the approaches described above demand human participation for view derivation. For the heuristic enumeration, even if highly relevant objects are recommended, the user needs to inspect the recommended objects and confirm them into the media view. Moreover, the semantics-based enumeration is based on the previously defined media views, whose definition in turn requires manual effort. For object algebra, since they are also applied on existing media views, manual effort is involved anyway. In fact, it is extremely difficult (if not impossible) to automatically derive media views, which are defined from application-dependent semantic perspectives. Fortunately, both of the proposed approaches are able to reuse the previously defined media views when deriving new media views, such that the view definition process becomes easier and more efficient.

9.4 Applications of *MediaView*

In this section, we demonstrate how the *MediaView* mechanism facilitates flexible retrieval and browsing of multimedia objects in the database. (More application cases of MediaView in the context of multi-modal information retrieval are given in [17].)Furthermore, we discuss its potentials for optimizing the physical storage and indexing based on the semantic correlation between objects indicated in the graph model.

MULTIMEDIA QUERIES

In a multimedia database, a key factor to success lies on its data model's capacity to support multimedia-flavored queries, which are not well supported by traditional data models mainly due to their failure to capture the semantics of multimedia. In contrast, we demonstrate that view operators (see Table 9-1), coupled with the object algebra operators (see table 9-2), provide a strong basis for formulating rather powerful and sophisticated multimedia queries. In this subsection, we do not attempt to come up with a formal description of the retrieval capability of MediaView (e.g., in terms of a complete query language); instead, we provide some query examples here to illustrate the usage and expressive power of these operators as "querying constructs". For expository purpose, our query examples are all from the application domain of e-publishing. The "e-magazine", "e-journal", and "e-Newspaper" used in the examples are all media views, which have been already defined.

Example 1 - Locating a subject object.

Consider the following query: *Find out which e-magazine has published an issue in 1990s that contains a speech by Richard Nixon when he visited China in 1972.*
This query can be expressed by a combination of the following MediaView operators:

> **Select** x.title, y.volume **from** e-magazine x
> **where** y **in components**(x) **and** y.year **within** [1990, 2000]
>   **and** z in components(y) and class(z) = speech
>   **and** z.year = 1972 **and** z.venue = "China"
>   **and** z.speaker = "Richard Nixon";

Example 2 - Frequency checking

Consider the following query: *Find out how many issues of the e-ReadersDigest magazine for the year of 2001 have enclosed a song related to sea.*
This query can be expressed by a combination of the following MediaView operators:

> **Select** x.volume **from** e-magazine x
> **where** x.title = "e-ReadersDigest" **and** x.year = 2001
>   **and** y in components(x) and type(y) = song
>   **and** y.lyrics **contains** {"sea", "ocean"};

In the above query, contains is a "syntactic sugar" used for checking containment between a textual document (in this case, the lyrics of a song) and any keyword from a given set (in this case, "sea" or "ocean", or both).

Example 3 - Commonality checking

Consider the following query: *Find out from the e-Newsweek and e-ReadersDigest any common objects (articles, photos, audio, or videos) in the issues of the year 1999.*
This query can be expressed by a combination of the following MediaView operators:

> **cross** ((**select** x **from** e-magazine x
>   **where** x.title = "e-ReadersDigest"),
>   (**select** y **from** e-magazine y
>   **where** y.title = "e-Newsweek"));

Example 4 - Content-based retrieval
Consider the following query: *Find and rank all the e journals whose cover pages are visually similar to that of DMKD ("Data Mining and Knowledge Discovery") by Kluwer Academic.*
This query can be expressed by a combination of the following MediaView operators:

> **Select** x.title **from** e-journal x, y
> **where** y.title = "Data Mining and Knowledge Discovery"
> **and** y.publisher = "Kluwer Academic"
> **order_by similarity**(x.coverpage, y.coverpage);

In the above query, similarity is a content-based function used for calculating the low-level feature similarity between two media objects (in this case, two cover page images). Depending on the base class of the objects, this function can be implemented using visual features such as color, shape, and texture, or other primitive features such as keywords (if the objects are documents) and pitch contour (if the objects are audios).

Example 5 - Associative search
Consider the following query: *Find all the pairs of e-Newspapers so that not only both have similar headlines, but also common objects (articles, photos etc.) inside; rank the pairs according to their levels of similarity.*
This query can be expressed by a combination of the following MediaView operators:

> **Select** x.title, y.title **from** e-Newspaper x, y
> **where i-overlap** (x, y) = true
> **order_by similarity**(x.headline, y headline);

Here, similarity is also a content-based function used for calculating the similarity between two sets of strings (in this case, the headlines). The i-overlap predicate is used for testing two instance objects to see if there is any common sub-object (cf. Table 9-1).

NAVIGATION IN "MEDIA MAP"

Besides the query operations, browsing and navigation are also among the most popular user behaviors in multimedia databases. As a part of the MediaView mechanism, a novel navigation paradigm—"media map"—is introduced based on the semantic graph model. The proposed paradigm helps the user traverse the database in a semantics-oriented manner, which offers better user experience compared with traditional paradigm for navigation.

A tourist map guides readers on the routes that they should follow to visit some of the sites. Our semantic graph model plays the similar role in a multimedia database: it defines the paths (semantic links) through which the user can travel from an object to other semantically related objects. In this regard, we treat the semantic graph as a "media map" and propose the following scenario for navigation. When the user navigates to a certain multimedia object, he will be also given a set of candidate objects that he may travel to from his current position. Typically, the candidates are the objects that are linked (directly or indirectly) to the object currently being visited.

In a graphical interface, for each candidate object we display a small icon as a hint of its content, with its correlation factor with the current object shown alongside. After deciding the next object to visit, the user clicks the corresponding icon from the list and the system will lead him to the selected object.

In addition to traversal of paths at object level, the semantic graph model also supports traversal of paths at view level based on the relationship between media views, which is defined by the v-overlap operator (see table 9-1) based on their common members. The scenario is similar to the object-level navigation: When the user is browsing the content of a certain media view, a list of related media views identified by applying v-overlap operator is also suggested to him. These candidate views can be ranked according to the degree of their overlap with the current view in terms of common members.

The proposed navigation paradigm makes sense to multimedia databases because it captures the semantic correlations between objects or object clusters (media views). This semantics-oriented navigation offers a smoother user experience than the traditional navigation paradigm, in which users browse the objects organized by their base classes. Suppose a user is fascinated with van Gogh's painting "Sunflower" and therefore is interested in reading his biography. Traditionally, he has to browse all the textual documents (as a biography is a kind of textual document) in order to find the desired biography, which is tedious and annoying. In the "media map" paradigm, when the user is looking at "Sunflower", all other relevant materials (including van Gogh's biography) are recommended to him by the system, so that he can navigate to the biography without any trouble.

OPTIMIZATION OF STORAGE AND INDEXING

As a by-product of our mechanism, after many media views have been defined, we can improve the physical storage and indexing of multimedia objects based on a better knowledge of their mutual semantic correlation. Specifically, the (n-level) correlation factor between two objects indicates that the two objects will be accessed together or successively, providng a guideline for the optimization ofstorage and indexing tasks. For example, if two objects have a high correlation factor, they should be stored closer to each other physically (e.g., in the same disk block) to allow efficient access from the disk. On the other hand, we may create an index for a pair or a chain of objects with high mutual correlation, so that they can be retrieved much more efficiently.

Practical algorithms for the optimization of storage and indexing need to consider more factors, such as the structure of the schema and characteristic user queries. This would be an interesting topic for further study, which is beyond the scope of the work presented in this chapter. Our intention here is to show the potential of MediaView in supporting such "back links" from the view level to internal level for improving physical storage and indexing.

9.5 Conclusion

In this chapter, we have discussed the issues of semantic modeling of multimedia data and reviewed the related techniques on semantic and multimedia data modeling.In particular, we have presented *MediaView* as a novel object-oriented view mechanism for multimedia databases. Specifically, it consists of a collection of media views providing application-specific contexts in which the elusive semantics of multimedia data can be properly interpreted. *MediaView* supports sophisticated multimedia-flavored queries, promotes a novel paradigm for convenient navigation, and provides hints for the the semantic correlation between objects for the optimization of the physical storage and indexing.

Due to the great potential of *MediaView*, in the future much interesting work can be done to extend its capability.We have attempted to define a set of "view algebras" with explicit semantic interpretations, in addition to the object algebra currently used. Since each media view corresponds to a semantic concept, it is reasonable to define view algebras from a purely semantic perspective, such as "synonymous", "antonymous", and "hyponym". If successfully defined, these view algebras will increase the capacity of *MediaView* in view derivation and querying. Other future work involves the efficiency aspect of the *MediaView* mechanism. Along this direction, new algorithms for indexing and physical storage of objects will be developed, in order to enhance the efficiency of object retrieval from media views. Schema structure and statistics of object access pattern have to be studied for this purpose.

References

[1] Abiteboul, S., and Bonner, A., "Objects and Views", in *Proc. of ACM SIGMOD Conf. on the Management of Data*, pp. 238—247, 1991.

[2] Adiba, M., "STORM: Structural and Temporal Object-Oriented Multimedia Database Systems," *IEEE Int. Workshop on Multimedia DBMS*, NY, USA, August, 1995.

[3] Albano, A., Bergamini, R., Ghelli, G., and Orsini, R., "An object data model with roles," in *Proc. of the Int. Conf. on Very Large Databases*, pp. 39--51, 1993.

[4] Bertino, E., Catania, B., and Ferrari, E. "Query Processing". In *Multimedia Databases in Prespective*, pp. 181-217, Springer, 1997.

[5] Brin, S., and Page, L., "The anatomy of a large-scale hypertextual web search engine", in *Proc. 7th Int. WWW Conf.*, 1998.

[6] Butterworth, P., Otis, A., and Stein, J., "The GemStone object database management system." *Comm. of the ACM*, 34(10):64--77, 1991.

[7] Dayal, U., "Queries and Views in an object-oriented data model." in *Proc. of 2nd Int. Workshop on Database Programming Languages*, pp. 90--102, Morgan Kaufmann, 1989.

[8] Deux, O., et. al. "The O2 System." *Comm. of the ACM*, 34(10): 34--48, Oct. 1991.

[9] Faloutsos, C. Indexing of Multimedia Data. Multimedia Databases in Prespective, pp. 219-245, Springer, 1997.

[10] Haas, L. M., "Supporting Multi-Media Object Management in a Relational Database Management System", Technical report. IBM Almaden Research Center, 1989.

[11] Heiler, S., and Zdonik, S. B., "Object views: Extending the vision", in *Proc. of IEEE Data Eng. Conf.,* pp. 86-93, Feb, 1990.

[12] Kim, W., Garza, J. F., Ballou, N., and Woelk, D., "Architecture of the ORION Next-Generation Database System," *IEEE Trans. on Knowledge and Data Engineering*, Vol. 2, No.1, 1990, pp. 109-124.

[13] Klas, W., Bool, S., and Lohr, M. "Integrated Database Services for Multimedia Presentation". In *Multimedia Information Storage and Management,* Kluwer Academic Publishers, USA, 1996.

[14] Kleinberg, J., "Authoritative sources in a hyperlinked environment." in *Journal of ACM,* 46:5:604-632, 1999.

[15] Kuno, H. A., Rundensteiner, E. A., "Materialized Object-Oriented Views in *MultiView*", in *Proc. RIDE-DOM,* IEEE CS Press, 1995.

[16] Li, Q., and Smith, J., "A conceptual model for dynamic clustering in object databases," in *Proc. of the Int. Conf. on Very Large Databases*, pp. 457--468, Aug, 1992.

[17] Li, Q., Yang, J. and Zhuang, Y., "*MediaView*: A Semantic View Mechanism for Multimedia Modeling," in *Proc. 3rd IEEE Pacific Rim Conference on Multimedia (PCM'2002)*, Dec. 2002.

[18] Oomoto, E., and Tanaka, K., "OVID: Design and Implementation of a Video-Object Database System", in *IEEE Trans. on Knowledge and Data Engineering*, vol.5, pp 629-641, 1993.

[19] Orenstein, J. A., "A Comparison of Spatial Query Processing Techniques for Native and Parameter Spaces", In *Proc. of ACM SIGMOD Conf.* pp 343-352, 1990.

[20] Ozsu, M. T., et. al. "TIGUKAT: A uniform behavioral objectbase management system." *VLDB Journal*, 4(3), VLDB Endowment, July 1995.

[21] Pernici, B., "Objects with roles." in *Proc. of ACM Conf. on Office Information Systems*, pp 205—215, ACM, 1990.

[22] Richardson, J., and Schwartz, P., "Aspects: Extending objects to support multiple, independent roles." in *Proc. of the ACM SIGMOD Int. Conf. on Management of Data*, pp. 298--307, 1991.

[23] Rundensteiner, E. A., "*MutiView*: A Methodology for Supporting Multiple Views in Object-Oriented Databases", in *Proc. of the 18th Conf. on Very Large Database*, Canada, 1992.

[24] Santini, S., Gupta, A., and Jain, R., "Emergent Semantics through Interaction in Image Databases", in *IEEE Transactions on Knowledge and Data Engineering*, vol. 13, No. 3, 2001.

[25] Scholl, M. H., Lassch, C., and Tresch, M., "Updateable Views in Object-Oriented Databases", in *Proc. of 2nd DOOD Conf.*, Germany, Dec. 1991.

[26] Sciore, E., "Object specialization." in *ACM Trans. on Information Systems*, 7(2): 103--122, April, 1989.

[27] Su, J., "Dynamic constraints and object migration." in *Proc. of the Int. Conf. on Very Large Data Bases*, pp 283-242, 1991.

[28] Tanaka, K., Yoshikawa, M., and Ishihara, K., "Schema Virtualization in Object-Oriented Databases", in *Proc. of IEEE Data Eng. Conf.*, pp. 23-30, Feb, 1988.

[29] Woelk, D., Kim, W., Luther, W., "An Object-Oriented Approach to Multimedia Databases", in *Proc. ACM SIGMOD Conference*, pp. 311-325, 1986.

10 FEATURE-BASED RETRIEVAL IN VISUAL DATABASE SYSTEMS

Prof. Aidong Zhang, Dr. Ramazan Savaş Aygün, and Prof. Yuqing Song

This chapter presents issues related to architecture, query processing, and indexing in visual database systems. The architectural issues for visual databases have more requirements than traditional databases. Metadata hierarchy, indexing using clusters and templates, and clustering using heterogeneous features are core issues in feature-based retrieval and play significant roles in efficiency and accuracy of query results. Querying, ranking, and merging heterogeneous features are explained from the perspective of the performance of the system and the satisfaction of the requirements of the users. Relevance feedback from the users increases accuracy in the retrieval.

10.1 Introduction

The need to manage and retrieval visual data in database systems is growing rapidly in many fields, such as Web browsing, geographical image systems (GIS), information retrieval, and distributed publishing. To meet these demands, database systems must have the ability to efficiently index and query visual data. Research on visual database systems brings about a variety of interesting and challenging issues, which are being studied by the visual database community.

Visual database architectures have been addressed at different levels, including system, schema, and functional levels. Various types of system architectures are possible for distributed environments. The system architecture is important to support global content-based query access to various visual databases over the distributed systems. An integrated metaserver consisting of a metadatabase, metasearch agent, and query manager can provide efficient access to remote databases built based on semantic heterogeneous clustering of data using templates. The architecture is important in reducing the amount of time and effort that users spend in finding information of interest. In addition, the metaserver's performance can further be refined based on user feedback.

Feature-based indexing and retrieval have been commonly used in visual database systems [20, 29]. Selection, evaluation, and indexing features affect the performance of querying in visual databases. The extraction of semantic features and semantic clustering play significant roles in both retrieval and indexing. The objects in the database can be clustered and each cluster can be represented with a template. Template matching increases the performance of a visual database while reducing the time for searching data.

Querying in visual databases requires retrieval of data using heterogeneous features. Clustering on heterogeneous features helps visual database systems distinguish semantically different visual data that are similar in a single feature space. The query results need to be ranked according to the querying object. The incorporation of the relevance feedback from the user can enhance the performance of the query manager.

This chapter is outlined as follows. In Section 10.2, the architectural issues for visual databases are presented. Metadata hierarchy, indexing using clusters and templates and clustering using heterogeneous features are explained in Section 10.3. In Section 10.4, querying, ranking, and merging heterogeneous features are covered. The chapter concludes with a discussion of further issues in visual databases.

10.2 Visual Database System Design

Visual database architectures have been addressed at different levels, including system, schema, and functional levels. Various types of system architectures have been proposed. In [13], a generic visual database architecture is defined. Three logically independent repositories are identified. In a distributed environment, various types of system architecture are possible. Examples of such systems include GLOSS (from Stanford) [11], WHOIS++ (from Bunyip Corporation), HARVEST (from University of Colorado) [2], WAIS (Wide Area Information Servers) [17], ImageRover [26], WebSeek [31], MetaSEEk [3], and NetView [36].

An integrated metaserver consisting of a metadatabase, metasearch agent, and query manager facilitates global content-based query access to various visual databases over the distributed systems. This kind of system significantly reduces the amount of time and effort that users spend in finding information of interest. It also can further refine the metaserver's performance based on user feedback.

Figure 10-1 illustrates the overall architecture of the system. This system contains three main parts: visual databases at remote sites, a metaserver, and a set of visual display applications at the client machines. We will focus on the design of the metaserver. The role of the metaserver is to accept user queries, extract the information in the query for suitable matching of the metadata, produce a ranking of the database sites, and distribute the queries to selected databases. Figure 10-2 shows the metaserver components, including the metadatabase, the metasearch agent, and the query manager [36].

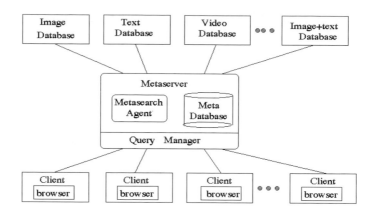

Figure 10-1 A distributed visual data retrieval system (©1998 IEEE)

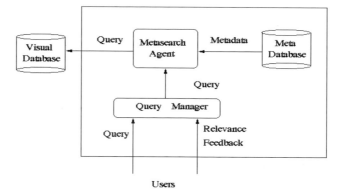

Figure 10-2 The metaserver architecture (©1998 IEEE)

The metadatabase houses both template and statistical metadata. The templates are feature vectors which are representatives of the feature vectors of component database images. The statistical metadata record the relationships between the templates and the database images. (See detailed discussion on metadata in the following section.) Three additional modules, the metadata collector, the template builder and the metadata refinement, are designed to support the interoperability and integrity of the metadata. The metadatabase, together with the metadata collector, template builder, and metadata refinement module, constitute the metadatabase management system [4], as illustrated in Figure 10-3. The collector gathers the metadata from the visual database, at the time when the database registers with the metaserver or at the time when the database resubmits its metadata. The template builder organizes and creates templates. The metadata refinement module periodically initiates the metadata update process by asking the database to resubmit its metadata.

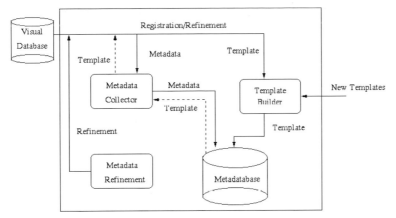

Figure 10-3 The metadatabase management system (©1998 IEEE)

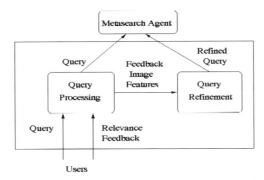

Figure 10-4 The query manager (©1998 IEEE)

The query manager consists of two modules: the query processing module and the query refinement module. The query processing module accepts user's query, extracts various feature vectors from the query, and submits the feature vectors to the metasearch agent. During the feedback step, the query processing module accepts images that are judged as relevant or irrelevant to the original query by the users. It extracts various feature vectors from the images, and forwards the feature vectors to the query refinement module. The query refinement module constructs a modified query from the image feature vectors sent by the query processing module. Figure 10-4 shows the functionalities of the query manager.

The metasearch agent invokes the site selection algorithm, which matches the query features to templates with the corresponding metadata of the databases to select the potentially relevant databases for the query. The query is then forwarded to the selected visual databases in an acceptable form. The searching mechanism of the local database searches its repository for possible answers to the posed query. The answer is then fed back to the client.

A prototype system, termed *NetView*, developed based on the above framework, consists of three major components: a central server, a remote database interface, and a user interface, as shown in Figure 10-5. *NetView* implements the metaserver functionalities in its central server. The central server interacts with the user interface through the socket interface to receive the user's query. The central server processes the query and returns a list of relevant database sites to the user. The user may choose the databases to search. The central server then forwards the query to the remote database interface for distributing the query to the chosen databases in an acceptable form.

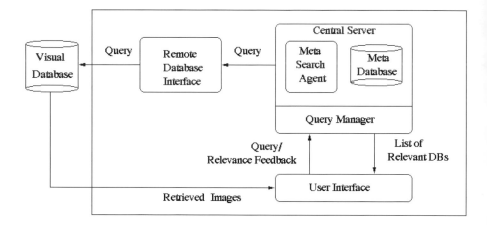

Figure 10-5 *NetView*: the web-based visual information system (©1998 IEEE)

Based on the metadatabase, the database sites can be ranked by the metasearch agent with regard to a particular visual query. The similarity between the feature vector of the query image and all the templates in the metadatabase is calculated to determine the most relevant template(s) for the query. The databases are ranked based on the statistical metadata associated with the template(s). In Figure 10-6, the table under the query icon shows the result of ranking the databases with the red bars indicating the percentage of the relevant images in each database. Once the ranked list of the databases with regard to a certain query has been generated, the metaserver will pose the query to the relevant databases selected by users. Users can choose the query images either from the databases or from the Internet using an image URL. The relevant images are then retrieved by the process described above.

10.3 Metadata Hierarchy

We will now outline the rationale of the metadata formulation that is used to select relevant databases for a given visual query. It may appear that such a selection can be done using the methods of text database discovery which maintain an index of relevant sites using text information associated with the database. Also, information such as monetary cost and latency of database sites can provide additional metadata

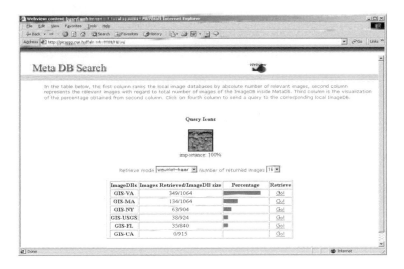

Figure 10-6 Database ranking result by metaserver

to enable early pruning of costly sites. However, for visual querying, text information can only perform a coarse pruning of the database sites. Further pruning must use the visual information in the query. But how can database relevance for visual queries be determined without a detailed examination of all images in the databases for possible matches? Clearly, it is not desirable to move the complete machinery of image content-based search used in the database engines to the distribution site to determine database relevance. Also, it is not possible to create off-line, an indexed set of database sites containing relevant images to queries, as that requires an anticipation of all possible visual queries. An intermediate approach, wherein the information in the database images is summarized and represented in a suitably abstracted form in the metadatabase, is utilized to address these problems.

Features such as texture, color and shape can represent information in a visual object. Computationally, the features of an image are typically represented by a set of numerical numbers, termed a feature vector. Visual queries can then be supported by matching the features of the query with the feature vectors in the database using similarity metrics. Many approaches have been developed to extract various features from visual data, including texture, color, and shape [14, 16]. Figure 10-7 shows two examples of visual queries and their matched images retrieved using the mechanisms presented in [24, 27, 34], texture and color queries as well as the matched segments within images are shown. The similarity attached to each image demonstrates that the similarity between the query and the image increases as their content becomes more similar. The semantic features about presence of objects in the images can enhance the indexing and retrieval since these features describe the content at the semantic level.

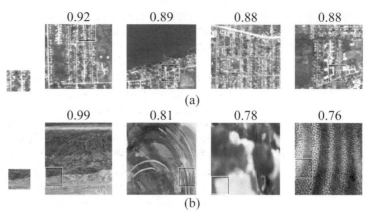

Figure 10-7 Visual queries and matched database images: (a) texture, (b) color

INDEXING OF FEATURES

Feature-based indexing and retrieval have been commonly used in visual database systems [20, 29]. The performance of the index structures is highly dependent on the extracted features. If the features are extracted based on the semantics and can be clustered using semantic clusters, the retrieval results can be improved significantly. Theories and approaches for querying and browsing images based on image semantics become increasingly critical as web-based information retrieval grows in popularity. However, effective and precise image retrieval by semantics still remains an open problem because of the extreme difficulty of fully characterizing images. J.P. Eakins [7] classified image features into three levels, ranging from the concrete to the very abstract. In this chapter, image features for visual databases are classified into the following three levels :

- *Primitive level:* Primitive features include color, texture, shape, and the spatial location of image elements.
- *Local semantic level*: Local semantic features describe the presence of individual objects in images. Two examples of queries by local semantic features are "find pictures with a bridge" (object of a given type) and "find pictures with sky and trees" (combination of objects).
- *Thematic level (or global semantic level)*: Thematic features describe the global meanings or topics of images. Two examples of queries at this level are "find pictures of a Chinese garden" and "find pictures of an earthquake." The thematic features of an image are based on all objects in the image along with their spatial relationships. High-level reasoning is needed to derive the global meaning of all objects in the scene and to determine the topic of the image.

Most existing feature extraction methods are based on primitive features [10, 19, 22, 30]. The difficulty of detection and recognition of general objects presents a significant challenge to the development of a visual database system that can extract general semantic features from images. However, techniques can be successfully applied to the simpler realm of specific domains.

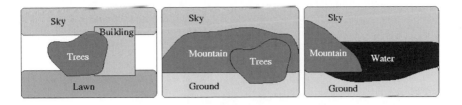

Figure 10-8 Some styles of scenery images

The domain of scenery images is a popular testbed for semantics extraction. There are several reasons why scenery images are easier to analyze than general images. Firstly, scenery images contain a limited range of object types. Common scenery object types include sky, tree, building, mountain, lawn, water, and snow. Secondly, shape features, which are difficult to characterize and match, are less important than other low-level features in analyzing scenery images. Finally, scenery images often fall into a small number of typical patterns, such as those shown in Figure 10-8. These typical patterns greatly simplify scene interpretation for scenery images.

Semantics-based image retrieval can be performed using the monotonic tree [32], a derivation of the contour tree for use with discrete data. The structural elements of an image are modeled as branches (or subtrees) of the monotonic tree of the image. These structural elements are classified and clustered on the basis of such properties as color, spatial location, harshness, and shape. Each cluster corresponds to some semantic feature. Following these steps, images can be automatically annotated with category keywords, including sky, building, tree, wave, lawn, water, snow, and ground, thus facilitating high-level (semantics-based) image querying and browsing.

The *SceneryAnalyzer* [33] is a scenery-querying system that enables semantic features extracted from scenes to be queried. The SceneryAnalyzer handles six types of scenery features: sky, building, tree, water wave, placid water, and ground. The visual properties of water wave and placid water are structurally different, and they are therefore treated as separate types. The ground feature in images can be further split into subtypes such as snow and lawn.

The process of generating the semantic keywords that support high-level scenery querying is shown in Figure 10-9. Feature extraction is the central and unique component of this system. The extraction of features is accomplished off-line; each image is processed to extract semantic features, which are then stored in a feature vector. These semantic features are automatically annotated onto a gray copy of the original image. All feature vectors are stored in a feature base.

CLUSTERS AND TEMPLATES

To support efficient retrieval, approaches that categorize the feature vectors in a database into clusters based on their features have been proposed [28, 35]. Each cluster can then be represented by a feature vector, denoted *template*, which is generally the centroid of the cluster. The cluster can be further classified into

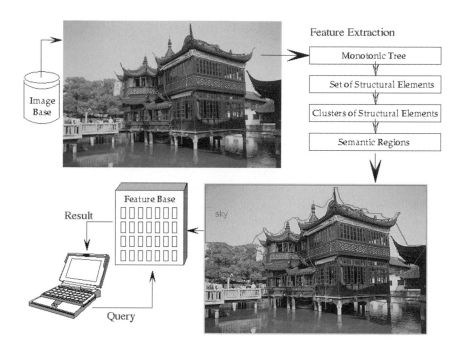

Figure 10-9 *SceneryAnalyzer* architecture.

subclusters, which can then be represented by their centroids. A template at a higher level represents the coarse features that contain all the features represented by its child templates. Such templates can be used to adequately represent the content of the database. Thus, templates are collected from the component databases as part of the metadata in the metadatabase.

To find the templates, sample images are selected from all local databases. Then using the hierarchical clustering method, a tree-like structure called a dendrogram [12, 28] is built. The dendrogram is cut at different levels resulting in different sets of clusters. Then the centroids of the resulting clusters are used as the templates. This process is applied to each feature class to find the corresponding templates. Thus these templates can concisely represent the images in all the local databases. Moreover, the hierarchical template structure can be realized to support an efficient search of the matched templates for a given query. Other methods may also be used to find a set of templates that can be used to accurately represent the content of the database. The metaserver can relate the content of the databases to the templates by calculating the statistical metadata for each database with respect to the templates. This statistical metadata will be used by the site selection mechanism. Users can view the templates and statistical metadata for each database. Figure 10-10 shows the templates (shown in the template images) and the statistical metadata (shown in the yellow box, for a certain template) in the database GIS-CA.

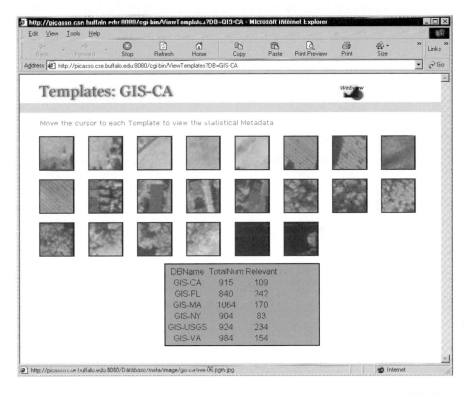

Figure 10-10 Visualization of templates and statistical metadata for the database GIS-CA

(a) Similar texture feature vectors.   (b) Similar color feature vectors.

Figure 10-11 Semantically different images with similar feature vectors

CLUSTERING ON HETEREGENEOUS FEATURES

In this section, we discuss the clustering of the database images based on the visual templates configured for each semantic cluster. The primitive features sometimes misrepresent the semantics of the images . Thus, the feature vectors of semantically irrelevant images may be located close together in the feature space. Figure 10-11(a) presents two images of wood and water where the distance between the texture feature vectors is very small, but these images are semantically dissimilar. Given a query whose feature vector is located in the neighborhood of the feature vectors of the two given images, it is highly probable that both images will be retrieved together in response to the query. Similarly, Figure 10-11(b) presents images of leaves and painting that have close color feature vectors but are not semantically related. Thus, indexing itself based on the closeness of feature vectors in the feature space sometimes may not provide satisfactory solutions.

Figure 10-12 demonstrates the intuition of a semantic clustering approach. Each semantic cluster is represented by two sets of images that belong to color and texture feature classes, respectively. For the color feature, the set of images in clusters C_1^c, C_2^c, and C_3^c are shown by solid lines. The sets drawn by dashed lines in C_1^t, C_2^t, and C_3^t represent the semantic clusters based on the texture feature class. A semantic cluster includes all the images within its scope. For each feature class, every semantic cluster is composed of a set of subclusters (not shown in the figure), and its scope is the union of the subcluster scopes. The scope of a semantic cluster based on both color and texture feature classes would be the intersection of the scopes of the clusters of the two feature classes. For example, the scope of the semantic cluster C_1 is $C_1^c \cap C_1^t$. An image will be assigned to a semantic cluster if it falls within the scope of all the heterogeneous clusters of the semantic cluster. For example, image q in Figure 10-12 will be assigned to the semantic cluster C_2, because it belongs to both C_2^t and C_2^c. However, since image p is not within the scope of C_2^t, it will not be assigned to C_2 [27].

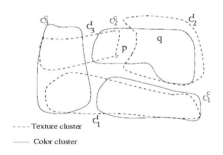

Figure 10-12 Set representation of image clusters based on color and texture feature classes

Let $F=\{f_1, f_2, ..., f_l\}$ be the set of feature classes, where $f_i \in \{texture, color, shape, ...\}$. The *scope of a subcluster* is the part of the feature space that contains the images belonging to the subcluster. In a two dimensional space, the scope of a subcluster is a circle with a radius obtained via the mean and variance of distances of the images in the subcluster to the template of the subcluster. One may also use other approaches such as the bounding box or convex hull to define the scope of the subcluster.

The *scope of the semantic cluster* in feature class f_i, denoted $S_k^{f_i}$, is defined as the union of the scopes of all subclusters of C_k in feature class f_i. The scope of the semantic cluster C_k based on the heterogeneous features, denoted S_k, is then defined as the intersection of the scopes of all feature classes. In practice, different approaches may be designed to accommodate various scopes in applications. A database image can be assigned to a subcluster if it falls within the scope of the subcluster, or if its distance to the border of the subcluster is less than a threshold.

A feature vector in feature class f_i that does not semantically belong to C_k may belong to $S_k^{f_i}$. To cluster the images under each feature class more precisely, heterogeneous features are used in determining the classification of the images in each $S_k^{f_i}$. An image is assigned to a semantic cluster only if all its features are represented in the cluster. Within each semantic cluster, the image is grouped into the subclusters represented by the templates matched by the image. Images that fall within the scope of more than one semantic cluster will be assigned to all the matched semantic clusters.

Outliers are the objects that are far away from all the templates and do not belong to any of the predefined semantic clusters; attempting to link the outliers with a cluster will cause inaccurate retrievals. Outliers are grouped into a special cluster, termed *other* cluster, denoted C_t, and are searched separately. An image V is assigned to cluster C_t if at least one of the feature vectors is not assigned to any semantic cluster.

We now consider the images shown in Figure 10-11(a). In this example, the image water is assigned to the cluster *water*, since it can be matched to at least one texture and one color template of the water cluster. Although the image wood can be matched to at least one texture template, it is not assigned to the semantic cluster *water*, since it does not match with any color templates of the water cluster. Similarly, in Figure 10-11(b), the image leaves is assigned to the cluster *leaves*, since it can be matched to at least one texture and one color templates of the leaves cluster. Although the image painting can be matched to at least one color template, it is not assigned to the cluster *leaves*, since it does not match with any texture templates of the leaves cluster.

The domain expert due to the changes in the database content may add new semantic clusters. In such cases, new images will be collected to generate the templates representing the new semantic cluster. Thus, given a set of existing semantic clusters $\{C_1,...,C_m\}$ and their corresponding scopes $\{S_1,...,S_m\}$, let S be the scope for the new semantic cluster C. The new semantic cluster C may be considered to be similar to the cluster C_i ($1 \leq i \leq m$) if the scope S overlaps with S_i. In such cases, the images that are grouped under cluster C_i and the images that are included in the *other* cluster may belong to the new semantic cluster C. These images must be regrouped following the steps outlined above.

Similarly, new templates may be added into a semantic cluster due to the additions of new images. In such cases, the scope of the semantic cluster will be changed. The new scope must be compared with the scopes of the other existing semantic clusters to determine similarity. Images that are grouped under similar semantic clusters and the images that are included in the *other* cluster must also be regrouped.

10.4 Querying on Heterogeneous Features

The query can be a still image or a video frame. An optional keyword, which describes the feature of interest, application domain, or semantics of the query, may also be given. The keyword can be used to narrow the scope of the features and images to be searched in the retrieving process.

If a keyword is attached to the query image and the keyword matches the name of a semantic cluster of an application, only images grouped under the specified semantic cluster will be considered in the retrieving process. Similarly, if only a feature class is specified, the retrieval can be based on the specified image features. Boolean combinations of the keywords can be supported. For example, given a query image containing its texture features and the keyword *floral*, the retrieval will be based on the texture features and the images within the floral cluster. If no keyword is attached to the query image, the retrieval will be based on the image features and the matched semantic clusters for the query.

Upon receiving a query q, a set of subqueries $\{q_1,...,q_n\}$ are generated from the query, with each subquery being a feature vector corresponding to a feature class in F. We then compare the subqueries to the templates, a set of matched templates T_{q_i} is selected for subquery q_i based on the approach specified in the previous section.

Let $T_{f_i}^{C_k} = \{ t_1^{k,f_i} ,..., t_2^{k,f_i} \}$ denote the set of templates representing cluster C_k ($1 \leq k \leq m$) in the feature class f_i. A semantic cluster C_k is chosen if, for every subquery q_i, at least one matched template can be found in C_k. The retrieval algorithm then searches the corresponding subclusters of the matched templates within the chosen semantic clusters to retrieve a ranked list of relevant images for the query.

Let q contain a set of subqueries $q_1,...,q_n$ and a set of chosen semantic clusters be $C_q = \{C_1,...,C_h\}$. Let each q_i match with multiple independent templates $t_1^{k,q_i},...,t_{m_i}^{k,q_i}$ within cluster C_k and $I_{t_j}^{k,q_i}$ be the set of relevant images retrieved from the subclusters represented by template t_j^{k,q_i}. The list of relevant images $r(q_i, C_k')$, within cluster C_k for subquery q_i, is defined as the union of all relevant images in $I_{t_j}^{k,q_i}$, $1 \leq j \leq m_i$:

$$r(q_i, C_k') = \bigcup_{j=1}^{m_i} I_{t_j}^{k,q_i} \tag{10-1}$$

Since the images sought contain features similar to all subqueries, the set of relevant images within cluster C_k for query q, denoted $r(q, C_k')$ is calculated as the intersection of $r(q_i, C_k')$, $i = 1,...,n$:

$$r(q, C_k') = \bigcap_{i=1}^{n} r(q_i, C_k') \tag{10-2}$$

The relevant images returned for each chosen semantic cluster are then unioned to become the final list of relevant images $\Re_q$:

$$\Re_q = \bigcup_{j=1}^{h} r(q, C_j') \tag{10-3}$$

The retrieved images in $\Re_q$ may be further ranked using the multi-layer neural network model to be introduced in the next section. Note that, if no semantic cluster is selected for the query, the special cluster *other* will be searched. In such cases, the search will be performed sequentially on the special *other* cluster.

Since each semantic cluster is formed based on the heterogeneous features, this approach can reduce the percentage of false-positives in the retrieval process. On the other hand, a small percentage of misses may be increased for some queries.

Figure 10-13 shows the relationships between a query, templates and database images. In this example, a subquery with texture features and a subquery with color features are extracted from the query image. Cluster *floral* is chosen for the query, since its subqueries match at least one texture template and one color template within the cluster. Thus, only images belong to the floral semantic cluster are considered in the retrieval process. Then the corresponding subclusters of the matched templates are searched. The images retrieved from the subclusters are then intersected to yield a list of relevant images within the floral cluster for the given query.

RANKING IMAGES USING HETEROGENEOUS FEATURES

Given a query image, a set of relevant images can be selected based on individual features, as discussed in the previous sections. However, a final ranked set of similar images to the query must be derived by merging the results obtained from individual features. In this section, we will discuss the issues related to the relative importance of individual heterogeneous feature classes and how the current methods consider them.

Various features are of differing importance in determining the ranking of database images. The ranking process must assign weights to the results obtained from the subqueries. In human perception, these feature classes do not have the same importance in distinguishing images. Thus, visual retrieval systems must consider the degree of importance of each feature class to determine the overall similarity of database images to the query image.

Most of the current image content-based retrieval systems such as Photobook [22], QBIC [10], Virage [1], and NETRA [19] use a weighted linear method to combine the similarity measurements of different feature classes. That is, given the similarity measurements $z_1,...,z_l$ of a database image to a query image with respect to feature

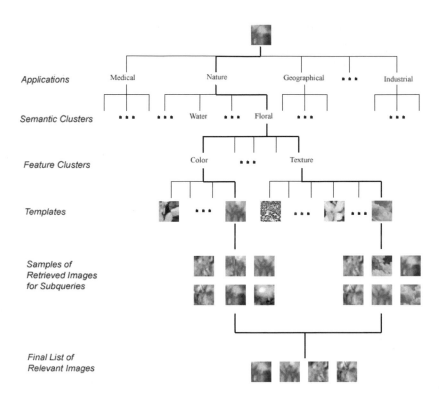

Applications

Semantic Clusters

Feature Clusters

Templates

Samples of
Retrieved Images
for Subqueries

Final List of
Relevant Images

Figure 10-13 Query, templates and database images (©1998 IEEE)

classes $f_1,...,f_l$ and the corresponding weights $w_1,...,w_l$, the overall similarity is calculated as $\sum_{i=1}^{l} w_i z_i$. Such a weighted linear combination of similarity measurements is called "*linear merging*". Studies have shown that various feature classes are not necessarily linearly-related. For example, the similarity measures of color and texture do not generally show a linear exchange. An important step in merging heterogeneous features is to apply a nonlinear transformation on each similarity measurement to make them more commensurate. One such transformation is the rank of retrieved images when ordered by the relevance score. Rather than merging scores (which differ in range, mean, and variance), ranks are merged to get new measures so that the retrieved images may be ordered accordingly [21]. Usually, the minimum, average, or maximum of the ranks are used. This method has achieved good results for document retrieval and has been applied to merge different texture feature classes [18]. Fagin provides an efficient algorithm to merge results when a monotonic function is used to merge individual matched values [8, 9]. In his method, the total number of retrieved objects in evaluating the query is sublinear in the database size. His algorithm uses fuzzy logic and merges the grades, or similarity values (fuzzy values) of the objects.

It is possible that for one feature class f_i, the images with close ranks have very similar feature vectors, whereas for another feature class f_j, the images with close ranks have very different feature vectors and look different with respect to that feature class. For example, the top retrieved images based on the color feature class might be very similar, but the top retrieved images with respect to texture might have very different feature vectors. Assume that image A has the ranks 5, 7 and image B has the ranks 10, 6 with respect to color and texture, respectively. In this case, images A and B are very similar to each other with respect to color (despite their difference in the rank), but since image B has a higher rank based on texture than image A (the lower the number, the higher the rank), it should be ranked higher than A based on both features. However, as Table 10-1 shows, using either the minimum, maximum, or average of ranks results in higher overall rank for image A. Therefore merging solely on the basis of rank may not guarantee the correct ranking of the images.

Image	Color Rank	Texture Rank	Overall Rank		
			Min	Max	Ave
A	5	7	5	7	6
B	10	6	6	10	8

Table 10-1 An example of ranks with respect to heterogeneous features

Another important issue is the weight of the feature classes. In many existing systems, the *user* would specify the weights. For example, the user would say "retrieve images using 50% of color feature class, 30% of texture feature class, and 20% of shape feature class". However, users do not intuitively sort images by similarity using this kind of language. In particular, as the number of feature classes increases, intuition about how to pick relative weightings among features is lost [23]. Also, since all the measurements of similarities are usually in the range of zero to one, the common practice is to normalize the distance measurements and convert them into similarity values. The normalization process for each feature class will be different, because of the difference in the feature spaces and distribution of feature vectors.

MERGING HETEROGENEOUS FEATURES

Neural networks have been used in many areas such as pattern recognition, computer vision, and control systems. Heterogeneous features can be merged using a multi-layer perceptron neural network. The input to the neural network is the set of measurements $\{z_1, z_2, ..., z_l\}$ between a query image q and a database image D with respect to the feature classes $f_1, f_2, ..., f_l$. If the features of images q and D are similar, the output of the neural network must be close to 1. However, if q and D are not similar, the output will be close to 0. The network implements an output function $o = F\{z_1, z_2, ..., z_l\}$. Cybenko [5] has proven that to approximate a particular set of output functions using multi-layer neural network, at most two hidden layers are needed. Arbitrary accuracy is obtainable given enough units per layer. It has also been proven that only one hidden layer is enough to approximate any continuous function [6, 15]. Consequently, this neural network model does not restrict the relationship between the feature classes to linear relationships.

To train the neural network and find the weights, a set of images that are visually similar (positive examples) and a set of images that are not similar (negative examples) are required. The system then finds the similarity (or dissimilarity) between images with respect to different feature classes and feeds these measurements to the neural network. Once the network has been trained, the feature classes will have the proper weights, so they can be used in merging heterogeneous features. In this approach, the user need not worry about assigning weights to feature classes. The weights are based on human perception. In comparison, in the neural network merging, each feature class can be measured in terms of similarity or distance independent of others, and the combination of these similarities and distances can be directly fed into the neural network. In this respect, the neural network model is more flexible than the previous approaches.

Given a query image q, let $\Re_q$ be the final list of relevant images. The goal is to rank the relevant images in $\Re_q$ with respect to all the heterogeneous features. Using the trained neural network, the similarity between the query image q and each image in $\Re_q$ can be found based on the heterogeneous features. In each step, the similarity measurements of heterogeneous feature classes of the query image q and an image in $\Re_q$ are fed into the neural network. The output value of the network will be the similarity between the two images. The images in $\Re_q$ are sorted and ranked based on the outputs of the neural network.

Note that the trained neural network can be used to determine the similarity between the query and all the images in the database. However, such an approach will result in a linear search of all images in the database, causing inefficient querying. By combining the semantics-based clustering with querying with the neural network model, both effectiveness and efficiency can be achieved.

INTEGRATION OF RELEVANCE FEEDBACK

Relevance feedback, first introduced in the mid-1960s, has been used extensively by the information retrieval community to improve a query's performance [1, 10]. This process modifies the query formulations based on the user's judgment of the initial retrieved documents [25]. The main idea consists of adding or subtracting terms that have been identified as relevant or irrelevant by the user, and altering term weights in a new query formulation. The system emphasizes terms included in the previously retrieved relevant documents and de-emphasizes terms included in the previously retrieved irrelevant documents. This query alteration process obtains an optimal query in anticipation of retrieving more relevant documents and fewer irrelevant documents in a later search.

Query refinement can be implemented using an interactive graphical display technique to establish communication between the system and users. After the initial search, a list of initial retrieved images are displayed to the user, and the mouse pointer can be used to designate images as relevant or irrelevant to the user's needs. The query process module extracts various features from each relevant or irrelevant image and forwards the features to the query refinement module.

Let $F=\{f_1,...,f_n\}$ be a set of features, where $f_i \in \{texture,color,shape,etc.\}$. Let $q=\{q_1,...,q_n\}$ be the original query and $q_i \in F$. Let I_r be the set of relevant images retrieved by the initial search and I_{ir} be the set of irrelevant images retrieved by the initial search as determined by the user. Given an image $g=\{g_1,...,g_n\}$, where $g \in I_r \cup I_{ir}$ and $g_i \in f_i$. Let $q_i^0 = q_i$. The query refinement module constructs a refined subquery q_i^k ($k \geq 1$) by adding features in the relevant images and subtracting features in the irrelevant images:

$$q_i^k = \alpha q_i^{k-1} + \beta \frac{1}{N_r} \sum_{g \in I_r} g_i - \gamma \frac{1}{N_{ir}} \sum_{g \in I_{ir}} g_i \,. \tag{10-4}$$

where N_r is the number of the known relevant images in I_r and N_{ir} is the number of the known irrelevant images in I_{ir}. The weights α, β, and γ must be determined experimentally.

The metasearch agent then takes the modified subqueries $q_1^k,...,q_n^k$ to produce a set of relevant database sites, D_i, for each subquery. The results are then merged to generate a final list of relevant databases. *Selection precision* measures the accuracy of database selection against manually verified relevant databases. The refinement process was applied to the nine queries with the lowest selection precision to evaluate the relevance feedback . The average selection precision of the nine queries before refinement was 0.47. The selected databases returned by the refinement process were compared with the results of the initial selection performed using the original query. Note that both initial and refinement selections were conducted against all databases. The changes in precision are used to evaluate the effect of the refinement process. For each query, the top 5 retrieved images returned by the top 4 selected databases (total of 20 images) were inspected and designated as relevant or irrelevant. In determining the relevancy of an image, similarity measures instead of user judgments were used to achieve consistent results. Images that had a similarity greater than or equal to 0.80 with respect to both texture and color subqueries of the given query were considered to be relevant to the query, the rest of the images were considered to be irrelevant. The thresholds α, β, and γ were set to 0.5, 1.0, and 0.5, respectively. After the refinement, the average selection precision of the 9 queries reached 0.58, which is a significant improvement with respect to the selection precision of 0.47 before the refinement. The result demonstrates that relevance feedback represents a powerful process for improving the output of the retrieval system.

10.5 Conclusion

This chapter has presented techniques for system design, indexing, and querying on semantic properties of visual data. The distributed architecture contains a metaserver, which is composed of the metadatabase, the metasearch agent, and the query manager. The extraction of semantic features as in scenes and semantic clustering provides a mechanism to retrieve query results that are compliant with the human perception. Semantics-based clustering can categorize the images into different clusters based on their heterogeneous features. Moreover, templates increase the efficiency of the retrieval process by reducing the number of databases to be searched. A neural network model is used to successfully rank the images retrieved based on individual features. By successfully combining the semantics-based and template-based clustering, retrieval results are improved in terms of efficiency and accuracy. In addition, there are also many other research issues related to the design of visual database systems, such as disk scheduling and allocation strategies, buffer management for prefetching and maximizing buffer utilization, multimedia synchronization, collaboration between clients and servers, and QoS metrics for continuity and synchronization specifications.

Acknowledgement

Some of the work conducted by the authors was supported by NSF grants.

References

[1]   J.R. Bach, C. Fuller, A. Gupta, A. Hampapur, B. Horowitz, R. Jain, and C.F. Shu. "The virage image search engine: An open framework for image management". In *Proceedings of SPIE, Storage and Retrieval for Still Image and Video Databases IV*, pages 76–87, San Jose, CA, USA, February 1996.

[2]   M. Bowman, P. Danzig, D. Hardy, U. Manber, and M. Scwartz. "Harvest: A scalable, customizable discovery and access system". *Technical Report CU-CS732-94, Department of Computer Science*, University of Colorado-Boulder, 1994.

[3]   S. Chang, J. Smith, M. Beigi, and A. Benitez. "Visual Information Retrieval from Large Distributed Online Repositories". *Communications of the ACM*, 40(12): 63–71, December 1997.

[4]   W. Chang and A. Zhang. "Metadata For Distributed Visual Database Access". In *Second IEEE Metadata Conference*, Silver Spring, MD, September 1997.

[5]   G. Cybenko. "Continous valued neural networks with two hidden layers are sufficient". *Technical report*, Department of Computer Science, Tufts University, Medford, MA, 1988.

[6]   G. Cybenko. "Approximation by superimposing of a sigmoidal function". *Mathematics of Control, Signals, and Systems*, 2: 303–314, 1989.

[7]   J. P. Eakins, "Automatic image content retrieval - are we getting anywhere", In *Proc. of Third International Conference on Electronic Library and Visual Information Research*, pp. 123-135, May 1996.

[8]   R. Fagin. "Fuzzy queries in multimedia database systems". In *Proc. 1998 ACM SIGACT-SIGMOD-SIGART Symposium on Principles of Database Systems*, 1998.

[9]   R. Fagin. "Combining fuzzy information from multiple systems". *Journal of Computer and System Sciences*, 58: 83–99, 1999.

[10]   M. Flickner, H. Sawhney, W. Niblack, J. Ashley, Q. Huang, B. Dom, and et al. "Query by Image and Video Content: The QBIC System". *IEEE Computer*, 28(9): 23–32, September 1995.

[11]   L. Gavarno, H. Garcia-Molina, and A. Tomasic. "The Effectiveness of Gloss for the Text Database Discovery Problems". In *Proceedings of the ACM SIGMOD'94*, pages 126–137, Minneapolis, May 1994.

[12] A. D. Gordon. Classification Methods for the Exploratory Analysis of Multivariate Data. Chapman and Hall, 1981.

[13] W. I. Grosky. "Multimedia Information Systems". *IEEE Multimedia*, 1(1): 12–24, 1994.

[14] V. N. Gudivada and V. V. Raghavan. Special Issue on Content-Based Image Retrieval Systems. *IEEE Computer*, 28(9), September 1995.

[15] K. Hornik, M. Stinchcombe, and H. White. "Multilayer feedforward networks are universal approximations". *Neural Networks*, 2: 359–366, 1989.

[16] R. Jain and S.N.J. Murthy. "Similarity Measures for Image Databases". In *Proceedings of the SPIE Conference on Storage and Retrieval of Image and Video Databases III*, pages 58–67, 1995.

[17] B. Kahle and A. Medlar. "An Information System for Corporate Users: Wide Area Information Servers". *ConneXions - The Interoperability Report*, 5(11): 2–9, November 1991. WAIS is accessible at http· //www.wais.com/newhomepages/techtalk.html.

[18] F. Liu and R. Picard. "Periodicity, directionality, and randomness: Wold features for image modeling and retrieval". *Technical Report 320*, MIT Media Laboratory Perceptual Computing, 1996.

[19] W. Y. Ma and B. S. Manjunath. "NETRA: A toolbox for navigating large image databases". In *IEEE International Conference on Image Processing*, 1997.

[20] B.S. Manjunath and W.Y. Ma. "Texture Features for Browsing and Retrieval of Image Data." *IEEE Transactions on Pattern Analysis and Machine Intelligence*, 18(8): 837–842, August 1996.

[21] T. Minka. An image database browser that learns from user interaction. *Master's thesis*, MIT, 1996.

[22] A. Pentland, R. Picard, and S. Sclaroff. "Photobook: Tools for Content-based Manipulation of Image Databases". In *Proceedings of the SPIE Conference on Storage and Retrieval of Image and Video Databases II*, pages 34–47, 1994.

[23] R. Picard. "A society of models for video and image libraries". *Technical Report 360*, MIT Media Laboratory Perceptual Computing, 1996.

[24] E. Remias, G. Sheikholeslami, A. Zhang, and T. F. Syeda-Mahmood. "Supporting Content-Based Retrieval in Large Image Database Systems". *The International Journal on Multimedia Tools and Applications*, 4(2): 153–170, March 1997.

[25] J. Rocchio. "Relevance Feedback in Information Retrieval". In *The Smart System - experiments in automatic document processing*, pages 313–323. Prentice Hall, Englewood Cliffs, NJ, 1971.

[26] S. Sclaroff, L. Taycher, and M. La Cascia. "ImageRover: A Content-based Image Browser for the World Wide Web". In *IEEE International Workshop on Content-based Access of Image and Video Libraries*, pages 2–9, 1997.

[27] G. Sheikholeslami, W. Chang, and A. Zhang. "SemQuery: Semantic Clustering and Querying on Heterogeneous Features for Visual Data". IEEE Transactions On Knowledge and Data Engineering, 14(5): 988–1002, Sep./Oct. 2002.

[28] G. Sheikholeslami, A. Zhang, and L. Bian. "Geographical Data Classification and Retrieval." In *Proceedings of the 5th ACM International Workshop on Geographic Information Systems*, pages 58–61, Las Vegas, Nevada, November 1997.

[29] J. R. Smith and S. Chang. "Transform Features For Texture Classification and Discrimination in Large Image Databases". In *Proceedings of the IEEE International Conference on Image Processing*, pages 407–411, 1994.

[30] J. R. Smith and S. Chang. "VisualSeek: a fully automated content-based image query system". In *Proceedings of ACM Multimedia 96*, pages 87–98, Boston MA USA, 1996.

[31] J. R. Smith and S. Chang. "Visually Searching the Web for Content". *IEEE Multimedia*, 4(3): 12–20, 1997.

[32] Y. Song and A. Zhang, "Monotonic Tree", In *the 10th International Conference on Discrete Geometry for Computer Imagery*, Bordeaux, France, April 3-5, 2002.

[33] Y. Song and A. Zhang, "Analyzing Scenery Images by Monotonic Tree". ACM Multimedia Systems Journal, Vol. 10, No.3, 2002.

[34] J. Wang, W. Yang, and R. Acharya. "Color Clustering Techniques for Color-Content-Based Image Retrieval". In *the Fourth IEEE International Conference on Multimedia Computing and Systems (ICMCS'97)*, pages 442–449, Ottawa, Canada, June 1997.

[35] W. Wang, J. Yang, and R. Muntz. "STING: A Statistical Information Grid Approach to Spatial Data Mining". In *Proceedings of the 23rd VLDB Conference*, pp. 186–195, Athens, Greece, 1997.

[36] A. Zhang, W. Chang, G. Sheikholeslami, and T. Syeda-Mahmood. "NetView: Integrating Large-Scale Distributed Visual Databases". *IEEE Multimedia*, 5(3): 47–59, July-September 1998.

Part II
Applications

11 DIGITAL LIBRARY

Dr. Michael S. Brown

In 1993, the United States National Office for High Performance Computing and Communications (HPCC) declared digital libraries a *National Challenge Application*. In response the National Science Foundation, together with other government agencies, launched the Digital Library Initiative (DLI) which funded six large-scale universities projects between 1994 and 1998. These projects targeted key digital library research challenges with particular emphasis on large scale, distributed collections. The DLI Phase-2 (DLI-2) is currently funding many more projects, including smaller assignments, to continue digital library research.

In this chapter, we give an overview of the Digital Library as an application and as a research challenge. We discuss three of the original DLI projects, their goals and research outcomes. We follow this with an overview of some of the current on-going DLI-2 projects and the challenges they address with respects to multimedia management and retrieval.

11.1 Digital Library Defined

The term *Digital Library* is synonymous with *Electronic Library*. Although the trend now has shifted towards using *digital library*, the term *electronic library* is still found in European literature and older published materials. There are many proposed definitions for digital libraries. The following definition from the Institute for Electronic Library Research [14] provides a nice consensus:

> An electronic library (known in the US as a 'digital library') is an organized and managed collection of mixed media materials in digital form, designed for the benefit of a particular user population, structured to facilitate access to its contents and equipped with aids to navigation of the global information network.

A slightly modified definition proposed by Nurnberg [19] categorized digital libraries into three broad components, resulting in the following definition:

Digital libraries consist of the following elements:

Data
A coherent entity of digital data that can be presented to the user

Metadata
Any information, inherent or derived, used to organize and manage data

Processes
Functions performed over library elements, including search, analysis, and visualization

As these definitions suggest, a digital library is not merely a repository of digital media, but consists of the *entire system* which stores and manages the repositories, including processes for indexing and harvesting useful information, along with the services needed to effectively navigate, analyze and visualize the data stored within (see Figure 11-1).

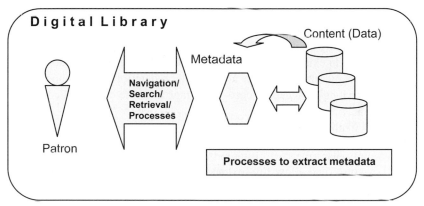

Figure 11-1 Digital library consists of more than the underlying content. The digital library involves the total system necessary to process, index, and disseminate the underlying content.

This idea of the digital library functioning as a system is comparable to our understanding of a physical library. A room full of books, microfilm, and other media does not constitute a library. It is the institution as a whole, including the organizational infrastructure, with the classification scheme, physical layout, loaning and fee policies, and so on, that make the actual library. The effectiveness of these institutions is judged by their ability to help patrons find and retrieve materials in a timely and relatively easy manner. The same calculation of effectiveness can also be applied to digital libraries. Digital libraries not only store the content, but also are designed to aid patrons in their access, search, and even understanding of the underlying information.

The last decade has seen a sharp increase in activity in the area of digital libraries. This is in part due to significant advances in networks, digital storage media, and the incredible growth and usage of the World Wide Web. As Schatz states, "[The] public awareness of the Net as a critical infrastructure in the 1990s has spurred a new revolution in the technology for information retrieval in digital libraries" [23].

Digital libraries use technology to create, organize, and navigate information in an electronic medium. Many materials are now electronically prepared and can be stored directly in a digital archive for reference, search, and distribution. This is quite common with technical articles which are created electronically and are made available by publishers in hardcopy and in their digital format. However, our libraries are far from all digital and significant efforts are underway to digitize existing physical materials currently not available in electronic format. One flagship effort to digitize a massive number of paper-based materials was performed by the University of Michigan's JSTOR (journal store) project. This project digitized older journals and used OCR to convert them to on-line documents that could be searched and retrieved. While pioneering for its day, this is now a common practice at many libraries and national institutions. The driving goals behind the JSTOR project are representative of many digital libraries. The goal is not to replace physical libraries but to augment traditional libraries collections. The JSTOR project goals included[16]:

- To provide a comprehensive digital archive of scholarly literature
- To dramatically improve access to this literature
- To fill gaps in existing library collection and old publications backfiles
- To address preservation issues and long-term deterioration of paper copy
- To reduce long-term capital costs of storage and care of journal collections

Text-Based and Multi-media Collections
On the surface, a digital library of documents seems a straightforward database application. Simply place the documents in the database, index on fields such as title, author, and keywords, and allow full-text searches. The user makes a query, and the database returns the documents satisfying the query. What is the challenge?

There are many questions to consider. What granularity of the documents elements can a patron search and retrieve? Perhaps only a table is desired and not the full document. There are non-textual semantics in the figures and tables. How can this non-textual data be indexed and searched? In addition, are there semantics beyond the words, such as concepts inherent to the collection that may not be encoded in the abstract and keywords?

Another consideration is how to make collections accessible to a large audience of people. For example, how do people without expertise in a particular discipline or understanding of the discipline's jargon effectively search the database ?

On a larger scale what are the fundamental challenges of accessing the digital library data? What are the most effective ways to remotely access content? How can distributed collections of content be access such that they appear as a single entity. How will this inter-operability be achieved given different content repositories use different query mechanism, with different document encoding, and are distributed all over the world?

Like their physical counterparts, digital libraries are not confined to text-based materials. Instead, they host a variety of mixed media, from images, to video clips, to geo-spatial items and materials that are exist only in digital form and have no analogue counterpart. However, there is a stark difference between the way physical and digital materials are managed. In a physical library, an indexer, usually a librarian, annotates incoming materials and organizes them based on a well-defined classification scheme. It is only after this "indexing" stage has occurred that the patron can gain access to the material. This manual annotation and indexing requires enormous amounts of human labor and is not feasible for digital library content, where data can be generated at incredible rates. Thus, one key challenge for multimedia digital libraries is the ability to automatically index multimedia content. In addition, traditional text-based queries may not be sufficient for retrieval. New interfaces and mechanisms for query, retrieval, and evaluation are necessary.

Thus there are many issues and challenges that must be addressed by digital libraries providers. Solutions to these challenges must be realized at all levels of the digital library research, from the underlying systems-level infrastructure, to content browsing and visualization. Not surprisingly, in 1993 the United States National Office for High Performance Computing and Communications (HPCC), declared digital libraries a *National Challenge Application*. A year later, the Digital Library Initiative was established to fund research to address digital library research challenges.

11.2 Digital Library Initiative (Phase I)

The Digital Library Initiative was a joint effort between three major US funding agencies, the National Science Foundation (NSF), Advanced Research Projects Agency (ARPA), and the National Aeronautics and Space Administration (NASA). Their goal was to fund six university sites to develop test beds to investigate research challenges in the digital library community. The idea behind this strategy was that these projects findings would help direct the future of digital libraries development, much like ARAPNET research helped shape today's Internet. Also, these projects were to establish test beds for use in further digital library research.

One of the themes of the initial DLI was the idea of seamless access and search, across *distributed collections* or the notion of federated heterogeneous repositories -- underlying archives, from different sources and in different formats would appear as one single entity to the accessing patron. With this challenge came the need for "deep-semantic interoperability" [18]. This implied that even very vastly different repositories could be tied together through technology that provided a deep understanding of content and a means to link derived metadata across heterogeneous sources.

The DLI funded six universities projects were as follows:

Institution	Project Title
• Carnegie Mellon University	"Full-content search and retrieval of video"
• University of California at Berkeley	"Work-centered digital information services"
• University of California at Santa Barbara	"Spatially-referenced map information"
• University of Illinois at Urbana-Champaign	"Federating repositories of scientific literature"
• University of Michigan	"Intelligent agents for information location"
• Stanford University	"Interoperation mechanisms among heterogeneous service"

These projects included a broad range of topics from video self-annotation to developing a comprehensive geo-spatial archive from heterogeneous sources. While all of the projects have made significant contributions to digital library research, for brevity, this chapter overviews three of the initial DLI projects. Addressed are: Carnegie Mellon University's (CMU) *Infomedia* project, which focused on issues concerning video metadata extraction, search and retrieval; the University of Illinois at Urbana-Champaign's (UIUC) project of federating repositories between different disciplines; and Stanford's Infobus project, which focused on providing a common interface for search and retrieval of distributed content via heterogeneous access protocols and modes of delivery. These three projects were selected because they provide insight into digital library problems at different operational levels. CMU is attacking problems inherent to the multimedia content. UIUC investigates deriving and linking deep semantics between different content, while the Infobus project examines the low level infrastructure of the underlying support systems providing content. We refer the reader to the DLI web-page for an overview of each project, their goals, and their contributions [12].

11.3 DLI Projects

11.3.1 Carnegie Mellon University
"Full-Content Search and Retrieval of Video"

The CMU project, called the *Infomedia Digital Video Library* project, established an on-line video digital library that allowed users to search, retrieve, and explore a repository of videos. The project used two main corpuses for content, "broadcast news" footage and "scientific documentaries". This consisted of not only the final content used in an actual broadcast, but also accompany raw footage from which the broadcast programs were produced. The research focused on means to efficiently search, retrieve, view, and understand video footage from the repository. One potential audience of this work was the entertainment and broadcast industry, where the ability to quickly find and explore old footage could greatly improve the time and effort needed to create new content. In this session we focus on two of their main challenges: (1) automatically deriving useful semantics from the video to provide intuitive and effective searching of this data, and (2) provide a useful way to present the results for evaluation by the user. (see Chapter 4).

EXTRACTING METADATA FROM VIDEO

Figure 11-2 gives an overview of the Infomedia's metadata extraction from video footage. Speech recognition, together with video processing, extracts useful information that can be queried by the user.

Segmentation of original footage

For most of the footage, the retrieval of the video in its entirety (such as a whole two hour documentary) was impractical. Therefore the video footage was segmented into *stories* that were on average only a few minutes in length. This granularity made the footage more manageable, and a simple hierarchical structure made it possible to obtain the full video from the segments.

Annotation via Speech Recognition

The Infomedia project observed that a typical video production generates about 50 to 100 times more video content than is actually used in a broadcast program [28] (e.g. 50-100 hours of video footage are used to create an hour of broadcast footage). While the resulting broadcast material is generally annotated with a transcript (closed-captioned), the rest of the raw footage is not annotated. Thus, an enormous amount of unused information remains imbedded in the audio track of the raw video footage.

To extract this data, speech recognition software is used to automatically derive the spoken words that accompanied the video. This extraction has to be performed amidst sound tracks, background noise, etc. The automatic text extraction was performed using the CMU Sphinx III speech recognition system. The CMU's Sphinx III speech recognition system was capable of transcribing spoken words with an error rate around 24% [27]. For some footage, such as newscasts, reorganization rates could be improved by incorporating information into the speech recognizer such as words used in the "news of the day", based on the probability that those words would appear in the audio track [28].

The automatic transcription proved a powerful technique to derive rich metadata to associate with the video. The transcribed text could be further processed to provide more semantics, such as segment keywords, or for assigning a "title".

Image Semantics, Key Frames, and Retrieval
While the audio provided textual information, the CMU project also looked at ways of deriving semantics from the video imagery. Their video was encoded using MPEG-1 at 30 frames per second with a resolution of 352x240 pixels. They used image processing techniques to extract a set of shots, or key-frames. These key-frames could then be used for queries and searches. Using image-similar techniques video segments with similar key-frames can be retrieved.

Finding Key frames
Key frames are found by bounding the video segment into regions where large changes occur. The CMU project used frame color histogram information metrics to determine where large frame changes took place (large changes in the histogram indicate a possible scene change). The key-frames are then taken to be the center of these bounded regions. The Infomedia project also used cinematic heuristics to determine better key frames. For instance they used optical flow to determine camera ego motion, selecting the key-frames that ended a pan sequence.

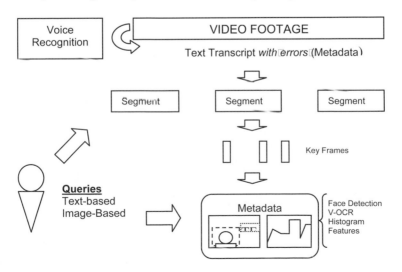

Figure 11-2 Infomedia automatic video process.

Key-frame Features

For non-textual retrieval, the users could apply searches that look for video clips containing faces. Key-frames were processed to see if they contained faces. This worked well for finding the news anchor who appeared very clear in well lit environments, but robust face recognition remains a challenging problem in computer vision for general imagery. However, even the ability to add simple recognition to the key-frames helps improve the semantics of the news clips.

In addition, color histograms were computed to allow the user to perform simple color matching image retrieval searches. A new color-clustering based histograms technique for retrieval was also developed [13], providing better performance than standard color histogram matching techniques.

Video OCR

Text appearing in the video can also be used to provide an insight into the video semantics. CMU developed a Video-OCR technique that could locate text regions in an MPEG-1 sequence [27]. This technique hypothesized that key frames with a cluster with sharp edges may contain text. Further processing filtered such probably regions and the extracted imagery was used in OCR software to determine the text. In news broadcast video, this technique could accurately identify 90 percent of the frames with text, and recognize 70 percent of the words in the identified frames. This provided additional textual information that could be used in text-based queries.

PRESENTATION

Video Skims

When a query is made relevant clips are returned based on keyword search and image feature search. As with text-based documents, the user needs to verify that these returned objects are relevant to the query. When retrieving well-structured documents, the user can generally look at document components, such as the abstract or figures and tables, to see if the data is relevant to their query. However, as video lacks this structure, making an evaluation of video footage may be a time consuming task, depending on the number of hits found from their query. For example, watching 1000 two minute segments is impractical, while fast-forwarding through a video makes it difficult for the viewer to comprehend content. In addition, useful information may be bypassed when intervals of frames are skipped while fast forwarding To address this problem, CMU developed what it termed *Video Skims*. These multimedia abstractions, used video and audio snippets along with text to create essentially a *video abstract* [15]. The skims contained only fragments of the original segment, using the extracted transcript to assign a texture head-line. The idea is that the user can watch the skim, which is significantly shorter than the original clip, to get the idea of the overall content. This can help reduce the time needed to evaluate the query's result.

SUMMARY
The CMU project provided an insight into how video metadata could be extracted and used. One key finding was that automatically derived textual transcriptions, even when erroneous, provided a powerful means of searching. They also began to focus on how to present such large quantifies of multimedia data to the user, developing novel approaches such as the video skims. In addition, they performed several user studies to get feed back on how video data could be presented and browsed. One finding was that users wanted an interface that allowed them to browse the entire collection, without having to specify any query. Such large scale video browsing remains an open challenge. The successful outcomes of this project led to its continued funding as a DLI-2 project, the Infomedia-II project.

11.3.2 University of Illinois at Urbana Champaign
"Federating Repositories of Scientific Literature"

The UIUC project addressed heterogeneous document search across multi-disciplined repositories produced by different publishers. The test bed supported full-text SGML documents, with associated articles, metadata, and bit-mapped figures. Their test bed corpus comprised five professional societies, including IEEE Computer Society and the American Institute of Physics Letters. The data was acquired directly from the publishers in SGML form. As of 1999, they had 63 journals containing 66,000 full-text articles [25].

Although the documents were SGML based, they were from different publishers and in significantly different technical areas. The goal of the project was to exploit the underlying SGML structure to provide a means for federated searching among the heterogeneous sources. One major task addressed by the UIUC project deep-semantic inter-operability. They were interested in ways that allowed experts in one area to effectively search for documents in another area, even though they were unfamiliar with the other disciplines' categories and terminology. In addition, this research investigated users' searching behaviors and how they used the results. This research helped determine ways future documents could be engineered for more effective publishing, usability, and retrieval.

Federated SGML repositories
SGML Granularity

SGML documents contain a rich markup that identifies the underlying document structure, which can be thought of as document components, such as title, abstract, text, tables, figures, and so on. The SGML encoding allows more granularity in the search and retrieval documents and the documents' components. For example, a search query could be performed only on tables, or figures, instead of the entire document. However, the tagging conventional can be differ among publishers. The UIUC project developed a canonical document structure which normalized these different tagging names for effective searching.

UIUC developed a search and retrieval software called "DELIVER" (Desktop Link to Virtual Engineering Resources) that searched the distributed repositories by text-queries. Returned articles contained active links between the bibliographies and other items in the test bed. The articles even contained links to article records outside the test bed, such as articles found in Inspec and Compendex databases.

Multiple-View Interface
UIUC also developed a Multiple-view interface, called IODYNE [23], to provide a novel search interface to the distributed repositories. Normal search and retrieval allow the user to ask a single query and then see the results. With a multiple-view interface, the user can integrate different types of queries from multiple sessions. This interface allows the user to drag-and-drop retrieved information from multiple queries to start other searches. In addition, IODYNE provides search suggestions from associated categories and from derived concept-spaces (explained below). This interface provides a much richer search and retrieval metaphor than current single query interfaces and may provided the basis for more intuitive and effective search engines.

Scalable Semantics, Concept-Spaces and Interspaces
Searching multi-discipline databases are challenging, not because of different publishers tagging markup, but also because of user domain expertise. *Applied physics experts* and a *computer scientist* may use different terms for a similar concept. These users need the ability to bridge the gap between concepts and jargon in different areas. This is inline with the deep-semantic inter-operability desired by the DLI.

The UIUC project addressed this problem with what they called the Interspaces and Scalable Semantics[23]. Figure 11-3 shows an overview of the Interspace idea. The Interspace provides a layout where each repositories represents a community's knowledge, with associated data, indexes and metadata. That is, a repository includes objects and indexes of a specific *user-knowledge space*. The challenge is to enable someone from outside the knowledge space to quickly understand it, by relating the knowledge-space to a more familiar space.

One way to approach this problem is to automatically derive concepts specific to a large collections of documents [8]. *Concept spaces* are computed by deriving patterns of nouns that appear frequently in documents [2, 3]. Concept spaces can be defined as the co-occurrence frequency between related terms within the documents of a collection.

Deriving concept spaces is a large task that requires an enormous amount of computing time to analyze the noun statistics in a large collection of documents. The UIUC project computed these concept-spaces using high-end supercomputers, with the computation taking days [23]. However, the final results are a derived set of deep semantics that can be associated with the collections.

The Interspace consists of concept spaces, user-domain categories and indexes, and the objects themselves (documents). UIUC provided an interface that allowedusers to navigate between knowledge-set by browsing the concept-space. Automatically linking different concepts is very difficult, so the navigation tools allowed the user to manually make connections between concepts [25]. This can serve to provide a vocabulary switching effect, where users can link up different terms that refer to the same concept between different subjects. One day ideally, this vocabulary switching will be performed automatically, linking together concepts of different knowledge sets to help facilitate deep understanding and searching among disciplines.

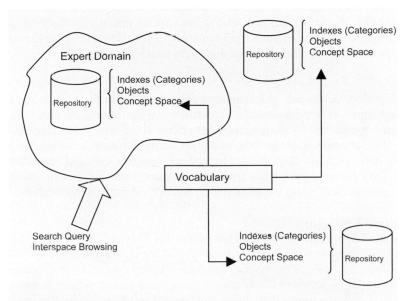

Figure 11-3 Overview of the Interspace idea. Content is index by categories and derived concept spaces. Concept spaces can be linked through vocabulary switching to provide powerful searches across expert domains.

Using Document Structure

Another part of the UICU project was to compare user habits to see how users searched for relevant documents, and how they used the retreived documents and its underlying components. [1]. Recall, that document components include elements such as abstracts, figures, equations, or bibliographies entries. The project compiled a short list of user habits in [23] as follows:

- To find articles of interest
- To determine relevance of the article before full retrieval
- To create a new document form sub-components of a full-text retrieval
- To find specific pieces of information (such as equations or fact)
- To find non-textual information such as figures and tables

They found that engineers had a common pattern that they used determined useful documents. They first read the tiles and abstracts, and then skimmed the section headings. They then examined components such a figures, summaries, and illustrations. Afterwards they would focus on key sections and then look at the references. However, after the initial skim, individual users had a very different approach to how they actually used the document. Users differed in how they selected bibliography entries, or generated new *surrogate documents* using key information from the original. These studies provided a useful insight to how future documents could be engineered, and with what sort of granularity the document should be structured. They also show a very interesting connection between the video skims and the idea of document skims, or document summaries.

Summary
The UCIC project developed a federated database of scientific knowledge that crossed disciplines. This work laid the foundation for the Interspace paradigm of semantic inter-operability. By deriving concept spaces along with multi-view search interfaces, they allowed users to drag-and-drop items between sessions. This is intended to help users quickly make connections between concepts in different disciplines. They envision a day when repository concept spaces can be automatically swapped to provide vocabulary switched search and retrieval across the database.

11.3.2 Stanford University
"Interoperation Mechanisms Among Heterogeneous Service"
Stanford's DLI project focuses on a distributed digital library infrastructure known as the *Infobus,* which addresses the issues of digital libraries content access, retrieval, and management at all levels of the system, ranging from user-interfaces to network transport. The Infobus assumes that clients and content service providers, create and deliver content independently. As a result, they use a variety of different protocols and delivery modes to their clients.

   The Infobus, which is built on the analogy of a hardware bus, tries to hide the underlying heterogeneous protocols and distributed complexity from the high-level clients. This project accomplishes this by building several plug-in services for processing distributed repositories of content. These services are provided using an object-based approach, where clients' interfaces talk directly to objects that provide uniform functionality, while dealing with complexities on the server's side. Figure 11-4 gives an overview of the infrastructure.

Objects and Services
The Infobus is composed of several objects and services. These communicate among one another to broker queries and information from clients to repositories. The following describes several of the key objects in the Infobus infrastructure.

Information Sources (IP)
These are the underlying content repositories, such as databases or Web-servers. These can be queried and return results. Note that they may have very different styles of interaction. HTTP web-engines get a query and return a full result. There is no notion of a true session taking place. That is, state is not maintained between the server answering the request and the low-level client asking for the information. Other services, such as Z39.50 (a standard information retrieval protocol) are stateful,

and establish a session between the low-level client and the information source. The Infobus infrastructure tries to hide such details from the high-level clients.

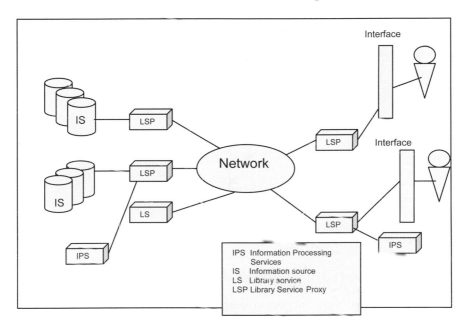

Figure 11-4 Stanford's Infobus architecture. IPS are "Information Processing Servives"; IS are "information Source"; LS are "Library Services"; LSP are "Library Service Proxy".

Library Service Proxy (LSP)
LSP wrap heterogeneous repositories such that they appear as a common source to the user. These define a clear set of search and retrieval methods that high-level user interface client can use. The LSP in turn connects to IS objects and hides low-level protocol details, such as stateful or stateless interactions with the actual information source. In addition, when a client uses a LSP to communicate with the Infobus, they are not aware if the LSP is providing an interface to a e-mail folder, to a multiple terabyte repository, or Web search engine [21]. Instead, they just see an object with a well-defined set of methods that can be invoked.

Library Services (LS)
These provide the necessary support functions to the other objects. This includes functions such as access rights management, metadata facilities, and query translations between the LSP and the different IS repositories.

Information Processing Services (IPS)

These are additional objects that can be invoked by the LSP to provide additional services, such as filtering the queried data, or serving as a cache to store information.

Example Query using the Infobus

Figure 11-5 shows an overview of a high-level client using the Infobus to retrieve

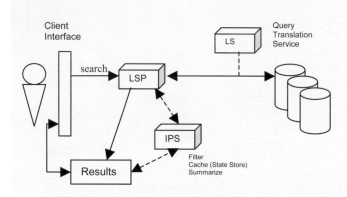

Figure 11-5 User query information flow using the Infobus

information. First the client interface creates a result collection and sends a pointer to this collection along with the text-based query to an LSP. The LSP then contacts the IS, possibly using a query translation services provided by an LS in the Infobus. The query results are then placed in the clients results object. An optional IPS can be allocated to help process the data. The IPS can act as a cache, to serve to filter data, or to provide other information processing services. As far as the client interface is concerned, it only needs to communicate with the LSP and examine results in the collection. All other complexities, such as the IS query, query translation, and the manner in which the data is processed are hidden. Providing this simple and uniform interface on top of a complicated underlying structure is the heart of the Infobus research.

Deployment of the Infobus

While the figure 11-5 provides a simplistic pictorial overview of object interaction in the Infobus, it does little justice to the massive undertaking in coding logistics and software development that was needed to realize the Infobus system. Distributing objects across wide-area-networks and is not easy and much of the technology the Infobus incorporated is still maturing. The Infobus system used a CORBA-based object system using Xerox PARC's Inter-Language Unification (ILU) implementation. Objects were implemented in C++, Python and Java. The initial project made significant progress developing the Infobus infrastructure and is being continued by the DLI-2 InterLib project, which is one of the largest DLI-2 projects.

SUMMARY

The Infobus project developed a set of well-defined objects and services to work together to hide the underlying complexity of distributed heteogenous repositories. These plug-in objects provide a modular approach to broker information between ever-evolving content providers and clients. Providing a common interface through which distrubted objects can help these collections appear as a single seamless entity and promises to make the access of digital content much easier to client applications.

11.4 DLI-Phase 2

After the success of the initial DLI projects, the NSF and other agencies, including the Library of Congress and the National Endowment of the Humanities, started a new round of funding called the Digital Library Initiative Phase 2 (DLI-2), which started in 1998 and is currently funding projects. The goal of the DLI-2 is to finance a broader range of projects and to address key issues in digital library research with the emphasis on more specialized collections. The DLI-2 project targeted three categories of Human-Centered, Content-Centered, and System-Center research, as well as funding for the establishment of digital libraries of test bed to support large research efforts.

This section will briefly describe the goals and example topics in each research category and describe projects addressing these issues. There are over 40 projects and not all are mentioned, furthermore, the ones listed may not fit into a single research category. Our classification may not reflect the views of the projects' investigators and is only intended to give the reader an overview of current research. We highly encourage interested readers to visit the DLI-2 web-site for more comprehensive information and direct access to each funded DLI-2 projects homepage [18].

11.4.1 Human-Centered Research

Human-Centered research focuses on ways digital libraries to will help users create, seek, and use information for technical research. Some of the key issues include:

- new methods and algorithms for information discovery and retrieval
- intelligent user-interfaces
- collaboration technologies
- human-computer interface issues
- economic and social impacts of the digital libraries

Projects addressing these issues include Arizona's *High-Performance Digital Library Classification Systems* which continues to focus on ways to automatically derive concept-spaces and other classification systems from large text-based collections. Stanford's *Encyclopedia of Philosophy,* which is designing a collaborative system to allow philosophy scholars a means to collaboratively write, maintain, and analyze newly published work in their area, while building an evolving working reference. Another much larger collaborative project, is the *Perseus Project*, part of Tuft University's *A Digital Library for the Humanities*. The Perseus Project has developed a scalable digital library for storing a wide-collection of scholarly literature about primary source materials. This work allows scholars from all over the world to collaboratively tag and cross-reference a huge collection of humanities materials.

Other projects include University of Washington's *Automatic Reference Librarians* for the World-Wide-Web which investigating intelligent software agents that can search the web to find high-quality information about complex technical topics.

11.4.2 Content and Collection-Based Research
The Content and Collection-Based category is intended to help mature the digitization and content management of new and novel collections. These projects focused on smaller, specific collections, that were currently not part of routine digitization efforts. Topics include:
- efficient acquisition for representation and preservation
- metadata encoding and preservation
- interoperability of content and objects
- domain-specific content
- development of new economic models addressing emerging electronic media

Current projects in this area include University of California's at Los Angeles *Cuneiform Digital Library Initiative*, which is developing a virtual collection of cuneiform items (stone tables). This work focused on creating novel media representations that integrates high-resolution images and 3D scans of cuneiforms with expert created markup annotation and cross-referencing. A similar undertaking is being performed by the University of Kentucky's *Digital Atheneum* project which is developing tools and techniques to provide old English scholars with novel ways to analyze the content of badly damaged manuscripts housed at the British Library.

Michigan State University's *A National Gallery of Spoken Word* is developing a content specific digital library of historically significant voice recordings and providing a fully searchable and free on-line collection. University of California Davis is developing a digital archive of Folk Literature using special textual transcription of spoken word narratives genres, including lyric poetry, folktales, and proverbs specific to Sephardim Jews. The University of Hawaii is developing the *Shuhai Wenyuan Classical Digital Database*, which is an open archive of Chinese classics, transcribed by experts, tied together with internet-based resources.

11.4.3 System-Centered Research
The system-centered category is focused on the underlying technologies for information systems that are able to cope with increasing high-volume of content. Some topics include:
- architectures for information environments
- intelligent agents
- middleware research for federating heterogeneous sources
- multimedia capture, representation and digitization
- evaluation and performance studies

This includes the Stanford's *Interlib* project, which is the continuation of the

Infobus work. Such work would also includes CMU's *Infomedia-II* project, which continues to focus on the archiving and automatic indexing of video collections.

Other projects include the University of Santa Barbara's *Alexandria Digital Earth Prototype*, which is another DLI-1 project. The Alexandria project is compiling an enormous digital library of geo-spatial information that ties together information from a range of disciplines, including the humanities, social science and biology through geo-spatial connections.

Oregon's Health Science University's *Tracking Footprints Through an Information Space* is undertaking a study to track ways in which medical experts gather technical information to solve given health care problems. By observing how these experts obtain and utilize information they hope to gain an insight into how to build new digital libraries for health-care workers.

11.4.4 Establishing Test beds
In addition to the listed categories, the DLI-2 is also funding projects to establish large test beds for further research. Many of the projects mentioned above are developing such test beds while performing other research. University of South Carolina at Columbia is building and maintaining a software and data library test bed for experiments and simulations used in the social and economic sciences. They call this a *"Web-Lab Library"* which can be used by other institutions to perform experiments on social and economic data. The Indiana University is deploying a *Digital Library of Music.* This test bed explores: how much content can be delivered, metadata standards, human-computer interactions, and ways that intellectual property can be managed.

11.4.5 Summary
The DLI-2 goals are to mature the standards, usability, and necessary infrastructures for digital libraries. While focused on the digital library at all levels of operation, from low-level systems support to their social and economic impact, the DLI-2 also targets content-specific digital libraries, encouraging new user-groups to develop innovative archives in their domains. In addition, the DLI-2 is funding projects to develop novel collaborative technologies that can link together scholars on a global scale and give users access to distributed repositories of intellectual information. The projects target a wide area of multimedia data, from audio, to video, to new medias that incorporate rich metadata tagging with existing media primitives. The finding and outcomes of these projects, along with other international efforts, will undoubtedly shape the way future digital archives of all types of media are developed, deployed, and utilized

11.5 Conclusion

Digital libraries are a reality. Many of us depend on them for our scholarly and technical research. However, more research is required to improve their usability, not for just content delivery, but for content discovery and information engineering. Challenges are present at all levels of digital library research, ranging from the development of cost-effect systems for storing, managing, and evolving enormous collections of digital information, to the creation of technologies that glue heterogeneous repositories together, across protocols, different queries and delivery mechanisms, and through deep semantic understanding of the underlying information.

   The works presented in this chapter have focused on DLI and DLI-2 projects that address some of these challenges. From automatically deriving metadata from video content, to providing novel video abstraction and allowing users to quickly understand the content contained within. We have seen the idea of using the *Interspace*, with derived concept-spaces, to give insight into ways multi-disciplinary knowledge can be linked together. While distributed objects provide the promise of hiding complexities and system specific protocols to provide a common and ubiquitous interface heterogeneous repositories. We have also seen scores of current DLI-2   projects that are addressing existing challenges while developing content specific test bed and archives.   While this chapter mainly focused on   projects specific to DLI and DLI-2, digital libraries and subsequent research is being performed world-wide, and promises to build a new area of information access and understanding.

References

[1] A. Bishop, "Digital Libraries and Knowledge Disaggregating: The Use of Journal Article Components," *In Proc. Third ACM Int'l Conf. Digital Libraries,* ACM Press, New York, 1998, pp. 29-39.

[2] H. Chen at al, " A Parallel Computing Approach to Creating Engineering Concept Spaces for Semantic Retrieval: The Illinois Digital Library Initiative Project, *IEEE Trans. Pattern Analysis and Machine Intelligence (PAMI),* 1996

[3] H. Chen et al, "Automatic Thesaurus Generation for an Elective Scientific Community", *J. American Soc. Information Science*, Vol 46, No. 3, Apr 1995, pp. 175-193

[4] T. Col and M. Kazmer, "SGML as a Component of the Digital Library," *Library Hi Tech*, Vol 13. No 4, 1995, pp. 75-90.

[5] H. Chen et al., "Automatic Thesaurus Construction for an Electronic Community System," *J. American Soc. Information Science,* Mar. 1995, pp. 175-193.

[6] http://www.canis.uiuc.edu/news/iodyne.html

[7] H. Chen et al., "Internet Browsing and Searching: User Evaluations of Category Map and Concept Space Techniques," *J. American Soc. Information Science,* July 1998, pp. 582-603.

[8] H. Chen et al., "Alleviating Search Uncertainty through Concept Associations: Automatic Indexing, Co-occurrence Analysis, and Parallel Computing," *J. American Soc. Information Science,* Mar. 1998, pp. 206-216.

[9] M. G. Christel, et al. "XSLT for Tailored Access to a Digital Video Library.", *Proc. IEEE Joint Conf. Digital Libraries* (Roanoke, VA, June 2001), ACM Press, 290-299.

[10] M.G.Christel "Visual Digests for News Video Libraries". *Proc. ACM Multimedia '99* (Orlando, FL, Nov. 1999), ACM Press, 303-311.

[11] DLI, NSF Digital Library Imitative (Phase 1), http://www.dli2.nsf.gov/dlione/

[12] DLI-2, NSF Digital Library Imitative Phase 2, http://www.dli2.nsf.gov

[13] Y.H. Gong, G. Proietti, and C. Faloutsos, "Image Indexing and Retrieval Based on Human Perceptual Color Clustering," *Proc. Computer Vision and Pattern Recognition*, IEEE CS Press, Los Alamitos, Calif., 1998, pp. 578-583.

[14] International Institute for Electronic Library Research. (Now the Digital Library Group), De Montfort University, Leicester LE1 5XY -- England

[15] R. Lienhart, S. Pfeiffer, and W. Effelsberg, "Video Abstracting," *Comm. ACM,* Dec. 1997, pp. 55-62.

[16] JSTOR, University of Michigan's Journal-Store Project, http://www.jstor.org

[17] NSF Research on Digital Libraries Announcement, NSF 93-141

[18] NSF Research on Digital Libraries Phase 2 Announcement, NSF 98-63

[19] P. J. Nurnberg, et al "Digital Libraries: Issues and Architectures", *Proc.First ACM Int'l Conf. Digital Libraries ,* Austin, Texas, June 11-13, 1995

[20] A. Paepcke et. al "Using Distributed Objects to Build the Stanford's Digital Library Infobus", *IEEE Computer*, 32, 2, Feb, 1999

[21] A. Paepcke et al. "Using Distributed Objects for Digital Library Interoperability", *IEEE Computer May,* 1996

[22] B. Schatz et al. "Digital Libraries: Technological advances and Social Impacts", *IEEE Computer*, 32, 2, Feb 1999

[23] B. Schatz et al. "Federating Diverse Collections of Scientific Literature", *IEEE Computer, May*, 1996, pg. 28-35

[24] B. Schatz, et al "Federated Search of Scientific Literate", *IEEE Computer*, 32, 2, Feb. 1999, 51-58

[25] B. Schatz et al., "Interactive Term Suggestion for Users of Digital Libraries: Using Subject Thesauri and Co occurrence Lists for Information Retrieval," *Proc. First ACM Int'l Conf. Digital Libraries,* ACM Press, New York, 1996, pp. 126-133.

[26] T. Sato et al., "Video OCR for Digital News Archive," *Proc. Workshop on Content-Based Access of Image and Video Databases*, IEEE CS Press, Los Alamitos, Calif., 1998, pp. 52-60.

[27] Wactlar, H., et al. "Lessons Learned from the Creation and Deployment of a Terabyte Digital Video Library". *IEEE Computer*, 32, 2, Feb. 1999, 66-73.

[28] Wactlar, H. et al "Intelligent Access to Digital Video: Infomedia Project", *IEEE Computer*, May 1996, 46-52

[29] M.J. Witbrock and A.G. Hauptmann, "Using Words and Phonetic Strings for Efficient Information Retrieval from Imperfectly Transcribed Spoken Documents," *Proc. Third ACM Int'l Conf. Digital Libraries*, ACM Press, New York, 1997, pp. 30-35

12 SCALABLE STORAGE FOR DIGITAL LIBRARIES

Prof. Edward A. Fox and Dr. Paul Mather

Digital libraries are voracious consumers of storage. As collections increase in size, the need for scalable storage becomes vitally important. Here, we survey the storage technology and research that is used in digital libraries. We first focus on issues related to performance, such as the limitation of devices and impact of workloads, and the characterisation and modelling done to design and tune storage systems to meet workload demands. Some tuning techniques include caching; physical data clustering; block allocation schemes; and log-structured approaches.

Disk arrays are a common way to aggregate individual disks to increase the total amount of storage whilst increasing the amount of I/O parallelism and fault tolerance. We describe disk array technology, focusing on the Redundant Array of Inexpensive/Independent Disks (RAID) approach and describe the most common RAID organisations and factors affecting RAID performance.

Finally, we describe networked storage systems and research. This includes the areas of intelligent disks; parallel file systems and parallel I/O; and distributed file systems. We end with a summary of how these many diverse strands relate to scalable storage for digital libraries.

12.1 Introduction

More than ever before, there is a vast amount of information available in digital form. Increasingly, this content is stored in large repositories accessed over wide-area networks. To address the voracious appetite for storage, research in that arena recently has turned towards the issue of *scalability*—in terms of hardware, software, and administration. Moreover, a current focus is the use of commodity off-the-shelf (COTS) hardware as a means of achieving scalability. In particular, approaches such as *clustering, networked attached storage*, and *storage area networks* are fast becoming industry standards as means for achieving manageable, scalable storage on the order of terabytes (2^{40} bytes) and upwards.

The bulk of storage research at the operating system level has looked at providing file system support and associated functions. Although there are several proposed architectures for digital libraries storage repositories, with the Kahn-Wilensky architecture [48] finding favour, all digital library implementations thus far have been built using databasesor directly or indirectly, atop traditional operating system file systems. Such file systems are relatively weak in terms of providing support for the rich metadata and file types typically used in digital libraries.

Hence, we aim to survey and reconcile diverse strands of scalable storage design, with support for multimedia digital library repository access semantics as the target. Particular emphasis will be placed on *scalability* in the storage realm, and on metadata support, as well as system support for different multimedia object types in the digital libraries realm.

12.2 Motivation

More data are available online than ever before. There is a dramatic increase in the amount of data available to the public served over wide-area networks, and the level of growth continues to increase. In some areas, data gathering has followed approximately Moore's law [49]: in astronomy, for example, the volume of data gathered is doubling roughly every 20 months [92].

Already, there are several terabyte-sized repositories of information of various kinds accessible to millions of users worldwide, and more are being proposed. The *TerraServer* [9] provides aerial, satellite, and topographic images of Earth delivered via the World Wide Web (WWW). As of February 2000, the database comprised over 1.5 terabytes of data. The system is configured to handle a maximum of 40 million hits per day, and 6,000 simultaneous users. The Sloan Digital Sky Survey (SDSS) intends to provide multi-terabyte astronomical observational data for interactive exploration and query via the Internet [92]. The SDSS expects to collect over 40 terabytes of raw data in the next five years.

Even larger datasets are being planned. The NASA Earth Observing System Data Information System (EOSDIS) project [51] seeks to provide access to a huge collection of remote sensing data collected over a 15 year time-span. It was predicted that raw data would be collected by 2002 at a rate of over 360 GB per day, and that the archive will contain over 3 petabytes (2^{50} bytes) of data across more than 260 data products. The system will utilise Distributed Active Archive Centers (DAACs) networked together via a high-speed backbone to collect and process the data, and the entire collection will be accessible to users via the WWW using a browser.

In addition to these large scale projects, ordinary production and consumption of data continues at a pace that has even *outpaced* Moore's Law. Decision support systems (DSS) are rabid consumers of data, with demand doubling roughly every 9–12 months, in contrast to the approximately 18-month doubling period of Moore's Law [49]. Of potentially even larger implication for storage is the growth in consumer still and video digital camera acquisition and use. Consider the implications of having a billion people a year each record 1000 images, each 3 megapixels in resolution, in addition to 100 hours of digital video!

The huge rate of increase in data suggests that massive-scale storage architectures need to be able to scale at a *faster* rate than the growth of processors and disks themselves: we cannot rely on hardware improvements alone to keep pace with the demand for data storage.

12.3 Subject Areas

Computer storage has long been the subject of research. With the rise of parallel and distributed systems came a plethora of associated storage systems. For our purposes, this activity can broadly be grouped in the areas of local file systems; parallel file systems; distributed file systems; and scalable networked storage. Within all these areas, attention has been paid to operating system and application programming support; layout, buffering, caching, and efficiency; reliability, fault-tolerance, and crash recovery; and security. Some of these goals are contradictory: high fault tolerance might lead to lower efficiency, for example.

12.4 Local File Systems

A *file system* is a means by which data is managed on secondary storage. At the user level, these data are organised using an abstraction known as a *file*. Files are a logical structuring mechanism that allows data to be organised as a byte-addressable sequence that may be randomly accessed. File systems mask the details of how this logical structure is stored on actual physical devices attached to the system. File systems not only manage the actual long-term storage of data—layout and structuring—but also manage the access to that data—name resolution, buffering, and caching. File systems employ metadata to assist in the organisation of files on storage. Examples of such metadata might be pointers locating blocks belonging to a given file on a disk, plus the length of that file. Another example might be a bitmap indicating the location of unused blocks on a disk.

We deem *local file systems* to be those whose storage is provided by directly-attached secondary storage. In the majority of cases, this storage is in the form of direct-access storage such as disk drives. The improvement in hard disk technology in terms of total storage (areal density) and transfer rate has improved roughly according to Moore's Law [42]. However, rotational speeds and, especially, seek times have increased at a much slower rate. Head seek times, in particular, are limited by the physics of inertia, and may not see vast improvement. To address this, file system research has explored the use of alternative layout schemes [72], aggregation [20], and aggressive caching and buffering [35, 73] amongst other techniques to ameliorate the effects of head seek times during reads to and writes from the disk.

12.5 Workload Studies

There have been several studies of the content and usage of local file systems [30, 40, 71, 75, 95]. Such studies characterise the *workload* of the file system. Workloads provide important information to file system architects, and can guide the design of layout schemes; buffering and caching strategies; migration policies; and performance tuning. Studies that have tracked the semantic content of files have found there is a good correlation between file type and file size [12, 30, 80].

One finding to arise repeatedly out of workload studies for typical engineering and office applications and environments is that most files are small: in Gibson and Miller's study [40] the majority of files are less than 4K in size, and in Douceur and Bolonsky [30] the median file size is 4K. The distribution of file sizes is heavy-tailed, with most bytes being concentrated in relatively few large files: most files are small, but large files use most disk capacity. Furthermore, a general trend is that most files are seldom accessed over the long term.

Traces of long-term file activity collected by Gibson and Miller [40] reveal that on a typical day, less than 3% of all files are used, and of that usage, file accesses account for over twice the number of either file creations or deletions, and the number of files modified is one third the number of creations or deletions. In addition, they found that files modified are about as likely to remain the same size as they are to grow, and that modified files rarely get smaller [40]. In fact, the amount of file growth during file modifications found in their study was usually less than 1K, regardless of the file's size [40]. Long-term activity of the file systems surveyed by Douceur and Bolonsky [30] determined the median file age to be 48 days, where file age is calculated as the time passed since the most recent of either the file's creation or last modification relative to when the file system contents snapshot was taken.

Traces capturing the short-term behaviour of file system workloads show that many files and accesses are ephemeral and bursty. The seminal Unix 4.2 BSD study of Ousterhout et al. [71] determined that files are almost always processed sequentially, and that more than 66% are whole-file transfers. Of the remaining non-whole file transfers, most are read sequentially after initially performing an *lseek*, i.e., to append. When measured over a short time scale, files tend to be very short-lived, or only open for a very short time.

Subsequent studies have confirmed the results of Ousterhout *et al.*, but their findings are tending towards extremes. Baker *et al.* [8] discovered the time between a file being created and subsequently either deleted or truncated to zero length to be less than 30 seconds for 65%–80% of all files, and those files tend to be small, accounting for only about 4%–27% of all new bytes deleted or overwritten within 30 seconds. Vogels [95] found that files are open for even shorter periods: 75% are open for less than 10 milliseconds; 55% of new files are deleted within 5 seconds; and 26% are overwritten within 4 milliseconds. Similarly, both Baker *et al.* [8] and Vogels [95] confirm that: most files accessed are short, and sequentially accessed (though shifting more towards random access); most bytes transferred belong to large files; and large files are getting larger (20% are 4 MB or larger).

Ramakrishnan *et al.* [75] collected relatively short-term traces of file system activity at eight different customer sites running VAX/VMS over several different time periods at various times of the day and night. The sites represented different workloads (interactive; timesharing and database activity; transaction processing; and batch data processing), but analysis showed that file system activity was consistent across different workload environments. The study revealed that for almost all workloads, most files—over 80%—are inactive (not opened, read, or written during the trace), and of those active files, a small percentage account for most of the file opens (in an airline transaction processing workload, less than 3% of active files account for over 80% of file opens). Over all workloads, over 50% of all active files were opened only once or twice, and 90% of active files were opened less than 10 times in a typical 9–12 hour "prime time" period.

Of those files active during the trace period, over 50% of them were read only once or twice, and 90% of active files in all but one workload had between 24 and 46 reads. In the case of writes, 31–46% of active files were not written at all, and 14–28% had only one write. Over 50% of the active files had fewer than two write operations. Once again, a small percentage of files often account for most writes, and skew the mean number of writes upwards. For the timesharing workloads studied, 1.3–2% of files accounted for 80–87% of writes. In the case of transaction processing workloads, 1.3–1.8% of files accounted for 95% of writes.

Analysis of the WWW by Adamic and Huberman [2] reveals that files and sites follow a power law with respect to many attributes, including file size, site size, popularity, and site linkage, thus confirming in a wider context the "heavy-tailed distribution" phenomenon common to all file system studies.

Some consequences of workload studies are that the *namespace* is dominated by small files, but that transfer bandwidth is dominated by large files. Furthermore, files exhibit a *generational* behaviour in that if they survive the short term then they will tend to persist—largely inactive—for the long term. The likely uncertain early lifetime makes write-back caching attractive in the anticipation that a file will subsequently be deleted or modified before actually being written to disk. Hence, the more this volatility that can be smoothed out by caching or buffering the better.

WORKLOAD CHARACTERISATION AND MODELLING

Workload characterisation has long been used in assessing the performance of systems [16]. In storage research, analysis of workload traces, as described above, features prominently. These workload traces capture I/O activity as it happens over a period of time, and contain important information about access requests and access patterns.

Because actual traces are often large and difficult to capture (because of technical and political reasons), researchers often have to resort to using synthetic traces to model workload. Generating synthetic traces that are representative of real-world activity is a significant problem [33]. Ganger and Patt [34] advocate a system-level approach to representing I/O workloads, and, in particular, that different classes of I/O request be treated with different importance to better model their effect on overall system performance. SynRGen [31] is a tool for generating workloads based upon algorithmic specifications of tasks, offering more realistic artificial traces of system activity.

Traces are primarily used as input to simulations of storage systems. Simulation is popular because of the flexibility with which data can be collected and operational parameters controlled. This allows "what-if" scenarios to be investigated to examine the effect of specific parameters within the storage system on global system performance, e.g., cache size, seek time, block size, bus speed, etc.

Ruemmler and Wilkes [78] describe a detailed disk simulation, which they apply to the HP C2200A and HP 97560 disk drives. Worthington *et al.* [100] describe the extraction of disk drive parameters directly by interrogation of SCSI drives, combined with empirical measurements, thereby improving modelling accuracy.

Ganger *et al.* [36] produced a highly-configurable disk simulator called DiskSim. DiskSim can model such storage aspects as device drivers, busses, controllers, adapters, and disk drives. DIXtrac [83] improves upon the extraction capabilities of [100] in that it can extract over 100 disk parameters in a fully automated fashion. DIXtrac can be used to provide input for disk drives simulated by DiskSim. Pantheon [97] is a toolkit that supports the development of storage simulations. It has been successfully used in such storage projects as TickerTAIP [18] and AutoRAID [98].

Although simulation is prevalent in storage research, it has the disadvantage of being costly in terms of time and resources needed to run simulations. Analytical modelling has the advantage of being a quicker and more direct answer for exploring alternative design spaces. The disadvantage of analytical modelling is its inaccuracy compared to simulation.

Uysal *et al.* [94] present an analytical model to predict throughput of a modern disk array that captures many of the complex optimisations found in such hardware. They validate their model against a real commercial array and obtain prediction accuracy within 32% in most cases, and to within 15% on average. Shriver [88] presents a detailed analytical model of a single disk drive. Barve *et al.* [10] extend this by analysing the performance of several drives on the same SCSI bus, as bus contention becomes a factor when large request sizes are used. Shriver *et al.* [89] present an analytical model of disk drives that support read-ahead and request reordering. Their prediction accuracy is to within 17% across a variety of disk drives and workloads. Borowsky *et al.* [14] present an analytical model of response time of a device subject to phased and correlated workloads. Their model is designed to determine whether a given quality of service bound is satisfied for the workloads presented.

Because analytical models lend themselves to direct computation, unlike long-running simulations, this makes them ideal for use in storage optimisation problems. Of recent interest, in particular, is the area of attribute-managed storage [41]. Attribute-managed storage is a constrained optimisation problem in which a set of workload units and devices are presented as input to a solver. The output is a mapping of workload units to devices such that the needs of the workload units are met. Shriver [87] describes the workload and device attributes and their mapping in detail. Alvarez *et al.* [3] describe a suite of tools called Minerva that can be used to design storage systems automatically. In a test, Minerva was able successfully to design a storage system to handle a decision-support workload that performs as well as a human-designed configuration. The Minerva-generated system used fewer disks than the human-designed system (16 vs. 30), and the design took only 10–13 minutes to produce. Minerva is limited in terms of its analytical models to the types of storage system it can design (currently, disk arrays).

12.6 Local File System Performance Issues

As mentioned previously, seek times dominate disk access times. A simple "break-even" point to make a seek worthwhile would be to spend as much time actually transferring data as seeking to find it. Obviously, though, the smaller the overall fraction of a total read or write is occupied by seeking, the better. Even COTS hard drives have relatively high sustained transfer rates, making the time lost seeking more troublesome. For example, a "desktop use" 25 GB EIDE drive with an average seek time of 9 ms supports a sustained transfer rate of 15.5 MB/s. A "server" 36 GB Ultra 160 SCSI drive with an average seek time of 4.9 ms supports a sustained transfer rate of 36 MB/s. For the 25 GB drive, just over 142 KB of data could be transferred in the time spent in an average seek (0.009 s at 15.5 MB/s); for the 36 GB drive, just over 180 KB could be transferred (0.0049 s at 36 MB/s). If only, say, 1 KB were transferred before seeking elsewhere, then less than 1% of the total time would be spent actually transferring useful data using the above mentioned drives as examples. Even for a larger block size — 4 KB — less than 3% of the time is spent usefully transferring data. Successful use of the available disk bandwidth means making larger transfers and fewer seeks.

CACHING

The impact of seeking on disk I/O is often addressed by caching. A large RAM-based cache can absorb the effects of multiple reads of the same disk block over time, and most modern operating systems employ such buffer caches. However, caches depend upon *locality of reference* to be successful: what happened before will happen again in the near future. As mentioned previously, workload studies indicate most reads are whole-file sequential reads. These reads have a negative impact on cache performance, as they can effectively stream through and flush the cache for large enough files, if action is not taken to mitigate such behaviour. Baker *et al.* report cache read miss rates of about 40% in their study, rising to 97% on machines processing large files [8].

Caches will cache the most commonly accessed disk blocks. In an engineering and office environment, the most commonly-used files will likely be application and system programs: word processors, compilers, editors, and the like. Across the entire user population, individual user data files will be accessed less often relative to the software that operates upon those data. Caching will likely be more successful for programs than data. The exception to this is perhaps the caching of writes.

A disk block that is read, modified, and written back will exhibit strong temporal locality of reference. Unfortunately, out of concern for safety, many applications will not update file data *in situ*, but, instead, often will make a temporary copy of the old file to which are applied the updates. After processing, the original file is either deleted or renamed as a backup copy, and the copy is renamed to the original. Baker *et al.* [8] report that only 10% of new data written is deleted or overwritten in the cache rather than actually being written to disk. However, they also report that the 30-second dirty block cleaning rule in their system accounted for over 70% of blocks written from the cache, indicating that cache content safety (flushing dirty blocks to disk to guard against loss due to hardware failure), not cache size, is a major factor on write-back cache performance.

CLUSTERING AND FRAGMENTATION

Another method of ameliorating the effects of seeks is to co-locate related data physically close together on the disk. In this way, only small seeks or a rotational delay must be incurred to read successive blocks of a file. Physical clustering is especially useful in the case of file system metadata. The Fast File System (FFS) for Unix greatly improved the performance of the previous file system design by spreading file metadata throughout the disk [62]. The new design divided the disk into cylinder groups, with each cylinder group having its own local metadata. In the previous design, *all* file system metadata was clustered at the beginning of the disk. Data blocks located on tracks away from the start of the disk thus incurred long seeks every time the metadata for the file stored there was accessed. McKusick [62] estimates that such a segregation of metadata, coupled with the small 512-byte block size, meant that the old Unix file system design was only able to use 3-5% of the disk bandwidth. But, by interspersing (clustering) metadata and data blocks nearby on the disk, and using a larger block size (4 K and up), disk bandwidth utilisation increases to 47% in the new FFS design [62]. Clustering in FFS also improves crash recovery, as file system metadata is not physically concentrated in one localised area of the disk, but is spread over it.

BLOCK SIZING AND ALLOCATION

A more general application of increasing the block size to lessen the effects of seeks and improve storage contiguity to benefit the large sequential transfers prevalent in file system traces is to adopt variable-sized blocks for files. In effect, the disk is treated as an area of storage from which variable-sized allocations are made. Such an approach is not new: efficient memory allocation and garbage collection has been long studied. Disks pose additional performance constraints because fragmentation seriously affects performance by translating into long seeks (excessive head movement), and garbage collection on disk is more costly than in main memory.

Wilson *et al.* [99] provide a comprehensive survey of the memory allocation literature. Memory allocation allowing variable-sized blocks has a disadvantage of extra metadata overhead for pointers to link allocated areas, and compaction suffers significant overheads. Knowlton [50] introduced the *buddy system* of allocation that is considerably more efficient when freeing blocks.

In buddy systems, only a fixed set of block sizes, $b_1<b_2<\cdots<b_k$, are available for allocation, so there is a potential for storage to be wasted if an object is placed in a bigger block than necessary. However, such waste is also a problem for traditional fixed-size allocation schemes (though it is strictly bounded). A standard block sizing scheme is to use increasing powers of 2 (the *exponential buddy system*). Hirschberg [45] describes a *Fibonacci buddy system* in which the block sizes follow the Fibonacci series. The use of the Fibonacci series permits a larger number of different sizes, and hence increases the probability of a good fit, thereby decreasing internal fragmentation.

Koch [52] designed a disk allocation scheme based upon the exponential buddy system. Files are stored in a bounded number of contiguous *extents*, and a reallocation algorithm runs periodically to improve the allocation of poorly arranged files. Koch reports a mean number of extents per file of 1.5 and an average internal fragmentation of less than 4%.

Ghandeharizadeh *et al.* [38] present and analyse several algorithms for managing files in a hierarchical storage environment with a focus on the tradeoff between contiguity of files and the amount of wasted space. Ghandeharizadeh *et al.* [37] describe algorithms for the layout of files on disk such that each n block file is stored in no more than $\lceil \lg n \rceil$ separate extents in order to minimise seeks.

Seltzer and Stonebreaker [86] examine several variable block size file allocation strategies and conclude that such extent-based approaches offer excellent performance in read-mostly environments because of seek time minimisation and large sequential reads.

LOG-STRUCTURED APPROACHES

A major trend in local file system design that took hold in the early 1990s was a move towards log-structured file systems to improve write performance [72, 77, 85]. Instead of the fixed layout schemes employed in traditional file systems such as FFS, log-structured file systems treat the blocks of the disk as a log to which data are appended. Unlike FFS-like file systems which incur seek overheads for each write, log-structured file systems instead batch all writes and then write all the changes to disk in a single large disk transfer operation. Typically, log-structured file systems divide the disk logically into *segments*, and write an entire segment at a time.

By performing large disk transfers, the impact of a seek is lessened. In addition, crash recovery is greatly assisted, because in the event of a crash, the file system software need only locate the last checkpoint and then perform a *roll forward* on the writes made after that point. This is in contrast to the time-consuming multi-pass consistency algorithm required in FFS-like file systems [63].

Because disks are finite in size, the log must *wrap* when it reaches the end of the disk. This necessitates special handling of free space in the file system. The most common approach is to employ a *cleaner* daemon that runs asynchronously alongside the mounted file system. The cleaner acts like a garbage collector, copying live blocks out of one or more segments containing deleted space and consolidating the live blocks together and writing them to clean segments in the log. The segments out of which the live blocks were copied are now free to be reused as clean segments for the log. Various policies can be enforced for segment cleaning, to balance the impact on the file system [60].

Another way to improve write performance, along the lines of log-structured file systems, is to log writes to a smaller, separate *logging medium* and then asynchronously commit the changes to the actual data disks of the file system. The logging medium could be non-volatile RAM [47,58], or a dedicated, separate log disk [22]. Non-volatile RAM also has been used to implement reliable buffer and file cache mechanisms to mitigate the impact of periodic writing of dirty blocks to ensure file system consistency. Prestoserve [66] uses non-volatile RAM to cache NFS writes. Rio [21] and Phoenix [32] implement reliable file caches, making writes to the file cache as permanent and as safe as files on disk, but with much greater speed. Rio, for example, performs 4–22 times as fast as a standard Unix file system that employs write-through caching, or 2–14 times as fast as a standard Unix file system using write-back caching. Even when using delayed metadata writes in a standard Unix file system [35], Rio is 1–3 times as fast, but has the reliability advantage of synchronous write behaviour.

12.7 Disk Arrays

File systems that reside on a single physical disk have an inherent bottleneck by virtue of there being only a single head assembly through which to perform all reads and writes. As we have seen, the time needed to move this head assembly over the disk surface is usually the major limiting factor in terms of disk I/O and hence file system performance.

One way to mitigate this bottleneck is to increase, in effect, the number of head assemblies. The easiest way to achieve this is to create a *virtual disk* out of an *array* of individual physical disks. Not only does this increase the potential parallelism of reading and writing—because different areas of the virtual disk can be read and written simultaneously—but it also increases the potential reliability, as a hardware failure now no longer renders the entire virtual disk unusable. Such virtual disk arrays are commonly called *disk arrays*. An important, and the most prevalent, class of disk arrays, in which performance and reliability are emphasised, is called Redundant Array of Independent/Inexpensive Disks (RAID) [20].

RAID employs *data striping* to improve performance, and *redundancy* to improve reliability. Striping is a way to distribute data transparently over disks in an array. The granularity of the stripe unit determines how many disks will be involved in a given I/O request. The larger the stripe unit, the fewer disks need be involved for small I/O requests, allowing multiple I/O requests to be serviced in parallel.

Redundancy in RAID involves using extra storage space to improve the reliability of the data stored in the RAID. The extra space holds *error correction* information for the actual file data stored on the RAID. The most prevalent error correction scheme used in RAID is *XOR parity*. In this scheme, a parity stripe unit is computed as the XOR of the corresponding stripe unit from each of the data disks. Such a scheme can tolerate the failure of a single disk, because the parity data can be used with the remaining good data to reconstitute the failed disk's data. There are other error correction schemes. For example, the *P+Q Redundancy* scheme uses *Reed-Solomon* codes to protect against the failure of up to two disks.

The major performance impact of using error correcting data in a RAID is that additional reads and writes are required when writing data, particularly for small writes that update only one data disk. This is because the parity data needs to be updated also on a write. In the case of a small write, a *read-modify-write* process must be used in which the old data is read before the new data is written, to determine how the new differs from the old, prior to applying the differences to the parity block. Thus, a small write actually requires two reads and two writes. Because parity is always written on a data write, if parity is kept on a single disk, that disk can easily become a bottleneck for the entire RAID.

The interleaving, or striping, of data and redundancy information over the disks of a RAID has a big impact on the performance and storage utilisation of the disks participating in the aggregate storage. There have emerged several characteristic organisations, or *levels* of RAID, briefly described below:

RAID 0: Data are striped across the disks comprising the RAID unit with no redundancy employed whatsoever. This gives maximal use of available disk space, and the best write performance, but with no fault tolerance. If a disk in the RAID fails, then the whole RAID fails (for all practical purposes).

RAID 1: Twice as many disks are employed in this RAID level as in level 0. Data are striped across the data disks, but, also, are *mirrored* on a corresponding redundant disk. So, for each data disk, there is another mirror disk. When data are written, both the data disk and its mirror are updated. When data are read, either of the data disk or the mirror can be read, as both contain identical information. Load balancing can be used to select the best disk for the read request, e.g., the disk with the shortest request queue, or the one whose head is currently closest to the data. If a disk in a RAID 1 fails, all reads and writes will be to and from its mirror. RAID 1 has very high read performance.

RAID 2: This RAID level uses bit-level striping and memory-style error-correcting codes to detect and correct disk failures. So, a RAID with n data disks using Hamming codes for redundancy will use log n additional disks to store parity. If a disk fails, the parity disks can, in combination, detect which disk must have failed, and correct the failed data. As the RAID set grows larger, the number of parity disks grows more slowly. However, because disk failures are *self-identifying*, the extra parity encoding needed to identify which disk failed is usually redundant, making this scheme little-used.

RAID 3: Another bit-level striping scheme, RAID 3 uses the fact that disk failures are self-identifying to eliminate the $O(\log n)$ parity disks, replacing them with a single disk per RAID set. When a disk fails, the parity disk can be used in conjunction with the remaining good disks to reconstruct the failed disk's data. In this bit-interleaved scheme, every disk participates in each read or write, and, in effect, all disks operate identically (the heads are synchronised). This simplifies implementation, and enables a high bandwidth to be delivered, making RAID 3 attractive for multimedia applications where quality of service bounds must be observed.

RAID 4: Similar to RAID 3, RAID 4 is a *block-interleaved* scheme that uses a single, dedicated parity disk. Read requests smaller than the striping unit need only access a single disk, increasing the I/O parallelism of this scheme. The drawback, however, is that all parity updates on writes go to a single disk, which can quickly become a bottleneck.

RAID 5: This organisation is similar to RAID 4, but improves upon that scheme by distributing the parity information across the data disks of the RAID, instead of having it reside on a separate disk. This *distributed parity* means that, now, all disks in the RAID are used to store data, but also, some blocks of each disk are used to store parity. The equivalent of a single disk's worth of storage is still consumed by parity storage, but it is spread across all the disks in the RAID. Although the layout is more complicated, it means that no single disk becomes a bottleneck for parity writes, and also has the advantage that all disks can participate in reads. The precise distribution of parity can affect performance [56]. RAID 5 generally has the best performance for small and large reads and large writes of any redundant disk array, but its biggest drawback is the read-modify-write penalty for small writes.

Those are the major RAID categories, but by no means all. For example, the aforementioned *P+Q redundancy* scheme, which can protect against up to two disk failures, is often classified as RAID level 6. In practice, the most commonly used RAID organisations are RAID levels 0, 1, 3, and 5.

Figure 12-1 shows the different layouts for an array of four disks organised according to some different RAID levels. Each box represents a stripe unit within a single disk, and is some multiple of one disk sector. The number in each box refers to the logical stripe unit of user data within the array. So, "5" designates the sixth logical user data stripe unit. (Numbering begins at zero.) The shaded boxes represent redundant data (used for mirrored user data or parity stripe units). The designation "Pn-m" refers to parity computed over the data in user data stripe units n to m respectively. Note that RAID level 3 is similar in organisation to RAID level 4 except that the granularity is that of a bit or byte. It can be seen from Figure 12-1 that RAID level 2 is the most redundant, and that in RAID level 4 all parity data is concentrated on a single disk, creating a potential bottleneck.

Although excellent performance can be obtained from a RAID with all disks operating, performance can be severely degraded if a disk fails. When a disk fails, the RAID is said to be operating in *degraded mode*. In this mode, *all* disks must participate in each read and write (except for RAID level 1), removing any I/O parallelism.

To counter this, a RAID can have *online* or *hot spare* disks—disks that are part of the RAID, but that are not used to store either data or parity in the course of normal operation. When a disk fails, its data is *reconstructed* onto a spare disk. This is done transparently, to ensure the RAID is not offline. If hot spares are not used, data reconstruction takes place when the failed disk is eventually replaced with a working unit. If *hot swappable* drives are used in the RAID, this can be accomplished without shutting down the entire RAID. *Distributed sparing* [64] and *parity sparing* [19] are two techniques that take advantage of online spare disks to improve the normal performance of the RAID, while still allowing fast reconstruction to begin.

Even in non-degraded mode, care must be taken to safeguard the integrity of the RAID to protect against system crashes. In particular, the RAID must keep track of not only which disks have failed, but also which logical sectors of a failed disk have been reconstructed, or which logical sectors are currently being updated. This is necessary to avoid reading *stale data* from a RAID. The information kept track of is referred to as the *metastate* of the RAID.

In addition, it is important to keep track of which parity sectors are *consistent* and which are *inconsistent* in the event of a system crash. Customarily, this means that before each write, the parity sector must be marked as inconsistent until new, valid parity is written. Upon a system crash, recovery mandates that all inconsistent parity must be regenerated. Inconsistent parity in the presence of a disk failure means that the data will not be able to be reconstructed correctly. However, because reconstructing a consistent parity sector also results in a consistent parity sector, it is

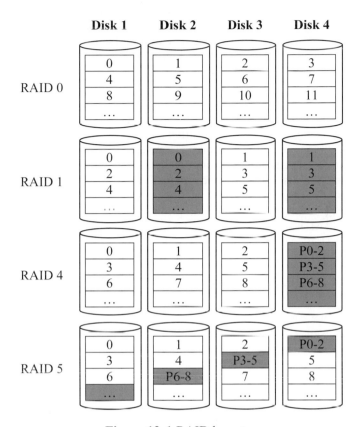

Figure 12-1 RAID layouts

permissable to trigger the regeneration of all parity sectors upon return from a system crash, instead of maintaining stable parity metastate information.

Worse than when individual disks fail, multiple disks can fail systematically. Typically, this is because the controller or bus to which they are connected fails. Thus, instead of a single disk failing, a *string* of disks fails. One way to combat this problem is to arrange error correction groups orthogonal to the actual hardware arrangement. Such a strategy is broadly called *orthogonal RAID* [67, 84].

RAID organisations are the subject of intense research, particularly in improving write efficiency and reconstruction performance. *Floating parity* [65] and *parity logging* [91] can improve small write performance. *Declustered parity* [46] can improve reconstruction performance by distributing the parity groups such that the load is balanced more evenly across the disk array in degraded mode.

Research has been undertaken into employing adaptive RAID organisations, to match workload with the best layout. The HP AutoRAID system [98] uses two levels of RAID between which files may migrate automatically depending upon how often they are updated. RAID 1 (mirroring) is used for frequently updated data, and RAID 5 for infrequently updated data. Initially, all but 10% of data are stored as RAID 5. In the course of normal operation, frequently-updated data in RAID 5 are promoted to mirroring.

One of the major problems with RAID schemes is that they are effectively *bus-based* solutions, and so have restricted scalability. Typically, the disks of a RAID array are connected directly to a *RAID controller* that mediates all disk I/O. The RAID controller may be a custom hardware controller, or, alternatively—and perhaps increasingly more common—a "software RAID" in which a RAID disk driver is available as part of the operating system. The RAID driver then can act as a "higher level" layer upon the native disk device drivers to create arbitrary RAID configurations using those disks (or partitions thereof).

The *RAIDframe* [26] software toolkit is a research vehicle for experimenting with different software RAID organisations. It has been ported to several variants of Unix, offering various levels of functionality: as a RAID simulator and a RAID device driver. For example, NetBSD 1.4 onwards includes a port of RAIDframe that enables, through a kernel configuration option, a RAID pseudo device driver. This driver allows RAIDs of levels 0, 1, 4, and 5 to be created using block devices or even using other RAIDs. So, it is possible to create a RAID 0 array out of several RAID 5 arrays, or to construct other arbitrarily complex hierarchies. The driver also supports hot spares, various parity declustering options, and so on.

Widely-available software RAID makes it fairly straightforward to build large disk arrays using COTS hardware. For example, a single PC clone with four wide SCSI host adapter cards, each driving fifteen 40 GB SCSI hard drives, could provide over 2 terabytes of RAID storage. However, in such a setup, the limiting factors are likely to be the memory bus bandwidth and SCSI controller capabilities.

12.8 Networked Storage

There is a vast literature on networked storage. Put simply, *networked storage* is storage available via a local-area or wide-area network. Typically, a *client/server* model is employed, in which the actual disk storage is provided by *servers*, and accessed over the network by *clients*. The division of labour between client and server can vary greatly. At one extreme, the server can act as a "black box," providing all functionality of authentication, name lookup, consistency, and data delivery (e.g., Network File System (NFS) [79]); at the other, it can provide simply a low-level storage abstraction that clients may read and write (e.g., the Swarm storage server [43]).

From a hardware perspective, networked storage can be divided into the following broad categories:

Server Attached Disks: Storage made available by the server consists of disks that are directly attached to the server host. An archetypal example is one or more hard disks connected to the server via a SCSI host adapter. Only the server host computer has direct access to the disks. All other access—by networked clients—must be via the server itself.

Network Attached Disks: This organisation has disks directly attached to the network and able to send data to clients via some high-level networking protocol such as TCP/IP. Usually, network attached disks require some form of lightweight aid in the form of file manager servers that mediate client requests to the disks themselves. The file manager interacts with the network attached disks via a private network to ensure integrity of data on the disks is maintained. Once authorised, network attached disks deliver data directly to the client making the request.

Storage Area Networks: Here, a high-speed, private network separates storage devices (disk, tape, etc.) and storage servers. The storage servers act as a front-end to the storage pool, and any storage server can communicate with any storage device on the storage area network (SAN). Clients make storage requests via any storage server. Data is transferred between client and storage devices via servers.

From a server/file manager point of view, networked storage can be divided into the following broad categories:

Central: All client requests are served by a single server, which carries the full burden of file management functionality. NFS [79] is the archetypal centralised server. Central server configurations are relatively straightforward in design and implementation, but lack scalability.

Distributed: Distributed servers cooperate with peer servers in order to satisfy client requests. Typically, distributed servers will partition the namespace and each will handle some portion of it. Examples of this category are the Andrew File System (AFS) [82] and the Sprite file system [70].

Merged client/server: In this category, clients can act as both clients and servers, acting as a server for locally-attached storage, and as a client when accessing non-local storage. Most of the file management burden is placed on the clients. Network of Workstations (NOW) [5] and xFS [96] are typical examples of this category.

There is a lack of standardised terminology in the area of large-scale, reliable, scalable storage. Devlin *et al.* [29] offer a taxonomy describing typical enterprise-scale storage organisations. Much terminology, however, is industry-driven.

Current emphasis is away from Small Computer Systems Interface (SCSI)-based disk systems to fibre channel-capable drives, which support the notion of network-attached storage. Fibre channel [11] is a high-speed serial interface offering speeds upwards of 100 MB/sec. It is more flexible than SCSI in that it supports three basic interconnection topologies that may be scaled up: point-to-point; fabric (switch); and arbitrated loop (ring). An arbitrated loop can support up to 126 devices without the need of a switch. This is more scalable than SCSI, which can support only up to 15 devices per bus. In addition, fibre channel can be driven over much larger distances—1 to 10 km, depending upon wiring—as opposed to a maximum of 25 metres with SCSI. Fibre channel integrates both storage and networking protocols, and is implemented using six protocol layers.

INTELLIGENT DISKS

Storage research continues the trend of migrating functionality into the disk itself. Instead of moving data to code, one approach is to move the code to the data and execute high-level functions on the drive itself [1, 49, 76]. In effect, make the disk an object-oriented device that responds to high-level method invocations from clients. In such a model, files become objects, the management of which becomes largely opaque to the outside world.

Keeton *et al.* [49] at Berkeley put forward a case for *Intelligent Disks*, or *IDISKs*, that provide higher-level application functionality on the disk itself. Gibson *et al.* [39] describe a disk-centric storage architecture called Network Attached Secure Disks (NASD) that also concentrates functionality on the disk itself. An extension of this work is that of *Active Disks* undertaken at Carnegie Mellon and at the University of California, Santa Barbara.

Riedel and Gibson [76] at Carnegie Mellon describe the issues and potential benefits of being able to execute code directly on NASD. Applications studied include database select and parallel sort. Their proposal of active disks localises activity on the disk, with relatively modest processing requirements and no communication between disks.

Acharya *et al.* [1] at UCSB also propose an active disk architecture that is a little more powerful in its scope. In it, *disklets* run on disks, dispatched and coordinated by a host. They outline several applications, including: SQL select and group-by; external sort; relational database datacube computation; image convolution; and generation of composite satellite images. Their simulation results indicate active disks outperform conventional disk architectures.

The IDISK architecture described in Keeton *et al.* [49] is probably the most powerful, in that it posits higher bandwidth disk-to-disk communication, allowing for more general-purpose parallelism.

All of the active disk architectures are predicated on increased technological sophistication of the disk drive itself. The processors used in current drives are typically one generation behind those used in the host computers driving them. Making a transition from 0.68 micron to a 0.35 micron processor would, for example, enable the integration of a 200 MHz StrongARM RISC core into the current ASIC die space, plus an additional 100,000 gates for specialised processing in addition to incorporating all existing formatting, servo, ECC, and SCSI functions [76]. Experts at Seagate anticipated that by 2001, drives would provide upwards of 100 MIPS in processing power, plus up to 64 MB of RAM and 2–4 MB of flash memory on the drive [4].

PARALLEL FILE SYSTEMS

In terms of software organisation, many networked storage architectures have been proposed. One important class is that of parallel file systems. Initially, these were relatively simple transpositions of the traditional Unix file system semantics that treated files as linear sequences of bytes [74]. However, one distinguishing feature of parallel file systems is that disk I/O is tightly coupled with the parallel computation. Thus, parallel I/O began to assume more importance in parallel computing, especially in those applications that are data-driven. The Message Passing Interface (MPI) working group is evolving standards for parallel I/O [23], as is the Scalable I/O Initiative [25].

Some parallel file systems recognise that files can be structured to mirror the application and workload under which they will be used. The Vesta Parallel File System [24] treats files as multiple disjoint partitions that can be accessed in parallel, and allows layout to be tuned to anticipated access patterns. The Hurricane File System [54] provides hierarchical building blocks that can be composed to custom-build access methods for files in a shared-memory multiprocessor environment. Application data files thus can be associated with precisely-tuned semantics appropriate to their chosen application and access patterns.

Disk-Directed I/O [53] is a scalable I/O approach for MIMD multiprocessor systems. Instead of compute nodes sending requests to I/O nodes independently, they instead *collectively* send a single request to all I/O nodes. The I/O nodes then control the flow of data back to the compute nodes to best maximise disk bandwidth. The Galley Parallel File System [68] implements a version of disk-directed I/O, although its I/O requests are not explicitly collective. The PARADISE system [15] integrates many theoretical parallel I/O techniques into an experimental file system, including disk-directed I/O; cooperative caching; flexible file structuring via metadata; and selectable file access methods. The TickerTAIP [17] parallel RAID architecture bridges across the parallel I/O and RAID models. River [7] is another approach to parallel I/O management. It provides a data-flow programming model and I/O layer for cluster environments. It uses a *distributed queue* to load balance the workload amongst data consumers in the system, and *graduated declustering* to do workload balancing amongst the data producers of the system. The goal is to provide persistent peak performance across workload perturbations.

The IEEE Storage System Standards Working Group has been working on a reference model for a large-scale, wide-area, scalable storage system designed for petabyte-scale repositories of potentially billions of datasets, each of which could be terabytes in size and spread across heterogeneous secondary and tertiary storage [27]. The model is designed to be widely dispersed, geographically, yet support an I/O throughput of gigabytes per second. The High Performance Storage System (HPSS) project is an implementation based upon the IEEE Mass Storage System Reference Model. The HPSS concentrates on highly-parallel, supercomputer, and cluster environments, and seeks to deliver data at upwards of 100 MB/sec [28]. One of the design criteria of the HPSS is to utilise existing and emerging standards for increased deployability. Another design goal is to provide API access at appropriate levels for use by separate digital library, object storage, and data management systems [28].

DISTRIBUTED FILE SYSTEMS
Distributed file systems are networked file systems that spread file system contents over several machines. In the simplest case, servers handle all file system responsibilities, including serving data; maintaining consistency; and handling locking. In more advanced cases, this workload is shared between clients and servers, and in some systems, clients may act as both clients and servers.

Although the NFS may be used to mimic a distributed file system, it is not really considered one. Sprite [70], developed at Berkeley, is a distributed file system in which the global file system namespace is partitioned across multiple domains. A file server may handle one or more domains. Sprite uses aggressive caching to improve performance. Clients cache files locally, to lessen server workloads. However, when a file becomes write-shared, the server becomes involved, sending call-back messages to all clients with the file open to disable caching for that file. The AFS [82] is similar to Sprite in that the file system is distributed across servers. Like Sprite, AFS maintains state, and performs callbacks to clients when cached shared file data is modified by another client. Consistency in AFS is at the whole file level, with the last file written being the version maintained by the server.

The biggest problem with AFS and Sprite is that they do not scale well to large numbers of clients. Availability is also a problem, should a given server become disconnected. The consistency granularity of AFS is also weaker than traditional Unix semantics. AFS was commercialised and developed as the Distributed File System (DFS). The DFS is the basis of the Open Software Foundation (OSF) Distributed Computing Environment (DCE) [69], and provides stronger consistency semantics over AFS. Coda ("COnstant Data Availability") [81] builds further upon AFS, and is intended to provide improved availability and even support disconnected operation for mobile users. In Coda, the global namespace is partitioned into volumes consisting of files and directories. A volume is the unit of replication. It may be explicitly replicated when it is created, and a *volume replication database* is used to keep track of all volumes. The volume replication database is itself replicated at every server.

Coda clients use more aggressive caching than in AFS in that they can proactively cache entire files for future disconnected operation. Disconnected operation may be the result of network inaccessibility to any Coda server replicating a volume used by the client, or a voluntary disconnection, such as a laptop computer being removed physically from the network. The local disk of a Coda client is treated as a file cache, and, during disconnected operation the client becomes a server, handling all requests out of its local cache. Upon reconnection, modifications are consolidated with the servers of the volumes used by the disconnected client. A repair tool is provided to allow the user to resolve inconsistencies that cannot be resolved by Coda automatically.

Berkeley's NOW project produced xFS, a *Serverless Network File System* [6]. In xFS, clients are both clients and servers: there is no centralised server (or small subset of machines designated as servers) in xFS. This greatly improves scalability. Metadata management is distributed across all nodes, instead of being partitioned onto canonical servers.

The Global File System (GFS) is a serverless file system in which a cluster of clients connects to network attached storage devices arranged in a network storage pool connected via a storage area network [90]. GFS uses a distributed lock manager controlled by the storage devices. Devices in the network storage pool are arranged into *subpools*, grouping devices with similar characteristics. These subpools are then exploited for efficiency. For example, file metadata is placed on low-latency subpools, whereas file data is placed on high-bandwidth subpools. Although GFS is designed for clusters, it can be extended to LAN and WAN environment by GFS clients exporting file systems via other protocols such as NFS and Hypertext Transfer Protocol (HTTP) [13].

Clusters are receiving a lot of attention as a way of producing low-cost, powerful, scalable computing facilities. Correspondingly, there has been significant cluster development resulting in mature products. Clusters are attractive because of their smaller administrative domain than wide-area distributed file systems. This allows for lower latency, higher stable bandwidth, and simpler security considerations. A distributed lock manager, for example, is easier to deploy with high performance in a SAN or cluster environment, than over a wide-area network.

VAXClusters and Tru64 Unix cluster provide high-performance, fault-tolerant cluster-level file systems [55,61]. Cluster members can share locally attached storage, or use NASD over a variety of interconnection fabrics. A hierarchical lock structure and distributed lock manager control file sharing. The lock manager can detect deadlocks, and the system features failure recovery. The Lustre cluster file system is similar in scope [59].

The Berkeley NOW work has sparked much cluster-level file system work. The development of Sprite [70] gave rise to Zebra [44], which provided striping across disks. Instead of striping files, Zebra uses a per-client log which is striped across the disks rather like a log-structured file system. The Swarm storage system, developed at the University of Arizona, is similar in that it provides a striped log abstraction on cluster storage nodes for clients [43]. DEC's Frangipani [93] is a cluster-level storage system based upon their Petal [57] virtual disks. Petal provides virtual disks, each of which has a 64-bit address space. A virtual disk is made up of a pool of physical disks that may be spread across multiple servers. Petal is a software RAID in many respects. It features component failure tolerance and chained declustered data access. Copy-on-write techniques enable Petal virtual disks to support efficient snapshots, for online backup purposes. Frangipani builds on Petal to provide a consistent view of the same set of files to all clients, and uses a distributed lock manager to ensure coherence.

12.9 Conclusion

There is a rich history and literature of research on storage, which has been partly surveyed here. When one focuses on ever-growing collections, the need for efficient and scalable storage is paramount.

Efficiency can be addressed in many ways. Primarily, efficiency is achieved when storage is tuned to its intended workload and to the limitations of device technology. In the case of disks, the time to perform long seeks is the most time consuming task. As a result, minimising seeks and performing large, sequential transfers can maximise disk drive performance and throughput. Different file system layouts can aid in this.

The performance effects of disk head movement may be mitigated by the use of disk arrays and networked storage, which offer greater possibilities for parallelism in data transfer. Not only do these allow the total amount of storage to scale quickly, but they also offer the possibility of increasing total reliability and fault tolerance through the use of redundancy.

With an increase in scale comes an attendant increase in storage management and administration.   The latest trend in storage research is to move towards self-management of data, and to migrate increasing functionality and responsibility for layout, access control, and administration onto the disk itself. Such a trend meshes well with the object-based paradigm of digital libraries collections, especially when multimedia content is involved. For scalable support of such multimedia collections, it seems wise to couple intelligent disks, parallel/distributed file systems, performance improvement techniques, and insights gleaned from workload studies. Many of these issues are considered further in the next chapter.

References

[1]   A. Acharya, M. Uysal, and J. Saltz. Active disks: Programming model, algorithms and evaluation. In *Proceedings of Architectural Support for Programming Languages and Operating Systems (ASPLOS) VIII*, pages 81–91. ACM, Oct. 1998.

[2]   L. A. Adamic and B. A. Huberman. The Web's hidden order. *Commun. ACM*, 44(9): 55–60, Sept. 2001.

[3]   G. A. Alvarez, E. Borowsky, S. Go, T. H. Romer, R. Becker-Szendy, R. Golding, A. Merchant, M. Spasojevic, A. Veitch, and J. Wilkes. Minerva: an automated resource provisioning tool for large-scale storage systems. *ACM Trans. Comput. Syst.*, 19(4): 483–518, Nov. 2001.

[4]   D. Anderson. Consideration for smarter storage devices. Presentation given at the National Storage Industry Consortium (NSIC) Network-Attached Storage Device working group meeting, June 1998. Available from http://www.nsic.org/nasd/.

[5]   T. E. Anderson, D. E. Culler, D. A. Patterson, and the NOW team. A case for NOW (Network Of Workstations). Technical report, University of California at Berkeley, 1994.

[6]   T. E. Anderson, M. D. Dahlin, J. M. Neefe, D. A. Patterson, D. S. Roselli, and R. Y. Wang. Serverless network file systems. Technical report, University of California at Berkeley, 1995.

[7]   R. H. Arpaci-Dusseau, E. Anderson, N. Treuhaft, D. E. Culler, J. M. Hellerstein, D. Patterson, and K. Yelick. Cluster I/O with River: Making the fast case common. In *Proceedings of IOPADS '99: Input/Output for Parallel and Distributed Systems*, Atlanta, Georgia, May 1999.

[8]   M. G. Baker, J. H. Hartman, M. D. Kupfer, K. W. Shirriff, and J. K. Ousterhout. Measurements of a distributed filesystem. In *Proceedings of Symposium on Operating Systems Principles (SOSP) 13*, pages 198–212, Oct. 1991.

[9]   T. Barclay, J. Gray, and D. Slutz. Microsoft TerraServer: A spatial data warehouse. In *Proceedings of the 2000 ACM SIGMOD International Conference on Management of Data*, Austin, Texas, USA, May 2000.

[10]   R. Barve, E. Shriver, P. B. Gibbons, B. K. Hillyer, Y. Matias, and J. S. Vitter. Modeling and optimizing I/O throughput of multiple disks on a bus. In *Proceedings of the 1999 Conference on Measurement and Modeling of Computer Systems (SIGMETRICS)*, pages 83–92, Altanta, Georgia, USA, May 1999.

[11]   A. Benner. *Fibre Channel: Gigabit I/O and Communications for Computer Networks*. McGraw-Hill, New York, NY, 1996.

[12]   J. M. Bennett, M. A. Bauer, and D. Kinchlea. Characteristics of files in NFS environments. In *ACM Symposium on Small Systems*, pages 33–40, 1991.

[13]   T. Berners-Lee, R. T. Fielding, and H. F. Nielsen. Hypertext Transfer Protocol – HTTP/1.0. RFC 1945, May 1996. http://www.ietf.org/rfc/rfc1945.txt.

[14]   E. Borowsky, R. Golding, P. Jacobson, A. Merchant, L. Schreier, M. Spasojevic, and J. Wilkes. Capacity planning with phased workloads. In *WOSP '98*, Santa Fe, New Mexico, USA, Oct. 1998.

[15]   M. Brodowicz and O. Johnson. PARADISE: An advanced featured parallel file system. Technical report, University of Houston, 1998.

[16]   M. Calzarossa and G. Serazzi. Workload characterization: A survey. *Proceedings of the IEEE*, 81(8): 1136–1150, Aug. 1993.

[17]   P. Cao, S. Lim, S. Venkataraman, and J. Wilkes. The TickerTAIP parallel RAID architecture. In *Proceedings of the 20th Symposium on Computer Architecture*, pages 52–63, May 1993.

[18]   P. Cao, S. B. Lim, S. Venkataraman, and J. Wilkes. The TickerTAIP parallel RAID architecture. *ACM Trans. Comput. Syst.*, 12(3): 236–269, Aug. 1994.

[19] J. Chandy and A. L. N. Reddy. Failure evaluation of disk array organizations. In *Proceedings of the International Conference on Distributed Computing Systems*, 1993.

[20] P. Chen, E. Lee, G. Gibson, R. Katz, and D. Patterson. RAID: High-performance, reliable secondary storage. *ACM Computing Surveys*, 26(2): 145–188, June 1994.

[21] P. M. Chen, W. T. Ng, S. Chandra, C. Aycock, G. Rajamani, and D. Lowell. The Rio file cache: Surviving operating system crashes. In *Proceedings of Architectural Support for Programming Languages and Operating Systems (ASPLOS) VII*, pages 74–83, Oct. 1996.

[22] T. Chiueh. Trail: a track-based logging disk architecture for zero-overhead writes. In *Proceedings of the 1993 IEEE International Conference on Computer Design (ICCD '93)*, pages 339–343, Cambridge, MA, USA, 3–6 Oct. 1993.

[23] P. Corbett, D. Feitelson, S. Fineberg, Y. Hsu, B. Nitzberg, J.-P. Prost, M. Snir, B. Traversat, and P. Wong. Overview of the MPI-IO parallel I/O interface. In R. Jain, J. Werth, and J. C. Browne, editors, *Input/Output in Parallel and Distributed Computer Systems*, volume 362 of *The Kluwer International Series in Engineering and Computer Science*. Kluwer, Amsterdam, 1996.

[24] P. F. Corbett and D. G. Feitelson. The Vesta parallel file system. *ACM Transactions on Computer Systems*, 14(4): 225–264, Aug. 1996.

[25] P. F. Corbett, J.-P. Prost, C. Demetriou, G. Gibson, E. Reidel, J. Zelenka, Y. Chen, E. Felten, K. Li, J. Hartman, L. Peterson, B. Bershad, A. Wolman, and R. Aydt. Proposal for a common parallel file system programming interface. Technical report, Department of Computer Science, University of Arizona, Tucson, Arizona, USA, 1996. Available via the WWW at http://www.cs.arizona.cdu/sio/api1.0.ps.

[26] W. V. Courtright, G. Gibson, M. Holland, and J. Zelenka. RAIDframe: Rapid prototyping for disk arrays. In *SIGMETRICS '96*, pages 268–269, May 1996.

[27] R. A. Coyne and H. Hulen. An introduction to the Storage System Reference Model, version 5. In *Proceedings of the Twelfth IEEE Symposium on Mass Storage*, Apr. 1993.

[28] R. A. Coyne, H. Hulen, and R. Watson. The High Performance Storage System. In *Proceedings of the Conference on Supercomputing '93*, pages 83–92, Portland, Oregon, USA, 15–19 Nov. 1993 Association for Computing Machinery.

[29] B. Devlin, J. Gray, B. Laing, and G. Spix. Scalability terminology: Farms, clones, partitions and packs: RACS and RAPS. Technical Report MS-TR-99-85, Microsoft Research, Dec. 1999.

[30] J. R. Douceur and W. J. Bolosky. A large-scale study of file-system contents. In *SIGMETRICS '99*, pages 59–70, Atlanta, Georgia, USA, 29 Apr.–6 May 1999.

[31] M. R. Ebling and M. Satyanarayanan. SynRGen: An extensible file reference generator. In *SIGMETRICS '94*, pages 108–117, Santa Clara, CA, May 1994.

[32] J. Gait. Phoenix: A safe in-memory file system. *Commun. ACM*, 33(1): 81–86, Jan. 1990.

[33] G. R. Ganger. Generating representative synthetic workloads: An unsolved problem. In *Proceedings of the Computer Measurement Group (CMG) Conference*, pages 1263–1269, Dec. 1995.

[34] G. R. Ganger and Y. N. Patt. The process-flow model: Examining I/O performance from the system's point of view. In *Proceedings of the 1993 Conference on Measurement and Modeling of Computer Systems (SIGMETRICS)*, pages 86–97, May 1993.

[35] G. R. Ganger and Y. N. Patt. Soft updates: A solution to the metadata update problem in file systems. Technical Report CSE-TR-254-95, University of Michigan, 1995.

[36] G. R. Ganger, B. L. Worthington, and Y. N. Patt. The DiskSim simulation environment version 1.0 reference manual. Technical Report CSE-TR-358-98, Department of Electrical Engineering and Computer Science, University of Michigan, Ann Arbor, MI, 27 Feb. 1998.

[37] S. Ghandeharizadeh and D. Ierardi. An algorithm for disk space management to minimize seeks. Technical Report 95–612, Department of Computer Science, University of Southern California, Los Angeles, California, 1995.

[38] S. Ghandeharizadeh, D. Ierardi, and R. Zimmermann. Management of space in hierarchical storage systems. Technical Report 94–598, Department of Computer Science, University of Southern California, Los Angeles, California, 1994.

[39] G. A. Gibson, D. F. Nagle, K. Amiri, J. Butler, F. W. Chang, H. Gobioff, C. Hardin, E. Riedel, D. Rochberg, and J. Zelenka. A cost-effective, high-bandwidth storage architecture. In *Proceedings of Architectural Support for Programming Languages and Operating Systems (ASPLOS) VIII*, pages 92–103. ACM, Oct. 1998.

[40] T. J. Gibson and E. L. Miller. Long term file activity patterns in a UNIX workstation environment. In *Proceedings of the Fifteenth IEEE Symposium on Mass Storage Systems*, pages 355–371, College Park, Maryland, USA, 23–26 Mar. 1998.

[41] R. Golding, E. Shriver, T. Sullivan, and J. Wilkes. Attribute-managed storage. In *Workshop on Modeling and Specification of I/O (MSIO)*, San Antonio, Texas, USA, 26 Oct. 1995.

[42] E. Grochowski and R. F. Hoyt. Future trends in hard disk drives. *IEEE Transactions on Magnetics*, 32(3): 1850–1854, May 1996.

[43] J. H. Hartman, I. Murdock, and T. Spalink. The Swarm scalable storage system. In *Proceedings of the 19th IEEE International Conference on Distributed Computing Systems*, pages 74–81, Austin, Texas, USA, 31 May–4 June 1999. IEEE Computer Society.

[44] J. H. Hartman and J. K. Ousterhout. The Zebra striped network file system. *ACM Trans. Comput. Syst.*, 13(3): 274–310, 1995.

[45] D. S. Hirschberg. A class of dynamic memory allocation algorithms. *Commun. ACM*, 16(10): 615–618, Oct. 1973.

[46] M. Holland and G. A. Gibson. Parity declustering for continuous operation in redundant disk arrays. In *Proceedings of Architectural Support for Programming Languages and Operating Systems (ASPLOS) V*, pages 23–35, 1992.

[47] Y. Hu and Q. Yang. DCD—Disk Caching Disk: A new approach for boosting I/O performance. In *Proceedings of the 23rd International Symposium on Computer Architecture*, pages 169–178, Philadelphia, PA, USA, May 1996.

[48] R. Kahn and R. Wilensky. A framework for distributed digital object services. Technical report, Corporation for National Research Initiatives, Reston, Virginia, USA, 13 May 1995. `urn:hdl:cnri.dlib/tn95-01`.

[49] K. Keeton, D. A. Patterson, and J. M. Hellerstein. A case for intelligent disks (IDISKs). *SIGMOD Record*, 27(3): 42–52, Aug. 1998.

[50] K. C. Knowlton. A fast storage allocator. *Commun. ACM*, 8(10): 623–625, Oct. 1965.

[51] B. Kobler and J. Berbert. NASA Earth Observing System Data Information System (EOSDIS). In *Proceedings of the Eleventh IEEE Symposium on Mass Storage Systems*, pages 18–19, Monterey, California, USA, 7–10 Oct. 1991.

[52] P. D. L. Koch. Disk file allocation based on the buddy system. *ACM Trans. Comput. Syst.*, 5(4): 352–370, Nov. 1987.

[53] D. Kotz. Disk-directed I/O for MIMD multiprocessors. *ACM Trans. Comput. Syst.*, 15(1): 41–74, Feb. 1997.

[54] O. Krieger and M. Stumm. HFS: A performance-oriented flexible file system based on building-block compositions. *ACM Trans. Comput. Syst.*, 15(3): 286–321, Aug. 1997.

[55] N. P. Kronenberg, H. M. Levy, and W. D. Strecker. VAXcluster: a closely-coupled distributed system. *ACM Trans. Comput. Syst.*, 4(2): 130–146, May 1986.

[56] E. K. Lee and R. H. Katz. Performance consequences of parity placement in disk arrays. In *Proceedings of Architectural Support for Programming Languages and Operating Systems (ASPLOS) IV*, pages 190–199, 1991.

[57] E. K. Lee and C. A. Thekkath. Petal: Distributed virtual disks. In *Proceedings of Architectural Support for Programming Languages and Operating Systems (ASPLOS) VII*, pages 84–92, Oct. 1996.

[58] B. Liskov, S. Ghemawat, R. Gruber, P. Johnson, L. Shrira, and M. Williams. Replication in the Harp file system. In *Proceedings of the 1991 Symposium on Operating Systems Principles (SOSP)*, pages 226–238, Oct. 1991.

[59] Lustre Developers. Lustre home. WWW page, 2000. `http://www.lustre.org`.

[60] J. N. Matthews, D. Roselli, A. M. Costello, R. Y. Wang, and T. E. Anderson. Improving the performance of log-structured file systems with adaptive methods. In *Proceedings of Symposium on Operating Systems Principles (SOSP) 16*, pages 238–251, Saint-Malo, France, Oct. 1997.

[61] K. McCoy. *VMS File System Internals*. Digital Press, 1990.

[62] M. K. McKusick. A fast file system for UNIX. *ACM Trans. Comput. Syst.*, 2(3): 181–197, 1984.

[63] M. K. McKusick and T. J. Kowalski. Fsck_ffs—the UNIX file system check program. In *BSD System Manager's Manual*, chapter 3. Computer Systems Research Group, University of California, Berkeley, July 1985.

[64] J. Menon, D. Mattson, and S. Ng. Distributed sparing for improved performance of disk arrays. Technical Report RJ 7943, IBM Almaden Research Center, 1991.

[65]   J. Menon, J. Roche, and J. Kasson. Floating parity and data disk arrays. *Journal of Parallel and Distributed Computing*, 17: 129–139, 1993.

[66]   J. Moran, R. Sandberg, D. Coleman, J. Kepecs, and B. Lyon. Breaking through the NFS performance barrier. In *Proceedings of the EUUG Spring 1990*, Apr. 1990.

[67]   S. W. Ng. Crosshatch disk array for improved reliability and performance. In *Proceedings of the 21st International Symposium on Computer Architecture*, pages 255–264, 1994.

[68]   N. Nieuwejaar and D. Kotz. Performance of the Galley parallel file system. In *Proceedings of IOPADS '96: Input/Output for Parallel and Distributed Systems*, pages 83–94, Philadelphia, Pennsylvania, USA, 1996.

[69]   Open Software Foundation. File systems in a distributed computing environment: A white paper. Technical report, Open Software Foundation, Cambridge, MA, 1991.

[70]   J. K. Ousterhout, A. Cherenson, F. Douglis, M. Nelson, and B. Welch. The Sprite network operating system. *IEEE Computer*, 21, pages 23–36, Feb. 1988.

[71]   J. K. Ousterhout, H. Da Costa, D. Harrison, J. A. Kunze, M. Kupfer, and J. G. Thompson. A trace-driven analysis of the UNIX 4.2 BSD file system. *Operating System Review*, 19(4): 15–24, 1985.

[72]   J. K. Ousterhout and F. Douglis. Beating the I/O bottleneck: A case for log-structured file systems. *Operating Systems Review*, 23(1): 11–27, 1989.

[73]   V. S. Pai, P. Druschel, and W. Zwaenepoel. IO-Lite: A unified I/O buffering and caching system. In *Proceedings of the Third Symposium on Operating Systems Design and Implementation (OSDI '99)*, pages 15–28, New Orleans, Louisiana, USA, 22–25 May 1999.

[74]   P. Pierce. A concurrent file system for a highly parallel mass storage system. In *Fourth Conference on Hypercube Concurrent Computers and Applications*, pages 155–160, 1989.

[75]   K. K. Ramakrishnan, P. Biswas, and R. Karedla. Analysis of file I/O traces in commercial computing environments. *Performance Evaluation Review*, 20(1): 78–90, June 1992.

[76]   E. Riedel and G. A. Gibson. Active disks - remote execution for network-attached storage. Technical Report CMU-CS-97-198, Carnegie Mellon University, Dec. 1997.

[77]   M. Rosenblum and J. K. Ousterhout. The design and implementation of a log-structured file system. In *Proceedings of the Thirteenth Symposium on Operating Systems Principles (SOSP)*, pages 1–15, Pacific Grove, California, USA, 13–16 Oct. 1991.

[78]   C. Ruemmler and J. Wilkes. An introduction to disk drive modeling. *IEEE Computer*, 27(3): 17–29, Mar. 1994.

[79]   R. Sandberg, D. Goldberg, S. Kleiman, D. Walsh, and B. Lyon. Design and implementation of the Sun Network File system. In *Proceedings of the Summer USENIX Conference*, pages 119–130, 1985.

[80]   M. Satyanarayanan. A study of file sizes and functional lifetimes. In *Proceedings of Symposium on Operating Systems Principles (SOSP) 8*, pages 96–108. Association for Computing Machinery, Dec. 1981.

[81]   M. Satyanarayanan. Coda: A highly available file system for a distributed workstation environment. In *Proceedings of the Second IEEE Workshop on Workstation Operating Systems*, Sept. 1989.

[82]   M. Satyanarayanan. Scalable, secure, and highly available distributed file access. *IEEE Computer*, 23, pages 9–20, May 1990.

[83]   J. Schindler and G. R. Ganger. Automated disk drive characterization. In *Proceedings of the 2000 Conference on Measurement and Modeling of Computer Systems (SIGMETRICS)*, pages 112–113, Santa Clara, California, USA, June 2000.

[84]   M. Schulze, G. Gibson, R. Katz, and D. Patterson. How reliable is a RAID? In *Proceedings of the 34th IEEE Computer Society International Conference*, pages 118–123, 1989.

[85]   M. Seltzer, K. Bostic, M. K. McKusic, and C. Staelin. An implementation of a log-structured file system for UNIX. In *1993 Winter USENIX Conference*, San Diego, CA, 25–29 Jan. 1993.

[86]   M. Seltzer and M. Stonebraker. Read optimized file system designs: A performance evaluation. In *Proceedings of the 7th International Conference on Data Engineering*, Kobe, Japan, 8–12 Apr. 1991.

[87]   E. Shriver. A formalization of the attribute mapping problem. Technical Report HPL-SSP-95-10 Rev D, Hewlett-Packard Laboratories, Palo Alto, California, USA, 15 July 1996.

[88]   E. Shriver. *Performance Modeling for Realistic Storage Devices*. PhD thesis, Department of Computer Science, New York University, New York, USA, May 1997.

[89]   E. Shriver, A. Merchant, and J. Wilkes. An analytical behaviour model for disk drives with readahead caches and request reordering. In *Proceedings of the 1998 Conference on Measurement and Modeling of Computer Systems (SIGMETRICS)*, pages 182–191, June 1998.

[90]   S. R. Soltis. *The Design and Implementation of a Distributed File System based on Shared Network Storage*. PhD thesis, University of Minnesota, Aug. 1997.

[91]   D. Stodolsky and G. A. Gibson. Parity logging: overcoming the small write problem in redundant disk arrays. In *Proceedings of the 1993 Symposium on Operating Systems Principles (SOSP)*, 1993.

[92]   A. S. Szalay, P. Kunszt, A. Thakar, J. Gray, D. Slutz, and R. J. Brunner. Designing and mining multi-terabyte astronomy archives: The Sloan Digital Sky Survey. In *Proceedings of the 2000 ACM SIGMOD International Conference on Management of Data*, Austin, Texas, USA, May 2000.

[93]   C. A. Thekkath, T. Mann, and E. K. Lee. Frangipani: A scalable distributed file system. In *Proceedings of Symposium on Operating Systems Principles (SOSP) 16*, pages 224–237, Oct. 1997.

[94]   M. Uysal, G. A. Alvarez, and A. Merchant. A modular, analytical throughput model for modern disk arrays. In *Ninth International Symposium on Modeling, Analysis and Simulation of Computer and Telecommunication Systems (MASCOTS 2001)*, Cincinnati, Ohio, USA, 15–18 Aug. 2001.

[95]   W. Vogels. File system usage in Windows NT 4.0. In *Proceedings of the 17th Symposium on Operating Systems Principles (SOSP)*, pages 93–109, Charleston, USA, 12–15 Dec. 1999. Association for Computing Machinery.

[96]   R. Y. Wang and T. E. Anderson. xFS: A wide area mass storage file system. Technical report, University of California at Berkeley, 1993.

[97]   J. Wilkes. The Pantheon storage-system simulator. Technical Report HPL-SSP-95-14, Hewlett-Packard Laboratories, Palo Alto, California, USA, May 1996.

[98]   J. Wilkes, R. Golding, C. Staelin, and T. Sullivan. The HP AutoRAID hierarchical storage system. In *Proceedings of Symposium on Operating Systems Principles (SOSP) 15*, pages 96–108, Dec. 1995.

[99]   P. R. Wilson, M. S. Johnstone, M. Neely, and D. Boles. Dynamic storage allocation: A survey and critical review. In *Proceedings of the 1995 International Workshop on Memory Management*, Kinross, Scotland, United Kingdom, 27–29 Sept. 1995.

[100] B. L. Worthington, G. R. Ganger, Y. N. Patt, and J. Wilkes. On-line extraction of SCSI disk drive parameters. In *Proceedings of the 1995 Conference on Measurement and Modeling of Computer Systems (SIGMETRICS)*, pages 146–156, Ottawa, Canada, May 1995.

13 OBJECT REPOSITORIES FOR DIGITAL LIBRARIES

Prof. Edward A. Fox and Dr. Paul Mather

Digital libraries encompass an enormous range of functionality. One fundamental task of a digital library is to store digital content. Storage management in digital libraries is delegated conceptually to the *repository* component. As the amount and complexity of content continues to grow, the issue of scalability grows more pressing. Like traditional file systems, digital library repositories must be capable of scaling to billions of objects and petabytes (2^{50} bytes) of total storage.

Here we address the issue of scalability in digital libraries object repositories, in particular with reference to existing approaches in storage systems and digital libraries research. The requirements of digital libraries object repositories are contrasted with those of file systems in order to identify the core needs of the former. The primary important characteristics of digital libraries objects are location-independent persistent identifiers for naming and an encapsulated object-based approach supporting diverse *metadata* for objects.

In this chapter we first differentiate digital objects from files. Then, we address the important area of naming and location of digital objects, and, in particular, survey many schemes that may be used to locate an object efficiently as the total number of objects grows dynamically to be very large. We then address the important area of metadata and the different schemes used to represent and encapsulate it. Of particular focus here is the *bucket* paradigm for representing digital objects. Finally, we address digital libraries object repositories, with an emphasis on the repository model introduced by Kahn and Wilensky and its derivatives.

13.1 Introduction

Although there has been extensive and intense research in the area of file and disk storage, there has been comparatively little undertaken in the area of storage for digital libraries. Most of what has been done is at the application level: data structures and file layouts for indices; protocols for search and data interoperation; file formats for archival storage; etc. Digital library metrics studies have focused on user interface and behavioural concerns, along with search session and query characterisation [35].

Although digital libraries are voracious consumers of storage, a tacit assumption in most cases is that the underlying operating system I/O substrate will be used for storage. An application-layer shim is applied to provide the more sophisticated I/O that is characteristic of digital libraries [48]. This may lead to a severe "impedance mismatch" between the lower and upper layers.

One notable situation, in which storage and access is given a great deal of attention, is in the multimedia area. There is a large volume of literature on the design and implementation of architectures for digital multimedia, with a particular focus on video servers. Although this is but a subset of the digital library universe, it indicates that an attention to low-level detail is warranted where performance is necessary.

Traditional file systems treat files abstractly, as simple linear sequences of bytes, with little or no additional semantics. Traditional file system files have limited *metadata* attached to files. What metadata exists is limited to information about the low-level storage of the file, and the ownership and access dispositions of the file. Almost no semantic information is associated with the file, reinforcing the "neutral linear byte stream" semantics prevalent in traditional file systems. There, the mantra is that applications provide semantics. Whereas that is true to an extent, great performance improvements can result from matching a file's storage and access at the low level to its intended application. File system studies that have tracked file types, consistently report a good correlation between file size and file type (semantics) [13].

In addition, a file's type may limit its access semantics. For example, compressed files are often not amenable to random access, because the decompression state of the current block depends upon that of all the previous blocks in the file. Such files, therefore, will almost certainly be accessed linear sequentially. (Yet, there are compressed file formats that have been designed to support nonlinear access, most notably many multimedia file formats.)

All this semantic information may be used to improve storage layout and efficiency. A file system that naively assumes all files will be accessed in the same way misses an opportunity to tune performance to file semantics.

13.2 Digital Objects vs. Files

A major differentiation between file systems and digital libraries is that file systems operate on files and digital libraries operate on *digital objects*. Digital objects are more sophisticated than files, and more abstract. Files, by comparison, are concrete and simple.

Digital objects have an underlying data representation of the digital object, and they also have associated metadata that contains information *about* the digital object. At a minimum, the metadata includes a unique *handle* [65] by which the digital object is known.

Files reside in file systems, whereas digital objects reside in *repositories* [23]. Repositories are responsible for storing digital objects and securing them according to *rights* associated with the digital object. These rights may include *terms and conditions* under which the digital object may be *disseminated* by the repository. A *repository access protocol* is used to effect appropriate access to digital objects stored in a repository, and to undertake the *deposit* of digital objects into a repository.

A digital object repository is *not* a digital library. A digital library is much more than a digital object repository, and includes functionality to facilitate user interaction; searching; browsing; workflow and content management; and more. A digital object repository is one *element* of a digital library, much in the same way a file system is part of a larger *operating system* infrastructure. A digital object repository can provide a functional substrate upon which digital libraries may be built.

Another characteristic difference between digital objects and files is that digital objects are typically longer-lived. An important facet of digital libraries—like that of conventional libraries—is *content management.* Digital libraries usually host *collections,* the contents of which are chosen carefully. There is cataloguing effort associated with introducing an item into a collection, and so to maximise the return on such investment, items are usually introduced with a long-term archival storage goal in mind.

Although collection items do degrade, go missing, or are otherwise deleted from conventional library collections, the frequency of deletion is small compared with the frequency of access in the collection as a whole. Unlike conventional file systems, which may contain very short-lived "temporary files," or files whose contents are updated frequently, digital objects are relatively *static,* or *read-mostly* in file system terms. Even then, updating a digital object is usually considered a significant event, and may result in different explicit *versions* of the digital object.

Digital objects also may be complex compound objects. For example, a digital object could be the result of running a given program on a given input file with given parameters.

Finally, a digital object may actually contain no digital data at all: it may consist entirely of metadata that refers to some non-digital archival object such as an *objet d'art* owned by a museum. Unlike file system metadata, which is relatively simple and focused on low-level storage concerns, digital object metadata often is very rich and it may not be uniform from digital object to digital object. In fact, a given digital object may be associated with varied encodings of the same underlying metadata, to make it inter-operable between several different cataloguing conventions.

13.3 Naming and Location

Naming is very important for digital libraries. Unlike local file systems, names for digital objects are usually intended to be *globally unique.* Part of this stems from conventional libraries, which deal with authoritative versions of objects, and part stems from the desire for federation.

Names in file systems denote specific bitstreams as they exist at the current time. Names for digital objects denote the underlying *intellectual property,* or digital object, as opposed to specific bitstreams or particular renditions of that digital object. A digital object may contain or support several possible *disseminations* of that digital object, but these are not generally considered to be separate digital objects in themselves. In file system terms, a digital object is ultimately more like a *directory* in that it is a *container* for all the digital artifacts (metadata and data) directly pertaining to that digital object. One could represent a digital object using a directory in which reside individual files containing the digital content (metadata and data), or even symbolic links to other directories representing digital objects. The directory would represent the digital object, and the directory pathname would serve as its name.

The important point to note about digital objects is that they are *containers* rather than simply individual flat entities. So, whereas a name refers to the digital object container, accessing the innards of that container is a more complicated operation, somewhat akin to accessing an *object* through its *methods* in the object oriented paradigm. In this respect, digital objects are closer to the object-based storage paradigm of IDISK [26], Active Disks [1, 59], and Network Attached Secure Disks (NASD) [18]. So, to access the content of a digital object might require qualifying the name with additional data (e.g., the "method"). The result of an access might be actual data, or other names to dereference.

UNIQUENESS AND LOCATION DEPENDENCE

There are two major classes of names: *locally unique* and *globally unique*. Locally unique names are unique to a particular server, that is, the same name on two different servers may refer to different objects. (For example, the file /etc/motd may contain two entirely different message-of-the-day contents on two different servers at the same instance in time, even though they have the same name: /etc/motd.)

File system file names are typically locally unique, but with a wide-area distributed file system they are usually globally unique (often made so by qualifying them with the global server name).

In addition, a name can be either *location dependent* or *location independent*. Names that are location dependent are tied to a particular server or server cluster, and incorporate the server name as some facet of the name itself. Location independent names incorporate no reference to any particular server holding the named object, and part of the name resolution process involves identifying a server storing the object to which the name refers.

Location dependent names are undesirable for at least three reasons: (1) they are tied to a "physical location," in that if an object is moved (not copied) to another server, the original name becomes invalid; (2) it is not immediately discernible whether two differently named objects are in fact the same thing; and (3) a set of location dependent names is harder to scale because the set is tied to a particular cluster.

A ubiquitous example of a location dependent globally unique name is the Uniform Resource Locator (URL) [RFC 1738] used in the World Wide Web (WWW). URLs explicitly encode the host and location within that host of a given resource. When accessing an object named by a URL, the host named in the URL is contacted. For popular objects, this can lead to severe load problems on the server host so named. There are techniques for alleviating such congestion. Hypertext Transfer Protocol (HTTP) redirects [RFC 1945], Active Names [70], Persistent Uniform Resource Locators (PURLs) [62], the Internet2 Distributed Storage Initiative [4], etc., seek to make location dependent names more location independent by interjecting a layer of indirection in the resolution process.

UNIFORM RESOURCE NAMES

A URL is a form of Uniform Resource Identifier (URI) [RFC 2396]. A Uniform Resource Name (URN) [RFC 1737] is also a form of URI, but one that is location independent. Unlike URLs, which have precise syntax and semantics, URNs have a precise syntax [RFC 2141] but currently rather loose semantics [RFC 2276].

An overview of work on URNs is given in [69]. Paskin [53] delineates many of the functional requirements and progress towards implementing URNs. The fundamental requirements of URNs include: global scope; global uniqueness; persistence; scalability; legacy support; extensibility; and independence [RFC 1737].

The basic syntactic form of a URN is *urn:nid:nss*, where *nid* is the *namespace identifier*, that identifies the particular type of URN, and *nss* is the *namespace-specific string* under the auspices of the *nid*. The semantic interpretation and syntactic format of the *nss* is local to, and defined by, the scheme implemented by the naming authority controlling the *nid* namespace.

There are several efforts currently underway to introduce a working URN scheme. The World Wide Web Consortium (W3C) has a working group looking at URNs [21]. Their effort concentrates on resolver discovery, and they propose adaptations to the existing Domain Name System (DNS) to enable a resolver to be located for a given URN [RFC 2483]. This approach adds a new type of record to DNS databases: the Naming Authority Pointer (NAPTR) [RFC 2915]. The NAPTR includes rewriting rules to enable a URI to locate an appropriate resolver for resolution [RFC 2915]. The W3C has proposed a simple mechanism for resolving URNs via HTTP [RFC 2169]. Powell [56] describes a simple technique that enables the resolution of URNs using the Squid WWW proxy cache.

Two major URN initiatives are the CNRI Handle system [65] and the Document Object Identifier (DOI) [52]. The DOI is a publisher-led effort to develop a URN scheme for persistent identification of intellectual property that includes a mechanism for resolving the DOI to some useful resource or service associated with that intellectual property. As well as identifying an item of intellectual property, a DOI includes basic metadata describing the target of the DOI [52].

The DOI actually uses the Handle system as its resolution mechanism. The Handle system is a global, decentralised, replicated URN system. It is comprised of a *Global Handle Registry* and many *Local Handle Services*. The general syntax of a handle is `urn:hdl:`*naming-authority*/*unique-identifier*. The *naming-authority* is the organisation that has authority over that particular portion of the global handle namespace. In DNS terms, it is akin to being authoritative for a domain. The *naming-authority* has administrative control over the *unique-identifier*s.

The Global Handle Registry is akin to the root nameservers in the DNS. Local Handle Services manage actual handles for one or more naming authorities. A naming authority may be *homed* at either the Global Handle Registry, or at a Local Handle Service. A naming authority can be homed at only one Local Handle Service (or, alternatively, at the Global Handle Registry). Both the Global Handle Registry and a Local Handle Service may actually be composed of several servers, for the purposes of fault tolerance and load balancing. The service at which a naming authority is homed has administrative control over the handles within that naming authority.

A client library is required for applications wishing to resolve handles. Handle servers feature caching, and handle resolution is somewhat akin to that of the DNS in that first the authoritative naming authority is sought, and then the handle is resolved via a handle server of that naming authority. Handles may be composed of a set of values. Each value has its own information, including a *type* entry. Handle resolution queries thus can specify that only values of a certain type, e.g., URL, be returned by a server. This allows a measure of *content negotiation* by the client.

The Handle system is described in more detail in [66]. The chief disadvantage of using handles as a ubiquitous naming scheme is the overhead of registering and resolving them in the global system. For short-lived, temporary objects, this is a significant additional overhead. However, handles are designed primarily to designate significant archival content: digital objects.

SCALABLE OBJECT LOCATION

Name space operations are characteristically lookup operations. An abstract way to think about name space operations is as a hash table in which the key is the name, and the value is the set of operations supported by the underlying object.

One way to scalably distribute such a hash table across the nodes of a cluster is to use a scalable distributed data structure (SDDS) [12, 28, 42-44, 71]. Born out of dynamic hashing [34], *linear hashing* [37] is one of the most common families of SDDS. Litwin, in particular, has proposed many extensions to the basic linear hashing approach, including variants featuring grouping [39], mirroring [40], security [41], scalable availability [38], Reed-Solomon codes [45], and a variant tuned for switched multicomputers [25].

SDDS have three basic design properties: (1) there is no central directory through which clients must access, avoiding "hot spots;" (2) expansion to new servers is gradual, and according to load; and (3) the access and maintenance primitives do not require atomic updates to multiple clients. These properties provide excellent incremental scalability.

SDDS hash tables are accessed via a primary key. For a digital library object repository, we can use the object's URN as a primary key. Because a URN is variable in length, we can canonicalise its format by hashing the URN string to a fixed signature, for example to the MD5 hash of the string. The MD5 hash will canonicalise any URN string to a 128-bit "object ID" with a low probability of collision (2^{-128}). It also will pseudo-randomise any URN, due to the one-way nature of the MD5 hash function. Another suitable candidate for the one-way hash function is the Secure Hash Algorithm (SHA) [17], which hashes an arbitrary length bit stream to a 160-bit string. SHA-1 is used in cryptographic applications for digital signatures.

Randomised placement approaches have been gaining favour in storage area networks because they offer statistical guarantees for access and provide good incremental scalability. The problem with approaches such as Redundant Array of Independent/Inexpensive Disks (RAID) and similar striping approaches is that they do not scale well with the addition of new disks. With high-efficiency parity layouts, often the entire layout must be recomputed with the addition of extra disks, necessitating a large movement of data. Randomised placement also has the advantage that it performs well for non-predictive access patterns.

The SICMA [7] and RIO [6] projects use randomised placement for multimedia servers. Both are resource efficient, and ensure an even distribution of requests amongst the disks in the system, even in the presence of non-predictive access patterns. RIO also employs partial replication of disk blocks to further evenly distribute the request load across the disks in the system [60]. It has been shown that even for predictive access patterns, RIO performs equivalently to that of normal striping approaches [61].

The PRESTO [5] project improves upon SICMA and RIO. PRESTO uses a parity approach rather than straightforward replication to improve space efficiency. PRESTO also has an efficient re-mapping technique that requires only $1/(n+1)$ virtual data blocks to be moved when adding an additional disk [5].

With the increasing popularity of peer-to-peer networks for cooperative wide-area storage and serving has come a focus on algorithms for scalable object location. The explosive growth of the WWW has necessitated a need for scalability to alleviate "hot spots" and ensure consistent and timely access to highly popular content. In particular, Karger *et al.* [24] introduced the concept of *consistent hashing* for locating trees of caches within a dynamically changing set of hosts. Consistent hashing has the desirable property that each node in the distributed hash table stores approximately the same number of keys, but there is relatively little movement of keys as nodes enter or leave the system. Consistent hashing also has the desirable property that not all clients need know about all the nodes participating in the distributed hash table, removing the need for global consistency and synchronisation. Chord [64] uses a variant of consistent hashing to effect distributed lookup in a peer-to-peer environment. It is used as the distributed hash layer of the Cooperative File System (CFS) [9], a peer-to-peer wide-area read-only storage system. Chord uses $O(\log N)$ messages to look up a key in a network of N nodes. The caching scheme of Karger *et al.*[24] has $O(1)$ lookup, but with more routing information stored at each node than Chord.

Plaxton *et al.* [55] describe a distributed data location technique designed to locate the closest copy of a data item to a client. Like Chord, it also uses $O(\log N)$ messages in the lookup, but has stronger guarantees about how far those messages travel within the network. The OceanStore [29] distributed storage system uses a variant of the algorithm of Plaxton *et al.*

Ratnasamy *et al.* use a d-dimensional Cartesian coordinate space for their Content-Addressable Network (CAN) [58]. The CAN provides $O(dN^{1/d})$ message key lookup, and requires $O(d)$ state maintained at each node. This keeps the per-node state fixed, but at an asymptotically worse lookup cost. Past [14] uses a protocol similar to Chord, but based on prefix routing.

The recent proliferation of schemes illustrates the necessity of a scalable lookup system for quickly and efficiently locating resources in a networked environment. The peer-to-peer schemes described above also tackle wider problems such as highly dynamic node participation in the network; inconsistent views; system security; anonymity; administration; and fault tolerance. In particular, the volatility of both the network membership and its topology is expected to be significantly lower in a storage cluster: nodes will join the network when new disks are added, and leave when they are removed for replacement or redeployment elsewhere.

13.4 Redundant Encoding for Reliability

Many schemes have been proposed for redundantly encoding data to improve reliability and tolerate faults. Several schemes have been proposed for the LH* family of SDDS, including mirroring [40], bit-level striping with parity [41], data with scalable parity [38], and Reed-Solomon encoding [45].

Other approaches are available for increasing reliability. Upfal and Wigderson [68] introduce a technique for replicating k copies of a datum across nodes in a distributed system to improve reliability in the presence of failures. Their technique requires that only a majority of the k copies be read or written in any given read or write of the datum.

Rabin [57] proposes a technique for "efficient dispersal of information" in which data can be encoded into n pieces such that only m out of n need be read to recreate the data. Alon and Luby [2] present a linear-time algorithm for an (n,c,l,r)-erasure-resilient code that can be used for encoding and decoding a n-bit string into a set of l-bit pieces such that when decoding, any subset of the l-bit pieces whose combined length is r is sufficient to recreate the original string. The encoded string is of size cn, and r need only be slightly larger than n.

The PRESTO project [5] employs a simple parity-based redundancy scheme. Each logical block is decomposed into k equal sized sub-blocks ($k>1$), plus an extra parity block the same size as a sub-block. When reading a logical block, any k of the $k+1$ sub-blocks can be read to reconstruct the original logical block. When writing, all $k+1$ blocks (sub-blocks plus parity sub-block) are updated. The sub-blocks are pseudo-randomly distributed throughout the system. The system can tolerate with high probability the failure of any single storage server. In PRESTO, $k=4$ for read-mostly objects, giving a 25% storage redundancy overhead [5].

13.5 Metadata

Metadata is, broadly speaking, "data about data." However, that simplification hides the rich complexity inherent in metadata. Metadata has widely varying uses, and that use may be subjective, meaning that metadata is interpreted in a specific *context*. A digital object may be put to different uses by different entities, such as publishers, archivists, and end-users. Such entities often will create and interpret metadata specific to their own intended purposes. Rather than requiring a global union of all possible metadata, it is more flexible to be able to *aggregate* diverse sets of metadata for a given object to enable interpretation of that object by a diverse set of end-users.

One approach that has attracted widespread support is to consider a digital object as a *container* into which is placed digital content and associated metadata. The Warwick Framework [31], which arose out of a metadata conference, describes this approach. It seeks to address some of the limitations of the Dublin Core [15, 67] metadata set, the chief one being that the Dublin Core seeks to be a global metadata set, and such generality is undesirable in some instances.

In the Warwick Framework, the metadata for an object is actually a container holding *packages* of metadata. Each package can either be, itself, a metadata container; a *metadata set* containing actual metadata; or an *indirect* reference to an external metadata object, referenced via a URN. A client may access the contents of a metadata container via an operation that returns a sequence of packages inside the container. The client may then skip over unknown packages—metadata that the client does not know how to deal with, or cannot access—and use only metadata that it "understands" for the application at hand [31].

The Resource Description Framework (RDF) is gaining popularity as a model for encoding metadata for digital objects (resources) [47]. The RDF uses XML as its encoding system, and provides a means for associating properties with resources. The properties may be assigned their own XML namespace, to disambiguate the notion of, say, "creator" in one scheme from "creator" in another scheme. In this way, the RDF supports the ability of being able to describe the same resource (digital object) with heterogenous metadata sets, as in the Warwick Framework.

BUCKETS

The theoretical framework for the Warwick Framework was laid down by Kahn and Wilensky [23]. The Warwick Framework was subsequently generalised by recognising that there is no real distinction between metadata and data, and so containers should store both. Daniel and Lagoze [10] describe such a generalisation, known as *Distributed Active Relationships*, later more commonly known as *buckets* [49]. One architecture based around this approach is called the Flexible and Extensible Digital Object Repository Architecture (FEDORA) [54].

The bucket paradigm treats digital objects as abstract data types whose contents are accessible through defined methods that allow a user to interrogate the contents of the digital object. In this way, more "intelligence" is invested in the digital object itself, and this makes the digital object less reliant on functionality built into the repository in which it is stored. This philosophy is termed Smart Objects, Dumb Archives (SODA) by Maly *et al.* [46]. There are interesting parallels between the philosophy of SODA and the disk-centric approach of IDISK and NASD, most notably that all emphasise self-management of objects.

Each bucket in the SODA model has a handle, and contains one or more typed *elements* and zero or more *packages*. Packages are used to aggregate elements. Typically, several different renditions of the same document, e.g., PostScript, PDF, and TeX, will form a single package, with each different format of the same document being a distinct element of that package. A package also can contain metadata, and terms and conditions for accessing the bucket, and even can be a pointer to a remote bucket, element, or package. Packages or elements in a bucket can, themselves, be buckets. The entire bucket has several primitive methods to allow controlled access to the contents of the bucket. The NCSTRL+ project [49] expanded upon the Networked Computer Science Technical Report Library (NCSTRL) distributed digital library [36] by using the bucket approach, along with other enhancements.

The FEDORA model is very similar, except that packages are typed byte streams called *DataStreams* that are accessed via *disseminators*. A *Primitive Disseminator* allows bucket-level access, while individual *Content-Type Disseminators* allow structured access to the individual DataStreams.

Figure 13-1 (adapted from [46]) illustrates conceptually a digital object in the bucket model. The bucket shown consists of three packages, representing, say, a technical report. One package in the bucket contains metadata for the digital object. Another package contains the report itself, and the third contains software described in the report. Each package contains multiple elements representing alternative encodings of the package contents. So, for example, the metadata package includes both Dublin Core and MARC metadata for the bucket. The report package includes TeX, PostScript, and PDF versions of the report's contents. The software package has versions of the software described in the technical report ported to both the C and Ruby programming languages. The content of the bucket is named and located via its handle. Access to the contents is mediated by access methods. Bucket-level access methods allow discovery and operation on the included packages. Package-level access methods allow discovery and dissemination of individual package contents.

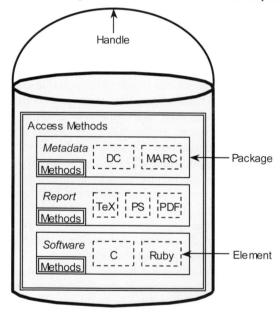

Figure 13-1 A digital object

TERMS AND CONDITIONS

An advantage of making digital objects self-contained, as in the bucket approach, is that it ties them less to a particular repository, and so makes replication and distribution easier in a heterogenous architectural environment. In particular, terms and conditions can be implanted within the digital object, instead of being wholly the responsibility of the repository. So, the move is to secure digital objects, rather than communication channels, which simplifies the superdistribution of digital objects.

Cryptolopes [19] adhere to this philosophy by encrypting the contents of a digital object, and including encrypted versions of the keys used to do so, along with digital certificates and checksums that may be used to verify the integrity and authenticity of the content. A third-party *clearinghouse* is used to enforce the terms and conditions of the bucket, facilitating the buying or obtaining of the appropriate keys to decrypt the content [27]. A similar technique is used in the DigiBox architecture [63].

The major disadvantage of many metadata encoding formats, from a low-level storage viewpoint, is that they are often free-format, and so relatively uneconomical and unwieldy in terms of storage management. Importantly, too, they are relatively expensive to parse when compared to typical file system metadata, which, for efficiency reasons, is necessarily compact and fixed in format. It is a significant challenge to extend the richness of digital object metadata to low-level repository storage.

13.6 Digital Object Repositories

Digital object repositories are akin to the file system layer of traditional operating systems. They are responsible for the storage, management, and access of digital objects. They provide low-level functionality upon which digital library functionality may be built.

Traditionally, a digital object repository is seen as a form of "middleware" that provides a shim to the operating system file system and network layers to allow low-level access to structured digital object content. This is a utilitarian stance, to leverage existing storage system functionality in a platform-independent fashion. It does, however, have the potential to divorce digital object storage from the actual low-level storage subsystem, enough to cause a significant loss of efficiency. This is especially problematic in the case of multimedia content.

There are two competing goals when designing a digital object repository: preservation versus efficiency. A repository designed with preservation as the primary goal seeks to make the storage as simple and transparent as possible, so that recovery of data can be done as easily as possible. This involves the use of *self describing* files and data structures; no deletions; and use of only disposable auxiliary structures (ones that can be faithfully recreated from scratch, if necessary). The major impact of such an approach is a significant decrease in performance; consequently, little work has been done in this area. Yet, the problem is important; Crespo and García-Molina [8] propose a layered digital archive architecture with long-term data preservation as its primary goal.

Instead of designing preservation into a digital object repository from the ground-up, it is usually assumed that the low-level storage will be supported by a disaster-recovery plan that allows it to be recovered after significant organisational loss. Allied to that is a *copy-forward* preservation approach, that migrates the low-level storage and organisation of digital objects into new, appropriate formats as system software undergoes generational change.

KAHN-WILENSKY AND ITS EXTENSIONS

A significant abstract semi-formal description of a digital object repository was introduced by Kahn and Wilensky [23]. They introduce many key concepts that have been built upon subsequently by researchers. Important amongst these concepts are the notion of a digital object and its handle, as discussed previously. They also describe the notion of *repository*, and a *repository access protocol (RAP)* which may be used to control the deposit of and access to digital objects in the repository.

A repository is a logical entity in that the name by which a repository is identified need not correspond to a unique server. Rather, it may correspond to a list of servers, each of which is responsible for the stewardship of digital objects stored in that repository. This necessarily makes the concept of a repository a distributed one.

In the Kahn-Wilensky approach, digital objects are accessed via their associated handle. Handle servers are used to locate a repository in which a digital object is stored. This repository then can be used to access the digital object by means of the RAP. Access to a digital object at a repository is by means of the digital object's handle, a service type, and possible additional parameters. Such an access is termed a *dissemination*.

Kahn and Wilensky [23] define only three service types for their RAP: ACCESS_DO, DEPOSIT_DO, and ACCESS_REF. The first two are for accessing and depositing a digital object, respectively. The third is to access the reference services of the repository. This allows a client to determine the contents of a repository.

Lagoze [30] describes an implementation of the Kahn-Wilensky repository approach called Interoperable Secure Object Stores (ISOS). ISOS uses the distributed object store Common Object Request Broker Architecture (CORBA) for its underlying implementation mechanism.

Dienst [11, 33] is another implementation of the Kahn-Wilensky approach. Dienst uses HTTP as its basic transport protocol, and partitions digital library functionality into several basic services. These services may be further refined by arguments to the service type when accessing digital objects within Dienst, analogous to the service type and parameter mode of repository access in the Kahn-Wilensky model. Dienst uses handles to name its objects.

NCSTRL [36] previously used Dienst for its digital library services. NCSTRL has been subject to active development, primarily to extend to it the notion of multiple collections or genres [49].

The Cornell Digital Library Research Group designed the Cornell Reference Architecture for Distributed Digital Libraries (CRADDL) infrastructure to support the collection concept [32]. CRADDL is a layered model, and uses CORBA for its implementation. The repository service layer of CRADDL is called FEDORA [54]. FEDORA is designed to support the bucket model of digital objects and metadata, as described in Section 13.5. The FEDORA approach invests much of the intelligence within the objects themselves. Each object is opaquely packaged and contains a *primitive disseminator* through which the actual contents of the object may be inspected and retrieved. The primitive disseminator can be thought of as basic *bootstrapping* primitive methods that allow a remote entity to access the particular content of a given object. The primitive disseminator guarantees a minimal set of services for handling an object. Handling, in this context, also means incorporating new content into a digital object [54].

Because much of the functionality resides within the objects themselves, the RAP in FEDORA is relatively simple, and involves mainly the creation and deletion of digital objects, and associating URNs with them. The repository also provides an environment in which content access and authorisation methods may be executed [54].

OTHER REPOSITORY APPROACHES

An alternative approach to digital object repositories is to view digital object stores as semi-structured databases. The metadata of a digital object is stored in fields in a database table, while the data of a digital object are treated as binary large objects (BLOBs) [20] stored in the database. A database provides a flexible method for searching and storing metadata, and the advances in distributed databases provides a ready platform upon which distributed digital object repositories may be deployed.

IBM Digital Library [22] employs the database approach. Its architecture consists of a *library server* and one or more *object servers.* The library server handles the metadata and searching, while the object servers handle storage and delivery of the actual digital content.

A client accessing the digital library gains access to a digital object via the library server. Once the library server has authenticated and located the object server storing the digital object the client wishes to access, the digital object is transferred directly between the object server in question and the client, requiring no further participation from the library server.

Although there may be multiple object servers, and those object servers may be attuned to the content they deliver (e.g., a streaming server for video), there is only a single library server per digital library. This creates a bottleneck in the architecture, especially in metadata-intensive applications. Because the library server is a database, it should be possible, though, to parallelise it, to allow it to scale up.

The Oracle Internet File System (*i*FS) is another database-driven file system [50]. The underlying database engine allows more flexible and extensible metadata to be attached to files than is supported by traditional file systems. It also supports flexible "views" of files within the file system. The *i*FS supports versioning of files, and check-in and check-out access for file update. The protection model is access control list-based, rather than permission-based as with many file systems.

Files in *i*FS are *parsed* and stored in a canonical internal format. *Renderers* are used to reconstruct parsed files in an application-specific way. One of the features of *i*FS is that it aims to be "protocol independent" when accessing files. The *i*FS can be accessed via SMB, HTTP, FTP, IMAP4, and SMTP. XML and Java are the technologies through which the *i*FS API is used.

Like the IBM Digital Library, *i*FS is a step towards *content management* rather than *file management.* The fields of content management and digital libraries share common goals and services. There is perhaps more emphasis on *workflow*, though, in the area of content management.

The SDSC Storage Resource Broker (SRB) architecture [3] is an API for managing access to widely distributed heterogenous data objects. One notable feature of SRB is that it allows access to objects stored in a variety of storage systems, such as databases, file systems, and tertiary storage, providing a simple, clean interface to the client. The Storage Resource Broker uses a metadata catalogue (MCAT) to keep track of resources belonging to collections managed by SRB.

The Stanford Infobus [51] is a CORBA-based digital library object repository mechanism for mediating access to distributed digital objects. The Infobus uses *library services* to perform various functions on objects accessed through the Infobus. *Library service proxies* mediate client access to the Infobus and its library services (and hence to the objects therein). The University of Michigan Digital Library [16] uses an agent-based approach to broker access to data stored within it.

Many of these systems are designed to handle widely distributed collections, and so do not focus on local, cluster-based repositories. This complicates the design requirements, placing stricter emphasis on authentication, heterogeneity, and network infrastructure. This, in turn, reduces the opportunities for optimisation.

13.7 Conclusion

This chapter described the relevant issues, research, and pertinent projects in storage systems and the special requirements of digital libraries as they pertain to storage system design. However, although there is a rich body of work in storage, there is a relative paucity when it comes to addressing directly the specific low-level needs of digital libraries and digital libraries object repositories. An ongoing project at Virginia Tech addresses this by proposing a cluster-based scalable digital libraries object repository.

Analysis of digital libraries object repository requirements and storage system research has led us to develop an architecture which meets the challenges and requirements of the former by applying the best appropriate practices of the latter.

Digital objects are naturally object-based, not block-based, and have rich hints available via explicitly catalogued metadata. By augmenting this explicit metadata with workload statistics, we can add a new, semantic-driven input to storage access and layout policies and mechanisms. This is not present in conventional storage systems, which, aside from some application-specific domains, treat files as neutral linear byte streams and thus miss a potentially rich source of performance optimisations.

Our target architecture is a cluster of intelligent disk nodes that supports a repository access protocol. Higher-level digital libraries functions are assumed to be provided elsewhere.

We propose to use scalable hashing with randomisation as a means of handling both the name space and workload distribution across the nodes of the cluster. Scalable hashing addresses the problem of quickly and scalably locating an object within the cluster, allowing for the growth or shrinkage of the cluster with minimal re-mapping of existing objects to other disks.

Disk seeks are an expensive commodity in any secondary storage system, because they are constrained by physical laws that severely limit performance improvement. However, most objects are small in relation to the collection, while most bytes are concentrated in relatively few objects. We thus propose to manage the tradeoff between minimising seeks as a threshold decision of when to stripe an object across more than one disk node. This serves two purposes: it allows even use of space across all disk nodes (and prevents large objects monopolising space on individual disk nodes), and it permits high-bandwidth transfers of large objects by involving more disk head assemblies. The size threshold at which to go "off node" is a matter for investigation. A simple solution is to have a fixed size for all objects. In a more complex implementation, it could be based upon the semantic type of the object, as identified by the object's metadata and access history.

Archival digital library collections are predominantly read-mostly. Such objects benefit from large transfer sizes, as they mostly are accessed in a whole-object sequential fashion. We propose to use a variable block size extent-based allocation scheme to improve I/O performance by making objects more sequential on disk. This will minimise seeking, and improve performance.

Our proposed architecture targets directly the needs of digital libraries object repositories. Beyond all else, it offers a platform for future experimentation tied directly to that important domain.

References

[1]   A. Acharya, M. Uysal, and J. Saltz. Active disks: Programming model, algorithms and evaluation. In *Proceedings of Architectural Support for Programming Languages and Operating Systems (ASPLOS) VIII*, pages 81–91. ACM, Oct. 1998.

[2]   N. Alon and M. Luby. A linear time erasure-resilient code with nearly optimal recovery. *IEEE Transactions on Information Theory*, 42(6): 1732–1736, Nov. 1996.

[3]   C. Baru, R. Moore, A. Rajasekar, and M. Wan. The SDSC Storage Resource Broker. In *Proceedings of CASCON '98*, Toronto, Canada, 30 Nov.–3 Dec. 1998. IBM Conference. Available via http://npaci.edu/DICE/Pubs/srb.ps.

[4]   M. Beck and T. Moore. The Internet2 Distributed Storage Infrastructure project: An architecture for Internet content channels. *Computer Networking and ISDN Systems*, 30(22-23): 2141–2148, 1998.

[5]   P. Berenbrink, A. Brinkmann, and C. Scheideler. Design of the PRESTO multimedia storage network. In *Proceedings of the Workshop on Communication and Data Management in Large Networks (INFORMATIK 99)*, Oct. 1999.

[6]   S. Berson, R. Muntz, and W. Wong. Randomized data allocation for real-time disks. In *COMPCON '96*, pages 286–290, 1996.

[7]   C. Brandt, G. Kyriakaki, W. Lamotte, R. Lüling, Y. Maragoudakis, Y. Mavraganis, K. Meyer, and N. Pappas. The SICMA multimedia server and virtual museum application. In *Proceedings of the Third European Conference on Multimedia Applications, Services and Techniques*, pages 83–96, 1998.

[8]   A. Crespo and H. García-Molina. Archival storage for digital libraries. In *Proceedings of Digital Libraries 98*, pages 69-78, Pittsburgh, PA, 1998. ACM.

[9]   F. Dabek, M. F. Kaashoek, D. Karger, R. Morris, and I. Stoica. Wide-area cooperative storage with CFS. In *Proceedings of the 2001 Symposium on Operating Systems Principles (SOSP)*, Banff, Canada, 21–24 Oct. 2001. Association for Computing Machinery.

[10]   R. Daniel, Jr. and C. Lagoze. Distributed Active Relationship in the Warwick Framework. In *Proceedings of the Second IEEE Metadata Conference*, Silver Spring, Maryland, USA, 16–17 Sept. 1997.

[11]   J. R. Davis and C. Lagoze. A protocol and server for a distributed digital technical report library. Technical report, Cornell University, June 1994. urn:hdl:ncstrl.cornell/TR94-1418.

[12]   R. Devine. Design and implementation of DDH: A distributed dynamic hashing algorithm. In *Proceedings of the 4th International Conference on Foundations of Data Organization and Algorithms (FODO)*, 1993.

[13]   J. R. Douceur and W. J. Bolosky. A large-scale study of file-system contents. In *SIGMETRICS '99*, pages 59–70, Atlanta, Georgia, USA, 29 Apr.–6 May 1999.

[14]   P. Druschel and A. Rowstron. Past: Persistent and anonymous storage in a peer-to-peer networking environment. In *Proceedings of the 8th IEEE Workshop on Hot Topics in Operating Systems (HotOS 2001)*, pages 65–70, Elmau/Oberbayern, Germany, May 2001.

[15]   Dublin Core Directorate. Dublin Core metadata initiative. WWW page, 2000. http://purl.org/dc.

[16] E. Durfee, D. Kiskis, and W. Birmingham. The agent architecture of the University of Michigan Digital Library. *IEE/British Computer Society Proceedings on Software Engineering*, 144(1), Feb. 1997.

[17] FIPS 180-1. *Secure Hash Standard*. U.S. Department of Commerce/NIST, Springfield, Virginia, USA, Apr. 1995.

[18] G. A. Gibson, D. F. Nagle, W. Courtright II, N. Lanza, P. Mazaitis, M. Unangst, and J. Zelenka. NASD scalable storage systems. In *Proceedings of USENIX 1999*, Monterey, California, USA, 9–11 June 1999.

[19] H. M. Gladney and J. B. Lotspiech. Safeguarding digital library contents and users: Assuring convenient security and data quality. *D-Lib Magazine*, May 1997. urn:hdl:cnri.dlib/may97-gladney.

[20] R. L. Haskin and R. A. Lorie. On extending the functions of a relational database system. In M. Schkolnick, editor, *Proceedings of the 1982 ACM SIGMOD International Conference on Management of Data*, pages 207–212, Orlando, Florida, USA, 2–4 June 1982. ACM Press.

[21] IETF Secretariat. Uniform Resource Names (URN) charter. World Wide Web page, Sept. 2000. http://www.ietf.org/html.charters/urn-charter.html.

[22] International Business Machines. IBM Digital Library, 1998. http://www.software.ibm.com/is/dig-lib/.

[23] R. Kahn and R. Wilensky. A framework for distributed digital object services. Technical report, Corporation for National Research Initiatives, Reston, Virginia, USA, 13 May 1995. urn:hdl:cnri.dlib/tn95-01.

[24] D. R. Karger, E. Lehman, F. T. Leighton, M. S. Levine, D. Lewin, and R. Panigrahy. Consistent hashing and random trees: Distributed caching protocols for relieving hot spots on the World Wide Web. In *Proceedings of the 29th Annual ACM Symposium on Theory of Computing*, pages 654–663, El Paso, Texas, USA, 4–6 May 1997.

[25] J. S. Karlsson, W. Litwin, and T. Risch. LH*lh: A scalable high performance data structure for switched multicomputers. IDA Technical Report 1995 LiTH-IDA-R-95-25, Department of Computer and Information Science, Linköping University, S-581 83 Linkping, Sweden, 1995. ISSN-0281-4250.

[26] K. Keeton, D. A. Patterson, and J. M. Hellerstein. A case for intelligent disks (IDISKs). *SIGMOD Record*, 27(3): 42–52, Aug. 1998.

[27] U. Kohl, J. B. Lotspiech, and M. A. Kaplan. Safeguarding digital library contents and users: Protecting documents rather than channels. *D-Lib Magazine*, Sept. 1997. urn:hdl:cnri.dlib/september97-lotspiech.

[28] B. Kröll and P. Widmayer. Distributing a search tree among a growing number of processors. In *Proceedings of the 1994 ACM SIGMOD International Conference on Management of Data*, pages 265–276, Minneapolis, Minnesota, USA, 24–27 May 1994.

[29] J. Kubiatowicz, D. Bindel, Y. Chen, P. Eaton, D. Geels, R. Gummadi, S. Rhea, H. Weatherspoon, W. Weimer, C. Wells, and B. Zhao. Oceanstore: An architecture for global-scale persistent storage. In *Proceedings of ACM ASPLOS*, Cambridge, Massachusetts, USA, Nov. 2000.

[30] C. Lagoze. A secure repository design for digital libraries. *D-Lib Magazine*, Dec. 1995. urn:hdl:cnri.dlib/december95-lagoze.

[31] C. Lagoze, C. A. Lynch, and R. Daniel, Jr. The Warwick framework: A container architecture for aggregating sets of metadata. Computer Science Technical Report TR96-1593, Cornell University, June 1996.

[32] C. Lagoze and S. Payette. An infrastructure for open-architecture digital libraries. Technical Report TR98-1690, Department of Computer Science, Cornell University, 1998. urn:hdl:ncstrl.cornell/TR98-1690.

[33] C. Lagoze, E. Shaw, J. R. Davis, and D. B. Krafft. Dienst: Implementation reference manual. Technical report, Cornell University, May 1995. urn:hdl:ncstrl.cornell/TR95-1514.

[34] P. Larson. Dynamic hash tables. *Communications of the ACM*, 31(4): 446–457, Apr. 1988.

[35] B. M. Leiner. Metrics and the digital library. *D-Lib Magazine*, July 1998. Guest editorial. urn:hdl:cnri.dlib/july98-editorial.

[36] B. M. Leiner. The NCSTRL approach to open architecture for the confederated digital library. *D-Lib Magazine*, Dec. 1998. urn:hdl:cnri.dlib/december98-leiner.

[37] W. Litwin. Linear hashing: A new tool for file and table addressing. In *Proceedings of Very Large Databases*, Montreal, Canada, 1980.

[38] W. Litwin, J. Menon, and T. Risch. LH* schemes with scalable availability. Research Report RJ 1021 (91937), IBM Almaden, May 1998.

[39] W. Litwin, J. Menon, T. Risch, and T. J. E. Schwarz. Design issues for scalable availability LH* schemes with record grouping. In *DIMACS Workshop on Distributed Data and Structures*, Princeton University, May 1999. Carleton Scientific.

[40] W. Litwin and M.-A. Neimat. High-availability LH* schemes with mirroring. In *Proceedings of the Conference on Cooperative Information Systems (COOPIS '96)*, Jan. 1996.

[41] W. Litwin, M.-A. Neimat, G. Levy, S. Ndiaye, and T. Seck. LH*s: A high-availability and high-security scalable distributed data structure. In *Proceedings of IEEE RIDE '97*, 1997.

[42] W. Litwin, M.-A. Neimat, and D. A. Schneider. LH*—linear hashing for distributed files. In *Proceedings of SIGMOD '93*, pages 327–336. ACM, May 1993.

[43] W. Litwin, M.-A. Neimat, and D. A. Schneider. RP*: A family of order-preserving scalable distributed data structures. In *Proceedings of Very Large Databases*, Sept. 1994.

[44] W. Litwin, M.-A. Neimat, and D. A. Schneider. LH*—a scalable, distributed data structure. *ACM Transactions on Database Systems*, 21(4): 480–525, Dec. 1996.

[45] W. Litwin and T. J. E. Schwarz. LH*rs: A high-availability scalable distributed data structure using Reed Solomon codes. In *Proceedings of the 2000 ACM SIGMOD International Conference on Management of Data*, Dallas, Texas, USA, 2000.

[46] K. Maly, M. L. Nelson, and M. Zubair. Smart Objects, Dumb Archives: A user-centric, layered digital library framework. *D-Lib Magazine*, 5(3), Mar. 1999. urn:doi:10.1045/march99-maly.

[47] E. Miller. An introduction to the resource description framework. *D-Lib Magazine*, May 1998. urn:hdl:cnri.dlib/may98-miller.

[48] R. Moore, C. Baru, A. Rajasekar, B. Ludaescher, R. Marciano, M. Wan, W. Schroeder, and A. Gupta. Collection-based persistent digital archives - part 1. *D-Lib Magazine*, 6(3), Mar. 2000. urn:doi:10.1045/march2000-moore-pt1.

[49] M. L. Nelson, K. Maly, S. N. T. Shen, and M. Zubair. NCSTRL+: Adding multi-discipline and multi-genre support to the Dienst protocol using clusters and buckets. In *Proceedings of the IEEE Forum on Research and Technology Advances in Digital Libraries, IEEE ADL '98*, pages 128–136, Santa Barbara, California, USA, 22–24 Apr. 1998. IEEE Computer Society. ISBN 0-8186-8464-X.

[50] Oracle Corporation. Oracle Internet File System (*i*FS). Product Overview, 2000. http://www.oracle.com/database/options/ifs/iFSFO.html.

[51] A. Paepcke, M. Q. W. Baldonado, C.-C. K. Chang, S. Cousins, and H. García-Molina. Using distributed objects to build the Stanford digital library Infobus. *IEEE Computer*, 32(2): 80–87, Feb. 1999.

[52] N. Paskin. DOI: Current status and outlook. *D-Lib Magazine*, May 1999. urn:doi:10.1045/may99-paskin.

[53] N. Paskin. Toward unique identifiers. *Proceedings of the IEEE*, 87(7), July 1999.

[54] S. Payette and C. Lagoze. Flexible and Extensible Digital Object and Repository Architecture (FEDORA). In *Second European Conference on Research and Advanced Technology for Digital Libraries, Lecture Notes in Computer Science v1513*, Heraklion, Crete, 21–23 Sep. 1998. Springer. http://www2.cs.cornell.edu/payette/papers/ECDL98/FEDORA.html.

[55] C. G. Plaxton, R. Rajaraman, and A. W. Richa. Accessing nearby copies of replicated objects in a distributed environment. In *ACM Symposium on Parallel Algorithms and Architectures*, pages 311–320, June 1997.

[56] A. Powell. Resolving DOI based URNs using Squid: An experimental system at UKOLN. *D-Lib Magazine*, June 1998. urn:hdl:cnri.dlib/june98-powell.

[57] M. O. Rabin. Efficient dispersal of information for security, load balancing, and fault tolerance. *J. ACM*, 36(2): 335–348, Apr. 1989.

[58] S. Ratnasamy, P. Francis, M. Handley, R. Karp, and S. Shenker. A scalable content-addressable network. In *SIGCOMM '01*, pages 161–172, San Diego, California, USA, 27–31 Aug. 2001. Association for Computing Machinery.

[59] E. Riedel and G. A. Gibson. Active disks - remote execution for network-attached storage. Technical Report CMU-CS-97-198, Carnegie Mellon University, Dec. 1997.

[60] J. R. Santos and R. Muntz. Performance analysis of the RIO multimedia storage system with heterogeneous disk configurations. In *Proceedings of the Sixth ACM International Conference on Multimedia*, pages 303–308, Bristol, United Kingdom, 13–16 Sept. 1998. ACM.

[61] J. R. Santos, R. Muntz, and S. Benson. A parallel disk storage system for real-time multimedia applications. In *Special Issue on Multimedia Computing Systems of the International Journal of Intelligent Systems*, 1998.

[62] K. Shafer, S. Weibel, E. Jul, and J. Fausey. Introduction to Persistent Uniform Resource Locators. WWW page, 1996. `http://purl.oclc.org/OCLC/PURL/INET96`.

[63] O. Sibert, D. Bernstein, and D. Van Wie. Securing the content, not the wire, for information commerce. Technical report, InterTrust Technologies Corp., 1996. `http://www.intertrust.com/architecture/stc.html`.

[64] I. Stoica, R. Morris, D. Karger, M. F. Kaashoek, and H. Balakrishnan. Chord: A scalable peer-to-peer lookup service for internet applications. In *SIGCOMM '01*, San Diego, California, USA, 27–31 Aug. 2001. Association for Computing Machinery.

[65] S. X. Sun and L. Lannom. Handle system overview. IETF Draft, Aug. 2000. `http://www.ietf.org/internet-drafts/draft-sun-handle-system-05.txt`.

[66] S. X. Sun, S. Reilly, and L. Lannom. Handle system namespace and service definition. IETF Draft, Aug. 2000. `http://www.ietf.org/internet-drafts/draft-sun-handle-system-def-03.txt`.

[67] H. Thiele. The Dublin Core and Warwick Framework. *D-Lib Magazine*, Jan. 1998. `urn:hdl:cnri.dlib/january98-thiele`.

[68] E. Upfal and A. Widgerson. How to share memory in a distributed system. *J. ACM*, 34(1): 116–127, Jan. 1987.

[69] URN Implementors. Uniform resource names: A progress report. *D-Lib Magazine*, Feb. 1996. `urn:hdl:cnri.dlib/february96-urn_implementors`.

[70] A. Vahdat, T. Anderson, and M. Dahlin. Active Names: Programmable location and transport of wide-area resources. Technical report, University of California at Berkeley, 1998.

[71] R. Vingralek, Y. Breitbart, and G. Weikum. Distributed file organization with scalable cost/performance. In *Proceedings of ACM-SIGMOD*, pages 253–264, May 1994.

14 INFORMATION DISCOVERY ON THE WORLD-WIDE-WEB

Dr. Ben Kao and Dr. David Cheung

This chapter discusses data-mining techniques for web-related data. In particular, we discuss techniques that can help information seekers locate relevant information on web. Two kinds of techniques, web-structure mining and web-log mining, are discussed. We also examine three techniques, *authorities and hubs* [10], *anchor points* [9], and *PageRank* [13] that examine the link structures of hypertext web pages. Since the web is huge and dynamic, it is not possible for any IR system to maintain a global view of the web. Recommendation of web information, therefore, has to be based on incomplete information. We discuss the idea of Internet *GlOSS* [2], which uses word statistics to make intelligent guess on the *topics of interest* of web sites. Also we discuss how the interest of web users can be abstracted in *user profiles*. Understanding both web users and web sites allows an effective matching of the two. Finally, we explain how mining web-log data can discover the topics of interest of web sites and user profiles.

14.1 Introduction

The Internet enables a computer user to be connected to virtually endless numbers of sites on the network. The World-Wide-Web (WWW) uses the Internet to distribute hypermedia documents among computer users located around the world. Large amounts of interesting and valuable information have been made available on the web for retrieval. In order to fully utilize the power of the WWW as an information source, it is essential to develop intelligent software systems to assist users to retrieve relevant documents.

Search engines are the most popular tools that people use to locate information on the web. A search engine works by traversing the web via the hyper-links that connect the web pages, performing text analysis on the pages it has encountered, and indexing the pages based on the keywords they contain [7]. A user who seeks information from the web formulates his information goal in terms of a few keywords composing a query. A search engine, on receiving a query, matches the query against its document index. All of the pages that match the user query are selected into an *answer set* and ranked according to how *relevant* the pages are with respect to the query. Relevancy here is usually based on the number of matching keywords that a page contains, with the "positions" of the matches taken into account. (For example, a matching keyword that appears in the title of a page is considered more relevant than an occurrence in the text body.)

Although search engines have been proven in practical use as indispensable tools for web information retrieval, they suffer from the following drawbacks [8, 9, 11]:

- large answer set,
- low precision,
- unable to preserve the hypertext structures of matching hyper-documents, and
- ineffective for general-concept queries.

Large answer set. Most search engines boast about their services, mentioning their extensive coverage of the web. Powered by high performance workstations, large storage servers, and sophisticated indexing techniques, these search engines handle user queries with reasonable response times. The quality of the responses, however, is sometimes questionable. The large number of pages indexed plus relatively loose matching criteria often result in large answer sets (sets of matching pages). Users are overloaded with the vast amounts of information returned from the search engines. Even though search engines rank the pages in the answer set by guessing how relevant they are with respect to the queries, the ranking systems are far from perfect given the limited expressive power of the keyword-based query interfaces. Table 14-1 illustrates this problem by showing the results obtained from querying four popular search engines with some sample queries.

Goal	Query	Search engine	No. of hits	No. of relevant pages in the first 30 hits	Rank of 1st relevant hit	No. of logical clusters in the first 30 hits
Find the site of Microsoft Windows	Microsoft Windows	Alta Vista	1,675,214	3	10th	4
Find riff specification	Riff specification	Lycos	Unknown	1	12th	26
Find NBA scoreboard	NBA score	Excite	71,118	27	3rd	2
Find general information about cricket	Cricket	Infoseek	40,990	1	22nd	20

Table 14-1 Example queries and results

In each row of the table, we show a search goal (*Goal*), the keyword-based query submitted (*Query*), the search engine that processed the query (*Search Engine*), the number of pages returned (*No. of hits*), the number of *relevant* pages that satisfied our goal in the first thirty hits (*No. of relevant pages in the first 30 hits*)[1], the rank of the highest ranked relevant page (*Rank of 1st relevant hit*), and the number of logical hypertext documents that contained the first thirty hits (*No. of logical clusters in the first 30 hits*). The numbers shown in the last column need further explanation. We observed that it is not unusual that a number of the pages returned are in fact parts of a logical cluster of a hypertext document. For example, if one submits the query "nba

---

[1] A page that simply contains the keywords as specified in the query but does not satisfy the goal is not considered a relevant hit.

scoreboard november" to a search engine, one would get a whole bunch of matching pages from a sports site, one for each basketball game that occurred on a November day in some year. One would find that all these pages share the same prefix in their URLs, and that they belong to a single logical cluster of a big hyper-document. The last column of Table 14-1 refers to the number of clusters into which the first thirty hits can be grouped.

From the table we see that the answer sets are huge and that the first relevant page may not be found until two dozen pages have been examined, many more if one is unlucky. Also, the first relevant page may not be the best page in the answer set. More screening is required if one would like to compare relevant hits looking for a better match.

Besides illustrating the large answer set problem, Table 14-1 also gives us a hint on how to avoid overwhelming the users with the numerous recommendations. The last column of the table suggests that the large number of pages can be grouped into a small number of logical clusters. Now, if a search engine were smart enough to identify the clusters and recommend to the users only one *representative* page per cluster, the users would be able to scroll through the suggested list much more efficiently. As an example, if many of the game summary pages that are returned for the query "nba scoreboard november" can be accessed by an index page, then returning that index to the user may be sufficient. In this chapter, we will discuss the idea of anchor points, authorities, and hubs, which can be considered as representative pages of an answer set.

Low precision. In addition to "information overload", low precision of the answer sets is sometimes another concern of the effectiveness of search engines. Previous studies have conducted experiments showing that relevant pages are often interspersed with irrelevant ones in the ranked query outputs [1]. The implication is that users cannot afford to examine only the first few, or any small subset, of the answer set. Table 14-1 illustrates this problem by showing the number of pages among the first 30 hits that are relevant to a search goal. We see that, for some queries, the results are not always useful.

The most attractive feature of the web is that it provides an on-line source of information which is structured as a huge network of web pages, each containing a certain piece of interesting knowledge. Users are free to access and navigate the web via hyper-links connecting related information units. Hypertext is especially useful for the "on-line presentation of large amounts of loosely structured information such as on-line documentation or computer-aided learning" [12]. It is therefore a useful concept that web document authors like to adopt. As an example, the *Usenet Hypertext Frequently Asked Questions Archive* stores FAQs on different subjects in hierarchical hypertext structures. A FAQ is usually broken up into a number of parts, each part containing a number of questions and their answers. Figure 14-1 shows a logical hypertext structure of a FAQ document.

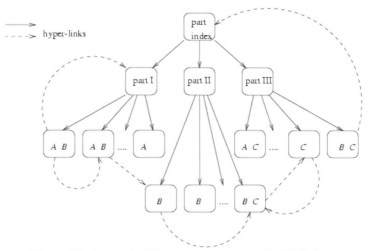

Figure 14-1 A typical hypertext structure of a FAQ document

In Figure 14-1, the FAQ is rooted at the part index page, which contains hyper-links to three second-level index pages: part I, part II, and part III. These index pages, in turn, point to a number of "leaf" pages containing information about certain subjects (A, B, and C in the figure). Besides the top-down hyper-links (solid arrows), there are cross references (dashed arrows) among the pages.

To the author of the FAQ, the whole structure constitutes a single document, and the part index represents the logical starting point of the document. Unfortunately, traditional search engines ignore the structural information embedded in the hyper-links and index the pages separately. If a user queries an engine with the keywords A, B, and C, none of the pages match all the keywords. The engines would rank these pages very lowly even though the hyper-document as a whole might provide all one needs to know about the subjects. On the other hand, if the author puts everything in one single web page, the document would match the user query very well. This reveals a drawback of the ranking mechanism of traditional search engines, namely, it works against documents that are presented in hypertext structures.

To deal with this ranking misjudgment, a search engine should recognize that the pages belong to a hyper-document (or a cluster) and that the document is relevant to the subjects that occur in the pages (A, B, C in our example). That is to say, the search engine should be able to (1) match queries against hypertext structures; and (2) return an *entry point* through which the constituent pages can be easily accessed. This observation suggests that the *structure* of web pages should be considered when a search engine is making a recommendation. We will discuss how this structural information can be used in web information retrieval in Section 14.2.

Destroying the hypertext structures of matching hyper-documents. Another inadequacy of traditional search engines again results from not preserving the hypertext structures of matching hyper-documents when making recommendations. Even if an engine could match a query with the pages of a hyper-document, the hypertext structure is flattened and the individual pages are returned out of order. As an example, we submitted the query "C80 faq" to *Alta Vista* in our search for information about a DSP chip named C80. *Alta Vista* successfully located a forum with hundreds of postings about the chip. The postings were organized as a hypertext with an index page pointing to the numerous postings, ordered by date. Each posting was contained in a separate web page. Although the pages returned by the search engine matched the query goal, all the postings listed in the answer set were out of order, and the index page was not listed in the first 100 hits returned. A user thus needs to reconstruct the logical reading path from the disorganized answer set. A better approach would be to recognize that the pages belong to the same cluster and to return a logical starting point (such as the index page).

General concept queries. Finally, while traditional search engines perform quite well for "specific" and "precise" queries (e.g., "find me the solution for solving the Rubik's cube"), they are not particularly effective in answering general concept queries. For example, someone may want to know about the sport cricket. Ideally, a site such as *CricInfo* that is dedicated to the sport would be a perfect match. Unfortunately, the query "cricket" is too general for traditional search engines with their simple keyword-matching systems. The result is that any web page that contains the keyword "cricket" matches the query, be it the start page of *CricInfo*, a page written by a fifth grader on his favorite sports, or a page reporting the scores of an international cricket tournament.

As the web develops, more and more information sources are dedicated to specific topics of interest. There are sites for *tennis*, sites for *salmon*, and even sites for *Big Foot*. If a user is looking for some general information about a topic, chances are that a specialized web site on the topic exists. Recommending web sites instead of individual web pages becomes more meaningful to this type of general concept queries. In Section 14.3, we will discuss how one could discover the nature of information held by a particular website, and how this knowledge can be used to enhance web information retrieval.

14.2 Web-structure Mining

In this section we discuss how the structure of the web can be used to enhance information retrieval. We refer to this process of analyzing web structure to extract knowledge *web-structure mining*. In the literature, it is also commonly called *link analysis*. There are a number of studies conducted on link analysis including *Authorities and Hubs* [10], *Trawling* [14], *Co-citation* [15], *Anchor Points* [9] and *PageRank* [13]. In this chapter, we briefly discuss authorities and hubs, anchor points, and PageRank.

AUTHORITIES AND HUBS

In [10], Kleinberg studied the concept of discovering authoritative pages from the web. He distinguishes two types of important web pages: authorities and hubs.

If a web page A contains a hyperlink that points to a web page B, we can consider such a referral as an endorsement of B from A. As an analogy, this book refers to a number of research articles on information retrieval techniques. In particular, we refer to Kleinberg's work [10]. Presumably, all these articles that are referred to by this book are important works on information retrieval, and thus in a sense, we endorse these works. A web page is an authority if it is linked to by a large number of other web pages. Again, as an analogy, if Kleinberg's work is referred to by many other documents, then his study must be a very important piece of work on the topic that it discusses.

A hub is a "dual" of an authority. A web page is a hub if it refers to many other useful web pages. A hub thus gives a good starting point from which related information sources are introduced and can be conveniently accessed. As an analogy, our book can be considered a hub for people to learn about information retrieval.

In general, a good hub is a web page that contains many links to many good authorities; and a good authority is a web page that is linked to by many good hubs. In [10], an algorithm called HITS (Hyperlink-Induced Topic Search) is introduced. Given a query, HITS tries to locate authorities and hubs that are relevant to the query. The algorithm works as follows.

Given a query Q, HITS first retrieves a number of web pages that match Q using a traditional search engine approach (i.e., based on the number of matching keywords that a web page has with respect to Q). These web pages are collected into a root set[2]. The root set is then expanded to a base set by collecting (1) all the web pages that contain links to at least one page in the root set, and (2) all the web pages that are linked to by at least one page in the root set.

Each page, p, in the base set is then assigned an authority weight, a_p, and a hub weight, h_p. Initially, all the authority weights and all the hub weights assume arbitrary numbers. Next, the algorithm enters an iterative phase in which authority weights and hub weights are updated by the following equations:

$$a_p = \sum_{q \to p} h_p, \ h_p = \sum_{q \leftarrow p} a_q$$

Here, we use the notation $q \to p$ to denote that page q contains a link that points to page p. The authority weights and the hub weights are normalized so that $\sum_p a_p^2 = \sum_p h_p^2 = 1$. This updating is iterated a number of times until the weights converge. Finally, HITS selects a small set of web pages from the base set that either have large hub weights or large authority weights. For more details about HITS, you are referred to [10].

The search engine Google [6], for example, uses a technique similar to HITS in retrieving (and ranking) relevant documents given a search query.

---

[2] In [10], Kleinberg uses Alta Vista to obtain the root set.

ANCHOR POINTS

We note that HITS is an on-line algorithm, that is, authorities and hubs are determined after a query is received. Not much pre-processing is done to speed up query processing. In [9], a similar technique that exploits link information is proposed. Given a user query and subsequently a set of matching web pages which form a number of clusters, the authors suggest that an IR system should recommend, instead of all the matching pages, a set of anchor points (possibly one for each cluster). An anchor point should possess the following properties:

it is a representative page of a cluster (i.e., by inspecting the anchor, a user can easily deduce what the pages in the corresponding cluster are about); and

it provides efficient and orderly accesses to the pages in the corresponding cluster.

In [9], the authors discuss the problem of *anchor point indexing*. In particular, the paper proposes an off-line algorithm for computing the *goodness* of a web page as an anchor point of a keyword. It also discusses how to retrieve anchor points given a user query. Here, we briefly mention the definition of anchor points, and how anchor point queries are processed.

We define the *distance* from a web page A to a web page B, denoted by $D(A,B)$, to be the minimum number of hyper-links that need to be traversed to reach page B starting from page A. If page B is unreachable from A, we have $D(A,B) = \infty$. We define the *k-neighborhood* of a web page A, denoted by $N_k(A)$, to be the set of all the pages that are within distance k or less from A. Figure 14-2 illustrates these terms. For example, the distance from page A to page J is 3 ($A \rightarrow E \rightarrow D \rightarrow J$) while the distance from page J to A is 4 ($J \rightarrow E \rightarrow D \rightarrow C \rightarrow A$). The 2-neighborhood of page A is the set of all the pages encircled by the dotted line. With our notations, we have $D(A,J) = 3$, $D(J,A) = 4$, and $N_2(A) = \{A,B,C,D,E,F,G,H\}$.

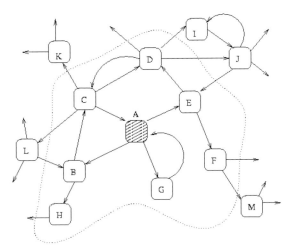

Figure 14-2 Distance and neighborhood

We assume a *scoring function f* exists such that given a keyword *a* and a web page *X*, *f(X,a)* measures the extent that page *X* matches the keyword *a*. We call the value of this function the *score* of page *X* with respect to *a*. There are many choices for such a function. One simple example would be:

$$f(X,a) = \begin{cases} 0 & \text{if X does not contain a;} \\ 1 & \text{if X contains a.} \end{cases}$$

Alternatively, *f(X,a)* could be the normalized occurrence frequency of *a* in *X*, as is done in term-frequency-inverse-document-frequency (*TFIDF*) [16]. A third example would be the probabilistic model as suggested in [17] in which the authors proposed a function that estimates the *probability* that a document *X* is relevant to a keyword *a*.

Recall that our goal is to recommend a starting page to a user from which he is likely to find relevant pages via *simple navigation*. Here, let us assume that by simple navigation, we mean that the user does not need to traverse more than *k* links away from the suggested starting point. In order to measure how well, potentially, a page *X* can lead a user to pages that match a keyword *a*, we define a *potential function Pₖ(X,a)* as follows:

$$P_k(X,a) = \sum_{Y \in N_k(X)} f(Y,a) \times \alpha^{D(X,Y)} \qquad (14\text{-}1)$$

where α is a constant parameter between 0 and 1.[3] In words, the potential function of a page *X* with respect to a keyword *a* gives the sum of all the scores of the pages in *X*'s *k*-neighborhood, with each page's contribution to the score scaled down exponentially with respect to its distance from page *X*. The constant α controls the pace of the scale-down (the smaller the value, the faster the pace).

Depending on the semantics of the function *f*, different quantitative interpretations can be associated with the potential function. For example, if *f(Y,a)* represents the *amount* of information that a page *Y* contains about the keyword *a*, and α is the probability that a user follows a hyper-link, then $\alpha^{D(X,Y)}$ gives the probability that page *Y* is visited if a user starts at page *X*, and *Pₖ(X,a)* gives the *expected* amount of information that a user would learn about keyword *a* if he starts from page *X* (assuming that the user does not take more than *k* hops away from *X*). As another example, if *f(Y,a)* measures the *probability* that page *Y* is relevant to keyword *a*, then $f(Y,a)\alpha^{D(X,Y)}$ is the probability that a user starting from page *X* would get a relevant page in *Y* about the keyword *a*. Hence, *Pₖ(X,a)* is equal to the *expected* number of pages that are relevant to *a* that a user would visit given that he starts at page *X*. In any case, intuitively, *Pₖ(X,a)* is an indicator of how much information about *a* one can get starting from page *X*. For the purpose of discussion, we will use the second interpretation of *f* (i.e., *f(Y,a)* measures the probability that page *Y* is relevant to *a*) for the rest of this section.

Given a query *Q* and a page *X*, we evaluate whether *X* is a good anchor point for *Q*

---

[3] We take $0^0 = 1$.

by estimating the number of relevant documents (w.r.t. Q) that are within distance k from X and that a user will visit if he starts from X. We call this estimate the *potential* of X with respect to Q (Potential(X,Q)). Anchor points are ranked based on their potentials.

Note that if $f(Y,a)$ is a measure of the probability that page Y is relevant to a keyword a, then $P_k(X,a)$ is the expected number of relevant pages (w.r.t. a) in X's k-neighborhood that a user would visit if he starts from X. So, for a single-keyword query $Q = a$, Potential(X,Q) is simply $P_k(X,a)$, as given in Equation 14-1.

As an example, Figure 14-3 shows the hypertext structure of a hypothetical FAQ document. If $\alpha = 0.8$, $k = 3$, and the scoring function $f(X,a)$ simply returns 1 if X contains a; 0 otherwise, then the tuple beside each page in the figure shows the values of Potential with respect to the keywords A, B, C, respectively. For example, we have

$$\text{Potential}(\text{``part I''},A) = P_k(\text{``part I''},A) = 2.4,$$
$$\text{Potential}(\text{``part I''},B) = P_k(\text{``part I''},B) = 2.752,$$
$$\text{Potential}(\text{``part I''},C) = P_k(\text{``part I''},C) = 2.512.$$

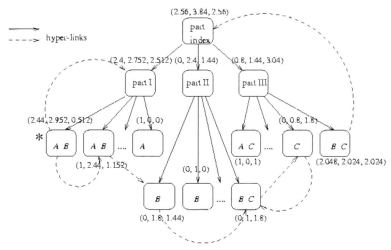

Figure 14-3 A FAQ document and the potential of each individual page

In this example, the anchor points for keywords A, B, and C are "part index", "part index", and "part III" respectively.

For a multiple-keyword query Q, Potential(X,Q) can also be estimated based on $P_k(X,a_i)$ where a_i's are the keywords in Q. Here, we distinguish two cases: conjunctive queries and disjunctive queries.

To simplify our discussion, let us define $Pr[Q]$ to be that, given a page in X's k-neighborhood that a user visits if he starts from X, the probability that the page is relevant to Q. Since the expected number of pages ($n_k(X)$) in X's k-neighborhood that a user would visit if he starts from page X is simply:

$$n_k(X) = \sum_{Y \in N_k(X)} \alpha^{D(X,Y)},$$

we have $Pr[a] = P_k(X,a) / n_k(X)$ for any keyword a.

Assuming that the occurrences of keywords in a document are independent from each other, we have, given a conjunctive query $Q = a_1 \wedge a_2 \wedge ... \wedge a_m$,

$$Pr[Q] = Pr[a_1 \wedge a_2 \wedge ... \wedge a_m]$$

$$= Pr[a_1] \cdot Pr[a_2] ... Pr[a_m]$$
$$= \frac{P_k(X,a_1)}{n_k(X)} \cdot \frac{P_k(X,a_2)}{n_k(X)} .... \frac{P_k(X,a_m)}{n_k(X)}$$
$$= \frac{\left(\prod_{i=1}^m P_k(X,a_i) \right)}{(n_k(X))^m}$$

Thus,

$$Potential(Q) = Pr[Q] \cdot n_k(X) = \left(\prod_{i=1}^m P_k(X,a_i) \right) \Big/ (n_k(X))^{m-1}. \qquad (14\text{-}2)$$

We remark that in practice the independent assumption may not hold. However, we would like to point out that the Potential function as defined above is used only to rank anchor points. It is not used to compute an accurate estimate of the number of relevant pages. In fact, this independent assumption is used in other information retrieval techniques, such as *GlOSS* [18, 19, 20].

The potential of a page with respect to a disjunctive query can be similarly estimated. Given a disjunctive query $Q = a_1 \vee a_2 \vee ... \vee a_m$, by the principle of inclusion and exclusion, we have:

$$Pr[Q] = Pr[a_1 \vee a_2 \vee ... \vee a_m]$$

$$= \sum_i Pr[a_i] - \sum_{i_1 \neq i_2} Pr[a_{i_1} \wedge a_{i_2}] + ... + (-1)^{m-1} \sum_{i_1 \neq i_2 \neq ... \neq i_m} Pr[a_{i_1} \wedge ... \wedge a_{i_m}]$$

$$= \sum_i \frac{P_k(X,a_i)}{n_k(X)} - \sum_{i_1 \neq i_2} \frac{P_k(X,a_{i_1})}{n_k(X)} \cdot \frac{P_k(X,a_{i_2})}{n_k(X)} + ... + (-1)^{m-1} \sum_{i_1 \neq i_2 \neq ... \neq i_m} \left(\prod_{j=1}^m \frac{P_k(X,a_{i_j})}{n_k(X)} \right).$$

Thus,

$$Potential(X,Q) = Pr[Q] \cdot n_k(X) \qquad (14\text{-}3)$$

can be estimated by the $P_k(X,a_i)$'s.

As an example, if we apply the above formula to the FAQ hypertext (Figure 14-3), the anchor points for the queries "$A \wedge B$" and "$A \vee C$" are the pages marked with an asterisk and "part index" respectively.

In [9], experiment results show that anchor point recommendation can significantly improve the performance of an IR system. Also, while the result set recommended by a traditional search engine changes with time, anchor points are relatively stable.

PAGERANK

The final link-analysis technique presented here is the *PageRank* scheme [13, 22]. It is a global ranking scheme, which assigns to each page a number according to their importance. Similar to the concept of a hub, a page that has more pages pointing at it is considered more important than a page with fewer pages pointing at it. The rank of a page *A* can therefore be defined as the number of pages that point at *A*. Furthermore, *A* attains a higher rank if the pages that point at it are themselves highly-ranked ones. Formally, we define the *rank* of a page *i* (*i* = 1 to *m*) by the following formula:

$$r(i) = \sum_{j \in B(i)} r(j) / N(j)$$

where *r(i)* is the rank of page *i*, *N(i)* is the number of outgoing links from page *i*, and *B(i)* is the set of pages that point to page *i*. Note that the definition of *PageRank* is recursive — the rank of a page both depends on and affects the rank of other pages. The divisor *N(j)* captures the notion that the rank of page *j* is evenly distributed among the pages that it points to.

To calculate *PageRank*, we note that the above formula can be written as $r = A^T r$, where *r* is an $m \times 1$ vector [*r(1)*, *r(2)*, ..., *r(m)*] and the elements a_{ij} of the matrix *A* is given by $a_{ij} - 1/N(i)$ if page *i* points to page *j*; $a_{ij} = 0$ otherwise. It can be shown that if a graph (with pages as nodes and links as edges) is strongly connected, then *r* is an eigenvector of A^T with eigenvalue 1, and *r* is unique. Techniques for handling more general graphs that are not strongly connected have also been proposed. Interested readers are referred to [13] and [22] for details.

14.3 Web-log Mining

Besides link analysis, another way to improve the effectiveness of a web IR system is to learn about the topics of interest to users, as well as those of web sites. The IR system thus acts as an agent that matches information providers and information consumers. The question is, how does a system obtain such knowledge without being too intrusive? In this section, we discuss a couple of techniques that mine a web log to learn about users' interest and web sites' content.

USER PROFILES

In [4], the authors devise a learning agent for discovering the topics of interest of web users, or their *user profiles*. A user profile is a set of *topic vectors*. Each topic vector is a collection of keywords and their weights. Essentially, a topic vector captures one topic of interest of the user. For example <(NBA, 0.6), (basketball,

0.55)> represents a topic vector that indicates that the user is interested in NBA basketball. The weights associated with the keywords in a topic vector measure the relative interestingness of the words. Knowing the user profiles allow a search engine to perform intelligent searching. For example, if one applies similar techniques to capture the profiles of web sites, one can perform automatic matching of user profiles and web site profiles. Web sites that match the interests of a user can thus be recommended to the user.

In the study, each user is associated with an access log that keeps track of the pages that have been read by the user. Each web page D is then converted into a *document vector* V_D. Similar to a topic vector, a document vector is a vector of terms and their associated weights. A term t is included in the document vector V_D if t appears in the document D.[4]After processing all the web pages accessed by a user, we have collected a set of document vectors for that user.

The next step in constructing the user profile is to cluster the document vectors. In [4], the *Leader algorithm* [21] was chosen as the clustering algorithm. The distance between any two document vectors is measured by their normalized inner product. When a new document vector is added to a pool of clusterized vectors, its distances from the centroids of all the clusters are calculated. The document vector will be added to the cluster whose centroid is the closest to the document vector and that the distance is smaller than a *cluster-distance* threshold value. If no such centroid can be found, the document vector forms a cluster by itself. After all the document vectors are considered, the *Reallocation method* [3] is applied to improve the clusters formed by the Leader algorithm. If too many clusters are formed, the clustering process is repeated with a smaller cluster-distance threshold.

After clusterization, the centroid of each cluster is extracted to derive a topic vector. Since a centroid in general may have many keywords and many of them may carry relatively small weights, a further selection of keywords within a centroid is performed. The centroids can be truncated with respect to a predefined length threshold or the keywords in it can be filtered against a weight percentage threshold. Finally, the set of topic vectors so extracted forms the user's profile.

INTERNET *GLOSS*

As we have explained in the introduction, traditional search engines suffer from a number of drawbacks. For example, the basic objects that a search engine indexes are individual web pages, which are innumerable. For many queries, this results in very large answer sets (sets of matching pages) with poor retrieval precision. As an example, if one tries to locate some general information about "Microsoft Windows" and queries a search engine with those two keywords, hundreds of thousand of pages are returned. Users could be disorientated by the number of documents. A more satisfactory answer to such a *general concept* query might be a recommendation to the Microsoft web site, or a web site that contains a FAQ of "Windows". Starting from the web site's home page, a user can search/navigate about the information source according to his own specific interest to pin-point relevant documents. We call this strategy of locating information on the web (first locate a relevant web site, then locate relevant information within the chosen web site) the *two-step approach*.

---

[4] Stop-words are first removed from D before the vector V_D is constructed.

In contrast, traditional search engines take a *one-step approach* in which potentially relevant documents are returned directly in reply to a user query.

As the web develops, more and more information sources are dedicated to specific topics of interest. To name a few, there are sites for *law reports* (Legal Data Search), *scientific research paper archives* (INSPEC), *sports digests* (ESPN), *news* (CNN), and even *music libraries*. If a user is looking for general information about a topic, chances are that a specialized web exists on that topic. Recommending web sites, instead of individual web pages, becomes more meaningful for this type of general concept queries. In [18], the problem of identifying relevant sources of documents is called the *text-source discovery problem*. In that study, it is assumed that information sources are static, and/or that they are fully co-operative in exporting metadata describing their collections. In [2], the authors investigate how the text-source discovery problem can be tackled in a *dynamic* and *uncooperative* web environment.

Before we discuss the techniques for discovering text sources, let us substantiate further the claim that the two-step approach to locating information on the web is, in many cases, preferable to the one-step approach of traditional search engines. First, the storage requirement of a traditional search engine is colossal. Alta Vista, for example, maintains more than 200 gigabytes of index data. It is conjectured that the web has a doubling period of under a year. Search engine indexes are expected to grow at a similar rate. Recommending a web site instead of individual web pages allows us to keep only *summaries* or statistical metadata of the information sources. As we shall see, these summaries only require a small fraction of the storage that is needed by a full-text index used in most search engines. Second, many information sources provide customized search interfaces in which queries are much more expressive than the simple keyword search adopted by most search engines. For example, a research paper archives could allow its users to search for articles based on attributes like "author", "title", "subject area", "keywords", and "references", etc. Directing a user to a relevant web site first often makes a more expressive query interface available to him/her. This results in more precise information retrieval. Third, many web sites nowadays are *dynamic* and *database driven*. By database driven, we mean that the information or documents maintained by the web site are stored in a database that is decoupled from the web server. This arrangement allows better maintainability and scalability of the information source. By dynamic, we mean that web pages are not statically created, instead, users are required to supply queries (e.g., via some pre-designed forms) and the web pages returned are composed on-the-fly based on the queries and the information stored in the back-end database. Essentially, the web only acts as an interface to the database. The information stored in the database is not directly put on the web (e.g., through hyper-links). Hence, "robots" sent by search engines would not be able to retrieve and index the information provided by the web sites.

One possible (and economical) solution to locating a web site that is relevant to a user query is to borrow the idea of *GlOSS* [18, 20]. *GlOSS* stands for *Glossary-of-Servers Server*. It is a system that, given a user query, recommends relevant document databases based on statistical information that it keeps about the document databases. The original study on *GlOSS* assumed that the document databases co-operated fully in exporting statistical information about their document collections. In order to apply *GlOSS* to a dynamic and uncooperative web environment, however, we need a mechanism for extracting statistical information from the web sites. We call the problem of extracting information from web sites to support the application of *GlOSS*, the *GlOSS* update problem. In the rest of this section we discuss *GlOSS* and briefly mention how to tackle the *GlOSS* update problem.

There are a number of options for tackling the text-source discovery problem. These options vary in their resource requirements and accuracy. As an example, a search engine with its full-text index on the documents stored in each web site has (almost) complete information to deduce the web site that is most relevant to a user query. The web sites could be ranked according to the number of *relevant* documents that they host. *Relevancy* here could be measured by the *similarity* between a document and the user query. There are a number of ways in which similarity is measured. Two popular models are the *Boolean model with conjunctive queries* and the *vector-space model* [20]. For simplicity, we only discuss the Boolean model in this chapter.

Under the Boolean model, a document is similar to and thus matches a query if the document contains all the keywords in the query. The similarity ($sim(Q,D)$) between a query, Q, and a document, D, can be defined as:

$$Sim(Q,D) = \begin{cases} 1 & \text{if D contains all the keywords in Q;} \\ 0 & \text{otherwise.} \end{cases}$$

The *goodness* of a document database db (a web site in our context) with respect to a user query Q, can be measured by simply counting the number of matching documents:

$$goodness(Q, db) = \sum_{D \in db} sim(Q, D)$$

With a full-text index, $goodness()$ can be easily calculated. A search engine is thus very "accurate" in ranking web sites given a relevancy and goodness measure. However, as we have discussed, the search engine approach is very costly, both in terms of the storage requirements as well as the network bandwidth requirements.

The idea of *GlOSS* is to recommend a document database based on the word statistics that the server keeps about the databases. Essentially, *GlOSS* tries to estimate *goodness()* using summary information. There are two versions of *GlOSS*: *bGlOSS* for the Boolean model [18], and *vGlOSS* for the vector-space model [20].

With the Boolean model, *GlOSS* keeps for each document database db the following statistical summaries: (1) the number of documents in db ($n(db)$); and (2) for each word t_j, the number of documents in db that contain t_j ($d_j(db)$). *GlOSS* assumes that keywords appear in the different documents of a database independently and uniformly.[5] Given a conjunctive query Q of k keywords $t_{i_1}, ..., t_{i_k}$, *GlOSS* estimates the goodness of a document database db with respect to Q by:

---

[5] We note that these assumptions are not realistic, and thus the estimates of *goodness()* are off. However, the estimates are only used to *rank* document databases, not to compute accurate values. In fact, it has been shown that such

$$goodness(Q,db) = \prod_{j=1}^{k} \frac{d_{i_j}(db)}{n(db)} \times n(db) = \frac{\prod_{j=1}^{k} d_{i_j}(db)}{(n(db))^{k-1}} \quad (14\text{-}4)$$

Taking a web site as a document database, we could apply the idea of *GlOSS* to locate a relevant web site given a user query. However, *GlOSS* assumes the availability of the statistical information about each web site. In practice, most web sites are dynamic and uncooperative. They do not actively export the word statistics of their collections. Even if such statistics are obtained, they become stale fast due to frequent updates on the web sites' contents. Maintaining these statistics is thus a very critical problem in applying *GlOSS* to solve the web site discovery problem.

As we have argued, it is not always possible to obtain the complete document collection of a web site from which a summary can be derived. Even if all the documents are accessible, retrieving all of them poses high overheads on the network bandwidth and storage. In many cases, however, obtaining the full set of documents from a web site is not necessary. This is particularly true when the goal of the system is to determine whether a web site contains a *non-trivial* number of documents that are relevant to a user query. As an example, the sports site ESPN is a good source of NBA basketball and contains numerous articles on that topic. To deduce that ESPN is a site relevant to the word NBA, it is sufficient for the system to examine only a fraction of the articles located at ESPN. Hence, instead of *actively* probing a web site for documents, an Internet *GlOSS* system can take a *passive* approach: it only examines and summarizes documents that a user community has ever retrieved from the web site within a certain period of time. Essentially, the Internet *GlOSS* server uses a *sample* of documents to project the information content of a web site. In [2], an Internet *GlOSS* prototype is proposed. Figure 14-4 shows the architecture of the system.

---

estimates are very effective in *ranking* databases. Readers are referred to [5] for a discussion on the efficacy of *GlOSS*.

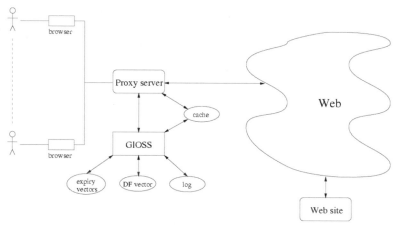

Figure 14-4 System architecture

In the prototype, it is assumed that users access the web via a Proxy server. Recently accessed web documents are cached at the Proxy. When a user issues a HTTP request, the request is forwarded to the Proxy. The Proxy then fetches the HTTP header of the document requested from the appropriate web site. The header usually carries the last-modified time of the document. This piece of information allows the Proxy to determine, if the document already exists in the cache, whether the cached copy is an update or not. If the cache does not contain the most updated document, the requested document will be downloaded from the web site; otherwise, the cached copy is returned to the user.

The Internet *GlOSS* server interacts closely with the proxy server. The proxy server notifies the *GlOSS* server on any modification to the cache. For each web site, *db*, that the *GlOSS* server monitors, the server maintains a *document frequency vector DF(db)* which models the document frequency statistics of the web documents that the system has ever retrieved from *db* in a certain period of time, say, in the past M days[6]. In words, if $\tilde{d}_b$ denotes the set of documents that the system has retrieved from *db* in the past M days, then the *j*-th component of the *DF(db)* vector represents the document frequency of the *j*-th word in $\tilde{d}_b$, or using our notation, $d_j(\tilde{d}_b)$. Besides the document frequency vector, the *GlOSS* server also keeps the size (the number of documents) of $\tilde{d}_b$, denoted by $n(\tilde{d}_b)$.

To process a query Q in the Boolean model, the goodness of *db* is estimated based on Equation 14-4 using the statistics of the sample, namely $d_j(\tilde{d}_b)$ and $n(\tilde{d}_b)$, in place of $d_j(db)$ and $n(db)$ respectively.

In [2], the authors propose algorithms for maintaining the *DF(db)* vector. Interested readers are referred to the original paper for detail. Experiment results show that the estimated document frequencies are fairly accurate. This is especially true for important keywords that occur frequently in a web site.

---

[6] In the prototype M is set to 400.

14.4 Conclusion

In this chapter we discussed a number of issues concerning information discovery on the World Wide Web. We pointed out a few shortcomings of traditional search engines in retrieving relevant information for an information seeker. We explained that web links and web logs are two valuable sources of knowledge that can significantly improve the effectiveness of a web IR system. We discussed the concept of authorities and hubs, anchor points, and PageRank in web-structure mining. We also discussed how user profiles can be extracted from web logs. Finally, we discussed the two-step approach to information discovery on the web, and mentioned the *GlOSS* update problem.

References

[1] V.N. Gudivada. Information Retrieval on the World Wide Web. *IEEE Internet Computing*, Vol. 1, No. 5, 1997, pp.58-68.

[2] C.Y. Ng, Ben Kao, David Cheung. Text-Source Discovery and *GlOSS* Update in a Dynamic Web. Proceedings of the Fourth Pacific-Asia Conference on Knowledge Discovery and Data Mining, 2000.

[3] W. Frakes, R. Baeza-Yates. Information Retrieval – Data Structures and Algorithms. Prentice-Hall, 1992.

[4] David Cheung, Ben Kao, Joseph Lee. Discovering User Access Patterns on the World-Wide-Web, in Knowledge Based Systems Journal, Elsevier Science, V10, N7, May 1998.

[5] A. Tomasic, L. Gravano, and H. Garcia-Molina. The Effectiveness of *GlOSS* For the Text-Database Discovery Problem, Proceedings of the 1994 ACM SIGMOD, 1994.

[6] http://www.google.com

[7] The Web Robots FAQ.
     http://info.webcrawler.com/mak/projects/robots/faq.html

[8] S. Feldman. Just the Answers, Please: Choosing a Web Search Service, The Magazine for Database Professionals, May 1997.

[9] Ben Kao, Joseph Lee, C.Y. Ng, and David Cheung. Anchor Point Indexing in Web Document Retrieval, in IEEE Transactions on Systems, Man, and Cybernetics Part C: Applications and Reviews, 30(3), pp. 364--373, 2000.

[10] J.M. Kleinberg. Authoritative Sources In a Hyperlinked Environment, Journal of the ACM, 46, 1999.

[11] B. Grossan. Search Engines: What They Are? How They Work?
     http://webreference.com/content/search/features.html

[12] J. Nielsen. The Art of Navigating Through Hypertext. *Communications of the ACM*, 33(3): 297-310, 1990.

[13] L. Page, S. Brin, R. Motwani, and T. Winograd. The PageRank Citation Ranking: Bringing Order to the Web. *Technical report, Computer Systems Laboratory, Stanford University*, 1998.

[14] R. Kumar, P. Raghavan, S. Rajagopalan, and A. Tomkins. Trawling the Web for Emerging Cyber-communities. *Proceedings of the Eighth International Conference on the Web-Wide-Web*, 1999.

[15] J. Dean and M.R. Henzinger. Finding Related Pages in the World-Wide-Web. *Proceedings of the Eighth International Conference on the Web-Wide-Web*, 1999.

[16] D.L. Lee et. al. Document ranking and the Vector-Space Model. *IEEE Software*, Vol. 14, No. 2, Mar/Apr 1997, 67-75.

[17] G. Salton. Automatic text processing: the transformation, analysis, and retrieval of information by computer. Mass: Add-Wesley, 1989.

[18] L. Gravano et. al. The Efficacy of *GlOSS* for the Text Database Discovery Problem. *ACM SIGMOD'94*, 1994.

[19] L. Gravano et. al. Precision and Recall of *GlOSS* Estimators for Database Discovery. *PDIS'94*, 1994.

[20] L. Gravano et. al. Generalizing *GlOSS* to Vector-Space Databases and Broker Hierarchies. *VLDB'95*, May 1995.

[21] J. Hartigan. Clustering Algorithms. Wiley, New York, 1975.

[22] A. Arasu, J. Cho, H. Garcia-Molina, A. Paepcke and S. Raghavan. Searching the Web. ACM Transactions on Internet Technology, Vol. 1, No. 1, Aug 2001, 2-43.

15 COOPERATIVE MULTIMEDIA INFORMATION SYSTEMS

Dr. Schahram Dustdar

Cooperative MMIS are increasingly gaining importance for multimedia information creation and dissemination in distributed teams. This chapter discusses relevant issues on cooperative aspects of Multimedia Information Systems (MMIS) and provides an insight into a variety of applications, technical constraints, and architectures enhancing efficient information dissemination and collaboration among people. Thus this chapter will offer breadth by reviewing the state of the art of commercial systems as well as research prototypes rather than details of one particular system.

15.1 Introduction

Communication, collaboration, and cooperation using multimedia information systems are essential ingredients for successful distributed teamwork. Recent advances in network technology (e.g. IP-Multicasting or Multicast Backbone or MBone for short) provide a highly scaleable network infrastructure for group work on the Internet. Relevant research streams in various disciplines within computer science are Computer Supported Cooperative Work (CSCW) [49, 50, 53, 61, 76,77, 15, 79, 26, 81], Multicasting and coordination issues [52, 58, 83, 2, 47, 6, 22, 23, 44 57, 72, 92, 91, 16, 24], Multimedia protocols [35, 37, 80], Compression Technologies [7, 33, 42], Intelligent Systems [27], Virtual Reality [29, 28, 45], Streaming [32, 7, 8], Software-Engineering [3, 17, 56, 64, 54], and mobile multimedia [39]. Cooperative Multimedia Information System examples are audio- and videoconferencing systems [1, 19, 20, 21, 25, 40, 41, 88, 43, 51, 67, 89], shared whiteboards [36, 66], application sharing systems [69], learning support systems [65, 24], workflow systems [90], groupware systems [31, 78, 70], and environments for designing and implementing cooperative multimedia systems [31, 87, 5, 63, 68]. These systems all aim at helping people to increase the efficiency of their collaborative work. Most collaborative work conducted on the Internet is asynchronous in nature, such as e-mail and Web pages [57]. More recent research on infrastructure (networks) suggests that IP multicasting offers a viable solution for supporting group work also on a large-scale on the Internet.

Cooperative Multimedia systems are inherently complex since they span many aspects requiring consideration when systems are being designed and implemented. The goal of this chapter therefore is to provide an overview and reference of the fundamentals and the current state of the art regarding research and prototype implementations.

COOPERATIVE MULTIMEDIA INFORMATION SYSTEMS DEFINITION

The most general definition of multimedia states that a multimedia system exploits a computer to combine text, data, graphics, animation, voice, and video into a single synchronised production or presentation [18]. Under that definition, TV is a multimedia device because it includes moving video, audio, text and animation in a single presentation system and uses computers in the process. This definition is not very precise and must be further clarified in terms of *interactivity* and *network infrastructure* to be useful in the context of this chapter.

Cooperative Multimedia Information Systems are interactive multimedia systems supporting human communications, collaboration, and cooperation by utilizing a networked infrastructure. We divide (multi)media into two classes: *continuous* and *discrete*. Continuous media (e.g. audio, video) change with time, whereas discrete media (e.g. text, data, images) are time independent.

TAXONOMIES OF MULTIMEDIA INFORMATION SYSTEMS

The purpose of this section is to provide an overview of several research streams of multimedia information systems and to provide a taxonomy for multimedia information systems. We will discuss important multimedia applications in each proposed category and give some examples on the usage. Whenever we communicate we use many senses simultaneously. Through each sense we interpret the external world using representations and organizations to accommodate that use. People communicate more effectively through multiple channels, such as gestures and other body movements that usually accompany speech or eye contact. Research on multimedia information systems is an interdisciplinary research area and consists of several research streams such as communication research, human-computer interaction, organization research, and computer science.

Each of these research streams of multimedia information systems has developed their own categorisation systems. However, a comprehensive and well-established categorisation of multimedia information systems - which cannot be found in the literature - is essential in order to generalise research findings across systems and research disciplines. Historically, communication research, which - in this context - investigates mainly Computer-Mediated Communication (CMC), was the first discipline to categorise media systems. Daft and Lengel [14] developed a taxonomy for assessing the "information richness" of media systems. They conceptualised richness as the "potential information-carrying capacity of data". In Daft and Lengel's system, face-to-face communication was highest in its "information richness". Next to face-to-face communication, in terms of information richness, came telephone communication. Daft and Lengel's taxonomy is often used as a classification system in organization research. However, research on the organizational implications of multimedia information systems [18] needs close and joint efforts across disciplines. Since multimedia information systems consist not only of communication systems enhancing human-human communication, a taxonomy, such as that developed by Johansen [53], which differentiates between two dimensions (e.g. synchronous/asynchronous communication and same place/different place) makes sense only to a limited number of multimedia information systems, in the literature termed "Computer Supported Cooperative Work" (CSCW) systems or "group multimedia information systems".

Therefore we propose a taxonomy of multimedia information systems consisting of two categories: *personal* and *cooperative* multimedia information systems. Table 15-1 summarises a proposed categorisation of multimedia information systems, provides system examples in each category and relates them to probable management issues. A detailed discussion of every system example used in the table and its main characteristics would go beyond the scope of this chapter, hence we will focus our efforts on the most important ones.

Categories	*Application examples*	*Management Issues*
Personal	Kiosk system Video-on-demand Music-on-demand Multimedia database Interactive-TV Personalised news Games	Change of focus on - mass information systems - external information systems
Cooperative	Point-to-point Conferencing Multipoint Conferencing Shared Workspaces Concurrent Engineering Multimedia Training Multimedia Broadcast	Redesign of - workplace - work content - processes - communication paths

Table 15-1 Multimedia information systems and management issues

15.2   Review of Cooperative Multimedia Information Systems

MULTIMEDIA CONFERENCING

Audio and videoconferencing are the main building blocks of cooperative multimedia information systems. It has been repeatedly hailed as on the brink of ubiquity and as a panacea for communications in distributed teams. The PicturePhone was presented at the 1964 World Fair and the Integrated Services Digital Network (ISDN) conferencing systems were introduced in the early 1980s. The 1990s marked the era of the emergence of desktop videoconferencing. However, videoconferencing never really caught on as well as it was expected to by technology analysts and scientists.

In this chapter we use the term "multimedia conferencing" instead of videoconferencing because the systems discussed in this section integrate multiple media formats into one system, not just video. The multimedia conferencing [47, 71] market is believed to be one of the key markets within the multimedia market segment. Recent developments in multimedia systems and networking technology show that using desktop multimedia conferencing for group decision-making on WANs such as the Internet is feasible [58]. Researchers have often discussed the failure of video to support interpersonal communication [11, 26, 81]. In the following section we review the design, hardware and software requirements as well as organizational issues in desktop multimedia conferencing systems.

As Johansen [53] shows, group work, and hence group decision-making, is a natural way of doing business. Early groupware systems and digital meeting systems lacked the ability to manipulate multiple media types such as audio, video and textual information in one integrated multimedia system. The merging of workstation technology and real-time computer conferencing has had a significant impact on CSCW and group decision-making and lead to the term "desktop conferencing". Research on early multimedia conferencing systems such as those developed at AT&T Bell Laboratories [1], Bellcore [77] and NEC [88] had as their aim the provision of the facilities found at face-to-face meetings with remote groups. It is generally accepted that computer-supported decision-making and communication results in many changes in communication patterns [25,40], greater task orientation and shorter meetings [38]. Regarding the video component, Ishii et al. [59, 50] point out the importance of gaze awareness - the ability to monitor the direction of someone's gaze and thus the focus of the attention. Heath and Luff [41] as well as Mantei et al. [61] found similar results. The main obstacle, we argue in this chapter, is that group problem solving and task accomplishments as well as organizational structure and - process support have never been addressed adequately. Research in communications studies showed [11, 26, 81] that voice is only a little slower than face-to-face communications. This might imply that video is not relevant for effective and efficient communications. Hence, studies provided evidence that the final outcome of any given task is not influenced positively by videoconferencing support, although people were happy to use videoconferencing [11]. Other research provides an interesting insight in this regard: video increases the "rate of social presence" [81] and therefore it makes a valuable contribution to negotiation, sales, and relationship building. However, support for social presence naturally includes a tighter integration with other corporate information systems such as workflow management systems (WfMS), groupware and project management systems.

Another promising research stream in videoconferencing research deals with so called "gaze-awareness" support. This research deals with the question of how to provide eye contact between videoconference participants. From a social perspective people who use frequent eye contact are perceived as more attentive, friendly, cooperative, confident, mature, and sincere than those who avoid it [30, 28]. The loss of gaze-awareness is one important contributing factor to the failure of videoconferencing as a mass tool. The remainder of this section however, deals with videoconferencing systems readily available. Videoconferencing systems supporting gaze-awareness are still in their infancy and mostly first research prototypes.

Cooperative multimedia information systems such as point-to-point or multipoint conferencing systems, are currently the driving force in the multimedia information systems arena and researchers in the CSCW and DSS domains are, and will be investigating multimedia elements of collaborative systems for business cooperati and e-learning [18, 30, 79, 84, 12, 57]. Corporate multimedia conferencing used t limited to boardrooms with built-in dedicated conferencing equipment c hundreds of thousands of dollars. Conferences had to be scheduled weeks in ? and run by experienced administrators. By contrast, desktop multimedia conf systems, which integrate multiple media formats, enable people to use desktop workstation for conferencing. Some industry observers and remain sceptical of any imminent or significant deployment of deskto conferencing in the organizational domain. Sceptics claim that organizations are able to function without seeing the person at the other end of the communication. The challenge to suppliers and researchers regarding desktop multimedia conferencing is to *de-emphasise* the technical aspects but to *emphasise* the qualitative new possibilities of collaborative work which desktop multimedia conferencing enables. We discuss the core functionality of the software regarding its usage for point-to-point and multipoint conferences. By *point-to-point* conference we mean the usage of desktop multimedia conferencing equipment between two people. By *multipoint* conference we mean the usage of desktop multimedia conferencing equipment between more than two people concurrently. The software we discuss in the following section allows both modes of conferencing. We start our review with *vat* (visual audio tool), which is a software tool that supports multiple audio channels between conference participants. It was developed by Van Jacobsen and Steve McCanne at Lawrence Berkeley Laboratory (LBL) [51]. The vat shows the active audio conference participants by providing their e-mail address. Clicking on a user brings up a new window displaying user statistics on receiving/sending audio packets. Clicking on the speaker or microphone button turns it on or mutes it. Finally, increasing or decreasing the volume is done by moving the buttons up or down. The vat tool can be used in point-to-point and multipoint conferences on LANs or WANs such as the Internet and uses TCP/IP (Transmission Control Protocol/Internet Protocol) as the transport protocol. The main disadvantage of this tool, like most other audio conferencing tools available today, is the inability to direct audio only to a selected user(s) in the audio conference. The video component is managed by *vic* [67], which allows users to transmit and receive real-time video over the Internet/Intranet. The video streams can be sent point-to-point or multipoint. The multipoint transmission mode sends to several destinations by using IP-multicasting and audio/video compression. Figure 15-1 shows a screenshot of a multi-point conference using *vic* on the Internet.

A small video of the active participants is shown together with their e-mail address. By clicking on the small picture one can enlarge the video as shown in Figure 15-1. Depending on the network infrastructure, users can choose their video transmission bandwidth up to 2 Mbps, the size of the displayed video and whether it should be color or greyscale. Finally one can select the "Stop sending" button, which stops the video transmission and "freezes" the sending picture on the receiving side of the videoconference.

For Internet videoconferences the TTL (time-to-live) value, which has to be set manually in advance, determines the scope of the videoconference. The value 16 transmits the video only on the LAN within the organization. A TTL value of more than 132 would have a worldwide scope and is used for some international conferences on the MBone which are of general interest, such as the WWW ferences. The ivs (INRIA videoconferencing system)[86] is a software tool that orts audio and videoconferences over LANs, WANs and ISDN networks in one re package. The ivs includes a software codec with an integrated dynamic-n-control mechanism and a protocol to manage the participants in a

Figure 15-1 A multi-point conference with vic

conference. It is based on ITU H.261. The main advantage of *ivs* is the possibility to use one integrated audio- and videoconferencing package on the desktop workstation. For document sharing we discuss the *wb* (whiteboard) utility, which is a collaborative software tool that supports a shared desktop whiteboard among a group of distributed users on a LAN or a WAN such as the Internet. The *wb* can be used to import Postscript or text documents and to share the content with other conference participants. Each user has the possibility to use wb's tools such as basic painting and word processing features. Whenever a conference participant modifies the shared document, the other conference participants show this updating process in real-time. At the end of a shared whiteboard session each user can save the document on a workstation for personal use.

Chang et al. [10] designed and implemented a **Web-based conference minute system** as a research prototype, which preserves relevant information produced during meetings (such as documents and URLs) and maintains hyperlinks to them. The system is based on a Java application-sharing framework and provides capabilities to record and replay the audio/video data of videoconferences and all operations of the shared applications used throughout the meeting. By providing these hyperlinks project participants can understand the flow of the meetings and by replaying the audio/video it is easy to know what actually happened. The authors of the conference minute system claim that this system increases the overall group productivity of their geographically dispersed members. Figure 15-2 depicts a screen shot of the conference minute system.

INTEGRATED TOOLSETS
JETS, shown in Figure 15-3 [82, 73], SCOOT [13], JASMINE (Java Application Sharing in Multi-user Interactive Environments) [73], JASBER (Java Shared Browser) [73] and Habanero [34] are well-known Java-based collaborative multimedia environments (research prototypes) on the Internet. Habenero, developed

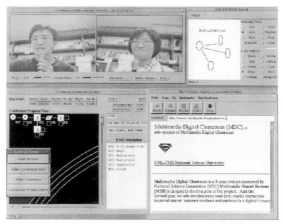

Figure 15-2 The conference minute system [10]

by NCSA, is an application suite as well as an environment that lets developers transform single-user applications into multi-user applications. JETS, is a research prototype and supports multiple users using any Java-enabled Web browser per session. It lets users share applications in form of Java applets in real-time. However, the systems don't support real-time audio and video on the Internet, nor scaleable communications using IP multicast.

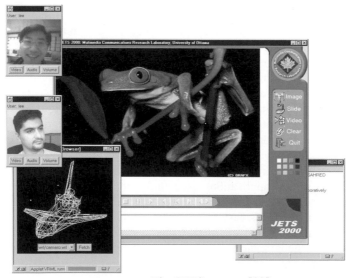

Figure 15-3 The JETS system [82]
Picture courtesy of Multimedia Communications Research Laboratory,
University of Ottawa

Groupkit [31, 78] is a software environment focusing on creating distributed Tcl-based applications. Although the initial intention was to create an overall software environment for group work, the toolkit is currently focused on groupware widgets, metaphors for session management, and programming abstractions for distributed systems. Groupkit also lacks support for real-time audio or video.

MSTAR [74] is a recent Java-based development from Sweden, commercially licensed to Marratech.com. It provides an integrated solution for generating, presenting, storing, and editing media in collaborative applications. It utilizes the IP multicast protocol to enable scaleable distribution of real-time audio and video as well as data among synchronously connected users (group members). Its strength lies in its integrated user interface for audio, video and shared whiteboard, chat, voting, and web-based presentations as Figure 15-4 shows. MSTAR utilizes an agent software architecture enabling developers to create new applications on top of the environment and to reuse existing ones. The architectural overview is described in Figure 15-6.

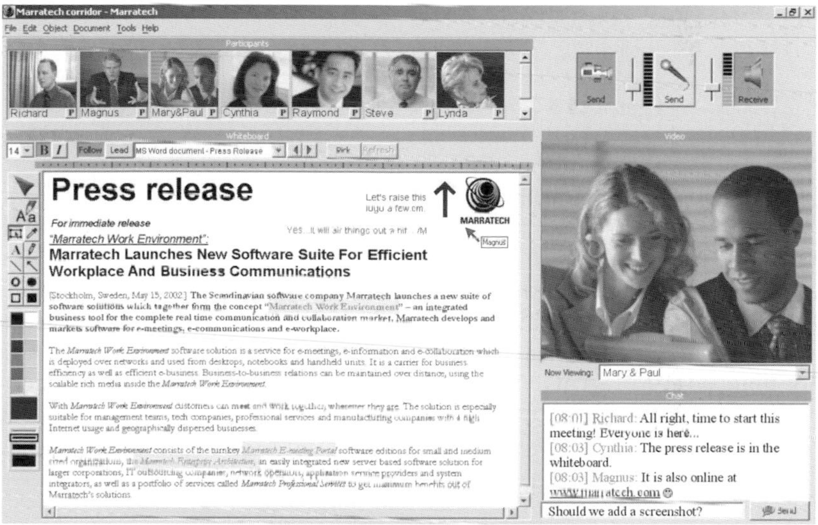

Figure 15-4 Marratech Pro Desktop
Picture courtesy provided by www.marratech.com

Groove [32] (founded by the creator of Lotus Notes) is a recently released peer-to-peer (P2P) system in the groupware area, with substantial monetary investment from Microsoft. It is listed here since it provides a toolset for cooperative work, which is relevant in the context discussed in this chapter. It consists of a P2P architecture consisting of transceivers and tools residing on each peer computer. Being a groupware system, it follows a *workspace metaphor*. Each peer (host) may create a workspace and invite other peers to join that workspace. Workspaces are typically based on a goal or project. The tools consist of audioconferencing facilities, calendar, chat, contact manager, discussion forum, file sharing, notepad, idea outliner, picture sharing, shared web-browsing, and a sketchpad. The toolset can be extended by third-party vendors or by Groove itself. Figure 15-5 shows a screen of a Groove workspace with a shared web-browser and ongoing audioconferencing (bottom left). Videoconferencing is currently not integrated.

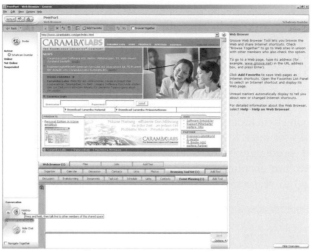

Figure 15-5 Groove peer-to-peer collaboration desktop

15.3 Architectural examples

MARRATECH

The Marrtech environment (previously called mSTAR) [74] is designed as an agent-architecture, as shown in Figure 15-6. An agent, in Marractech's terms, is a component that resides within an application and is responsible for specific tasks. It is not obligatory for agents to provide graphical user interfaces. Integration of the system with enterprise information system is done via Java Server pages and Servlets.

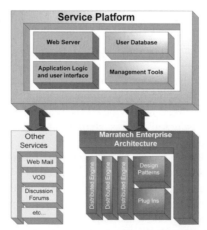

Figure 15-6 Marratech Enterprise Architecture
Picture courtesy provided by www.marratech.com

MULTIMEDIA INTERACTIVE TELELEARNING SYSTEM (MITS)

MITS [65] is a web-based multimedia interactive telelearning system with an emphasis on metadata and media content management mechanisms. The metadata model supports multimedia courseware creation, management, delivery and learners' evaluation. The proposed multimedia document architecture describes the multimedia content, which represents the course material. Both, metadata and media contents are generated, stored, and utilized to facilitate the search and on-demand presentation of learning objects. The students' preference, performance, and progress information is maintained by a profile database. One of the design goals was to create highly reusable multimedia learning objects.

Figure 15-7 [65] illustrates the MITS three-tier client/server system architecture. It consists of user service tier, business Logic tier, and data service tier. The clients interact with the system using a web-browser. The data service tier hosts the Courseware database, media content database, profile database and optionally an email server. A detailed discussion of the architecture would be beyond of this chapter. The number of multimedia objects and resources available on the Internet is growing exponentially and it is of paramount importance to develop new tools and methods to these resources to build efficient and effective collaborative multimedia teaching and learning systems. The MITS research prototype provides effective metadata management mechanisms and therefore enables efficient searching, evaluation, and utilization of multimedia resources.

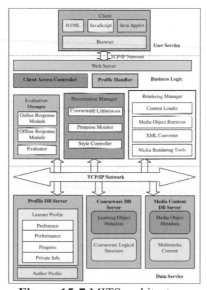

Figure 15-7 MITS architecture

CONFERENCE MINUTE SYSTEM

The **Conference minute system architecture** by Chang et al. [10] is depicted in Figure 15-8. The four-layer architecture consists of the graphical user interface, the progress flow/description/template managers, the minute object manager, the media recorder, and the player. The *minute object manager* maintains all existing conference minutes and creates a new minute with its respective ID in response to the template manager. The *template manager* gets the content of the last meeting by this ID, whenever a meeting is established, and generates an appropriate template for the minute of this meeting through the primitives provided by the minute object manager. During meetings, the *description manager* or the *progress manager* is executed depending on the operations requested by the user (*graphical user interface*). Brief object descriptions can be added to the objects. Based on the type of operations (recording or replaying), the minute object manager handles the media objects. In the recording process, the *media recorder* connects to the share manager and to the audio/video conference system to intercept input events (shared applications and/or audio/video). To summarize: The Conference minute system has the goal to capture all formal and informal information needed to maintain a project organizational memory. Information such as current project status, resolutions reached at meetings, and informal communications carried out using audio/video conferencing are stored by the system. The overall goal of the conference minute system is to increase group productivity of distributed groups.

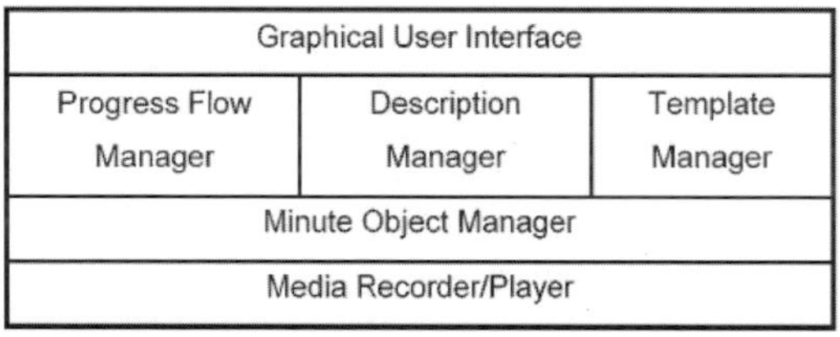

Graphical User Interface		
Progress Flow Manager	Description Manager	Template Manager
Minute Object Manager		
Media Recorder/Player		

Figure 15-8 Conference minute system architecture [10]

TEAMSMART

TeamSmart [85], a research prototype developed by NTT Multimedia Communications Labs, is discussed in this section. TeamSmart is based on the Inquiry Cycle model of collaborative document creation [75]. Documents, such as design specifications, represent relevant deliverables in many projects. Although team members have different styles to work on documents, all members still meet and synchronize their work and conduct peer review of the entire documents. The overall goal of TeamSmart is to maintain complete consistency and traceablility of contributions to the document creation process made by team members. The Inquiry Cycle model involves three activities: expression, discussion, and commitment. Expression deals with preparing and presenting ideas to be documented. Discussion involves the discussion of documents and sharing of ideas, annotations or any form of comments. Commitment includes planning changes to the documents based on results from the discussion and also execution of the plan. In order to represent and keep track of relationships among pieces of information generated throughout this cycle, metalevel links are used [85], as shown in Figure 15-9.

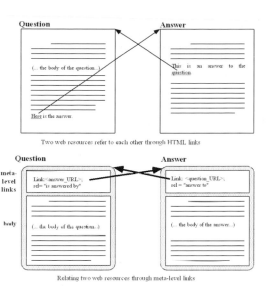

Figure 15-9 Metalevel links [85]

Current systems mostly support distinct project activities (e.g. e-mail for exchanging documents, document management tools for version control, and videoconferencing for peer reviews). Many misunderstandings are due to the lack of integration and traceability support. TeamSmart's goal is to address these problems by integrating the Mbone toolset (vat, rat, vic, wb) with their own collaborative document creation client named TQ and document visualization tool named TeamViewer. The persistence for this cooperative multimedia system is an enhanced HTTP server. The overall system architecture is depicted in Figure 15-10.

The TeamVCR [9] is part of NTT's TeamSmart suite and provides a multimedia note taking service. During a collaborative session using the TeamSmart suite, TeamVCR records the conference and automatically indexes its contents. Upon request TeamVCR plays back the audio/video data using the Mbone tools. The TeamVCR system itself consists of clients, server, and an index agent as depicted in Figure 15-11. The server component consists of a session management server and VCR servers. The TeamVCR clients handle interactions with users and communicate with session management server via the session initiation protocol (SIP) [37] and the Real-Time Streaming Protocol (RTSP) [80]. The session management server creates and controls the VCR server. It also initiates, monitors, and controls the recording and playing sessions, while the VCR server records and replays the audio/video streams.

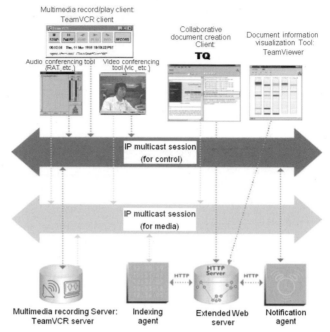

Figure 15-10 TeamSmart architecture [85]

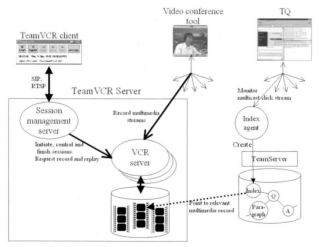

Figure 15-11 TeamVCR architecture [85]

15.4 Management Issues

Today organizations that benefit most from new and emerging technologies will be those that take advantage of new multimedia technologies to integrate the work of people within organizations, to establish new links with organizations, to build "virtual organizations", and use knowledge, services and "economies of space" more efficiently. Malone and Rockart [59] show that the changes take place in three categories: substitution, increased use, and new structures. Regarding organizational use of cooperative multimedia information systems, we are witnessing changes within organizations in these three dimensions as well. A large body of literature documents the interest in IT at the organizational level. However, a clear consensus on how organizations are evolving in response to changing technology is missing. Markus and Robey [62] reviewed the current literature on the organizational impact of IT from three differing viewpoints (imperatives) as depicted in Figure 15-12: the technological, the organizational and the interactionist perspective. The *technological perspective* postulates IT as an exogenous variable, which determines the behaviour of individuals and organizations. The *organizational perspective* takes the stand that organizational needs determine the type of IT [18]. IT is treated as a dependent variable. The ramification of this perspective is that there exists a one-to-one mapping between organizational information needs and the type of IT. The *interactionist perspective* makes no assumption about a causal relationship between IT and organizational change. On the contrary, it postulates that the type of IT that an organization adopts is the result of the interaction of organizational variables [18]. The organizational and technological perspectives are deterministic, whereas the interactional perspective is probabilistic. Rather than viewing the organizational impacts of multimedia information systems as being inherently positive or negative, we view multimedia information systems as neutral regarding their organizational impacts [18].

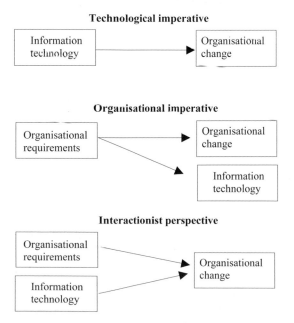

Figure 15-12 Research approaches for cooperative MMIS implications

The nature of an organization's use of IT - and therefore the use of multimedia information systems as well - emerges through complex interactions among the intentions of key actors, attributes of the technology involved and dynamic organizational processes. We view multimedia information systems as an enabling factor which, by providing qualitatively new capabilities in storing, presenting and processing information, makes certain organizational changes possible. Hence, we believe that for research on multimedia information systems it is essential to investigate first the new capabilities that multimedia information systems offer and then to study the relationship to organizational change. Huber [46] argues that organizations face two important aspects of change - changes in organizational environments and changes in organizational processes. Thus, he concludes that the level of environmental complexity, the level of turbulence and their absolute growth rate will be significantly greater in the future than in the past. Therefore decisions will be made more frequently and more rapidly. According to our understanding of the implications of cooperative MMIS on organization design, use and implementation, cooperative MMIS in the organizational context need considerable planning. The organizational implications will differ depending on the systems investigated.

Personal MMIS (see Table 15-1) enhance the development of mass information systems directed to (potential) users outside the organization. This category of information systems is, regarding its reach and scope, relatively new to most organizations. A cultural adaptation process of organization members and system developers, to the requirements of an implementation process is required. From the perspective of users, the amount of available information will rise. Consequences of retrieval and storage of multimedia information will enhance organizational learning processes. The application of interactive TV and the delivery of services to residential and organizational users produce more than alpha or beta changes. It provides opportunities for redesigning the organization (e.g. through redesigning its delivery channel).

Cooperative MMIS such as multimedia conferencing systems and multimedia groupware have the potential to fundamentally redesign organizational communication patterns. For example, some studies show that the use of desktop multimedia conferencing leads to higher efficiency in organizational communications [18]. Is efficiency the only change enabled by cooperative MMIS, or does it enable qualitatively new communication possibilities? Cooperative MMIS enhance a "networked" or "virtual" organizational structure. The advent of desktop multimedia conferencing systems for collaborative work raises new research questions on design issues of inter- and intra-organizational communication, organizational design and organizational computing [14]. Prospective users of cooperative MMIS should take into account the volume and ambiguity of the "information environments" as well as the cultural prerequisites of their organization. Most of the technological problems in cooperative MMIS are due to a lack of network bandwidth of existing networks and lack of integration with voice networks. The social and organizational implications such as structural and cultural changes in building virtual organizations, will be even more challenging.

Cooperative MMIS aim to support people working together in geographically dispersed distributed teams. The media for cooperation may be audio, video, and data. Currently a huge challenging domain is the question of multimedia information delivery to a multitude of devices. These devices may range from desktop computers to smart phones, personal digital assistants (PDAs), mobile phones, new devices inside in cars, trains, or airplanes, just to name a few. Today we witness a trend towards device independence and different modes of connectivity (mobility). We suggest distinguishing between three modes: fixed, mobile, or ad-hoc. We speak of fixed connectivity when users work on computers permanently connected to a network. For example in an office where each employee has a personal computer connected to the company-wide network or a wide-area-network (WAN). Mobile connectivity essentially describes a mode where people are "on the move" but access data and applications located on their remote network. The ad-hoc mode allows users to establish a "virtual" group of users on the fly. Participants in ad-hoc groups may have network connectivity either permanently or sporadically. The mobility of participants also offers new ways of distributed collaboration: processes are no longer bound to locations of resources (such as participants or artifacts) but can consider several availability modes.

15.5 Conclusion

Cooperative MMIS' goal is to provide support for efficient information creation and dissemination of multimedia data among distributed team members. In other words, cooperative MMIS provides a vehicle to qualitatively increase communications and collaboration among people, by providing a comprehensive framework for co-operation. The challenges of cooperative MMIS are manifold. Technical constraints, as discussed in this chapter, limit the possibilities of cooperation compared to face-to-face meetings. On the other hand, cooperative MMIS provide features not found in real-world face-to-face meetings, and therefore potentially increase productivity and efficiency of people considerably. Furthermore, organizational and management issues are vital to the success of cooperative MMIS in organizations, as can be seen in the last section of this chapter.

References

[1] S. R. Ahuja, J. R. Ensor, and D. N. Horn, "The Rapport Multimedia Conferencing System", Proc. Conf. on Office Information Systems, Palo Alto CA, pp. 1-8, Mar. 1988.

[2] E. Amir, S. McCanne, and R. Katz, "Receiver-Driven Bandwidth Adaptation for Light-Weight Sessions, " Proceedings ACM Multimedia, ACM Press, New York, pp. 415-426, 1997.

[3] K. Almeroth and M. Ammar, "The Interactive Multimedia Jukebox (IMJ): A New Paradigm for the On-Demand Delivery of Audio/Video," Proc. Seventh International World Wide Web Conference (WWW7), Brisbane, Australia, Apr. 1998.

[4] M. C Anglides and S. "Dustdar, Multimedia Information Systems," Kluwer Academic Publishers, Boston 1997.

[5] B. Bailey *et al*, "Nsync - A Toolkit for Building Interactive Multimedia Presentations," Proc. ACM Multimedia 98, pp. 257-266, Bristol UK, Sep. 1998.

[6] S. Bradner, A. Mankin, A. Romanow, and V. Paxson, "IETF criteria for evaluating reliable multicast transport and application protocols," Internet Draft Internet Engineering Task Force, May 1998. (Work in progress)

[7]   J. A. Brotherton, J. R. Bhalodia, and G. D. Abowd, "Automated Capture, Integration, and Visualization of Multiple Media Streams," Proc of IEEE Multimedia '98, July, 1998.

[8]   M. C. Buchanan and P. T. Zellweger, "Automatically Generating Consistent Schedules for Multimedia Applications," Multimedia Systems, Vol. 1, No. 2, pp. 55-67, 1993.

[9]   F. Cao, J. Smith, and K. Takahashi, "An architecture of distributed media servers for supporting guaranteed QoS and media indexing," Proceedings of the IEEE International Conference on Multimedia Computing and Systems, Vol.: 2, pp. 1 –5, 1999.

[10]   I. C. Chang *et al*, "A Multimedia World Wide Web Based Conference Minute System for Group Collaboration", Multimedia Tools and Applications 9(3): 199-226; Nov. 1999.

[11]   A. Chapanis *et al*, "Studies in Interactive Communication: The Effects of Fours Communication Modes on the Behaviour of Teams During Cooperative Problem-Solving," Human Factors, 14(6), pp. 487-509, 1972.

[12]   P. Chiu *et al*, "Room with a View: Meeting Capture in a Multimedia Conference Room", IEEE Multimedia, 7(4), pp. 48-54, 2000.

[13]   E. Craighill, *et al*, "SCOOT: An object-oriented toolkit for multimedia collaboration," Proc. ACM Multimedia 94, pp. 41-49,San Francisco CA, Oct. 1994.

[14]   R. L. Daft, and R. H. Lengel, "Information richness: a new approach to managerial information processing and organization design," in L.L. Cummings and B.M. Staw, Eds., Research in organizational behavior, Vol. 6, pp. 191-234, JAI Press: Greenwhich, CT, 1984.

[15]   R. C. Davis, J. A. Landay, V. Chen, J. Huang, R. B. Lee, F. Li, J. Lin, C. B. Morrey III, B. Schleimer, M. N. Price, and B. N. Schilit, "NotePals: Lightweight Note Sharing by the Group, for the Group," Human Factors in Computing Systems: CHI 99 Conference Proceedings, Pittsburgh, PA, May 1999.

[16]   H. P. Dommel and J. J. Garcia-Luna-Aceves, "A Coordination Architecture for Internet Groupwork," Proceedings of the 26th Euromicro Conference, 2000., Vol. 2, pp. 183 –190, 2000.

[17]   D. J. Duke and I. Herman, "A Standard for Multimedia Middleware," Proc. ACM Multimedia 98, Bristol UK, pp. 381-390, Sep. 1998.

[18]   S. Dustdar and M. C. Angelides, "Organizational impacts of multimedia information systems," Journal of Information Technology, 12(1), pp. 33-43, 1997.

[19]   S. Dustdar and R. Huber, "Group decision making on urban planning using desktop multimedia conferencing," Journal of Multimedia Tools and Applications, 6(1), pp. 1-14, 1998.

[20]   S. Dustdar, "Critical Issues using MBone-Videoconferencing on the Internet", Journal of Computing and Information Technology, 6(,3), pp. 273-283, 1998

[21]   S. Dustdar and G. Hofstede, "Videoconferencing accross cultures - a conceptual framework for floor control issues," Proceedings of the European Conference on Information Systems 1998, Aix-en-Provence, 1998.

[22]   R. Finlayson, "liveCaster: Multicast your data throughout the Internet!," http://www.live.com/-liveCaster/, Jan. 1998.

[23]   S. Floyd, V. Jacobson, C. Liu, S. McCanne, and L. Zhang, "A Reliable Multicast Framework for Light-weight Sessions and Application Level Framing," Proceedings ACM SIGCOMM 95, pp. 342-356, Aug. 1995.

[24]   G. Fortino, L. Nigro, F. Pupo, "An MBone-based on-demand system for cooperative off-line learning," Proceedings. 27th Euromicro Conference, pp. 336 –344, 2001.

[25]   W. Gaver, T. Moran, A. MacLean, L. Lovstrand, P. Dourish, P. K. Carter, and K. W. Buxton, "Realizing video environment: EuroPARC's RAVE system," in Proc. CHI '92, Conference on Human Factors in Computing Systems, pp. 27-35, 1992.

[26]   S. Gale, "Adding Audio and Video to an Office Environment," Studies in Computer Supported Cooperative Work, in. J.M. Bowers and S.D. Benford (eds.), Elsevier Science Publishers, New York, pp. 49-62, 1991.

[27]   J. Gabbe, A. Ginsberg, and B. Robinson, "Towards Intelligent Recognition of Multimedia Episodes in Real-Time Applications." Proc. ACM Multimedia 94, San Francisco CA, Oct. 1994.

[28]   S.J. Gibbs, C. Arapis, and C. J. Breiteneder, "Teleport-Toward Immersive Copresence," Multimedia Systems, 7(3), pp. 214-221, 1999.

[29]   A. Ginsberg and S. Ahuja, "Automating Envisionment of Virtual Meeting Room Histories," Proc. ACM Multimedia 95, San Francisco CA, Nov. 1995.

[30]   J. Gemmell *et al*, "Gaze Awareness for Video-conferencing: A Software Approach," IEEE Multimedia, Vol. 7(4), pp. 26-35, 2000.

[31]   Groupkit, http://www.cpsc.ucalgary.ca/grouplab/groupkit

[32]   Groove, http://www.groove.net

[33]   B.G. Haskell, *et al*, "Image and Video Coding – Emerging Standards and Beyond," IEEE Trans. On Circuits and Systems for Video Technology, Vol. 6, pp. 814-837, Nov. 1997.

[34]   Habanero, http://www.ncsa.uiuc.edu/

[35]   M. Handley and I. Wakeman, "CCCP: Conference Control Channel Protocol - A Scalable Base for Building Conference Control Applications," Proc. SIGCOMM 95, Aug. 1995.

[36]   M. Handley and J. Crowcroft, "Network Text Editor (NTE): A scaleable shared text editor for the Mbone, Proceedings of SIGCOMM 1997.

[37]   M. Handley *et al*, "SIP: Session Initiation Protocol", RFC2543, Internet Engineering Task Force (IETF), 1999.

[38]   R.C. Harkness, and P. G Burke, "Estimating teleconferencing travel substitution potential in large business organizations," in L.A. Parker and C.H. Olgren, Eds., The teleconference resource book: a guide to applications and planning, pp. 256-264, Elsevier Science Publishers: Amsterdam, 1984.

[39]   S. Hartwig *et al*, "Mobile Multimedia – Challenges and Opportunities (invited paper)", IEEE Transactions on Consumer Electronics, 46(4), pp.1167-1178, Nov. 2000.

[40]   M. Hatcher, "A video conferencing system for the United States Army: Group decision making in a geographically distributed environment," Decision Support Systems, Vol. 8, pp. 181-190, 1992.

[41]   C. Heath and P. Luff, "Disembodied conduct: communication through video in a multimedia office environment," in Proc. of CHI '91, Conference on Human Factors in Computing Systems, pp. 99-103, 1991.

[42]   J. L. Herlocker and J. A. Konstan, "Commands as Media: Design and Implementation of a Command Stream," Proceedings ACM Multimedia 95, San Francisco CA, Nov. 1995.

[43]   T. Hodes, *et al*, "Shared Remote Control of a Video Conferencing Application: Motivation, Design, and Implementation," Multimedia Computing and Networking 1999, Proc. IS&T/SPIE Symposium on Electronic Imaging: Science & Technology, pp. 17-28, San Jose, CA, Jan. 1999.

[44]   W. Holfeder, "MBone VCR --- Video Conference Recording on the Mbone," http: // www.icsi. berkeley.edu/mbone-vcr/, 1995.

[45]   M. Hosseini and N. D. Georganas, "Collaborative Environments for Training," Proceedings of the ACM Multimedia, pp. 621-622, 2001.

[46]   G. P. Huber, "A theory of the effects of advanced information technologies on organizational design, intelligence, and decision making," Academy of Managment Review, 15(1), pp. 47-71, 1990.

[47]   International Multimedia Teleconferencing Consortium, http://www.imtc.org/.

[48]   IP Multicast Initiative, http://www.ipmulticast.com/.

[49]   H. Ishii, M. Kobayashi, and K. Arita, "Iterative design of seamless collaboration media," Communications of the ACM, Vol. 37, pp. 83-97, 1995.

[50]   H. Ishii, M. Kobayashi, and J. Grudin,, "Integration of inter-personal space and shared workspace: clearboard design and experiments," in Proc. of CSCW 92, Conference on Computer Supported Cooperative Work, pp. 33-42, 2001.

[51]   V. Jacobson and S. McCanne, "Visual Audio Tool (vat)," Software Online at ftp://ftp. ee.lbl.gov/conferencing/vat

[52]   V. Jacobson, "Multimedia Conferencing on the Internet," in Proc. of ACM SIGCOMM '94 Conference Tutorial, London, England, 1994.

[53]   R. Johansen, "Groupware: computer support for business teams", The Free Press: New York, 1988.

[54]   D.H. Kim *et al*, "Collaborative Multimedia Middleware and Advanced Internet Call Center," Proceedings of the 15th International Conference on Information Networking, pp. 246 –250, 2001.

[56]   J. F. Koegel-Buford, "Middleware System Services Architecture," chapter in Multimedia Systems J. F. Koegel-Buford (Editor), Addison-Wesley, 1994.

[57]   T. Liao, "Webcanal: a Multicast Web Application," in 6th International WWW Conference, Santa Clara, CA, April 1997.

[58]   M. R. Macedonia, and D. P. Brutzman, "MBone provides audio and video across the Internet," IEEE Computer, 27, pp. 30-36, 1994.

[59]   T. W. Malone and J. F. Rockart, "How will information technology reshape organizations? Computers as Coordination Technology'," in S. P. Bradley, J. Hausman and R. L. Nolan, (eds.), Globalization, Technology, and Competition: The Fusion of Computers and Tele-communications in the 1990s, Boston, MA: Harvard Business School Press, 1993.

[60]   R. Malpani and L. A. Rowe, "Floor Control for Large-Scale Mbone Seminars," Proc. of ACM Multimedia, Seattle WA, pp. 155-163, Nov. 1997.

[61]   M. M. Mantei, R. M. Baecker, A. J. Sellen, W. A. S. Buxton, and T. Milligan, "Experiences in the use of media space," in Proc. of CHI '91, Conference on Human Factors in Computing Systems, pp. 203-208, 1991.

[62]   M. L. Markus, and D. Robey, "Information technology and organizational change: causal structure in theory and research", Management Science, 34(5), pp. 583-598, 1988.

[63]   K. Mayer-Patel and L. A. Rowe, "Design and Performance of the Berkeley Continuous Media Toolkit," in Multimedia Computing and Networking, Proc. IS&T/SPIE Symposium on Electronic Imaging: Science & Technology, pp. 194-206 San Jose CA, Jan. 1997.

[64]   S. McCanne, et al, "Toward a Common Infrastructure for Multimedia-Networking Middleware," Proc. NOSSDAV 97, St. Louis MO, May 1997.

[65]   O. Megzari, L. Yuan, and A. Karmouch "Meta-Data and Media Management in a Multimedia Interactive Telelearning Systems," Multimedia Tools and Applications, Vol. 16, pp.137-160, 2002.

[66]   S. McCanne, "A distributed whiteboard for network conferencing," U.C. Berkeley: CS268 Computer Networks term project and paper, May 1992.

[67]   S. McCanne and V. Jacobson, "vic: a Flexible Framework for Packet Video", In Proceedings of the ACM Multimedia 1995.

[68]   R. F. Mines, J. A. Friesen, and C. L. Yang, "DAVE: A plug and play model for distributed multimedia application development," Proc. ACM Multimedia 94, pp. 59-66, San Francisco CA, Oct. 1994

[69]   S. Minneman, et al, "A Confederation of Tools for Capturing and Accessing Collaborative Activity," Proc. ACM Multimedia 95, pp. 523-534, San Francisco CA, Nov. 1995.

[70]   D. Miras, V. Hardman, and A. Steed, "A New Approach to Collaborative Real-Time Internet Multimedia Applications," IEE Image Processing and its Applications Conference, pp.326-330, 1999.

[71]   Multimedia Communications Forum, http://www.mmcf.org/.

[72]   K. Obraczka, "Multicast Transport Mechanisms: A Survey and Taxonomy", IEEE Communications Magazine, Vol. 36 No. 1, Jan. 1998.

[73]   J. C. de Oliveira, et al, "Java Multimedia Telecollaboration," Proceedings of the ACM Multimedia pp.607-608, 2001.

[74]   P. Parnes, K. Synnes, and D. Schefström, "mSTAR: Enabling Collaborative Applications on the Internet," IEEE Internet Computing, pp. 32-39, Sep.-Oct. 2000.

[75]   C. Potts, K. Takahashi, and A. I. Anton, "Inquiry-Based Requirement Analysis," IEEE Software, 11(2), pp. 21-32, Mar. 1994.

[76]   T. Rodden, "Technological support for cooperation," in D. Diaper and C. Sanger, Eds., CSCW in practice: an introduction and case studies, Springer: New York, pp. 1-22 ,1993.

[77]   R. W. Root, "Design of a multi-media vehicle for social browsing," In Proc. of CSCW, pp. 25-38, 1988.

[78]   M. Roseman and S. Greenberg, "Building Real-Time Groupware with GroupKit," ACM Transactions on Computer-Human Interaction, 1(3), pp. 66-106, 1996.

[79]   E. Schooler, "Conferencing and Collaborative Computing," ACM Multimedia Systems Journal, Vol. 4, pp. 210-225, 1996.

[80]   H. Schulzrinne, A. Rao, R. Lanphier, "Real-Time Streaming Protocol (RTSP), RFC2326, Internet Engineering Task Force (IETF), 1998.

[81]   A, Sellen, "Remote Conversation: The Effects of Mediating Talk with Technology," Journal of Human-Computer Interaction, 10(4), pp. 410-444, 1995.

[82]   S. Shirmmohammadi et al, "Web-Based Multimedia Tools for Sharing Educational Resources", Journal on Educational Resources in Computing (ACM JERIC), 1(1), Article No. 9, March 2001.

[83]   D. Sisalem and H. Schulzrinne, "A Floor Control Application for Light-Weight Multicast Conferences," Proceedings International Conference on Telecommunications, Thessaloniki, Greece, pp. 130-134, Jun. 1998.

[84]   M. Sohlenkamp, et al, "Supporting the Distributed German Government with POLITeam", Multimedia Tools and Applications, Vol. 12, pp. 39-58, 2000.

[85]   K. Takahashi, E. Yana, "A Hypermedia Environment for Global Collaboration, " IEEE Multmiedia, 7(4), pp. 36-47, 2000.

[86]   T. Turletti "INRIA Video Conferencing System (ivs). Institut National de Rechereche en Informatique et an Automatique. http://www.inria.fr/rodeo/ivs.html

[87]   University of Berkeley OpenMash project, http://www.openmash.org

[88]   K. Watabe, S. Sakata, K. Maeno, K. Fukuoka, and T. Ohmori, "Distributed multiparty desktop conferencing system: MERMAID," in Proc. of CSCW, pp. 27-38, 1990.

[89]   A. Watson and M. A. Sasse, "The Good, the Bad, and the Muffled: The Impact of Different Degradiations on Internet Speech," Proceedings of the ACM Multimedia, pp.269-276, 2000.

[90]   M. Weber, *et al*, "Combining multimedia collaboration and workflow management," Proceedings of the 23rd Euromicro Conference, pp. 114–119, 1997.

[91]   L. K. Wright, S. McCanne and J. Lepreau, "A Reliable Webcast Protocol for Multimedia Collaboration and Caching," Proceedings of the ACM Multimedia, pp.21-30, 2000.

[92]   D. Wu, A. Swan, and L. A. Rowe, "An Internet Mbone Broadcast Management System," Multimedia Computing and Networking 1999, Proc. IS&T/SPIE Symposium on Electronic Imaging: Science & Technology, San Jose, CA, Jan. 1999.

16 PACS, IMAGE MANAGEMENT, AND IMAGING INFORMATICS

Prof. H. K. Huang

The research and development of Medical Imaging started in early 1970s, it has gradually matured as a science discipline during the past 30 years. This field requires the knowledge of physical sciences including engineering, physics, mathematics, computer science, as well as biomedical sciences including medicine and biology. Some significant contributions of medical imaging in healthcare are the development of various medical imaging modalities, 3-D rendering, image processing, image-aided diagnosis, and image-fusion. These contributions are important in the sense of small-scale applications. The development of picture archiving and communication system (PACS) in the 1990s provides very large image databases allowing the exploration of large-scale medical imaging applications. The method of the exploration is generally called medical imaging informatics.

In this Chapter, we first summarize the development of PACS and terminology used in this field. The medical imaging informatics infrastructure consisting of several major components (including large image databases; image processing, visualization, graphic user interface, data security, and computer networking; database and knowledge base management; and application servers and application software) are then described. To conclude, we give an application in Neuro-surgical Command Module as an example of medical imaging informatics.

16.1 Picture Archiving and Communication System (PACS) Fundamentals

PACS COMPONENTS
The concept of Picture Archiving and Communication System (PACS) that originated at the SPIE Medical Imaging Conference in Newport Beach, CA, February 1982, has just passed its 20th birthday [3]. PACS is a system integration of medical images originally designed for facilitating radiologists in interpreting images more efficiently. It has evolved over the years to become a cornerstone of the modern healthcare delivery system. PACS consists of several major components as shown in Figure 16-1. This chapter concentrates on the image data management aspect of PACS, but for completeness, the other components are described briefly in the following paragraphs [5].

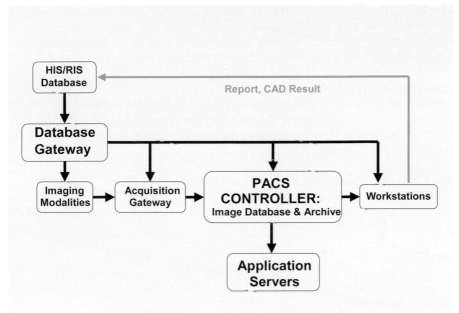

Figure 16-1 PACS Components & Data Flow

HIS/RIS Database
Modern hospitals or clinics use Hospital Information System (HIS) or Clinical Management System (CMS) to manage their operations. HIS/CMS are large information system consisting of many subsystems developed during the life time of the HIS/CMS, some of which are legacy subsystems. HIS/CMS has many databases, among them are various clinical information databases, and RIS (Radiology Information System). RIS handles radiology operations including patient registration, examination scheduling, acknowledgement of examinations completed, diagnosis reports, etc. HIS/CMS collects patient related information including demographic, test results, diagnostic results from imaging examinations, and hospital/clinic management data; and distributes them strategically to facilitate the hospital/clinic operation. Neither the HIS nor CMS consists of an imaging subsystem.

Database Gateway
PACS requires patient information from HIS/CMS and RIS to direct the radiology operation effectively and efficiently. In order to assure that patient information is preserved and to avoid human errors, a direct connection from these databases to the imaging modalities is necessary. A database gateway (GW) is used for this connectivity. The database GW transmits a work list of the patient from HIS/CMS or RIS directly to the imaging modalities and PACS.

Imaging Modalities
Imaging modalities used in PACS include magnetic resonance (MR), computed tomography (CT), ultrasonic (US), computed and digital radiography (CR and DR),

digital subtraction angiography (DSA), nuclear medicine, and endoscopy images. After an image or a series of images has been generated from a modality, it is converted to DICOM (digital imaging and communication in medicine) format and transmitted to other components using the DICOM communication protocol [2].

Acquisition Gateway (GW)
For those modalities which do not convert images to DICOM standard, the images are routed to an acquisition gateway where they are converted to DICOM standard. The GW is also used to stage images from different modalities as a buffer before they are sent to the PACS controller.

PACS Controller: Image Database and Archive
Images from the acquisition GW, are sent to the PACS controller consisting of an iamge database and an archive server. . Images are archived according to the DICOM data model and distributed to workstations.

Workstations (WS)
There are three types of WS, diagnostic (very high resolution, with 2.5K x 2K display), clinical (medium resolution with 2K x 1.6K display), and desktop (from 1K to standard 800 line display)

Application Servers
Selected images from the PACS controller can be sent to various application servers. These servers can be a standard server or Web-based server.

MEDICAL IMAGES
Different imaging modalities generate images of various sizes. Figure 16-2 shows the sizes of some common medical images: a US hospital of 500 beds, performing 200,000 radiological examinations per year, accumulates about 10 gigabytes/day or 3-5 terabytes/year. These images after acquired from modalities have to be organized in a standardized DCIOM data model, archived, and distributed to: 10-20 diagnostic workstations (WS) each with multiple 2K high resolution display monitors within radiology departments; 20 or more clinical WSs with multiple 1K display monitors at hospital wards; and hundreds of desktop WSs through out the hospitals. Figure 16-3 shows the DCIOM worklist (left) and CT images (right).

	One Image (bits)	# of Images/Exam	One Examination
Nuclear medicine (NM)	128x128x12	30-60	1-2MB
Magnetic resonance imaging (MRI)	256x256x12	60-2000	8-500MB
Ultrasound (US)*	512x512x8(24)	20-230	5-60MB
Digital subt. Angiography (DSA)	512x512x8	15-40	4-10MB
Digital microscopy	512x512x8	1	0.26MB
Digital color microscopy	512x512x24	1	0.79MB
Computed tomography (CT)	512x512x12	40-1000	20-500MB
Computed radiography (CR)	2048x2048x12	2	16MB
Digitized x-rays	2048x2048x12	2	16MB
Digital mammography	4000x5000x12	4	160MB

* Doppler US with 24 bit color images

Figure 16-2 Size of Some
Common Medical Images

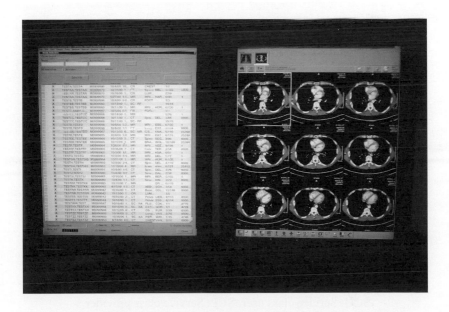

Figure 16-3 Worklist and Images Display.

TERMINOLOGY AND STANDARDS

Among many terminologies used in medical imaging, the ICD (International Classification of Diseases) is most important. ICD has the following standard format:

ICD-10 xxx.x

means version 10, the disease identifier is xxx.x, e.g., 484.3 means pneumonia in whooping cough.

There are two major standards used in PACS and in medical data, DICOM 3.0 and HL-7 (health level 7) [4]. DICOM has two specifications one for image and associated data communication, and the other for image and data format. After an image is generated by an image modality, it is formatted to DICOM, and transmitted to other components of PACS with the DICOM communication protocol which is a higher level TCP/IP (transfer control protocol/Internet protocol). Figure 16.4 shows the image data communication between two DICOM compliant components in the PACS based on the concept of DIMSE (DICOM Message Services Elements), SCU (Service Class User) and SCP (Service Class Provider). When image data is transmitted within the same device, it is called DICOM Service; whereas data transmitted between two devices is called DICOM Protocol.

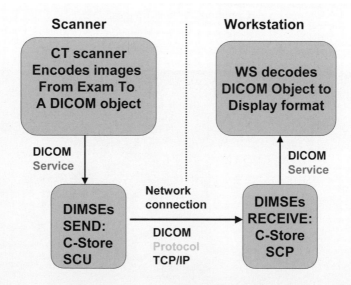

Figure 16-4 Transmitting a set of CT images from the Scanner to WS
DIMSE: DICOM message services elements
SCU: Service class user, SCP: Service class provider

IHE (INTERGRATING THE HEALTHCARE ENTERPRISE) AND PACS DATA FLOW

With image and medical information standards established, image and patient information can be transmitted between components in PACS. To transmit this information through the hospital effectively and efficiently, careful data flow analysis of the hospital is needed. In order to do that, understanding the concept of Integrating of the Healthcare Enterprise (IHE) is essential. IHE are sets of protocols, which require all scheduled workflow adhere to certain terminology and standards so that data generated by devices from different manufacturers can be integrated [12]. These protocols define, among others, the following scheduled workflow:

- Admission
- Order
- Schedule
- Data acquisition
- Notification of completed steps
- Diagnosis
- Distribution

The Radiological Society of North America (RSNA) together with HIMSS (Healthcare Information and Management Systems Society) Working Groups have collaborated to define these scheduled workflows, and invited manufacturers to participate in live-demonstrations of workflow connectivity during their Annual Meetings, since 1999. Successful results in these demonstrations stimulate open architecture in the manufacturers' products for easier system integration, create competitive product lines and lower the costs, and streamline scheduled workflow in diagnostic imaging. These translate to directly benefiting the patients and improving healthcare delivery.

HIS/CMS (HOSPITAL INFORMATION SYSTEM/CLINICAL MANAGEMENT SYSTEM) AND ePR (ELECTRONIC PATIENT RECORD)

Both the HIS and CMS described earlier are hospital- or clinic-based, they are not designed for an individual patient. A recent trend is to replace HIS/CMS by the Electronic Patient Record (ePR) or Electronic Medical Record (eMR) system, a more patient-centered database system [8].

The Electronic Patient Record is an emerging concept which replaces or supplements the hospital- or clinic-based healthcare information system. The concept of ePR is that when a patient requires clinical services, the patient data accompanies the patient everywhere in the hospital. The major functions of ePR are:

- Accepting direct digital input of patient data.
- Analyzing across patients and providers
- Providing clinical decision support and suggest courses of treatment
- Performing outcome analysis, and patient and physician profiling
- Distributing information across different platforms and health information systems

Figure 16-5 depicts the relationship between the connectivity of databases in HIS/CMS and PACS in the VistA Imaging system developed by the US VA Medical Center [1]. This connectivity allows its ePR data model to have the output shown in Figure 16-6. This figure illustrates a page of the GUI at the ePR workstation showing relevant patient information including data and images. (Courtesy of Dr. H. Rutherford).

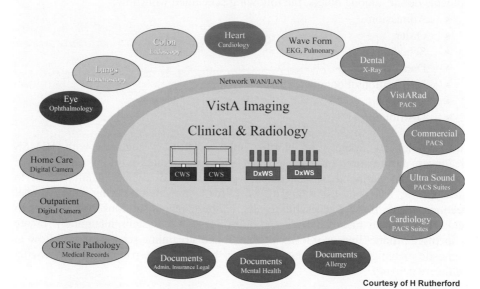

Courtesy of H Rutherford

Figure 16-5 VistA Imaging: Input from Image and Data Sources

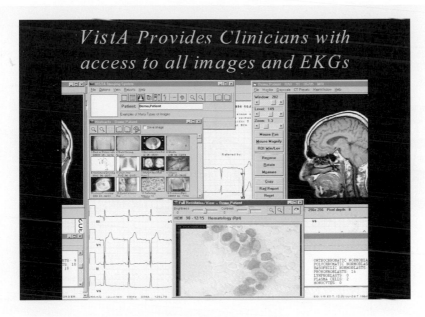

Figure 16-6 VistA Provides access to all images and EKGs
(Courtesy of H Rutherford) [1]

16.2 Image Data Communication and Management

IMAGE COMMUNICATION

PACS requires high speed networks to transmit large image files between components (see Figure 16-2). In the case of the intranet, gigabit/sec switches and mbits/sec drops to WSs are almost standard in most hospital network infrastructure. The transmission rate, even for a large image size, is acceptable for clinical operation. However, in the case of the Internet, where images have to be transmitted between hospitals through wide area networks (WAN), high speed WAN is still very expensive for image transmission.

Internet 2, founded in the summer of 1996, is a high speed network infrastructure for high speed data transmission. It is a consortium of more than 180 research universities, known as UCAID (The University Corporation for Advanced Internet Development). Currently Internet 2 has three backbones: vBNS (very high speed backbone network service, Figure 16-7), Abilene, and CalREN (California Research and Education Network, Figure 16-8). Internet 2 sites connect to the backbones through regional GigaPoP (point-of-presence). Figure 16-9 shows how to connect to the Internet 2 backbone from the Image Processing and Informatics Laboratory at the Childrens Hospitals Los Angeles. Currently, using the Internet 2 can achieve about 5 Mbits/sec from San Francisco or Los Angeles to Washington DC. The routes are shown in Figure 16.7, RED (UCSF – NIH, 4), and Green (USC – NIH, 4), and the performance is shown in Figure 16-10. [9]

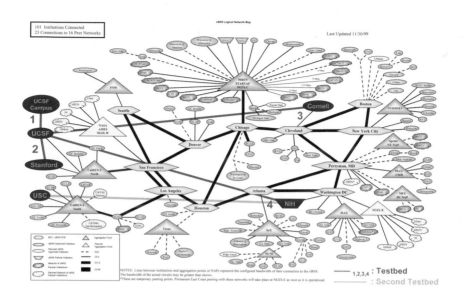

Figure 16-7 vBNS Backbone networks and its connectivity (Courtesy of vBNS)

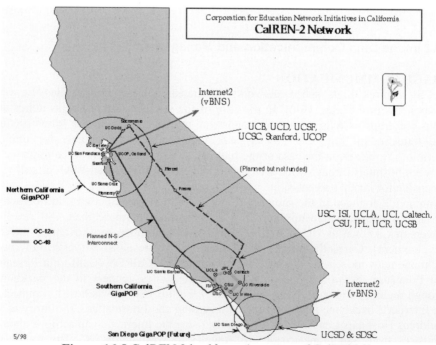

Figure 16-8 CalREN 2 backbone (courtesy of CalREN 2)

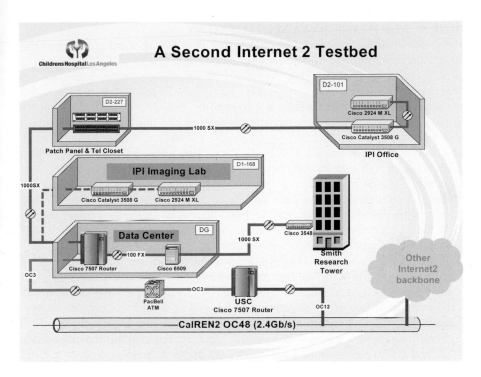

Figure 16-9 Connection from the Childrens Hospital Los Angeles
to the CalREN 2 , an Internet 2 backbone

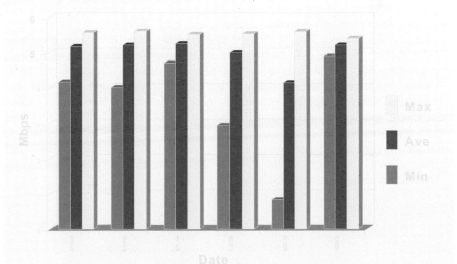

Figure 16.10 Internet 2 Throughput Summary: SF - NLM, NIH

LARGE SCALE IMAGE DATA MANAGEMENT
Three topics of importance to large scale image data management are hierarchical archive, data migration, and off-site archive.

Hierarchical Archive
Large scale image data management is governed by two criteria: no loss of images once received by the PACS from imaging modalities, and minimizing image access time at workstations.

To ensure no loss of images, PACS always retains more than two copies of an individual image on separate temporary disks until the image has been archived successfully to a permanent storage device (e.g., digital tape library). A second permanent copy should also be kept off-site as a back-up. The off-site back-up archive will be discussed in the next Section. Figure 16-11 shows various storage units in the PACS components. This multiple copy scheme is achieved via the PACS inter-component communication as follows:

- *At the imaging modality.* Images are not deleted from the imaging modality's local storage unless the radiographer/technologist has verified the successful archiving of individual images through the PACS software acknowledgment. In the event of failure of the acquisition process or of the archive process, images can be resent from these imaging devices to the PACS permanent archive.

- *At the acquisition gateway computer.* Images acquired at the acquisition gateway (GW) remain in its local disks until the permanent storage system has acknowledged a successful archive has been completed. These images are then deleted from the GW disks.

- *At the PACS central archive system.* The PACS central archive system has two components, a RAID for short storage, and a DLT (digital linear tape) library for permanent archive. Images arriving in the central archive system are first stored in the RAID which supports queries from the WS for the most recent examinations. Images from RAID are also copied to the DLT. After they have been copied, they are deleted from the RAID using an aging first-in first-out algorithm. The central archive system has two identical databases for fault-tolerance. For image data management, the archive server provides the following functions: image receiving, stacking, routing, archiving, retrieving, studies grouping, archive management, RIS interfacing, PACS database updating, and image pre-fetching.

- *At the workstation.* Images stored in the designated WS will remain there until the patient is discharged or transferred. Images in the PACS archive can be queried/retrieved from any WS.

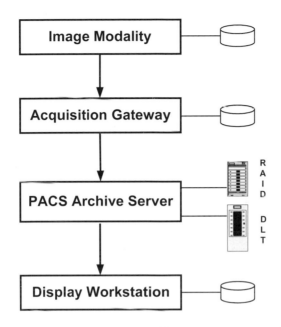

Figure 16-11 Hierarchical Storage In Pacs

Back-up Archive
To build a fault tolerance in the PACS server to assure no loss of images, a back-up permanent archive system can be used. The simplest method is to copy the image from the permanent archive to an off-site second permanent storage. An alternative method is to use an off-site ASP (Application Service Provider) model.

The off-site ASP back-up archive model can provide instantaneous, automatic backup of acquired PACS image data and instantaneous recovery of stored PACS image data. The ASP model is a third party service with a data center which consists of a fault-tolerant back-up archive server with RAID and DLT library. Image data sent to the PACS permanent archive in the hospital is also sent to this ASP server in parallel. In case the on-site permanent archive is down, images can be sent from the off-site back-up to the hospital PACS server. Figure 16.12 shows such an off-site back-up permanent archive.

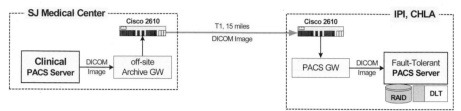

Figure 16-12 The DICOM compliant off-site ASP
(Application Service Provider) archive server at the IPI Laboratory,
Childrens Hospital Los Angeles for St. John's Medical Center PACS.

Large Scale Archive Upgrade and Data Migration

Archive upgrade and data migration are two problems always confronting the PACS designer. In the case of archive upgrades, the PACS archive accumulates images at a very fast rate, especially when new imaging modalities are introduced. For example, multi-sliced detector CT scanner can comprise of over 600 images (300MB of data) per examination. In a 300 bed community hospital, for example, with these new image modalities, the data accumulation rate is about 10 GB/day or 3.5 TB/year. For this reason, regardless how careful the planning was on the archive requirement in the beginning, chances are that the archive will run out of storage capacity. In the case of data migration, even though storage technology is always improving, oneday, say in three years, it will be necessary to replace the old storage technology with a new one, e.g., MOD by DLT. When this happens, data migration of a very large image data volume becomes inevitable. To move a 10 TB image data from clinical environments running 24 hours a day, 7 days a week is a very challenging task.

As an example, at Saint John's Health Center, Santa Monica, California, PACS data volumes have increased dramatically since the hospital became filmless in April of 1999. This is due in part to continuous image accumulation, and the integration of a new multi-slice detector CT scanner into PACS. The original PACS archive would not be able to handle the distribution and archiving, load and capacity in the near future. Furthermore, there is no secondary backup of all the archived PACS image data for disaster recovery purposes. A clinical and technical process template to upgrade and expand the PACS archive, migrate existing PACS image data to the new archive, and provide a back-up and disaster recovery function was designed and implemented. The server hardware configuration was upgraded from a 2.5 TB MOD Library to a 10 TB DLT library, and an off-site secondary backup implemented for disaster recovery based on an ASP model. The upgrade included new software versions and database reconfiguration. Upon completion, all PACS image data from the original MOD library was migrated to the new tape library and verified. The migration was performed in the background with the clinical operation running continuously. Once the data migration was completed the MOD library was removed. Some of the pitfalls and challenges during this archive upgrade and data migration process included hardware reconfiguration for the original archive server, clinical downtime during the upgrade, data migration planning to minimize the impact on clinical workflow, and the design of a contingency plan to minimize the impact of clinical downtime [9].

16.3 Imaging Informatics

MEDICAL IMAGING INFORMATICS INFRASTRUCTURE
CONCEPT OF MIII

Medical image informatics infrastructure (MIII) is an emerging field focused on taking advantage of existing PACS resources and their images and related data for large-scale horizontal and longitudinal clinical, research, and education applications which could not be performed before due to insufficient data [11].

MIII Architecture and Components

MIII is composed of the following components: medical images and associated data (including the PACS database), image processing and analysis, data/knowledge base management, visualization, graphic user interface, communication networking, image data security, system integration, and application server and user application software. These components are logically related, and are depicted in Figure 16-13, and their functions are summarized as follows.

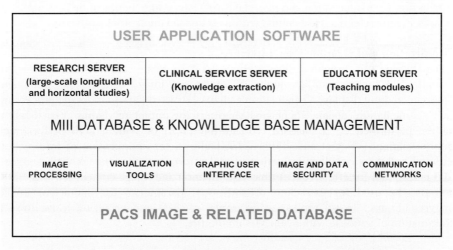

Figure 16.13 Medical Image Informatics Infrastructure (MIII) components and their logical relationship

PACS and Related Data

The PACS databases and other related health information systems containing examination accession numbers, patient demographic data, case histories, medical images and corresponding diagnostic reports, and laboratory test results constitute the data source in MIII. These data are organized and archived with DICOM and HL7 standards. In addition, controlled health vocabulary like the ICD-10 can be used for medical identifiers, codes, and messages proposed by the American Medical Informatics Association.

Image Processing and Analysis Tools
Image processing and analysis software tools allow for setting up data extraction, image content indexing, and data retrieval mechanism. Their functions include segmentation, region of interest determination, texture analysis, content analysis, morphological operations, and image registration. The output from image processing and analysis can be a new image or some features describing certain characteristics of the image. Image processing and analysis functions can be performed automatically or interactively from the PACS image database extracted by an application server. Data extracted by the image processing and analysis tools can be appended to the image data file.

Database and Knowledge Base Management
The database and knowledge base management component software has several functions. First, it integrates and organizes PACS images and related data, extracts image features and keywords from image processing, and derives medical heuristics rules and guidelines into a coherent multimedia data model. Second, it supports on-line database management, content-based indexing and retrieval, formatting and distribution for visualization and manipulation. This component can be developed on top of a commercial database engine with user add-on application software.

Visualization and Graphic User Interface
Visualization and graphic user interface (GUI) relate to workstation design. Visualization includes surface and 3-D rendering, image data fusion, static and dynamic imaging display. Visualization utilizes extracted data from image processing (i.e., segmentation, enhancement, and shading) for output rendering. Visualization can be performed on a standard workstation (WS) or with a high performance graphic engine. Efficient and effective GUI allows for optimization of information retrieval and data visualization with minimal effort from the user. A well-designed GUI is essential for effective real-time visualization and image content retrieval. GUI can also be used for extraction of additional parameters for interactive image analysis. Figure 16-14 shows an example of 3-D rendering of a mciro CT examination of a mouse with a 500 MB image file.

Communication Networks
Communication networks include network hardware and communication protocols required to connect MIII components together. MIII communication networks can have two architectures: a network of its own with a connection to the PACS's networks, or it can share the communication networks with the PACS. In the former, the connection between the MIII and the PACS networks should be transparent to the users and provide the necessary high speed throughput for the MIII to request PACS images and related data and to distribute results to user's workstation. In the latter,

the MIII should have a logical segment isolated from the PACS networks so that it would not interfere with the PACS daily clinical functions.

Image Data Security

Security includes data authenticity, privacy, and integrity. Authenticity is to authenticate the sender. Privacy considers who can access what type of data and when. Both issues are normally taking care of by the network authority. Data integrity is the responsibility of the user who generates the image data. Detailed encryption methods needed to be used to guarantee that the data has not been altered during image transmission. Digital signatures and digital envelopes derived from image data are methods that can be used for this purpose [11].

System Integration

System integration includes system interface, and shared data and workspace software. System interface software utilizes existing communication networks and protocols to connect all infrastructure components into an integrated information system. Shared data and workspace software allocates and distributes resources including data, storage space, and workstation to the on-line users.

Once the aforementioned components are implemented in the MIII, application oriented software can be designed and developed to integrate necessary components for a specific clinical, research, or education application. This provides rapid prototyping and reduces costs required for the development of every application. It is in the application server layer and the user application software layer (see Figure 16-13) that the user encounters the advantages and power of the MIII.

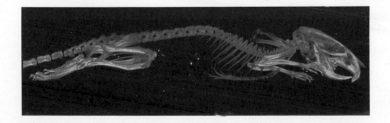

Courtesy of ORNL

Figure 16-14 3-D Rendering of a Micro CT Mouse Exam - 50 micro pixel
1000 slices (500MB) (ORNL)

NEURO SURGICAL COMMAND MODULE (NCM) - MANAGEMENT OF NEUROSURGICAL PATIENT DATA

Concept of the NCM

An application using imaging informatics and PACS for lower back spinal surgery is the Neurosurgery Command Module (NCM) [6]. The NCM is a very small version of an electronic patient record (ePR) system with a database managing patients scheduled for, or/and have gone through, lower back spinal surgery. It consists of a server with extracted images and related data of patients from PACS who require lower back spinal surgery, and WSs to be used during surgery and post-surgery conferencing. Before the surgery, the server collects patient's CT, MR and other pertinent images including 3-D rendering pre-surgery plan from PACS and sent them to workstations. During pre-surgery, the WS is wheeled to the surgical suite to connect with the C-arm X-ray and endoscopy. This will allow for on-line imaging examination and real-time image display of all types of images during the surgery. After the surgery, the workstation down loads all information obtained during the surgery to the server. The informatics software in the NCM server combines all relevant images, related patient's information, and documented surgical procedure and results systematically into the patient soft folder based on an ePR data model for future review and use.

NCM Workflow

The NCM server can be considered as one application server shown in Figure 16-1. The ePR design concept allows the image and data of a patient to be staged at the server and delivered to the WS for review before, during, or after surgery. Remote WS can be connected to the on-site WS by a high-speed network where a second opinion and remote teaching can take place. A more detailed version of Figure 16-1 for NCM application is shown in Figure 16-15.

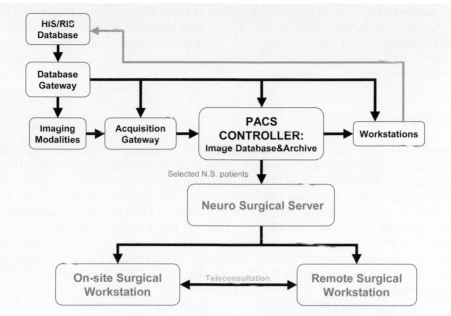

Fig.16-15 PACS Workflow & the Neuro Surgical Server

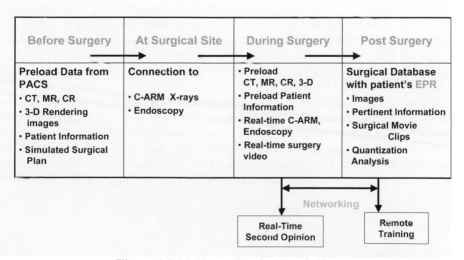

Figure 16-16 Neuro Surgical Work Flow

Figure 16-16 shows the minimally invasive spinal surgery workflow before surgery, preparation required at the surgical site, during surgery, and post surgery. The two rectangular boxes at the bottom right represent the two WSs shown in Figure 16-15. The concept of NCM is based on PACS, ePR, and workflow analysis as an aid for spinal surgery. NCM can be used as a stand-alone imaging-guided tool during surgery or as a remote teaching resource during or post surgery. NCM is being developed in research laboratories.

16.4 Summary

In order to understand PACS, it is necessary to be familiar with PACS various components, terminology and standards. Integrating the healthcare enterprise (IHE) are protocols of image data workflow allowing connectivity of components in PACS from various vendors based on existing standards. The information system used in hospitals is called HIS (Hospital Information System) or CMS (Clinical Management System) which consists of many clinical databases, like RIS (Radiology Information System). These databases are operation oriented designed for special clinical services. The new trend in healthcare information system is ePR (Electronic Patient Record) which is patient oriented, i.e., the data goes where the patient goes.

A medical image data file is usually very large- tens and hundreds of megabytes. For this reason, high-speed communication technologies are required for rapid data transmission. In the intranet environment, current gigabits/sec technology is adequate. However, it is not the case in the Internet where high-speed networks are still expensive to use. The new technology is the Internet 2. It allows for high-speed transmission with acceptable costs. The popularity of Internet 2 is still in its infancy because of the difficulty in the last mile connection from the network backbone to the research or clinical laboratories.

Hierarchical Archive in PACS is important for two reasons. Firstly, a well designed hierarchical storage system can improve the image data flow resulting in a more efficient healthcare delivery system. Secondly, back-up archive is necessary to assure no loss of image data. An archive system requires periodic upgrading and data migration, these two issues require careful planning since PACS is a clinical system running 24 hours a day, 7 days a week.

Medical image informatics infrastructure (MIII) is an emerging field focused to take advantage of existing PACS resources and its image and related data for large-scale horizontal and longitudinal clinical, research, and education applications which could not be performed before due to insufficient data. An example is given in neurosurgical command module (NCM), an ePR based server to facilitate the data management of patient required spinal surgery. The development of the NCM is in progress.

16.5 Conclusion

The works presented in this chapter have focused on PACS fundamentals and medical imaging informatics. PACS is a large system integrating many medical imaging modalities with storage devices, and display workstations linking together by high speed networks and image management software. PACS is one of the basic ingredients in a digital healthcare delivery system. PACS also provides image databases for performing large-scale research, education, and clinical service. The method of manipulating and managing large-scale medical image databases is called medical imaging informatics.

Medical imaging informatics requires an infrastructure design, which consists of five levels of major components. The first level is large image databases. The second level consists of tools including image processing, visualization, graphic user interface, data security, and computer networking. The third level is database and knowledge base management. The fourth and fifth levels are application servers and application software, respectively. These components are arranged in a hierarchical level design. Medical Imaging informatics is a key technology required in a cost effective healthcare delivery system.

Acknowledgement

This work is partially supported by a contract from TATRC, US Army Medical Research and Materiel Command: DAMD17-99-P-3732, and National Institute of Biomedical Imaging and Bioengineering, NIH: R01-EB 00298

References

[1] R. E. Dayhoff, K. Meldrum, and P. M. Kuzmak, Experience Providing a Complete Online Multimedia Patient Record. Session 38. *Healthcare Information and Management Systems Society,* 2001 Annual Conference and Exhibition, Feb.4-8, 2001.

[2] DICOM: Digital Imaging and Communication in Medicine. National Electrical Manufacturers' Association. Rosslyn, VA: NEMA, 1996.

[3] A. J. Duerincks, Picture Archiving and Communication System (PACS). *Proc. SPIE for Medical Applications.* Vol. 318. Newport Beach, CA, 1982.

[4] HL7: Health Level Seven. An Application Protocol for Electronic Data Exchange in Health Care Environments. Version 2.1. Ann Arbor, MI: Health Level Seven, Inc., 1991.

[5] H. K. Huang, Picture Archiving and Communication Systems: Principles and Applications. Wiley & Sons, NY, p.521, 1999.

[6] H. K. Huang PACS, Informatics, and the Neurosurgery Command Module. *J. Mini Invasive Spinal Technique.* Vol 1, 62-67, 2001.

[7] H. K. Huang, S. T. C. Wong, E. Pietka, Medical Image Informatics Infrastructure Design and Applications. *Medcial Informatics ,* Vol 22, No 4, 279-289, 1997.

[8] C. J. McDonald,The Barrier to Electronic Medical Record Systems and How to Overcome them. *J Amer Med Informatics Assoc* Vol 4, May/June, 213-221, 1997.

[9] B. J. Liu, L. Documet, D. Sarti, H. K. Huang, J. Donnelly, PACS Archive Upgrade and Data Migration: Clinical Experiences. Proceedings SPIE Medical Imaging, Saqn Diego, CA, 2002.

[10] F. Yu, K. Hwang, M. Gill, and H. K. Huang, Some Connectivity and Security Issues of NGI in Medical Imaging applications. J High Speed Networks. 9, 3-13, 2000.

[11] X. Zhou and H. K. Huang, Authenticity and Integrity of Digital Mammography Image. *IEEE Trans. Medical Imaging,* Vol 20, No. 8, 784-791, 2001.

[12] www.rsna.org/IHE

17 CONTENT-BASED RETRIEVAL FOR MEDICAL DATA

Dr. Tom Weidong Cai, Prof. David Dagan Feng, and Dr. Roger Fulton

The recent information explosion has led to massively increased demand for multimedia data storage in integrated database systems. Content-based retrieval is an important alternative and complement to traditional keyword-based searching for multimedia data, and can greatly enhance information management. However, current content-based image retrieval techniques have some deficiencies when applied in the medical imaging domain. Many of the proposed techniques for content-based retrieval of medical data use features or patterns specific to medical images. In this chapter, we address content-based retrieval techniques for the following types of medical data: one-dimensional ECG signals (Section 17.2); two-dimensional X-ray projection images (Section 17.3); three-dimensional CT / MRI volume images (Section 17.4); and four-dimensional PET / SPECT dynamic images (Section 17.5). Finally, a summary is given in Section 17.6.

17.1   Introduction

The increasing use of information and telecommunications technologies, including Internet technologies, has laid the foundation for a revolution in traditional healthcare, and resulted in the development of Electronic Health (e-health), to improve the efficiency and effectiveness of current healthcare systems [1-2]. E-health includes the use of digital multimedia medical data transmitted electronically – for clinical, educational and administrative applications – both locally and at a distance. Healthcare services today rely heavily on digital multimedia medical data, including one-dimensional electrocardiogram (ECG) signals, two-dimensional X-ray projection images, three-dimensional computed tomography (CT) and magnetic resonance imaging (MRI) volume images, and four-dimensional positron emission tomography (PET) and single photon emission computed tomography (SPECT) dynamic images. The volume of medical data generated in the clinical environment has been rapidly growing due to an increase in both the usage and the range of digital medical diagnostic modalities. To illustrate the amount of data generated in a typical examination consider the example of a dynamic PET study of the brain which involves the reconstruction of 30 or more cross-sectional images, each of 128×128 pixels, at 22 time points. The resulting four-dimensional data set requires upwards of 22 megabytes of storage. Dynamic whole body PET studies generate even more data due to the need to scan at several positions to cover the axial length of the body. The amount of data has not only made clinical interpretation and examination more difficult, but also hinders classification and management. Most medical multimedia databases can retrieve data based on textual or numerical fields (e.g. image file name, patient hospital identification number, examination date, or pathological description). However, purely text-based methods pose significant limitations on multimedia data retrieval. Some visual properties or features of images, such as shape and texture, are extremely difficult to describe in text. The description limits the scope of the search to that predetermined by the author of the system or the current application, and leaves no means for using the data beyond that scope. Description can be subjective, and different people may give quite different descriptions for the same image. Since descriptive text is entered manually, the indexing process is slow.

In contrast to traditional text-based methods, content-based image retrieval usually uses color [3-7], texture [8-10] and shape [11-13]. In addition to these three major visual features, some other features such as icons [14], and geometrical (Radon) features [15] may also be included. Color is one of the most widely used visual features. It is relatively unaffected by the background and independent of image size and orientation. As one of the major visual features, texture plays an important role in human perception. It involves visual patterns with homogeneity properties that cannot be provided by color and intensity alone. Shape includes boundary-based and region-based shape representations, such as the Fourier Descriptor and Moment Invariants. All of the above visual features are extracted automatically in the indexing process when images are entered into a multimedia database. Queries and retrievals can be based directly on the visual properties of the images and return results ranked by the degree of content matching. In recent years, various content-based image retrieval systems have been developed in research prototypes and commercial systems [16-24]. Obviously, content-based data retrieval has become a major technique for overcoming the drawbacks of existing retrieval methods. Content-based retrieval of medical data could potentially open up interesting new avenues for research. However, existing content-based image retrieval approaches may not be applicable to medical images due to their unique characteristics. For example, color is not normally captured in the medical imaging process and medical images are usually acquired and displayed in grayscale, or pseudo-color. Therefore, the color index is likely to be of lesser importance in medical images. Texture may be confounded by the noise in many medical images, especially in nuclear medicine. There is a need for efficient techniques for multimedia medical data browsing, searching, and retrieval. Previous work in this area based on features and patterns specific to medical images is described in the following sections.

17.2   Content-based Retrieval of One-dimensional Medical Data

The electrocardiogram (ECG) is a one-dimensional recording of cardiac electrical potentials [25-26]. The electrical signal originates in the cardiac muscle fibers. The ECG may roughly be divided into the phases of depolarization and repolarization. The depolarization phases correspond to the P-wave (atrial depolarization) and QRS-wave (ventricular depolarization). The repolarization phase signifies the T-wave (ventricular repolarization). Figure 17-1 shows a typical ECG waveform. The ECG is measured by placing ten electrodes on selected spots on the human body surface. Six electrodes are placed on the chest, and four electrodes are placed on the extremities.

For regular ECG recordings, the variations in electrical potentials in 12 different directions out of the ten electrodes are measured.

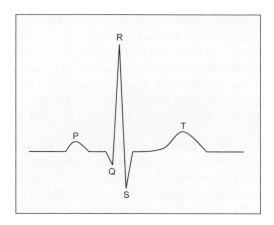

Figure 17-1 A typical ECG waveform. The points P, Q, R, S and T are known as the P-wave, Q-wave, R-wave, S-wave and T-wave respectively.

The ECG is used to identify cardiac abnormalities, such as abnormal rhythms (arrhythmias) [27], sleep apnea [28-29] and myocardial infarction [30-31]. Large amounts of ECG data are generated in the primary care, home and hospital environments which often need to be processed, evaluated and shared among clinicians. The classification of ECG data into different physiological disease categories is a complex pattern recognition task. Computer based classification of the ECG can achieve high accuracy and offers the potential of affordable mass screening for cardiac abnormalities. Successful classification is achieved by finding patterns in the ECG data that discriminate effectively between the required diagnostic categories. Conventionally, a typical heart beat is identified from the ECG data, and the QRS-, T- and possibly P- waves are characterized using measurements such as magnitude, duration and area. The ECG classification is then performed on the basis of these measurements. The content-based search for ECG data can then be worked out based on such a classification.

Several classification algorithms with alternative representations of the diagnostic content of the ECG have been investigated [27-35]. Chazal et al. [32] derived a standard cardiology feature set from standard QRS features. For each QRS detection, the associated QRS onset and offset was determined. The features were derived from the scalar lead parameters abstracted from the P-waves, QRS complexes and T waves; vector loops; and 3D loop e.g. planarity of the QRS plane. Bousseljot et al. [33] presented a technique which was not based on the measurement and extraction of individual features of the ECG data. With comparison of the signal patterns of 12-channel ECG of unknown diagnosis with signal waveforms of an ECG database, those ECGs whose waveforms were most similar to those of the unknown ECG are identified. The result of each reference ECG comparison was a 12-dimensional vector whose elements give the correlation values for the respective lead. Chazal et al. [34] later showed that it is possible to classify with features extracted from the wavelet transform (WT) of ECG data and achieve comparable diagnostic accuracy to the standard cardiology features. The advantage of this representation is that the approximate QRS detection point is the only characteristic point required. By eliminating the need to find other characteristic points a significant amount of computation is saved. Bozzola et al. [30] presented an approach for the automated ECG classification and searching based on a hybrid Neuro-Fuzzy model. In this approach the classification power of the connectionist paradigm has been coupled with the ability of the Fuzzy Set formalism to treat natural language in a quantitative way, which allows clinicians to build up a system capable of both a good classification accuracy and to give meaningful explanations of the diagnoses for cardiac abnormalities, such as myocardial infarction. Recently, Bousseljot et al. [35] proposed a method for the interpretation and comparison of 12-lead ECG data without feature extraction, using a modified algorithm of cross correlation for normalized multi-lead waveforms described in [33], and demonstrated this ECG multi-lead waveform recognition with 10,000 ECGs. Reference cases are selected from the ECG database which best match the signal patterns of the unknown ECG.

17.3 Content-based Retrieval of Two-dimensional Medical Data

Traditional X-ray projection imaging provides a map of tissue density that reflects the composition of the human body in two-dimensional domain. The radiograph is classically a negative image on photographic film made by exposure to X-rays that have passed through the body. Radiographically the breast mainly consists of two component tissues: fibroglandular tissue and fat. Fibroglandular tissue is a mixture of fibrous connective tissue (the stroma) and the functional (or glandular) epithelial cells that line the ducts of the breast (the parenchyma). The remainder of the breast is fat. In terms of X-ray attenuation, fat is more radiolucent than fibroglandular tissue; thus, regions of fat appear darker on a radiograph of the breast. Regions of brightness associated with fibroglandular tissue are referred to as mammographic density.

The detection of mass lesions on mammograms can be a difficult task for clinicians or machines. The potential variability and heterogeneity of normal breast tissue often produces localized findings that may simulate mass lesions or create distractions during the search process. Several multi-scale methods based on the wavelet have been introduced [36-40]. In [38], tumour detection is directly accomplished within the transformation domain, relying on a thresholding of the wavelet coefficients to produce a detection or no-detection result. Karsseme [41] proposed a method to incorporate a priori knowledge within the analysis phase and to exploit structural geometric knowledge which mainly relies upon Bayesian techniques. Recently Balestrieri et al. [42] developed a retrieval method based on a hierarchical entropy-based representation (HER) [43], which transforms two-dimensional visual signals such as contour and texture into a one-dimensional representation with a number of invariance properties: rotation, reflection, translation, contrast, luminosity, and optical zoom.

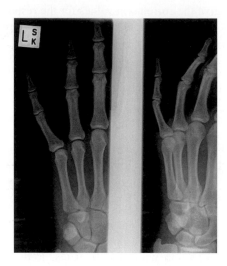

Figure 17-2 An example of hand-bone X-ray image

A number of techniques have also been developed to handle content-based retrieval for other types of X-ray images, such as dental and hand-bone X-rays. Figure 17-2 shows an example of hand-bone X-ray image. Zhang et al. [44] developed a prototype clinical dental radiograph image database to be indexed by image content. The underlying content-based search engine is based on an efficient modal shape description method, embodying two main techniques -- the finite element method (FEM) and eigen-decomposition, for characterizing the shape of a two-dimensional image region. Pietka et al. [45-46] developed content indices for hand-bone X-ray image retrieval. Since the X-ray images of the phalanges have very well-defined edges and lend themselves to automatic image processing routines, this method assesses and estimates bone age using the following steps. It first separates the third finger image from the image, then measures the lengths of the distal, middle and proximal phalanxes. After converting the measurement into an age estimate using the standard phalangeal length table, it finally compares the estimated age with the patient's age. This method has been used as a feature extraction tool in KMeD, a knowledge-based multimedia medical distributed database system [47].

17.4 Content-based Retrieval of Three-dimensional Medical Data

The most widely used three-dimensional medical imaging techniques are magnetic resonance imaging (MRI) and computed tomography (CT). MRI and CT provide more detailed, comprehensive information on skeletal structures than traditional X-rays, and also image soft tissue more effectively. Magnetic resonance is a phenomenon that relies on the "magnetic moment" present in a wide variety of organic and inorganic materials [48]. The most important source of magnetic moment is hydrogen nucleus. Thus, the gradient differences in magnetic resonance signals rely mainly on the amount of hydrogen in various tissue types. For example, as shown in Figure 17-3(a), in magnetic resonance signals, bone and air, which contain minimum amounts of hydrogen, produce dark pixels, while fat and water, which contain large amounts of hydrogen, produce bright pixels. MRI can be used to obtain exquisitely detailed images of anatomical and pathological structures. A typical MRI is 256 × 256 pixels per slice, and the number of slices can vary between 16 and 124 on a standard clinical protocol. Like MRI, CT is capable of producing three-dimensional images. CT images are based on density within the imaged object. The denser the tissue is the brighter the pixel. For example, bone produces very bright pixels in CT, whereas air produces very dark pixels, as shown in Figure 17-3(b). CT images typically consist of 512 × 512 pixels per slice, and the total number of slices can vary between 44 and 64.

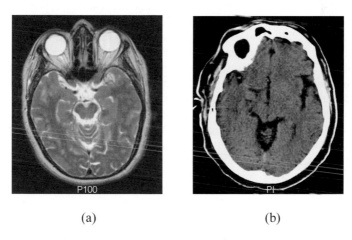

(a) (b)

Figure 17-3 (a) a brain MRI; (b) a CT brain image

Content-based retrieval for MRI and CT images has been investigated previously [43, 49-56]. The capacity to retrieve images containing objects with shapes similar to a query shape is desirable in medical image databases. Robinson et al. [49] proposed a similarity measure and an indexing mechanism for non-rigid comparison of shape which added this capability to an MRI database. The similarity measure is based on the following observations. Firstly, the geometry of the same organ in different subjects is not related by a strictly rigid transformation. Secondly, the orientation of the organ plays a key role in comparing shape. A KD-tree was constructed for indexing shapes and an algorithm for using the similarity measure along with the KD-tree for efficient retrieval was proposed. The indexing strategy changes with similarity measure and as such does not require recomputation of the indexing tree? (This sentence doesn't make sense) Hou et al. [50] proposed a content-based medical image indexing technique based on spatial features derived from the relative spatial relationships among internal image entities. The similarity measurement is based on causality (probability) that indicates the degree of similarity between a user's query and images. The technique was demonstrated on MRI chest images. Orphanoudakis et al. [51-52] developed a network of servers which provided content-based similarity search for medical images. The representation of image content consists of geometric properties and texture descriptors of selected ROIs. Such features, differentiated with respect to their relative clinical significance, play an important role in comparisons of medical images routinely carried out by clinicians. For each physiological ROI extracted from images, the following set of features was computed: location, shape, size and a set of texture descriptors. Shape and size properties include roundness, compactness, area, and orientation. Chu et al. [47, 53] developed a knowledge-based approach to retrieve medical images by feature and content with spatial and temporal constructs. Selected objects of interest in a medical image are segmented, and contours are generated from these objects. A knowledge-based semantic image model is proposed that consists of four layers (raw data layer, feature and content layer, schema layer and knowledge layer) to represent the various aspects of an image object's characteristics. Such a four-layered integrated spatial and temporal data model characterizes low-level image features (such as raw image data, and contours), abstract semantic image representations (including image objects and streams), and generic domain knowledge. This approach also supports automatic feature analysis and classification for knowledge-based query answering. Recently, Liu et al. [54] proposed a content-based approach for volumetric pathological CT neuroimage retrieval. A set of novel image features were computed to quantify the statistical distributions of approximate bilateral asymmetry of normal and pathological human brains. They applied a memory-based learning method to find the most-discriminative feature subset through image classification according to predefined semantic categories. This selected feature subset is used as indexing features to retrieve medically similar images under a semantic-based image retrieval framework.

17.5 Content-based Retrieval of Four-dimensional Medical Data

Biomedical functional images obtained from positron emission tomography (PET), single photon emission computed tomography (SPECT) and other nuclear medicine imaging modalities play an important role in modern biomedical research and clinical diagnosis. Unlike X-ray, CT and MRI imaging, which primarily provide structural information, PET and SPECT are dynamic functional imaging techniques that allow the in vivo study of physiological processes. Physiological function can be estimated by observing the behaviour of a small quantity of an administered substance 'tagged' with radioactive atoms. Images are formed by the external detection of gamma rays emitted from the patient when the radioactive atoms decay. Because they allow observation of the effects of physiological processes, functional imaging techniques can provide unique diagnostic information [56-57]. Figure 17-4 illustrates the classical approach to processing and analysis of functional images, and subsequent generation of brain parametric images using PET with the glucose tracer 18F-fluoro-deoxyglucose (FDG).

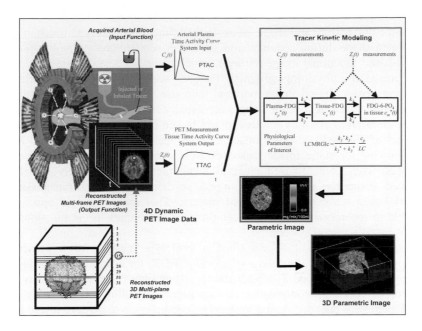

Figure 17-4 Quantitative estimation of regional glucose metabolic rate with PET

After intra-venous injection of the radiotracer (FDG) the time course of the regional radio-tracer concentration in the brain is obtained by acquiring a series of images. At the same time the input function is obtained from a series of blood samples. The physiological parameter of interest, in this case the local cerebral metabolic rate of glucose, is estimated by fitting a compartmental model to the data. Four-dimensional dynamic data (three dimensions in space and one in time) are required to construct the three-dimensional parametric image, which depicts regional glucose metabolism quantitatively in quantitative units of mg/100g/min. To estimate physiological parameters and form parametric images, the PET scanner acquires a series of scans at a pre-determined rate (not necessarily constant), typically for 20–60 minutes. Each scan, when reconstructed, represents the 3-D radiotracer distribution at a particular point in time during the study. From these data a tissue time-activity curve (TTAC) can be plotted for each voxel, and the physiological parameter value for that voxel calculated by the application of a tracer kinetic model to the TTAC. The parametric image is obtained by displaying the parameter estimates for each voxel. In dynamic functional imaging many different physiological parametric images can be obtained. Glucose metabolism, oxygen utilisation, and blood flow in the brain and heart can be measured with compounds labelled with carbon (^{11}C), fluorine (^{18}F), nitrogen (^{13}N), and oxygen (^{15}O), which are the major elemental constituents of the body.

Existing content-based medical image retrieval approaches may not be suitable when applied to functional images since quantitative physiological information in the functional image is unlikely to be recognized by common image retrieval techniques based on color, texture and shape index. Indeed function is likely to result in changes in apparent shape during acquisition as the tracer redistributes. A new approach for content-based functional image retrieval, which could support indexing and retrieval based on specific physiological kinetic features, seems to be required. Furthermore, in current content-based image retrieval database systems, the textual annotations that describe the content of the images are usually stored in a standard database. The images themselves are only referenced and linked by text strings or pointers, rather than part of the database. Such image archival techniques are only suited to small, slow growing, or static image databases. When applied to large functional image databases, such content-based retrieval techniques would severely tax the most advanced computer systems, and impede efficient archival, transmission, and network-based retrieval of the stored images. Therefore, it would appear highly desirable to incorporate image data compression techniques in the functional image content-based retrieval system.

Recently, Cai et al. [58] described a prototype design for a content-based functional image retrieval database system (FICBDS) for dynamic PET images based on a previously described 3-step functional image data compression technique that can achieve very high compression ratios without degrading image quality [59-60]. Pixel kinetics are encoded during image data compression to achieve image indexing and compression simultaneously. The proposed functional image retrieval system not only supports efficient content-based retrieval based on physiological kinetic features of the stored functional images, but also greatly reduces image storage requirements.

TRACER KINETIC FEATURE EXTRACTION AND DATA COMPRESSION
Tracer kinetic modeling techniques are widely applied in PET to extract physiological information about dynamic processes in the human body. Generally, this information is defined in terms of a mathematical model $\mu(t|p)$ (where $t = 1, 2,\ldots,$ T are discrete sampling times of the measurements and p is a set of the model parameters), whose parameters describe the delivery, transport and biochemical transformation of the tracer. The input function for the model is the plasma time activity curve (PTAC) obtained from serial blood samples. Reconstructed PET images provide the output function in the form of a tissue time-activity curve (TTAC), denoted by $Z_i(t)$, and $i = 1, 2,\ldots,$ I corresponds to the i-th pixel in the imaging region. Application of the model on a pixel-by-pixel basis to measured PTAC and TTAC data using certain rapid parameter estimation algorithms [59-62], yields physiological parametric images.

TTAC curves extracted pixel-by-pixel can be used to build a content-based functional image retrieval system. However, there are some potential problems. Firstly, since pixel TTACs are obtained from a sequence of twenty or more image frames acquired with a conventional sampling schedule (CSS), the high dimensional TTAC feature vectors (twenty or more dimensions) need to be indexed for the retrieval system, making similarity measurement much more complex and inefficient. Secondly, due to the high level of statistical noise in dynamic PET image data, TTACs are inherently noisy. Thirdly, when large numbers of extracted TTACs are stored in the database, content-based image retrieval and network-based retrieval may be inefficient. Fortunately, these problems can be overcome by applying a three

-step dynamic image data compression technique [59-60], as illustrated in Figure 17-5.

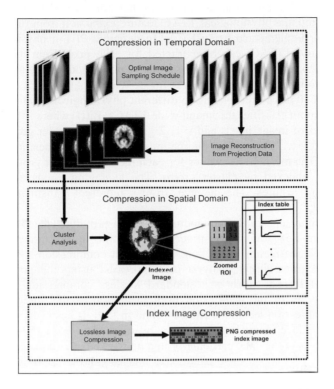

Figure 17-5 The three-step dynamic data compression method which supports fast content-based retrieval

In conventional PET image-wide parameter estimation, TTACs are obtained from a sequence of twenty or more image frames acquired with sampling schedules that have been largely empirically developed and may not be optimal for the extraction of accurate physiological parameters. Recently, Feng et al. [63] have developed an optimal image sampling schedule (OISS) which greatly reduces image storage requirements, as well as improving the signal-to-noise ratio for individual frames. For a five-parameter FDG model, only five image frames are needed [64]. The first step of the data compression technique is to apply OISS to the acquired images which removes temporal redundancies and minimizes the number of temporal frames while preserving data quality and fidelity. OISS can also be used to reduce the number of TTAC feature dimensions. Since only five image frames are needed, the number of extracted TTAC feature dimensions can be reduced to five, considerably reducing the computation required for similarity measurement. The OISS design has been integrated into a prototype image retrieval database system as a pre-processing routine for dimension reduction of physiological TTAC feature vectors.

In the second step, compression in the spatial domain exploits spatial redundancies in the image data. Using a knowledge-based cluster analysis (KCA) technique [59], the reduced set of temporal frames can be further compressed to a single indexed image. In this step, the physiological TTAC feature extraction from functional image contents can be implemented using KCA. In dynamic PET a physiological tissue time-activity curve (TTAC) can be extracted for each pixel from the sequence of image frames. Pixels in physiologically similar regions should have similar kinetics. The KCA method automatically classifies image-wide TTACs into a certain number of typical TTAC types corresponding to different physiological kinetic patterns. The clustering algorithm classifies the image-wide TTACs, $Z_i(t)$ (where $i=1, 2, ..., R$, R is the total number of image pixels and $t=1, 2, ..., 5$), into S cluster groups C_j (where $S \ll R$, and $j=1, 2, ..., S$) based on the magnitude of natural association (a similarity measure) [59]. The mean TTAC values for each identified cluster group are stored in an index table, indexed by cluster group. The averaging yields a set of highly smoothed TTAC features. The average TTACs represent j distinct physiological behaviours.

The KCA algorithm has been integrated into a prototype database and image retrieval system [58] as a pre-processing routine to facilitate storage and physiological kinetic feature extraction. In the final compression step, the indexed image obtained from cluster analysis is compressed using the portable network graphics (PNG) format which is a standard lossless compression file format. A detailed description of the theory and implementation of this three-step dynamic image data compression technique can be found in [59]. It has been shown that this technique can reduce storage requirements by more than 95% [59-60], providing much faster access to the data, particularly in a networked environment.

Raw dynamic PET images and parametric images can be rapidly recovered from the compressed data [59]. Firstly the indexed image can be decompressed. The perfectly reconstructed index image can then be used to recover the dynamic PET image sequence and generate the parametric image. Using the cluster TTAC's defined in the index table, parameter estimates for the tracer kinetic model are obtained by fitting the model parameters to the cluster TTACs. Finally, the parameter estimates and calculated physiological parameters for each cluster TTAC to their respective pixel locations are mapped by reference to the indexed image to obtain parametric images.

FOUR-DIMENSIONAL FUNCTIONAL IMAGE RETRIEVAL SYSTEM

A prototype content-based functional image retrieval database system (FICBDS) has been developed on a SUN Ultra-2 workstation running Solaris 2.5, using IDL and the C programming language. Figure 17-6 shows the overall architecture of the prototype retrieval system. Key components are: (1) the functional image processing engine; (2) the image database engine; (3) the dynamic image archival system; and (4) the graphical user interface.

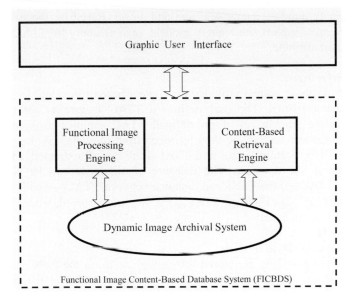

Figure 17-6 The prototype database system for four-dimensional functional image retrieval

Functional Image Processing Engine

The role of the functional image processing engine, which is based on the functional image processing system (FIPS) [65] with a set of special image processing tools, is to manipulate and process dynamic image data. It includes several special image processing routines: (1) A routine to resample CSS data to conform to the optimal PET image sampling schedule (OISS) [66]; (2) a knowledge-based cluster analysis (KCA) tool, which identifies specific physiological kinetic features within the functional image content; (3) a number of fast algorithms for the generation of functional images, such as the Patlak graphic approach (PGA) [67], the linear least squares method (LLS), and the generalized linear least squares algorithm (GLLS) [68], which can rapidly calculate functional parameters from the database; (4) tools for the statistical analysis of dynamic image, such as a tissue time activity curve (TTAC) and plasma time activity curve (PTAC) viewer, dynamic image player, profile, histogram, and surface plotting routines, and (5) image display functions, including a 3D parametric image rendering routine.

Image Database Engine

The FICBDS database engine manages functional image data with a two-level data model and supports information retrieval by image content.

Image data modeling

The image data model includes two levels: (i) physiological features or attributes, such as the tissue time activity curve (TTAC); and (ii) textual patient attributes, e.g., patient ID, study number, name, sex, age and so on. Based on the two-level data model, the engine supports some query models, such as query by TTAC feature and query by textual attribute.

Content-based retrieval

The FICBDS database system provides for content-based functional image retrieval on TTAC characteristics. The user defines a TTAC feature vector or selects a querying example from a set of pre-defined TTAC samples. Images containing similar TTAC feature vectors are then retrieved from the database. Let $TTAC_{EX}$ be the example TTAC feature vector and $TTAC_{DB}$ be the all extracted TTAC feature vectors stored in the database. The measure of similarity can be computed by $D_E(TTAC_{EX}, TTAC_{DB})$, the Euclidean distance between $TTAC_{EX}$ and $TTAC_{DB}$. All $TTAC_{DB}$ satisfying $\{ D_E(TTAC_{EX}, TTAC_{DB}) \leq M \}$ are retrieved, where M is a user-defined TTAC matched degree.

Dynamic Image Archival System

Dynamic functional images are stored for each plane using the three-step technique for dynamic image data compression. There are three tables in the database: (1) Image TTAC Feature Vector Index Table which contains the study number, patient ID, plane number, cluster number and TTAC vector (activity concentration in different time points); (2) TTAC Index-Map Table which stores the TTAC cluster index map of each image plane. A conventional approach would store the entire image data set prefixed with a header in the image table and would require a large amount of storage space. However, in this functional retrieval system, the use of the three-step functional image data compression means that only a cluster index table needs to be stored for each image plane; (3) Patient Information Table, which includes textual information about the patient, such as patient ID, name, sex, age, height, weight, study number, PTAC input function, injected dose, physician name, physical examination date, medical history, pathological results and so on.

Graphical User Interface

Figure 17-7 shows the graphical user interface of the FICBDS system. There are three individual query entry windows. With the "Query by TTAC Feature" option, the user can use a slider control to define a TTAC sample curve for querying. Weighting parameters can be added to different time points. The sample curve is plotted in the "TTAC Sample" window. The "Query by Activity Concentration" entry window allows the user to retrieve TTAC, associated with a certain activity range of measurement. The "Query by Textual Attribute" allows the user to enter textual information, such as patient ID, patient name, sex, age range, etc. The system also supports combinations of these different query methods using the "Query Model Select" option. Studies containing planes that match the search criteria are listed, together with their matching planes. The numbers of matched planes for a highlighted list are shown in the "Matched Plane List" window. Some utilities, such as matched plane viewing and frame viewing are also available.

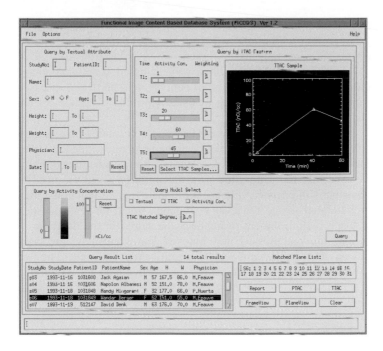

Figure 17-7 The graphical user interface of the database system
with different query methods

CASE STUDIES

To assess the feasibility and practicality of this content-based functional image retrieval database design, a set of 15 dynamic clinical FDG PET studies was used to create a prototype database. The studies were acquired by a SIEMENS ECAT 951R PET scanner, at the PET and Nuclear Medicine Department, Royal Prince Alfred Hospital, Sydney. The number of cross-sectional image planes was 31. A typical conventional sampling schedule (CSS) consisting of 22 temporal frames was used to acquire the PET projection data. The PET scanning schedule was 6×10.0 second scans, 4×30.0 second scans, 1×120.0 second scans, and 11×300 second scans. The dynamic PET data were corrected for attenuation, decay-corrected to the time of injection and then reconstructed using filtered back-projection with a Hanning filter. The reconstructed images were 128×128 with a pixel size of 2mm×2mm. The 22-frame CSS-format image data were re-sampled into a 5-frame OISS-format, using the dynamic image re-sampling tool of the functional image processing engine. The OISS consisted of the following five scanning intervals: 1×41, 1×136, 1×567, 1×1711, and 1×1145 seconds. The prototype database contained 465 images (15 studies, 31 planes for each study).

Figure 17-8 shows a user-defined TTAC query. The sample TTAC associated with a high metabolic rate is shown in Figure 17-8(a). The results are shown in Figure 17-8(b). Images in the first row are tissue regions with similar TTAC features to the sample TTAC. Their cluster index images and original temporal images are shown in the second and third rows, respectively. The results show the query successfully identified brain regions with a relatively high metabolic rate. It is possible to identify and retrieve images containing tissue regions which exhibit physiological behaviour similar to a pre-defined pattern. Such a content-based functional image database system could facilitate the identification of patients with particular disease states, and potentially increase the understanding of underlying disease processes and improve specificity in diagnosing disease.

The content-based functional image database retrieves index images corresponding to specific physiological kinetic features. A dynamic image sequence representing the original kinetic information and image data are then reconstructed from the index image. Kinetic feature indexing allows our database to avoid the need to store the raw image data but still enables rapid access to the dynamic functional image data. In contrast, conventional image retrieval techniques based on features such as color, texture and shape require the entire image to be stored in the database. Also, since dynamic images are compressed and indexed simultaneously, the complexity of database design and implementation is significantly reduced. The reduction in database size speeds network-based access to dynamic data for interpretation by the physician.

17.6 Conclusion

Current content-based image retrieval techniques may not be applicable in the medical imaging domain. In this chapter, we have summarized the major content-based retrieval techniques for medical data which include: one-dimensional ECG signals; two-dimensional X-ray projection images; three-dimensional CT / MRI volume images; and four-dimensional PET / SPECT dynamic images. These content-based medical data retrieval techniques may offer potential advantages in clinical decision making, retrospective studies, surgical planning, radiation therapy and telemedicine.

Acknowledgment

The authors would like to thank Z. Chen for her valuable assistance in preparing this chapter. This research is supported by the ARC, UGC and CMSP Grants.

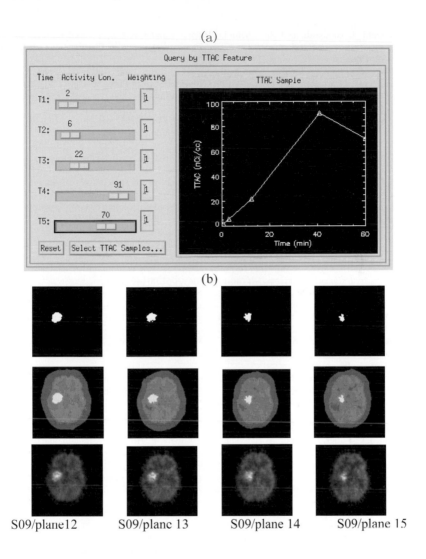

(a)

(b)

S09/plane12   S09/plane 13   S09/plane 14   S09/plane 15

Figure 17-8 Some query results by using an user-defined TTAC sample

References

[1]   R. C. Coile, "E-Health: Reinventing Healthcare in the Information Age", *Journal of Healthcare Management*, Vol. 45, No.3, pp206-210, May/Jun, 2000.

[2]   J. Mitchell, *From Telehealth to E-Health: The Unstoppable Rise of E-Health*, The Commonwelth Department of Communications, Information Technology and the Arts (DOCITA), Canberra, 1999, available at http://www.dcita.gov.au.

[3]   E. Binaghi, I. Gagliardi, and R. Schettini, "Indexing and fuzzy logic-based retrieval of colour images", *Proc. IFIP Working Conf. On Visual Database Systems II,* Elsevier Science Publishers, pp79-92, 1992.

[4]   A. Nagasaka and Y. Tanaka, "Automatic video indexing and full-video search for object appearances", *Visual Database System, II, IFIP,* Elsevier Science Publishers, pp113-127, 1992.

[5]   S. W. Smoliar and H. J. Zhang, "Content-based video indexing and retrieval", *IEEE Multimedia,* vol.1, pp62-72, 1994.

[6]   M. J. Swain, "Interactive indexing into image database", *Proc. SPIE,* vol.1908, pp193-197, 1993.

[7]   J. R. Smith and S. F. Chang, "Tools and techniques for color image query", *Proc. SPIE,* vol. 2670, 1995.

[8]   J. Weszka, C. Dyer and A. Rosenfeld, "A comparative study of texture measures for terrain classification", *IEEE Trans Sys. Man and Cyb.,* SMC-6 (4), 1976.

[9]   P. P. Ohanian and R. C. Dubes, "Performance evaluation for four classes of texture features", *Pattern Recognition,* vol.25, no.8, pp819-833, 1992.

[10]   H. Tamura, S. Mori and T. Yamawaki, "Texture features corresponding to visual perception", *IEEE Trans. Sys. Man and Cyb.,* SMC- (6), 1978.

[11]   W. Niblack, R. Barber, W. Equitz, M. Flickner, E. Glasman, D. Petkovic, P. Yanker, C. Faloutsos and G. Taubin, "The QBIC project: querying images by content using color, texture and shape", *Proc. SPIE,* vol.1908, pp173-187, 1993.

[12]   K. Tanabe and J. Ohya, "A similarity retrieval method for line drawing image database", *Progress in Image Analysis and Processing,* Chichester: Wiley, pp138-146, 1989.

[13]   B. M. Mehtre, M. Kankanhalli and W. F. Lee, "Shape measures for content based image retrieval: A comparison", *Information Processing and Management,* vol.33, no.3, 1997.

[14]   F. Rabitti and P. Stanchev, "GRIM-DBMS: a graphical image database management system", *Proc. IFIP Working Conf. On Visual Database Systems,* pp.415-430, 1989.

[15]   H. Wang, F. Guo, D. Feng and J. Jin, "A signature for content-based image retrieval using a geometrical transform", *Proc. of the 6th ACM Int. Multimedia Conf. MM'98,* pp.229-234, Bristol, UK, Sept., 1998.

[16]   M. Flickner, H. Sawhney, W. Niblack, J. Ashley, Q. Huang, B. Dom, M. Gorkani, J. Hafner, D. Lee, D. Petkovic, D. Steele, and P. Yanker, "Query by image and video content: the QBIC system", *IEEE Computer,* vol. 28, no. 9, pp. 23-32, 1995.

[17]   A. Pentland, R. W. Picard, S. Sclaroff, "Photobook: tools for content-based manipulation of image databases", *Proc. SPIE,* vol. 2185, pp. 34-47, 1994.

[18]   J.R. Bach, C. Fuller, A. Gupta, A. Hampaur, B. Horowitz, R. Humphrey, R. Jain, and C. Shu, "The Virage image search engine: an open framework for image management", *SPIE Digital Image Storage and Archiving Systems,* vol. 2670, pp. 76-87, Feb, 1996.

[19]   W. Y. Ma, "Netra: A toolbox for navigating large image databases", Ph.D. Dissertation. Department of Electrical and Computer Engineering, University of California, Santa Barbara, 1997. (http://vivaldi.ece.ucsb.edu/Netra)

[20]   Y. Rui, T. S. Huang, and S. Mehrotra, "Content-based image retrieval with relevance feedback in MARS", *Proc. of IEEE Int. Conf. on Image Processing '97,* October 26-29, 1997, Santa Barbara, California, USA, ppII815-818. (http://jadzia.ifp.uiuc.edu:8000)

[21]   P. M. Kelly, T. M. Cannon, "Efficiency issues related to probability density function comparison", *Proc. IS&T/SPIE Symposium on Electronic Imaging,* vol. 2670, pp. 42-49, 1996.

[22]   V. E. Ogle, M. Stonebraker, "Chabot: retrieval from a relational database of images", *IEEE Computer,* vol. 28, no. 9, pp. 40-48, 1995.

[23]   J. S. Jin, H. Greenfield and R. Kurniawati, "CBIR-VU: a new scheme of processing visual data in multimedia systems", *Lecture Notes in Computer Science: Visual Information Systems,* C H C Leung (Ed.), pp.40-65, 1997.

[24]   V. N. Gudivada and V. V. Raghavan, "Content-based image retrieval systems", *IEEE Computer,* vol. 28, no. 9, pp. 18-22, 1995.

[25]   A. L. Goldberger and E. Goldberger, *Clinical Electrocardiography A Simplified Approach,* 5th ed. St. Louis, MO: Mosby, 1994.

[26]   J. G. Webster, *Medical Instrumentation, Application and Design,* 2nd ed. New York: Wiley, 1995.

[27]   G. Clifford, L. Tarassenko and N. Townsend, "Fusing Conventional ECG QRS Detection Algorithms with an Auto-associative Neural Network for the Detection of Ectopic Beats", *Proceedings of ICSP2000,* pp.1623-1628, 2000.

[28]   T. Penzel, G. B. Moody, R. G. Mark, A. L. Goldberger and J. H. Peter, "The Apnea-ECG Database", *Computers in Cardiology*, vol. 27, pp.255-258, 2000.

[29]   P. de Chazal, C. Heneghan, E. Sheridan, R. Reilly, P. Nolan and M. O'Malley, "Automatic Classification of Sleep Apnea Epochs Using the Electrocardiogram", *Computers in Cardiology*, vol. 27, pp.745-748, 2000.

[30]   P. Bozzola, G. Bortolan, C. Combi, F. Pinciroli and C. Brohet, "A Hybrid Neuro-Fuzzy System for ECG Classification of Myocardial Infarction", *Computers in Cardiology*, vol. 27, pp.241-244, 1996.

[31]   J. H. van Bemmel, A. M. van Ginneken, H. Stam, D. Assanelli, P. W. MacFarlane, N. Maglaveras, P. Rubel, C. Zeelenberg and Chr Zywietz, "Integration and Communication for the Continuity of Cardiac Care (I4C)", *J. Electrocardiology*, vol. 31(supp):60-68, 1998.

[32]   P. de Chazal and B. G. Celler, "Selection of Optimal Parameters for ECG Diagnostic Classification", In: Computers in Cardiology. Piscataway: IEEE Computer Society Press. 1997:13-16.

[33]   R. Bousseljot and D. Kreiseler, "ECG Signal Analysis by Pattern Comparison", *Computers in Cardiology*, vol. 25, pp.349-352, 1998.

[34]   P. de Chazal, B. G. Celler and R. Reilly, "Using Wavelet Coefficients for the Classification of the Electrocardiogram", *Proceedings of World Congress on Medical Physics and Biomedical Engineering*, 2000:4pages.

[35]   R. Bousseljot and D. Kreiseler, "Waveform Recognition with 10,000 ECGs", *Computers in Cardiology*, vol. 27, pp.331-334, 2000.

[36]   S. Liu, C.F. Babbs and E.J. Delp, "Multiresolution Detection of Spiculated Lesions in Digital Mammograms", *IEEE Transactions on Image Processing*, vol.10, no.6, pp874-884, 2001.

[37]   J. M. Dinten et al, "A Global Approach for Localization and Characterization of Microcalcification in Mammograms", *Digital Mammography* (1996), Elsevier: 235-238.

[38]   Y. Chitre et al, "Adaptive Wavelet Analysis and Classification of Mammographic Calcification", *Digital Mammography* (1996), Elsevier:323-326.

[39]   R. N. Strickland et al, "Wavelet Transform for Detecting Microcalcification in Mammograms", *IEEE Transaction on Medical Imaging*, pp.218-229, 1996.

[40]   T. C. Wang et al, "Detection of Microcalcification in Digital Mammograms Using Wavelet", *IEEE Transactiosn on Medical Imaging*, pp.498-509, 1998.

[41]   I. J. Karssemc et al, "Adaptive Noise Equalization and Recognition of Microcalcification Cluster in Mammograms", *Journal of Pattern Recognition and Artificial Intell.*, no. 7, pp.263-274, 1993.

[42]   Angelo Balestrieri, Antonella Balestrieri, A. Barone, A. Casanova and M. Fraschini, "Information Retrieval from Medical Database", *Lecture Notes in Computer Science 2184*, pp.42-52, 2001.

[43]   R. Distasi, S. Vitulano et al, "A Hierarchical Representation for Content-based Image Retrieval", *Journal of Visual Language and Computing*, no. 11, pp.369-382, 2000.

[44]   W. Zhang, S. Dickinson, S. Sclaroff and J. Feldman, "Shape-Based Indexing in a Medical Image Database", *Workshop on Biomedical Image Analysis*, pp.221-230, June, 1998.

[45]   E. Pietka, L. Kaabi, M. L. Kuo and H. K. Huang, "Feature Extraction in Carpal Bone Analysis", *IEEE Computer Graphics and Applications*, 1991.

[46]   E. Pietka, M. F. McNitt-Gray, M. L. Kuo and H. K. Huang, "Computer Assisted Phalangeal Analysis in Skeletal Age Assessment", *IEEE Transactions on Medical Imaging*, vol. 10, no. 4, pp.616-620, 1991.

[47]   W. W. Chu, A. F. Cardenas and R. K. Taira, "KMeD: A Knowledge-based Multimedia Medical Distributed Database System", *Information Systems*, vol. 20, no. 2, pp.75-96, 1995.

[48]   S. Napel, G. D. Rubin, C. F. Beaulieu, R. B. Jeffrey and V. Argiro, "Perspective Volume Rendering of Cross-sectional Images for Simulated Endoscopy and Intraparenchymal Viewing", *Proceedings SPIE: Medical Image*, Newport Beach, CA, pp.1-12, 1996.

[49]   G. P. Robinson, H. D. Tagare, J. S. Duncan and C. C. Jaffe, "Medical Image Collection Indexing: Shape-based Retrieval Using KD-Trees", *Computerized Medical Imaging and Graphics*, vol. 20, no. 4, pp.209-217, 1996.

[50]   T Y Hou, P. Liu, A. Hsu and M-Y Chiu, "Medical Image Retrieval by Spatial Features", *IEEE Inter. Conf. Systems, Man and Cybernetics*, vol. 2, pp.1364-69, 1992.

[51]   S. C. Orphanoudakis, C. Chronaki and S. Kostomanolakis, "I²C: A System for the Indexing, Storage, and Retrieval of Medical Images by Content", *Medical Informatics*, vol. 19, no. 2, pp.109-122, 1994.

[52]   S. C. Orphanoudakis, C. Chronaki and D. Vamvaka, "I²Cnet: Content-Based Similarity Search in Geographically Distributed Repositories of Medical Images", *Computerized Medical Imaging and Graphics*, vol. 20, no. 4, pp.193-207, 1996.

[53]   W. W. Chu, C-C Hsu, A. F. Cardenas and R. K. Taira, "Knowledge-Based Image Retrieval with Spatial and Temporal Constructs", *IEEE Transactions on Knowledge and Data Engineering*, vol. 10, no. 6, pp.872-888, 1998.

[54]   Y. Liu, F. Dellaert, W. E. Rothfus, A. Moore, J. Schneider and T. Kanade, "Classification-Driven Pathological Neuroimage Retrieval Using Statistical Asymmetry Measures", *Lecture Notes in Computer Science*, vol. 2208, pp.655-665, 2001.

[55]   G. Potamias, "Content-Based Similarity Assessment in Multi-segmented Medical Image Data Bases", *Lecture Notes in Computer Science*, vol. 2123, pp.347-361, 2001.

[56]   S. C. Huang and M. E. Phelps, "Principles of tracer kinetic modelling in positron emission tomography and autoradiography", In: *Positron emission tomography and autoradiography. Principles and applications for the brain and heart*. New York: Raven Press, pp.287-346, 1986.

[57]   D. Feng, D. Ho, H. Iida & K. Chen, "Techniques for Functional Imaging", an invited chapter contributing to *Medical Imaging Techniques and Applications*, Edited by: C. T. Leondes, in 'Gordon and Breach International Series in Engineering, Technology and Applied Science', *Gordon and Breach Science Publishers*, pp85-145, 1997.

[58]   W. Cai, D. Feng and R. Fulton, "Content-Based Retrieval of Dynamic PET Functional Images", *IEEE Transactions on Information Technology in Biomedicine*, vol. 4, no. 2, pp.152-158, 2000.

[59]   D. Ho, D. Feng and K. Chen, "Dynamic image data compression in spatial and temporal domains: Theory and algorithm", *IEEE Trans. Info. Tech. Biomed.*, vol. 1, no. 4, pp. 219-228, 1997.

[60]   W. Cai, D. Feng and R. Fulton, "Clinical investigation of a knowledge-based data compression algorithm for dynamic neurologic FDG-PET images", *Proceedings of the 20th Annual International Conference of the IEEE Engineering in Medicine and Biology Society (EMBS)*, vol. 20, part 3, pp. 1270-1273, Hong Kong, Oct 29 – Nov 1, 1998.

[61]   S. C. Huang, M. E. Phelps, E. J. Hoffman, K. Sideris, C. Selin, and D. E. Kuhl, "Non-invasive determination of local cerebral metabolic rate of glucose in man", *Amer. J. Physiol.*, vol. 238, pp. E69-E82, 1980.

[62]   D. Feng, D. Ho, K. Chen, L. C. Wu, J. K. Wang, R. S. Liu, and S. H. Yeh, "An evaluation of the algorithms for determining local cerebral metabolic rates of glucose using positron emission tomography dynamic data," *IEEE Trans. Med. Imag.*, vol. 14, pp. 697-710, 1995.

[63]   D. Feng, W. Cai and R. Fulton, "An optimal image sampling schedule design for cerebral blood volume and partial volume correction in neurologic FDG-PET studies", *Australia and New Zealand Journal of Medicine*, vol. 28, no.3, pp. 361, 1998.

[64]   X. Li, D. Feng, and K. Chen, "Optimal image sampling schedule: A new effective way to reduce dynamic image storage space and functional image processing time", *IEEE Trans. Med. Imag.*, vol. 15, pp. 710-718, 1996.

[65]   D. Feng, W. Cai and R. Fulton, "FIPS: A functional image processing system for PET dynamic studies", *Proceedings of the Ivth International Conference on Quantification of Brain Function with PET (BrainPET'99)*, pp772, Copenhagen, Denmark, Jun 13-17, 1999.

[66]   D. Ho, D. Feng and L.C. Wu, "An assessment of optimal image sampling schedule design in dynamic PET-FDG studies", pp. 315-320, in *Quantitative Functional Brain Imaging with Positron Emission Tomography*, Herscovitch, Publisher Academic Press, 1998.

[67]   C. S. Patlak, R. G. Blasberg, and J. Fenstermacher, "Graphical evaluation of blood to brain transfer constants from multiple-time uptake data," *J. Cereb. Blood Flow Metab.*, vol. 3, pp. 1-7, 1983.

[68]   D. Feng, S. C. Huang, Z. Wang, and D. Ho, "An unbiased parametric imaging algorithm for non-uniformly sampled biomedical system parameter estimation," *IEEE Trans. Med. Imag.*, vol. 15, pp. 512-518, 1996.

18 BIOMETRICS FEATURE RETRIEVAL USING PALMPRINT IMAGES

Dr. David Zhang and Dr. Jane You

This chapter presents a new approach to palmprint image retrieval for personal identification. Three key issues in image retrieval are considered - feature selection, similarity measurement and dynamic search for the best match of the sample in the image database. A statistically based feature selection scheme is introduced to guide the dynamic extraction of palmprint features for hierarchical retrieval. A texture feature measurement which is characterized by its high convergence of inner-palm similarities and good dispersion of inter-palm discrimination is used to facilitate the fast search for the appropriate matching candidates in the database at a coarse level. A wide range of palmprint images have been tested to demonstrate the feasibility and effectiveness of the proposed method.

18.1 Introduction

Automatic personal identification is becoming a significant component of the security systems with many challenges and practical applications in today's networked world. Biometrics, which deals with the identification of individuals based on their biological or behavioral characteristics, has been emerging as a new and effective identification technology that achieves accurate and reliable identification results. Although no single biometric is expected to satisfy all identification requirements, the use of such unique, reliable and stable personal features has invoked increasing interest in the development of biometrics-based identification systems for civilian, military and forensic applications.

Currently, a number of biometrics-based technologies are commercially available, and many more are being proposed, researched and evaluated worldwide [11, 14]. Each biometric has different strengths and limitations in different applications. So far, there are eight main, different biometrics (face, fingerprint, hand geometry, iris, retinal pattern, signature, voice-print, and facial thermograms) used for identification. Other biometrics such as hand vein, keystroke dynamics, odor, DNA, gait and ear are currently under investigation.

The selection of a biometric for a given identification task is application oriented, depending on the requirements of the specific application, the characteristics of the application and the properties of the biometrics. As stated in [11], the following aspects need to be considered and studied carefully at the design stage: (1) the nature of the application (identification or authentication); (2) the degree of automation for operation (semi-automatic or fully automatic); (3) the capture of biometrics data from users (habituated and feasibly measured); (4) the access to biometrics data (covert or overt); (5) the quality of biometric data (cooperative or non-cooperative of the subjects); (6) the storage requirement constraints (the size of the internal representation for the chosen biometrics); (7) the performance requirement constraints (accuracy and reliability); and (8) the acceptability of a biometrics from users (users' preference and willingness for data collection).

The general problem of personal identification involves a number of important and difficult research issues such as design, evaluation, integration and circumvention. For a biometrics-based identification system, the design issue could be related to a pattern recognition system which deals with data acquisition and representation, feature extraction, matching criteria, search methods, database organization and system scalability. In order to be widely accepted by the public for practical applications, a biometrics-based identification system should be characterized by its easy of operation, short response time and high accuracy. The investigation of fingerprint identification and speech recognition has drawn considerable attention over the last 25 years. Recently, face recognition and iris-based verification issues have been studied extensively, resulting in the successful development of biometric systems for commercial applications. However, limited work has been reported on palmprint identification and verification.

Palmprint refers to the principal lines, wrinkles and ridges on the palm. Like fingerprints, the palmprint has been used in law enforcement for criminal identification because of its stability and uniqueness. The rationale for choosing hand features as a base for identity verification originates from its user friendliness, environment flexibility and discriminating ability. Normally people do not feel uneasy about having their hand images and prints taken for testing. More importantly, these hand features are stable and uniquely represent each individual's identity. Consequently, it is essential to develop an effective method for automating palmprint identification and verification for security reasons.

Matching plays an important role in identity verification and identification. Verification refers to the comparison of a claimant's biometrics feature against the bonafide person's sample which has been stored in the verification system. In other words, verification is regarded as a one-to-one match. Identification is concerned with the search for the best match between the input sample and the templates in the database, which is termed one-to-many matching. Palmprint verification consists of determining whether two palmprints (test sample from input and the template in the database) are from the same palm. Therefore, a comparison must be made between the template feature set and the feature set derived from the hand image at the time of verification. The selection of features and similarity measures are two fundamental issues to be solved. A feature with good discriminating ability should exhibit a large variance between individuals and small variance between samples from the same person.

The advances in image feature extraction/representation and similarity measures play an essential role in image identification and classification. Thus, dynamic image feature extraction, guided search and knowledge-based hierarchical matching have significant potential, enabling both image identification and image classification to be performed more effectively. However, most of the existing systems and research adopt a fixed image feature selection scheme for image feature extraction and search.

To achieve flexibility and multiple feature integration, we extend a statistical approach to construct a decision tree for the personal identification task, where a statistically based feature selection criterion is used to guide the selection and integration of multiple palmprint features for matching in a hierarchical structure.

At a first glance, palmprint and fingerprint patterns appear to resemble each other as both consist of a large amount of ridges. Although the minutiae based matching which utilizes terminations and bifurcations of the ridges is powerful for fingerprint verification and identification, such an approach is not suitable for palmprint patterns due to the change of orientations. Zhang and Shu [23] proposed using datum point invariance and line feature matching for palmprint feature extraction and verification. They introduced a directional projection algorithm to localize the datum points for matching. However, their algorithm is subject to the following limitations: (1) The localization of datum points is based on the detection of distinct principal lines in palmprint patterns. The approach is not suitable for the patterns without clearly dominant principal lines. (2) A set of fixed masks are used to detect feature lines for matching. (3) The link of feature line segments for palmprint line feature representation requires a high level of computation resources and lacks accuracy. (4) The algorithm is not suitable for dealing with partially occluded palmprint patterns. Recently Jain et al. investigated the feasibility of matching palmprints based on feature points [7]. Instead of extracting feature lines explicitly as in [23], they applied the following six steps to extract isolated feature points that lie along palm lines: (1) palm image smoothing; (2) image binarization by interactively thresholding the smoothed gray palm image; (3) successive morphological erosions, dilations and subtractions for feature point detection; (4) location adjustment for the feature points; (5) calculation of the orientation of each feature point (the orientation of the palm line which a feature point is associated with); and (6) removal of redundant points. They used their previous point matching technique to determine whether the two sets of feature points/orientations were the same by computing a matching score. Although this approach has the advantages of simplicity and reasonable accuracy, it lacks robustness in the following aspects: (1) the detection of feature points relies on the selection of a threshold value for image binarization and the consequent successive morphological operations; (2) the feature point matching is based on the traditional exhaustive comparison method, which is very time consuming and may not meet the real-time requirement for on-line matching from a large collection of palmprint patterns.

Unlike these existing techniques, we propose a dynamic selection scheme to facilitate the coarse-to-fine palmprint pattern matching for personal identification, which combines global and local palmprint features in a hierarchical manner. The global texture energy (GTE) is introduced to represent the global palmprint feature, which is characterized by a high convergence of inner-palm similarities and a good dispersion of inter-palm discrimination. Such a global feature is used to guide the dynamic selection of a small set of similar candidates from the database at a coarse level for further matching. In addition, a fractional discrimination function is proposed to enhance image feature points in conjunction with image decomposition and contextual filtering for the final identification by feature point based image matching at a fine level. To handle the case of occluded palmprint matching, the Hausdorff distance measurement is used to determine the similarity between two palmprint feature point sets. To speed up the search for the best matching, a multi-step image matching procedure is applied associated with different orders of the proposed fractional discrimination function.

This chapter is organized as follows: Section 18.2 presents an overview of the detection of palmprint features. Section 18.3 introduces a dynamic feature selection scheme which embraces a statistically based feature selection criterion and a fractional discrimination function for palmprint feature extraction. The combination of multiple features for matching measurement and a hierarchical search for the best match is described in Section 18.4. Section 18.5 reports the experimental results and Section 18.7 summarizes the chapter.

Palmprint Feature Extraction

A key issue in palmprint identification involves the search for the best matching of the test sample from input and the templates in the palmprint database. The selection of features and similarity measures are two fundamental problems to be solved. A feature with good discriminating ability should exhibit a large variance between individuals and small variance between samples from the same person. Principal lines and datum points are regarded as useful palmprint features and have been successfully used for verification [23]. In addition, there are many other features associated with a palmprint, such as geometry features, wrinkle features, delta point features and minutiae features [1]. All of these features are concerned with the local attributes based on points or line segments. The lack of global feature representation resulted in the high computation demand for matching which measures the degree of similarity between two sample sets. This section summarizes the existing palmprint feature descriptions and introduces a texture based palmprint feature measurement.

The Description of a Palmprint

The palm is the inner surface of a hand between the wrist and the fingers. The palmprint refers to principal lines, wrinkles and ridges on the palm. There are many features present on a palm. There are three principal lines caused by flexing the hand and wrist , which are named the heart-line, head-line and life-line. Figure 18-1 shows the layout of a palm, where the palm has been divided into three regions, namely the finger-root region (I), the inside region (II) and the outside region (III). The three marked curves, 1, 2 and 3 represent the three principal lines the heart-line, head-line and life-line respectively. The two endpoints, a and b, are determined by the intersections of the life-line (curve 3) and heart-line (curve 1) with the sides of a palm. Due to the stability of the principal lines, the locations of endpoints and their midpoint o in a palm remain unchanged in respect to rotation of the hand and the change of time. Therefore, these feature lines are regarded as reliable and stable features to distinguish a person from other people.

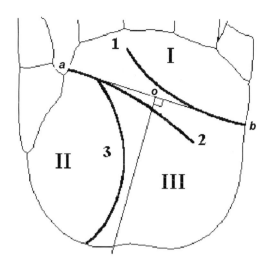

Figure 18-1 The layout of a palmprint
Regions: I, finger-root region; II, inside region; and III, outside region
Principal Lines: 1, heart-line; 2, head-line and 3, life-line
Datum Points: a, b endpoint; o, midpoint

In addition to the principal lines and datum points (two endpoints and midpoint) as described above, many other features such as wrinkles and ridges are associated with a palmprint. The following lists these features, and the relevant line patterns are illustrated in Figure 18-2.

- Geometry Features: width, length and area are the geometry features in accordance with a palm's shape.
- Principal Line Features: very important physiological characteristics to distinguish different individuals because of their stability and uniqueness.
- Wrinkle Features: thin and irregular lines and curves different from principal lines.
- Delta Point Feature: the center of a delta-like region in a palmprint, which is normally located in finger-root region and outside region.
- Minutiae Features: significant feature measurement of ridges existing in a palm.

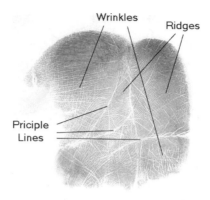

Figure 18-2 The line pattern of a palmprint

Principal Line Detection and Datum Point Localization

Principal lines and datum points are used to describe palmprint features for some palmprint identification and verification systems [23]. In [23], Zhang and Shu proposed using the directional projection algorithm for the detection of principal lines and their endpoints. The following highlights the major steps of this technique and more details are given in [23].

- Heart Line Detection by Horizontal Projection Algorithm.
- A pixel with suitable offsets to the edges of outside and topside of a palm is considered in a palmprint. All pixels belonging to heart line are located by the horizontal projection algorithm.
- Life Line Detection by Vertical Projection Algorithm.
- Another sub-image close to the wrist is processed by the vertical projection algorithm for the detection of points along the life-line.
- Endpoint Localization.
- The location of endpoint can be determined based on the peculiarities of the heart-line by the horizontal projection algorithm.
- End Point Localization by Principal Line Tracing.
- Midpoint Location.

PALMPRINT FEATURE POINTS

In [7], palmprint feature points are defined as the pixels which lie on the prominent palm lines. The orientation of the associated palm line for such feature points is also regarded as an important palmprint feature measurement. The following summarizes the feature extraction algorithm proposed by Jain et. al. for palmprint matching:

- Palm image smoothing by a 4-neighbour averaging for noise and thin palm line removal.

- Palm image binarization by thresholding.
- Compact region removal by successive morphological operation for feature point detection.
- Feature point alignment to the medial axis by using a 4x4 neighbourhood for location adjustment.
- Computing the orientation of the corresponding palm line with which each feature point is associated.
- Spurious feature point removal based on contrast thresholding.

Line Feature Extraction and Representation

Line features include both curved and straight lines. So far there have been many methods proposed to detect lines [5, 22]. However, most of these algorithms are not suitable for palmprint images because a palmprint consists of lots of ridges and fine wrinkles of varying lengths and thickness, running in different directions. The pyramid edge detection method used a stack filter to detect strip-like line segments in an image and is useful for extracting the principal lines of a palmprint. Unfortunately, this method fails to detect thin line segments in a palmprint image. Although some nonlinear filters can detect thin vertical lines and the extension of these filters can extract both thick and horizontal lines, these filters are not able to identify other line segments such as wrinkles and ridges with diverse orientations. An improvement in this approach is reported in [23]. The method aims to repeatedly detect short line segments at each individual orientation and combine them with a post-process at the final stage.

Texture Feature Measurement

A palmprint consists of many thin and short line segments represented by wrinkles and ridges. Such a pattern can be well characterized by texture. Thus, we use a texture feature measurement as a powerful technique for palmprint feature extraction. Figure 18-3 shows different palmprint texture patterns.

Texture provides a high-order description of the local image content. The analysis of texture requires the identification of texture attributes that can be used for segmentation, discrimination, recognition, or shape computation. Historically, structural and statistical approaches have been adopted for texture feature extraction [9]. The structural approach assumes the texture is characterized by some primitives following a placement rule. In the statistical approach, texture is regarded as a sample from a probability distribution on the image space and is defined by a stochastic model or characterized by a set of statistical features.

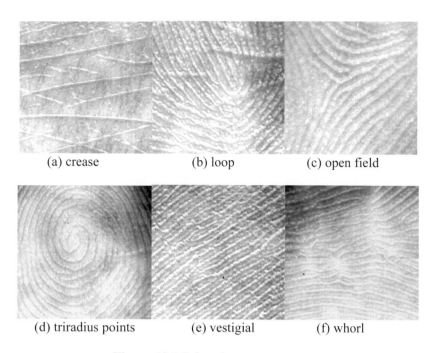

(a) crease   (b) loop   (c) open field

(d) triradius points   (e) vestigial   (f) whorl

Figure 18-3 Palmprint texture patterns

A special characteristic of our proposed method for palmprint identification is determination of a feature, the texture energy computed using Laws' convolution masks [12], which is capable of functioning directly function as a classifier. This statistical approach pioneered, by Laws, is notable for its computational simplicity. He introduced the notion of a single parameter, the local "texture energy" (E) evaluated at each pixel location (i, j) in the convolved image over a large window size 15*15 as the measure of the texture features in the spatial domains, which is defined as the mean deviation within the given window and given by:

$$E(i,j) = \frac{1}{(2n+1)^2} \sum_{k=i-n}^{i+n} \sum_{l=j-n}^{j+n} |F(k,l)| \qquad (18\text{-}1)$$

where

$$F(i,j) = A(i,j) * I(i,j) = \sum_{k}^{a} {}_{=-a} \sum_{l}^{a} {}_{-a} A(k,l) I(i+k, j+l) \text{ represents}$$

the original image and A refers to the selected Laws' convolution mask.

Although individually Laws' masks can be used in isolation as a texture classifier, they are subject to a number of limitations due to the lack of robustness with respect to the fixed masks. One feasible approach to improve the performance is to replace the fixed masks with adaptive masks "tuned" to be robust to the classification tasks. This process involves the determination of the texture class "tuned" mask which, when applied to a textured image, smoothes out regions of common texture so that the variance of the convolved image, typically over a 15x15 local window, is (a) reasonably constant over all locations within a region of uniform texture, so that such a region is essentially converted into a region of uniform gray scale, (b) markedly different in value between regions of different texture. The approach reported here is an extension of [2] and our previous work [19]. For the fast selection of a small set of similar palmprint patterns from the database, we applied four "tuned" masks to capture the global palmprint texture features which are sensitive to horizontal lines, vertical lines, 45° lines and –45° lines (listed in order of decreasing sensitivity). Figureure18-4 lists the four of our modified "tuned" masks. In our approach the local variance after convolution is well-approximated by the sum of squared values of the convolved image within the test window, which is expressed as:

$$TE(i, j) = \frac{\sum_{W_x} \sum_{W_y} (I * A)_{RS}^2}{P^2 W_x W_y} \tag{18-2}$$

where the rs sum is over all pixels within a square window W of size $W_x * W_y$ centered on the pixel at i,j,; A is a zero sum "tuned" 5*5 convolution mask; and P is the parameter normalizer $P^2 = \sum_{i,j} (A_{i,j})^2$. Such a texture energy measurement for global palmprint feature extraction has the following characteristics: (a) insensitive to noise; (b) insensitive to shift changes; (c) easy to compute; and (d) high convergence within the group and good dispersion between groups.

−1	−2	−4	−2	−1		−1	0	2	0	−1
0	0	0	0	0		−2	0	4	0	−2
2	4	8	4	2		−4	0	8	0	−4
0	0	0	0	0		−2	0	4	0	−2
−1	−2	−4	−2	−1		−1	0	2	0	−1

Horizontal Line   Vertical Line

0	−1	−4	0	2		2	0	−4	−1	0
−1	−6	0	8	0		0	8	0	−6	−1
−4	0	12	0	−4		−4	0	12	0	−4
0	8	0	−6	−1		−1	−6	0	8	0
2	0	−4	−1	0		0	−1	−4	0	2

45^0 Line -45^0 Line

Figure 18-4 The four "tuned" masks for global palmprint texture feature extraction

18.2 Dynamic Feature Selection

Feature extraction is a key issue for pattern recognition and palmprint consists of very complicated patterns. It is very difficult, if not impossible, to use one feature model for palmprint matching with high performance in terms of accuracy, efficiency and robustness. Although the research reported in [7] resulted in a more reliable approach to palmprint feature point detection than the line matching algorithm detailed in [23], the issues of efficiency and robustness remain untackled. The technique detailed in [7] involves one-to-one feature point based image matching, which requires high computation resource for a large palmprint database. In addition, the matching lacks flexibility because only one similarity measurement is applied. To speed up the search process for the best match with reliable features and flexible matching criteria, we introduce a statistically based feature selection scheme to rank different features for dynamic feature selection and flexible similarity measurement. Instead of using a fixed feature extraction mechanism and a single matching criterion as in [7, 23], we apply a mask tuning scheme to extract global palmprint texture features to identify a small set of the most relevant candidates at a coarse level. A multi-step feature point based image matching is performed at a fine level. Unlike the matching scheme described in [7], our fine matching is conducted in a hierarchical manner based on the use of a fractional discrimination function. Furthermore, the Hausdorff distance measurement is used as the matching criterion to deal with the cases of occluded images. This section briefly describes the feature selection criteria and the discrimination function.

Feature Selection Criteria
When constructing a decision tree for pattern recognition, it is important to identify the important data features for the classification and rank them in order for feature integration and similarity measurement. The use of a feature selection criterion to determine the ranking of the input attributes as initially proposed by Carter and Catlett in 1987. As summarized in [18], the proposal is based on the following observations:

- It is difficult to define a suitable feature selection criterion for groups of variables taken at a time.
- A combination of variables at a node will lead to an increased number of possible children nodes.
- Since decision trees are constructed by examining all possible tests at each stage, increasing the forms to be explored would lead to a significant increase in computational cost.
- The decision tree may be much more difficult to understand if each test is a complex function.

ID3 algorithm [16] and CART (Classification And Regression Trees) approach [4] are examples of induction algorithms that construct decisions in a systematic way. These algorithms adopt a divide-and-conquer strategy and use a feature selection criterion to guide the generation of the decision tree. However, there are a number of problems associated with these such as their restriction to one variable, structure complexity, high computational demands, sensitivity to noise and changes of data, inflexibility of rules, and difficulties in handling continuous features and knowledge discovery under uncertainty [18, 24]. Thus, it is essential to apply some heuristic which will facilitate the fast construction of efficient, near-optimal decision trees. It is expected that the feature selection criteria should cover the following important issues [24]:

- the direct handling of multi-valued features that leads to the construction of multi-branching decision trees;
- the use of a Boolean combination of features;
- the use of a linear combination of features;
- the extension of decision trees as probabilistic classifiers;
- dealing with a mixture of discrete (categorical) and continuous features;
- pre-pruning during tree building.

Recently feature selection criteria based on statistical measures have been extensively studied. These statistically based criteria include the Chi-square criterion, Asymmetrical Tau and Symmetrical Tau [24]. A comprehensive description and comparison of these criteria is given in [18]. Chi-square statistic provides a test of significance with regard to the independence between variables, however, it does not give a measure of the degree of association between the variables. Goodman and Kruskal proposed their measure of association, the Asymmetrical Tau [8], to overcome the limitations of the Chi-square criterion for cross-classification tasks in the statistical area. The Asymmetrical Tau is a measure of the relative usefulness of one variable in improving the ability to predict the classifications of members of the population with respect to a second variable. Nevertheless, when this approach was used directly as a feature selection criterion for constructing decision trees, it tended to favor features with more values. To improve this approach, Zhou and Dillon defined their Symmetrical Tau criterion [24] by combining a proportional-reduction-in-error (PRE) measure and a cost-complexity heuristic to obtain a balanced statistical-heuristic criterion for building multi-branching decision trees. We extend the use of the Symmetrical Tau criterion to guide the selection and integration of the appropriate features from multiple palmprint features for similar palmprint pattern grouping. The major steps are summarized as below:

Step 1: Identify all of the individual features to be used for matching and obtain their feature vectors. For n features f_i $(i = 0, 1, ..., n-1)$, there will be n individual feature vectors V_i $(i = 0, 1, ..., n-1)$.

Step 2: Apply Gaussian normalization to the above feature vectors.

Step 3: Initialize a set of weights a_i $(i = 0, 1, ..., n-1)$ and obtain its corresponding combined feature vector $V_c = \sum_{i=1}^{n-1} a_i V_i$

Step 4: Calculate the corresponding Tau using the following formula:

$$Tau = \frac{\sum_{j=1}^{J}\sum_{i=1}^{I}\frac{P(i,j)^2}{P(+j)} + \sum_{i=1}^{I}\sum_{j=1}^{J}\frac{P(i,j)^2}{P(i+)} - \sum_{i=1}^{I}P(i+)^2 - \sum_{j=1}^{J}P(+j)^2}{2 - \sum_{i=1}^{I}P(i+)^2 - \sum_{j=1}^{J}P(+j)^2}$$

(18-3)

where

- the contingency table has I rows and J columns;
- P(ij) = probability that a variable belongs both to row category and to column category j;
- P(i+) and P(+j) are the marginal probabilities in row category i and column category j, respectively.

Step 5:   Adjust the set of weights, and obtain a new combined feature vector V'$_c$ and calculate the corresponding Tau.

Step 6:   Repeat Step 5 for all of the given adjustment weight sets.

Step 7:   Find the maximum value of Tau from the sequences of Tau obtained in the previous stage.

Step 8:   Choose the combined feature with the maximum Tau value.

18.3 Fractional Discrimination Function

The central problem for feature point based image matching is to find an effective approach to extract feature points invariant of scales and orientations. Conventional edge detection techniques based on detection of discontinuities in pixel properties are not suitable for distinguishing boundaries between differently textured regions because texture is characterized by its local features over some neighborhood rather than in pixel scales. Therefore, an operator sensitive to different textures is required to enhance some texture components and to suppress others for texture edge detection. In contrast to the traditional methods, our solution involves the introduction of fractional functions as discrimination functions to perform robust (in the presence of noise), selective (band limited) and contextual feature extraction. A general framework is defined to encompass the existing functions and masks for feature extraction. The use of fractional discrimination for the extraction of embedded features within random textures provides a systematic and iterative method to search for dynamic features for image analysis [20].

The idea behind this method is to enhance the feature points by means of image decomposition and contextual filtering on the basis of the proposed fractional function. In addition, such a function is described in a unified form with three-parameters, where the function performs differentiation with inherent aggregation. Consequently, unlike the existing approaches which mostly rely on different individual operators, our study simplifies the feature extraction operation in a general format based on the perceptual/scale-space properties of the visual receptive fields, which provides an effective approach to feature extraction and is suitable for many vision tasks.

The concept of fractional derivatives is the approximation of the non-integer derivatives to a continuous domain with convergence to the first-, second- and finite-order derivatives. The fractional derivative of a constant is not always equal to zero [15]. Based on the different approaches to define such fractional derivatives, different fractional discrimination functions will be developed.

Let $f(\varepsilon)$ represent any function in the domain of ε, and $D^{\nu}f(\varepsilon)$ be referred to as the fractional differentiation of function $f(\varepsilon)$ at the order of ν, where ν is any real positive number. If the standard Gamma function is used, we have:

$$D^{-\nu}f(\xi) = \frac{1}{\Gamma(\nu)} \int_{c}^{\xi} (\xi - t)^{\nu-1} f(t)dt \qquad (18\text{-}4)$$

where c and ε are the lower and upper limits of the integral respectively. Accordingly, if $\nu = m - \mu > 0$ and $\mu > 0$, we have:

$$D^{\mu}f(\xi) = D^{m}[D^{-\nu}f(\xi)] \qquad where\ \xi > 0 \qquad (18\text{-}5)$$

Consequently, we can define the fractional discrimination functions by combining the aggregation function $\alpha(\xi)$ (is this correct?) with the fractional discrete derivatives.

Let D^{μ} denote the fractional discrete derivative operator (FDD). The fractional discrete derivative is performed by convolving the coefficients of FDD with the given image. The FDD coefficients for the order from 0.1 to 1.0 can be determined based on a Maclaurin series expansion of $(1-x)^{p}$, $p < 1.0$. Higher orders of FDD coefficients can be derived by repeated convolutions of the lower orders of FDDs. The fractional discrimination functions gradually phases out the contrasts and eventually results in a peak or zero-crossing map for the 1st and 2nd order operation. As a result, subtle features in an image will be extracted at different resolutions in terms of various fractional-orders, which is very powerful to detect boundaries of different regions for image matching.

18.4 Hierarchical Palmprint Matching

Coarse Level Similar Palmprint Pattern Grouping

It is very important to group similar palmprint patterns from the given palmprint database for further identification and verification. This grouping task can be viewed as a decision-making process which allocates an input palmprint sample to those categories with similar measurements in the database. In traditional classification problems, various *ad hoc* means are used to derive feature sets, and this process is clearly demarcated as the feature stage [6]. What is involved in the general classification stage is the abstraction of features of the classes of entities being considered. For the classification to be useful, it must be reliable and computationally attractive. This means the choice of the textural features or the models must be as compact as possible, and yet as discriminating as possible[19]. In other words, the texture energy measurements of palmprint samples with distinct texture primitives should exhibit large variances while the measurements of the similar patterns should possess very small diversity. Therefore, such a global feature is characterized with high convergence of inner-palm similarities and good dispersion of inter-palm discrimination. Figure 18-5 shows four palmprint samples from four different individuals with distinctive texture features and Figure 18-6 demonstrates the distribution of global palmprint texture energy measurements.

strong principal lines   full of wrinkles   less wrinkles   strong wrinkles

Figure 18-5 Sample of different palmprint pattern with distinctive texture feature

Although different individuals have different palmprint patterns, some of these patterns are so similar that it is very difficult, if not impossible, to classify them based on the global texture features only. Figure 18-7 shows the samples of the similar palmprint patterns from different groups. To tackle such a problem, we propose a dynamic selection scheme to group a small set of the most similar candidates in the database for further identification by image matching at a fine level. The idea behind this is to eliminate those candidates with large GTE differences and generate a list of the very similar candidates with very close GTEs. The following summarizes the main steps for implementation:

Step 1:   Convolve the sample palmprint image I_{sample} with the four tuned masks A_i, $i=1,2,3,4$ and obtain the corresponding global texture energy $GTE_{sample}(i)$, $i=1,2,3,4$.

Step 2:   Compare the sample with the candidate in the database in terms of GTE and calculate their difference d, where d is given by:

$$d = \sum_{i=1}^{4} \mid GTE_{sample}(i) - GTE_{candidate}(i) \mid, \, i=1,2,3,4.$$

Step 3:   If d is smaller than the pre-defined threshold value, this candidate is added to the list for further matching.

Step 4:   Go to Step 2 and repeat the same procedure until all of the candidates are considered.

Step 5: Provide the final list of candidates to guide the search for the best matching at fine level.

Fine Level Multi-step Image Matching

A guided image matching scheme is developed to determine the degree of similarities between the template pattern and every possible sample pattern selected throughout the texture classification procedure which is detailed in Section18.3. Unlike the traditional image matching methods which are based on the detection of edge points, we propose to detect interesting points in textured images to achieve high performance. To avoid blind searching for the best fit between the given samples, a guided search strategy is essential to reduce computation burden. We applied a dynamic selection of interesting points to search for the best matching in a hierarchical structure.

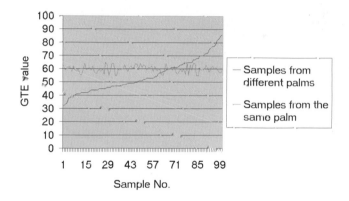

Figure 18-6 The comparison of palmprint GTE distribution inter-palm dispersion vs. inner-palm convergence

Given two images, a matching algorithm determines the location of the template image on the target image and places a value on their similarity at this point. The value determines the degree of similarity. Based on Huttenlocher et al. [10], we are able to use the Hausdorff distance algorithm to search for portions, or partial hidden objects. This feature also allows us to partition the target image into a number of sub-images, and then carry out the matching process simultaneously on these sub-images in order to accelerate the process. Instead of using edge points, we use interesting points as a basis for computing the Hausdorff distance. A comprehensive discussion about interesting points is given in [21. It aims to reduce the computation required for a reliable match. Also by using a matching scheme such as the Hausdorff distance we are able to find partially hidden objects while keeping the amount of computation minimal.

Group A: Similar palmprint samples

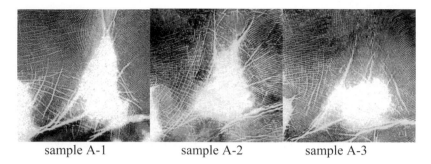

sample A-1   sample A-2   sample A-3

Group B: Similar palmprint samples

sample B-1   sample B-2   sample B-3

Figure 18-7 Samples of similar palmprint patterns from different groups

The hierarchical Chamfer image matching scheme was first proposed by Borgefors [3] in order to reduce the computation required to match two images. This section details our extension of this scheme by introducing a guided search strategy to avoid blind searching for the best fit between the given patterns. In order to avoid blind searching for the best fit between the given patterns, a guided search strategy is essential to reduce computation burden. Our extension of the hierarchical image matching scheme (H.I.M.S) was based on a guided searching algorithm that searches from low level, coarse grained images, to high level, fine grained images. To do this we needed to obtain a Hausdorff distance approximation for each possible window combination of the template and target the image at the lowest resolution. Those that returned a Hausdorff distance approximation equal to the lowest Hausdorff distance for those images were investigated at the higher resolution. The use of interesting points for hierarchical image matching is reported in our recent work [21]. To strengthen the robustness of the guided search, in the work reported in this chapter, we apply the fractional discrimination function to detect palmprint feature points and create a feature point hierarchy in terms of different fractional orders for the multi-step feature point based image matching.

18.5 Experiment Results

The palmprint image samples used for the testing are 232 * 232 size with the resolution of 125 dpi and 256 grayscales. In our palmprint image database, a total of 200 images from 100 individuals were stored. These palmprint samples were collected from both female and male adults ranging in age from 18 to 50. Although there are some electronic sensors available to get digitized palmprint images on-line, such devices are more application dependent and sensitive to noise and unexpected disturbance such as the movement of hands, lighting settings, etc. In our tests reported in this chapter, the palmprints were printed on paper by coloring palms with washable ink, using a similar method as what is described in [17]. The palmprint images marked on paper were digitized using a scanner at 125 dpi with 256 grayscales. Samples of the palmprint images are shown in Figure 18-8. A series of experiments were carried out to verify the high performance of the proposed algorithms.

Laws' texture energy concept was extended by the use of four "tuned" masks to extract the global palmprint texture features sensitive to the horizontal lines, vertical lines, 45° lines and -45° lines. Such a global texture energy measurement was used to guide the fast selection of a small set of the most similar palmprint patterns for fine matching. For a given test palmprint sample, on the average, 91% of the candidates in the database were classified as distinctive from the input data and filtered out at the coarse classification stage. In the worst case, the elimination rate of the candidates was 72% and only 28% of the samples remained for further identification at a fine level by image matching. Figure 18-8 demonstrates the effectiveness of this selection scheme by global palmprint texture feature extraction.

Feature point based image matching is performed for the final confirmation at a fine level by the proposed hierarchical structure. Figure 18-10 shows the comparison of interesting points and edge points for palmprint feature point detection for fine matching. The average accuracy rate was 95%. Since the majority of samples were filtered out from by the coarse classification, the execution speed of fine identification was increased significantly.

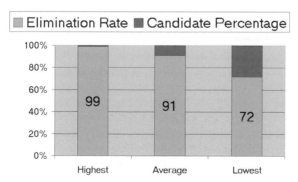

Figure 18-8 The performance of selection scheme: elimination rate vs. candidate percentage

Compared with the current existing techniques for palmprint classification and identification, our approach integrates multiple palmprint features and adopts a flexible matching criterion for hierarchical matching. Table 18-1 summarizes the major features of our method and the other techniques [11, 23] with respect to database size, feature selection, matching criteria, search scheme and accuracy. The experimental results presented above demonstrate the improvement and advantages of our approach. For future work, a dynamic indexing will be developed to handle the expansion of the palmprint database with robustness and flexibility.

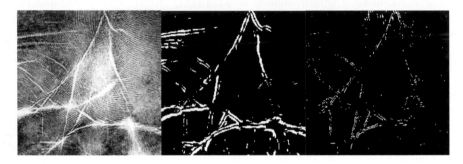

Figure 18-9 The comparison of palmprint feature point detection: edge points vs. interesting points

	Feature Point Based Matching [7]	Line Based Matching [23]	Hierarchical Matching (Proposed Approach)
Database Size	30 samples	200 samples	200 samples
Feature Extraction	feature points (single feature extraction)	lines (single feature)	texture &feature points (multiple features)
Matching Criteria	distance measurement (fixed measurement)	Euclidian distance (fixed measurement)	energy difference & Hausdorff distance (flexible measurement)
Search Method	one-to-one comparison (sequential)	one-to-one comparison (sequential)	guided search (hierarchical)
Accuracy	good	limited	good

Table 18-1 Comparison of different palmprint matching methods

18.6 Conclusion

Palmprint is regarded as one of the most unique, reliable and stable personal characteristics and palmprint verification provides a powerful means to authenticate individuals for many security systems. Palmprint feature extraction and matching are two key issues in palmprint identification and verification. In contrast to traditional techniques, we proposed a statistically based feature selection scheme to integrate multiple palmprint features for hierarchical palmprint matching. The feature selection criterion, the Symmetrical Tau, offers a general guideline for dynamic feature selection and multiple feature integration. The combination of global texture measurement via mask tuning and local feature points detection by fractional discrimination function provides the base for hierarchical matching. Our comparative study of palmprint feature extraction shows that palmprint patterns can be well described by textures, and the texture energy measurement possesses a large variance between different classes while retaining a high level of compactness within the class. The coarse level classification by global texture features is effective and essential to reduce the number of samples for further processing at a fine level. The guided search for the best match based on interesting points further improves the system efficiency. The use of Hausdorff distance as the matching criterion can handle partially occluded palmprint patterns, which makes our system more robust. The experimental results provides the basis for the further development of a fully automated palmprint-based security system with high performance in terms of effectiveness, accuracy, robustness and efficiency.

Acknowledgement

Some of the work conducted by the authors is supported by both CRC/UGC grant (3-ZB31) and CERG grant (B-Q380), Hong Kong Government.

References

[1]   P. Baltscheffsky and P. Anderson, "The palmprint project: Automatic identity verification by hand geometry," Proc. 1986 International Carnahan Conference on Security Technology, Gothenburg, Sweden, pp. 229-234, 1986.

[2]   K.K. Benke, D.R. Skinner and C.J. Woodruff, "Convolution operators as a basis for objective correlates for texture perception," IEEE Trans. Syst., Man and Cybern.,} vol. 18, pp. 158-163, 1988.

[3]   G. Borgefors, "Hierarchical Chamfer matching: a parametric edge matching algorithm," IEEE Trans. Patt. Anal. Machine Intell., vol. 10, pp. 849-865, 1988.

[4]   L. Breiman, J.H. Friedman, R.A. Olshen and C.J. Stone, Classification and Regression Trees, Wadsworth International Group, Belmont, California, 1984.

[5]   J.B. Burns, A.R. Hanson and E.M. Riseman, "Extracting straight lines," IEEE Trans. Pattern Analysis and Machine Intelligence, vol. PAMI-8, pp. 425-455, 1986.

[6]   T.M. Caelli and D. Reye, "On the classification of image regions by color, texture and shape," Pattern Recognition, vol. 26, pp. 461-470, 1993.

[7]   N. Duta, A.K. Jain and K.V. Mardia, "Matching of palmprints," Paper under review in IEEE Trans. Pattern Anal. Machine Intell., currently downloadable from http://biometric.cse.msu.edu.

[8]   L.A. Goodman and W.H. Kruskal, "Measures of association for cross-classifications," J. Amer. Statist. Assoc., vol. 49, pp. 732-764, 1954.

[9]   R.M. Haralick, "Statistical and structural approaches to texture, Proc. IEEE, vol. 67, pp. 786-804, 1979.

[10]   D.P. Huttenlocher, G.A. Klanderman and W.J. Rucklidge, "Comparing images using the Hausdorff distance," IEEE Trans. Patt. Anal. Machine Intell., vol. 15, pp. 850-863, 1993.

[11]   A. Jain, R. Bolle and S. Pankanti, Biometrics: Personal Identification in Networked Society, Kluwer Academic Publishers, 1999.

[12]   K.I. Laws, Textured Image Segmentation, Ph.D. Thesis, University of Southern California, 1980.

[13]   D. Marr and E. Hildreth, "Theory of edge detection," Proc. R. Soc. Lond. B, vol.207, pp. 187-217, 1980.

[14]   B. Miller, "Vital signs of identity," IEEE Spectrum, vol. 32, no. 2, pp. 22-30, 1994.
[15]   K.S. Miller and B. Ross, "An introduction to fractional calculus and fractional differential equations, John Wiley and Sons, New York, 1994.
[16]   J.R. Quinlan, Learning Decision Trees, Technical Report 87.5, School of Computing Sciences, NSWIT, Sydney, Australia, 1987.
[17]   T. Reed and R. Meier, "Taking dermatoglyphic prints: A self-instruction manual," Supplement to the Newsletter of the American Dermatolglyphics Association, pp. 1-45, 1990.
[18]   S. Sestito and T.S. Dillon, Automated Knowledge Acquisition, Sydney, Prentice Hall, 1994.
[19]   J. You and H.A. Cohen, "Classification and segmentation of rotated and scaled textured images using texture `tuned' masks," Pattern Recognition, vol. 26, pp. 245 - 258, 1993.
[20]   J. You, S. Hungenahally and A. Sattar, "Fractional discrimination for texture image segmentation," Proc. 1997 IEEE International Conference on Image Processing (ICIP'97), vol. 1, Santa Barbara, CA. USA, Oct. 26-29, 1997, pp. 220-223.
[21]   J. You and P. Bhattacharya, ``A Wavelet-based coarse-to-fine image matching scheme in a parallel virtual machine environment, IEEE Trans. Image Processing, vol. 9, no. 9, pp. 1547-1559, 2000.
[22]   P.S. Wu and M. Li, "Pyramid edge detection based on stack filter," Pattern Recognition Letters, vol. 18, pp. 239-248, 1997.
[23]   D. Zhang and W. Shu, "Two novel characteristics in palmprint verification: datum point invariance and line feature matching," Pattern Recognition}, vol. 33, no. 4, pp. 691-702, 1999.
[24]   X. Zhou and T.S. Dillon, "A statistical-heuristic feature selection criterion for decision tree induction," IEEE Trans. Pattern Anal. Machine Intell., vol. PAMI-13, no. 8, pp. 834-841, 1991.

19 FINDING HUMAN FACES IN A FACE DATABASE

Dr. Kin-Man Lam, Kenneth

In this chapter, we introduce the technologies used in human face recognition. The different parts of a human face recognition system will be described, namely, locating human faces, extracting facial features, face recognition, and searching for faces from a database. Eigenface is a useful technique for face representation and recognition, which is reviewed in Section 19.2. The algorithms and techniques for detecting human faces in a complex background are described in Section 19.3. After detecting a human face, the techniques for extracting the respective facial features are presented in Section 19.4. Section 19.5 describes a method for human face recognition, which is efficient at searching a large face database.

19.1 Human Face Recognition: Overview and Challenges

Face recognition is an important social skill, and we acquire this skill in early childhood. We remember hundreds or even thousands of faces during our life, and can easily identify a face irrespective of the perspective, illumination, facial expression, age etc.. The development of computer based face recognition systems started more than 20 years ago, but significant advancements have been made in the last 10 years. However, computers still have difficulty and require a lot of computation to identify and locate human faces amongst a complex background, locate the important facial features, and identify detected faces from large databases. The face recognition algorithms must be able to identify a person under the following conditions:

- different expressions of the face;
- change of scale and lighting conditions;
- presence of details such as moustache, glasses, etc.;
- different perspectives and orientations of the face; and
- noisy images: presence of Gaussian noise, partially occluded images.

As the face images as captured and used by computers are 2-D projections of the 3-D human face, the above requirements make face recognition by computer a challenging task.

APPLICATIONS OF AUTOMATIC FACE RECOGNITION

Face recognition is a useful for law enforcement, and may be applied to a wide range of commercial applications. Although many possible approaches such as speech recognition, iris scanning, signature verification and fingerprint scanning can be used to identify a person, face recognition has the advantage of being non-intrusive, and requiring little co-operation from the person to be identified, to collect useful data for recognition. Face recognition applications [1-2] range from static matching of controlled format photographs in mugshot matching to real-time matching of surveillance video images. Different applications of face recognition will have to process face images of differing quality, resolution, etc. This poses a difficult technical challenge to the process of face recognition. Possible applications of face recognition to law enforcement and commercial situations include credit cards, driver's licenses, passports and personal identification, and mugshot matching; bank/store security; crowd surveillance; expert identification; witness face reconstruction; electronic mugshot books; electronic line-ups; reconstruction of faces from remains; and computerized aging.

Face recognition technology can also be applied to man-machine interfaces that control or communicate with a computer via face position and facial gesturing. Computer systems have been developed [3] to track the eyes and the nose of a subject and to compute the direction of the face. Face direction and movement can then be used to control the cursor and select a desired menu item on screen. The system can enhance both the work and the living environment of disabled persons. Another application related to face recognition is the image compression of facial images. In video phone or video conferences, the human face is the major component of the communicated image, and the capability of video compression algorithms for dealing with the human face will affect the video quality significantly. This kind of coding technique is called model-based image coding, MBIC [4]. For this application, the accuracy and efficiency of detecting and extracting the facial features have a direct effect on the performance of the coding system.

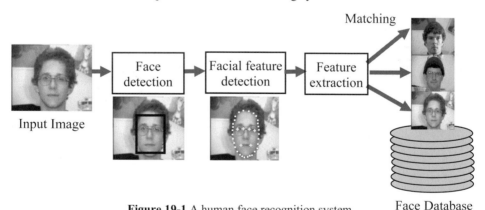

Figure 19-1 A human face recognition system.

STRUCTURE OF A HUMAN FACE RECOGNITION SYSTEM
An automatic human face recognition system can be divided into two parts: facial feature detection and face recognition. The facial feature detection includes the procedures to detect the existence and position of the human face, and then to extract the important facial features. This is an important step in the sense that the accuracy in detecting the facial features will significantly affect the performance of the next step - face recognition. Usually, the two faces to be compared will be normalized to a specific size and aligned to each other based on the position of the facial features. The input face will be compared to the faces stored in a face database. This will be a

time-consuming process if the size of the database is very large. Therefore, to design a practical and reliable automatic face recognition system, the algorithms used to detect human faces either from an image file or a video camera should be fast, accurate, and insensitive to lighting conditions, while the features used to represent human faces should be simple and effective. Figure 19-1 shows the structure of an automatic human face recognition system.

In an automatic human face recognition system, a face region is extracted and then normalized based on the position of the two eyes. The normalized human face is aligned with the human faces in a database, and they are then compared. In other words, the accuracy of face detection and facial feature extraction will affect the performance of an automatic human face recognition system significantly.

19.2 Eigenface

Eigenfaces [5-6] are the principal components of a number of human face images, so they have a good representation of human faces. In this chapter, the eigenfaces are used for both face detection and face recognition. A face image is represented by a 2D pixel array $\mathbf{F}(x, y)$, where $0 \le x \le M$ and $0 \le y \le N$, and $M \times N$ represents the resolution of the image. The values of $\mathbf{F}(x, y)$ represent its gray-level intensities. To obtain an efficient representation of face images, principal component analysis can be applied on a set of training face images represented as $\mathbf{F}_1, \mathbf{F}_2, \mathbf{F}_{3,...}, \mathbf{F}_k$, where k is the number of training face images. A face vector Γ_i, $i = 1, 2, \ldots, k$, is formed by concatenating one row of pixels followed by the next row of $\mathbf{F}(x, y)$ in an order from top to bottom. The dimension of each face vector Γ is therefore $1 \times NM$. The average face, Ψ, and difference face, Φ, are defined as follows:

$$\Psi = \frac{1}{k} \sum_{n=1}^{k} \Gamma_n \quad \text{and} \quad \Phi_i = \Gamma_i - \Psi \tag{19-1}$$

Then, the correlation matrix $\mathbf{C}$, which is of dimension $NM \times NM$, is computed as follows.

$$\mathbf{C} = \Phi \cdot \Phi^{\mathrm{T}} \tag{19-2}$$

The eigenvectors of this matrix $\mathbf{C}$ represent the principal components of the training face images, which exhibit a large magnitude at the important facial features such as the eyes, mouth, eyebrow, etc. As $\mathbf{C}$ has a dimension of $(NM)^2$, determining the $(NM)^2$ eigenvectors and eigenvalues is an intractable task. However, as k is much less than the dimension of $\mathbf{C}$, $\mathbf{C}$ is singular and cannot be an order greater than k. The analysis is simplified by considering the eigenvectors e_i of $\Phi^{\mathrm{T}}\Phi$ such that

$$\Phi^{\mathrm{T}}\Phi e_i = \lambda_i e_i \tag{19-3}$$

where λ_i is the eigenvalue of the eigenvector e_i. Premultiplying both sides by Φ,

$$\Phi\Phi^{\mathrm{T}}\Phi e_i = \lambda_i \Phi e_i \tag{19-4}$$

From which, Φe_i are the eigenvectors of $\mathbf{C} = \Phi\Phi^{\mathrm{T}}$. These vectors determine the linear combination of the k training face images to form the eigenvectors e_i' as follows:

$$e_i' = \sum_{j=1}^{k} e_{ij} \Phi_j \tag{19-5}$$

With this analysis, the calculations are greatly reduced from the order of the number of pixels in the image to the order of the number of images in the training set. The associated eigenvalues are used to rank the eigenvectors according to their usefulness in characterising the variation among the images. Figure 19-2 shows the first eight eigenfaces of the faces in the MIT face database according to the magnitude of their eigenvalues λ. It can be seen that these eigenfaces have a significant magnitude at the important facial features and at the face contour.

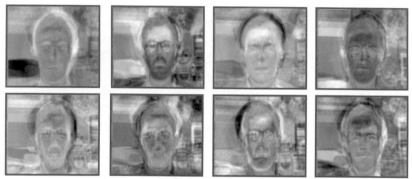

Figure 19-2 The first eight eigenfaces of MIT face database.

19.3   Human Face Detection in a Complex Background

Human face detection is a challenging task because the human face is highly variable. The detection performance may be affected by the presence of glasses, different skin colors, genders, facial hairs, facial expressions, etc. Furthermore, a complex background will make this task even more complicated. Different approaches [7] have been devised for the detection of human faces in gray-level images. These include shape analysis [8-9], template-based [10], feature-based [11-14], neural network based [15], example-based [16], and more often, a combination of all of these. The computational complexity of these methods is usually too high for real-time applications.

In order to improve detection time and detection performance, the color-based approach [17-20] has become a new direction for human face segmentation. The basic idea of this approach is that although skin color differs from person to person and race to race, it is distributed in a very small region in the color space. It implies that the possible face-like region can be segmented by using simple thresholding techniques. As a result, the search space in an unconstrained image is greatly reduced. For example, in [17], the skin-like regions are extracted by using both the normalized RGB color model and HSV color model, the possible face regions are then verified by means of wavelet decomposition. In [18], the face-like regions are segmented by using the hue and saturation color information, then the face region is determined by detecting whether the shape of the extracted region is elliptical or not. In [19-20], the chrominance information in YCrCb color space is used for the segmentation of skin-like regions. These methods can achieve a high detection rate if the images are under good lighting conditions. However, the human face is a 3D object, so the reflection of intense light from and shadow over the face regions will result in some skin pixels not being located. As a result, the performance of these segmentation methods will be degraded in real situations.

In this section, a robust method that considers the lighting effect on the color of human face for segmenting the skin-like regions in an image is described. This is particularly important for live detection of human faces using a video camera as we cannot control the lighting conditions.

COLOR SPACES

The selection of a color model to represent the human skin color is important for face detection in a color image. Examples of commonly used color spaces include RGB, hue-saturation-value (HSV), YCrCb, etc. RGB color space stands for the three primary colors: red, green, and blue, and the range of each component is from 0 to 255. The hue-saturation-value (HSV) model is commonly recognized to be similar to the human perception of color. The hue (H) is a measure of the spectral composition of a color and is represented as an angle from 0 to 360°, while saturation (S) refers to purity of the color and its value ranges from 0 to 1. Value (V) refers to the darkness of a color, which ranges also from 0 to 1. The YCrCb is a hardware-oriented color model. Y represents the luminance component, while Cr and Cb represent the chrominance components of an image. For real-time applications, the effectiveness of a color model for segmentation must be considered. In RGB color space, brightness and color information are coupled together, so it is not suitable for color segmentation under unknown lighting conditions. The performance of using HSV color space to model skin color is good, but it may not be suitable for real-time applications because the output format used by digital cameras or capturing equipment is usually either YCrCb or RGB. Therefore, in order to avoid the extra computation required in conversion, the face detection method described in this chapter uses the YCrCb color model.

SEGMENTATION OF FACE-LIKE REGIONS

The performance of digital cameras and the characteristics of human skin color are important considerations for a color-based face detection approach. However, the response of camera sensors is non-linear to the spectral power input, and the light reflected or absorbed by human skin is also non-linear to the illumination [21]. This means that the measured intensity on each color channel of skin color varies under different lighting conditions. As a result, the traditional thresholding technique for skin-color extraction may not work under poor or intense lighting conditions. To investigate the distributions of skin color under different lighting conditions, more than 220,000 face skin pixels are used. The face images used are captured from a digital video camera or downloaded from the Internet. It is found that the distributions of the Cb and Cr components change with the illumination over a face region. Therefore, the skin color distribution is divided into six groups of different luminance, as shown in Figure 19-3. The color space for skin color with the luminance value ranging from 113 to 190 has a larger area than that under intense or poor lighting conditions. Furthermore, the distributions of skin color on the Cr-Cb plane under different illumination are not the same. Nevertheless, the distributions change slightly when luminance is between 113 and 190. This implies that the distributions of skin color are affected by the luminance values. Based on the distribution of Cr and Cb in each of the six groups, six skin-color chromatic maps, denoted as Map_m, $m =0$, ..., 5, are generated. The x-axis and y-axis of the chromatic maps represent the intensity levels of Cr and Cb, respectively. Based on these chromatic maps, a pixel in an image will be identified as a possible skin pixel as follows:

$$P_1(x, y) = \begin{cases} 1, & \text{if } [\ Map_m(a,b) = 1] \\ 0, & \text{otherwise,} \end{cases} \qquad (19\text{-}6)$$

$$\text{where } m = \begin{cases} 0, & \text{if } [Y(x, y) \leq 87] \\ 1, & \text{if } [112 \leq Y(x, y) \leq 88] \\ 2, & \text{if } [113 \leq Y(x, y) \leq 190] \\ 3, & \text{if } [191 \leq Y(x, y) \leq 203] \\ 4, & \text{if } [204 \leq Y(x, y) \leq 219] \\ 5, & \text{if } [Y(x, y) \geq 220] \end{cases}$$

a, b and Y are the intensity levels of the Cr, Cb, and the luminance value of the input image at location (x,y), while m is the number of chromatic maps. The results of skin color segmentation proposed in [17] and [19], and based on this method are illustrated in Figure 19-4(b), (c), and (d), respectively. The black color in the figure represents the non-skin regions, while others are the segmented possible face regions. The result of skin-like region detection using the HSV color space is illustrated in Figure 19-4(b), which works well over the strong light regions but fails to detect those skin regions under shadow. Figure 19-4(c) shows the results using a simple thresholding technique with the YCrCb color space, which does not work properly for those face regions under intense lighting conditions. In contrast, using the 6 chromatic maps can detect most of the face regions irrespective of the lighting conditions.

In order to improve the reliability level for face detection, the noise and small holes in the segmented results due to the facial features such as the eye and mouth regions are considered. Morphological operations (dilation and erosion) [22] are applied to the segmentation output $P_1(x, y)$ for removing the noise and holes. The purpose of dilation is to fill in small holes in the facial areas, while erosion is to remove small objects in the background. The equation is defined as follows:

$$P_2(x, y) = P_1(x, y) \bullet B \qquad (19\text{-}7)$$

where $\bullet$ represents the closing operation, and B is the structuring element. The closing operation involves the application of dilation followed by erosion. Therefore, small holes in the segmentation output $P_1(x, y)$ will be filled up after this closing operation. The structure element B is a 5×5 square. The input image is then divided into a number of blocks of 8×8 pixels. Since the human face is a 3D object, the luminance throughout the facial region is non-uniform and the background region tends to have a more even distribution of brightness. Therefore, the standard deviation, denoted as σ, of the Y component of each block is calculated to verify whether it belongs to a face region or not. If a block has its σ below 1.5, it will be considered as a background region. The output of this stage, denoted as P_{face}, is defined as below:

$$P_{face}(x, y) = \begin{cases} 1, & \text{if } [\sigma(u, v) > 1.5] \\ 0, & \text{otherwise,} \end{cases} \qquad (19\text{-}8)$$

where $\sigma(u,v)$ is the standard deviation of the 8×8 image block at location (x,y). The segmentation result of Figure 19-4(a) is illustrated in Figure 19-4(e). With this scheme, the search space for face detection can be greatly reduced. The segmented results are then input to the second stage for the detection of eye regions.

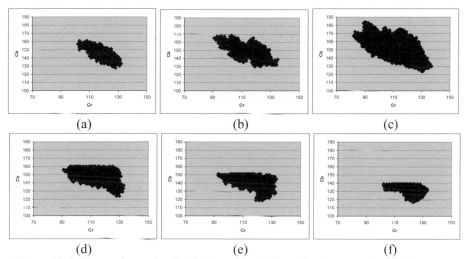

(a)   (b)   (c)

(d)   (e)   (f)

Figure 19-3 Human face color distribution under different luminance values: (a) Lower or equal to 87, (b) between 88 and 112, (c) between 113 and 190, (d) between 191 and 203, and (e) between 204 and 219, and (f) higher or equal to 220.

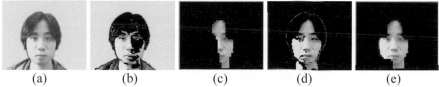

(a)   (b)   (c)   (d)   (e)

Figure 19-4 Color segmentation results: (a) Original image, (b) results based on HSV color space [17], (c) results based on YCrCb color space with simple thresholding [19], (d) results of using multiple chromatic maps, and (e) results of (d) after removing the noise and holes.

Possible Eye Candidate Detection in Color Images

In the second stage, the possible eye candidates will be detected and then used to form the possible face regions in an input image. With color images, the search space for possible eye candidates can be restricted to skin color regions. The Y component is used for the detection of the possible eye regions within the skin color regions. Possible eye candidates are located by detecting valley points within the segmented regions from stage one. Based on the low gray-level intensity characteristics of the iris, a pixel at (x,y) will be declared to be a possible eye candidate if P_{eye} is equal to one.

$$P_{eye}(x,y) = \begin{cases} 1, & \text{if } [P_{face}(x,y) = 1] \cap [F(x,y) < T_{eye}] \cap [\Phi_v(x,y) > T_v] \\ 0, & \text{otherwise} \end{cases} \quad (19\text{-}9)$$

where $F(x,y)$ represents the gray level intensity of an input image at location (x,y), Φ_v is the valley field, T_{eye} and T_v are pre-defined thresholds. The valley field, Φ_v, can be extracted using morphological operators. The equation for valley field extraction is

$$\Phi_v = F(x,y) \bullet B - F(x,y) \quad (19\text{-}10)$$

where $F(x,y)$ is the image and B is the structuring element. The valley image is obtained by performing a closing operation, which is then subtracted by the original image. Figure 19-5(a) and (b) show an input image from stage one and its corresponding possible eye regions, respectively.

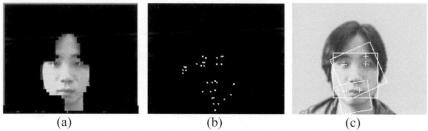

(a)   (b)   (c)

Figure 19-5 Eye candidate detection: (a) extracted face-like region from stage one, (b) possible eye candidates, and (c) possible face candidates.

Since the size of a human face is proportional to the distance between its two eyes, a possible face region containing the eyebrows, eyes, nose, and mouth can be formed based on this relationship. A square block is used to represent possible face candidates. The size of the square block is determined by means of the head model in [23]. Based on the head model and the location of the possible eye pairs, a number of possible face regions at different locations, of different sizes, and orientations are generated. A possible face region is formed if the following criteria are satisfied:

$$T_{\theta 1} < \theta < T_{\theta 2},$$
$$T_{size1} < F_{size} < T_{size2}, \quad (19\text{-}11)$$
$$Total_{seg} > T_{seg},$$

where θ represents the angle of a line passing through a possible eye pairs, F_{size} is the face length, and $Total_{seg}$ is the sum of P_{face} over the total number of pixels inside the selected region. The range of the orientation angle of a human head is assumed to be between $T_{\theta 1}$ and $T_{\theta 2}$, while the face size is assumed to vary from T_{size1} to T_{size2}. The threshold T_{seg} is used to make sure that most parts of the selected region are of skin-color pixels. Figure 19-5(c) shows those possible face regions satisfying Equation 19-11. In order to improve the detection reliability and accuracy, each of the selected possible face regions is then processed to compensate for non-uniform lighting on the face region. After the normalization process, the image will be passed to the final stage for further verification.

Face verification using Eigenmask

To determine whether a normalized face candidate is a face or not, the similarity between the face candidate and a face template is measured. In our approach, the face template is a gray-level image and is obtained by calculating the average of a set of pre-processed training face images. Let the face template, denoted as $F_{template}$, be a two-dimensional $N \times N$ array of 8-bit intensity levels. The $F_{template}$ is calculated as follows:

$$F_{template}(x, y) = \frac{1}{n} \sum_{i=0}^{n} F_i(x, y),\qquad(19\text{-}12)$$

where n and $F_i(x,y)$ are the number of training face images and ith training face image, respectively. The training set contains face images of different races, genders, ages, with and without glasses and a moustache. The face template derived from the training images is shown in Figure 19-6(a).

The distance between a possible face region and the face template can be measured by means of the Euclidean distance. However, the eyes, nose and mouth regions are important features for a human face. The detection performance will be improved if the distance measure is weighted according to the importance of the corresponding regions. An eigenmask, denoted as E_{mask}, is used as a weighting function in measuring the distance. The eigenmask is generated from the first eigenface of the training face images. The eigenmask used in the distance measure is generated based on the absolute value of the first eigenvector. The following equation shows the generation of the normalized eigenmask:

$$E_{mask}(x, y) = \begin{cases} 255 & \text{if } [m(x, y) > 255] \\ m(x, y) & \text{otherwise,} \end{cases}\qquad(19\text{-}13)$$

where

$$m(x, y) = \frac{255}{\max(e_{nor}) - \min(e_{nor})} \cdot e_{nor}(x, y) + V,$$

$$e_{nor}(x, y) = |e_1(x, y)|,$$

V is a constant and e_1 is the first eigenvector. The dynamic range is stretched and the minimum value of the mask is controlled by the constant V as every part of the face region should be considered in calculating the distance. The constant V is obtained by measuring the number of points that $m(x,y)$ exceeds the peak value, i.e. 255. The value of V used in generating the mask is determined by increasing its magnitude until 5% of the points have their magnitudes over 255. Figure 19-6(b) shows the eignmask generated and used in the distance measure. As shown in the figure, the magnitudes of the mask at important features, such as the eyebrows, eyes, nose, and mouth, are larger.

(a)   (b)

Figure 19-6 (a) The face template, and (b) The eigenmask.

The use of the eigenmask can increase the distance when comparing a non-face candidate with the face template, and so reduce false alarms. The distance measure between an input candidate and the template is given as follows:

$$\varepsilon = \sum_{y=0}^{N}\sum_{x=0}^{N} E_{mask}(x,y) \cdot \left(F(x,y) - F_{template}(x,y)\right)^2, \qquad (19\text{-}14)$$

where F is the input candidate. As a result, a true face will be declared if the value of ε is smaller than a threshold T_{face}. For overlapping regions, the one with the lowest value of ε is chosen as the face region.

Detection Results

The detection performance of this scheme was tested using 227 color images. Some of the images were captured by a digital camera and some of them were downloaded from the Internet. These test images consist of both indoor and outdoor scenes under different lighting conditions. The image size is 176×144 pixels and the face size varies from 30×30 to 80×80 pixels. The detection performance of our approach with and without using the eigenmask is shown in Table 19-1. A face is considered to be fairly detected if the detected position of the two eyes is more than 2 and less than 5 pixels from the correct eye position.

Detection method	Correctly detected	Fairly detected	False alarm	Missed faces	Detection rate (%)
Eigenmask enabled	212	4	7	15	93.39
Eigenmask disabled	198	10	24	29	87.22

Table 19-1 Detection performance.

Table 19-1 shows that the method can achieve an overall hit rate of 93.39% when the eigenmask was considered in calculating the distance. Without using the eigenmask, the overall hit rate was reduced to 87.22%. Also, the numbers of false alarms and missed faces increased from 11 to 34 and 15 to 29, respectively. This means that the eigenmask can greatly improve detection performance. Some of the detection results are shown in Figure 19-7. The experiments were conducted on a Pentium III 500 MHz computer. The average processing time for locating faces in a picture was about 1.6 s.

Figure 19-7 Some face detection results.

19.4 Facial Feature Detection and Extraction

In the previous section, a method based on skin color and eigenmask was described for human face detection in a complex background. In this method, the position of the two eyes of a face is detected, and enabling the face region to be determined. However, to ensure accuracy in face recognition, the position of the two eyes must be precise. In addition, for many methods, other feature points such as the face contour, position of the mouth and nose, etc. are required. In this section, a brief overview of the techniques for extracting the face contour and the important facial features will be described.

EXTRACTING THE FACE CONTOUR
To locate a boundary, classical edge detection algorithms provide local evidence for edges, but it is very hard to organise this low-level information into a sensible contour for an object. Snake, which is an active contour model, can be used to locate the head/face boundary. In this section, an efficient algorithm for snake will be described to locate a boundary.

An Active Contour Model - Snake
Snake [24] is an active contour model for representing image contours. It is an energy-minimizing spline, which can be operated under the influence of internal contour forces, image forces, and external constraint forces. A snake is represented as a parametric curve $\mathbf{v}(s) = [x(s), y(s)]$, where the arc length s is a parameter. The snake starts at the point where $s = 0$ and ends where $s = 1$. An energy functional of the snake is defined as

$$E_{snake}^{*} = \int_{0}^{1} E_{internal}[\mathbf{v}(s)] + E_{image}[\mathbf{v}(s)] + E_{constraint}[\mathbf{v}(s)]ds \qquad (19\text{-}15)$$

where $E_{internal}$ represents the internal energy of the contour due to bending or discontinuities, E_{image} refers to the image forces, and $E_{constraint}$ is the external constraint forces. The location of the snake corresponds to the local minima of the energy functional.

In order to locate the boundary of an object in an image, an accurate estimation of the initial position and the number of points of the snake is necessary. As the approximate position of the face has been identified as described in Section 19.3, the initial face boundary can be set easily. A snake is represented by a number of points which are labeled in a sequential order. The initial snake is set slightly larger and outside the detected face region, as shown in Figure 19-8(a). The initial snake is then drawn by image forces to the object's boundary. However, the image force is effective only when the snake is placed near the edge. To attract the snake from a fairly large distance away from the edges, the image force is blurred by a 2-D Gaussian smoothing operator, $G(x, y)$, with a large window size. The edge image and the smoothed edge image are illustrated in Figure 19-8(b) and (c), respectively.

$$G(x, y) = \frac{1}{2\pi\sigma^2} \exp\left(-\frac{x^2 + y^2}{2\sigma^2}\right) \tag{19-16}$$

This allows the snake to move to the blurry energy-functional and the blurring is then slowly reduced when the snake is close to the image.

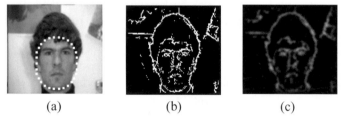

| (a) | (b) | (c) |

Figure 19-8 (a) The initial snake, (b) the edge image, and (c) the smoothed edge image.

The energy-functional to be minimised for the snake points is

$$E = \int \alpha(s)E_{continuity} + \beta(s)E_{curvature} + \gamma(s)E_{image} \; ds \tag{19-17}$$

The optimal position of the snake points is where the value of this energy-functional is a minimum. An efficient method for this purpose is the greedy algorithm [25-26], which is a fast iteration method. The energy function is computed at v_i and at each of its eight neighbours as illustrated in Figure 19-9. The points before and after it on the contour are used in computing the continuity constraints. The location having the smallest value is chosen as the new position of v_i.

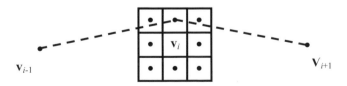

Figure 19-9 Neighbours searched in the greedy algorithm.

Continuity force, $E_{continuity}$:

A term which encourages even spacing of points and satisfies the first-order continuity is used. The difference between the average distance between points, d, and the distance between the two points under consideration: $|d - |\mathbf{v}_I - \mathbf{v}_{i-1}||$ is used. Thus, points having a distance near the average will have the minimum value. The value is normalised by dividing it by the largest value in the neighbourhood to which the point may move. At the end of each iteration a new value of d is computed.

Curvature force, $E_{curvature}$:
Since the formulation of the continuity term causes the points to be relatively evenly spaced, $|\mathbf{v}_{i-1} - 2\mathbf{v}_i + \mathbf{v}_{i+1}|^2$ gives a reasonable and quick estimate of curvature. This term is also normalised by the largest value in the neighbourhood.

Image force, E_{image} :
E_{image} is the gradient magnitude which is also to be normalised. For a point with magnitude *mag*, the image force is defined as $(min - mag)/(max - min)$, where *max* and *min* are the maximum and minimum gradient in each neighbourhood.

At the end of each iteration, the curvature at each point on the new contour is determined. If the value is larger than a threshold, β_i is set to zero for the next iteration. This allows for the formation of corners on the contour.

Terminating Criteria
Possible stopping criteria for the iteration process is when the number of points moved is less than a threshold or the number of iterations has reached a certain quantity. It is observed that, sometimes, the points of the snake shift along the boundary, and the contour only moves very slightly. However, as most of the points are moved, the iteration process will be continued. A better terminating criterion is the contour area criterion (CA-criterion) [27], which makes use of the normalized total area to determine the convergence of the process.

Suppose that A^k is the area of the snake at the kth iteration. Then, the CA-criterion, η^k, is defined as follows:

$$\eta^k = \left| \frac{A^k - A^{k-1}}{A^k} \right| \tag{19-18}$$

The area of the snake A^k can be calculated by using the equation as follows:

$$A^k = \frac{1}{2} \sum_{i=1}^{n} \begin{vmatrix} x_i^k & y_i^k \\ x_{i+1}^k & y_{i+1}^k \end{vmatrix} \tag{19-19}$$

where $\left(x_{n+1}^k, y_{n+1}^k \right)$ is equal to $\left(x_1^k, y_1^k \right)$ for a closed snake and n is the total number of snake points. The value of η^k converges quickly and is close to zero when the object boundary is located by the snake. Figure 19-10 shows the rate of convergence with the CA-criterion. The CA-criterion exhibits a smooth and fast convergence, and also has good stability in convergence.

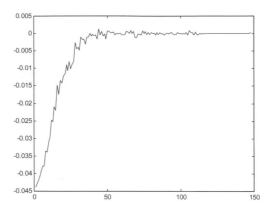

Figure 19-10 Plot of CA-criterion η^k against the iteration.

Using the relations of the area and the perimeter of a closed contour, an inequality is expressed as follows:

$$\frac{Area}{Perimeter^2} \leq \frac{1}{4\pi} \text{ or } Area \leq \frac{1}{4\pi} Perimeter^2 \tag{19-20}$$

It can be observed that the area of the object is bounded provided that its perimeter is convergent. If this is the case, then the area of the object will converge faster. Therefore, the use of the area as a stopping criterion can give a better convergence. Figure 19-11 shows the final contours for different head-and-shoulders images.

Figure 19-11 Final face contours.

DEFORMABLE TEMPLATES

A facial feature extractor can detect the locations of the different facial features first, then the exact features are extracted. Usually, the feature extractor can automatically extract the eyebrows, eyes, nostrils and mouth from a human face. Among the facial features, it has been found that the upper part of the face is more useful for face recognition than the lower part. A lot of research has been conducted to extract the eye features. To locate the different facial features, various methods rely on the fact that almost every face has bilateral symmetry, with two eyebrows, two eyes, one nose and one mouth with a very similar layout.

The deformable template [28] is a common method used in extracting the contour of the eye and mouth. Conventional edge detectors are unable to find the boundary of the eye reliably because the local evidence of the edges cannot be organised into a sensible global percept. Furthermore, due to the low gray level contrast around the eye and mouth boundaries, the edges of these two facial features may not be detectable. The method of the deformable templates makes use of global information, and hence improves the reliability of locating the contour. This template is specified by a set of parameters, which allows *a priori* knowledge of the expected shape of the features. This knowledge of the shape, in turn, guides the detection process.

Figure 19-12 illustrates the eye template and the corresponding parameters used. In order to determine the parameters of the templates, appropriate energy functions are defined using the information about the edges, valleys, peaks and the intensity of the image. The templates are flexible enough to be able to change their size and other parameter values, so as to minimise the energy function and match themselves to the input feature. The minimisation is done by the steepest descent of the energy function in the parameter space. The final values of these parameters can be used to describe the features. The disadvantages of the template are that it involves a large number of parameters whose characteristics are different to each other and the optimization process is slow. The template must be started at or below the eye; if it is started above the eye, the valley force from the eyebrows may cause problems. Furthermore, during the optimisation process, changing the weights of the energy terms from one epoch to the next is critical to the successful extraction of the facial features.

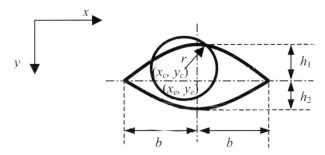

Figure 19-12 The deformable eye template.

An Efficient Eye Model and its Representation

In this section, the shape of the eye is extracted by a new scheme [29] based on the ideas of the snake and the deformable template. In this new scheme, the two boundaries of the eye are represented by four points, while the iris is described by one point and one parameter to represent its centre and radius, respectively. This representation is in a form similar to snakes, but the respective energy-functionals to be minimised are based on some new energy terms and those terms used by the deformable template. This new scheme, which is matched to the image model, combines both the flexibility of the snake and the ability of the deformable template to include *a priori* knowledge concerning the expected shape of the facial feature to guide the detection process. A fast algorithm based on the greedy algorithm for active contour modelling is presented for the optimisation step. Furthermore, the weights of the energy terms need not change in the course of optimisation.

As shown in Figure 19-13, the upper and lower boundaries together are represented by four points, and the iris is described by one point, as its position, and one parameter, as its radius. Here, three points form a parabola which is used to represent a boundary. For example, the upper boundary of the eye is represented by $\mathbf{X}_0$, $\mathbf{X}_1$, and $\mathbf{X}_2$. One advantage of this new model is that it matches the image model. The image model is composed of pixels, and the points of the new model move along the pixels during the matching. When a point has moved to the desired position, it will not, or only slightly, be disturbed by the movement of the other points. In the case of the deformable template, when one of the parameters is changed, some of the other parameters must also be adjusted in the optimisation process. Other important advantages of this new method are that the weights of the energy terms do not need to change and that it is possible to use a fast algorithm for the optimisation process.

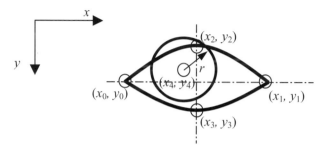

Figure 19-13 New eye model.

This eye model has a representation similar to snake in terms of its form, but the energy functions to be minimised are similar to those used in the deformable template. The peak forces Ψ_p, which are extracted by a morphological operator as in the deformable template, together with the valley forces Ψ_v and edge forces Ψ_e are blurred to give Φ_p, Φ_v and Φ_e, respectively. This facilitates the initial points of the models being attracted to their correct positions on the eye.

In order to extract the eye effectively, a number of energy functions are defined that capture all the relevant eye features. For different points of the eye model, different sets of the energy functions are used in the minimisation process. The energy functions are defined as follows:

1. The corner potential, E_c, is a measure of the fitness of a point to be an eye corner. The potential is defined as being equal to the region dissimilarity of the corner at the point:

$$E_c(\mathbf{X}_i) = -\left|D(R_1, R_2)\right|$$

where $\mathbf{X}_i$ is the corner position, R_1 and R_2 are the two regions in the window separated by the two parabolae $para(\mathbf{X}_0, \mathbf{X}_1, \mathbf{X}_2)$ and $para(\mathbf{X}_0, \mathbf{X}_1, \mathbf{X}_3)$.

2. The valley energy term, E_v, is given as the integral of Φ_v over the interior of the circle divided by the area of the circle:

$$E_v = -\frac{1}{|R_c|}\int_{R_c}\Phi_v(x)dA$$

where R_c represents a circle with centre $\mathbf{X}_i$ and radius r, and $|R_c|$ is the area of the circle. This term measures the size and position correlation of the iris to the valley field of the image.

3. The edge potential, E_e, is a measure of the edge intensities along a parabola:

$$E_e(\mathbf{X}_i) = -\frac{1}{L_{para(\mathbf{X}_0,\mathbf{X}_1,\mathbf{X}_2)}} \int_{para(\mathbf{X}_0,\mathbf{X}_1,\mathbf{X}_2)} \Phi_e(x)ds$$

4. The edge potential of the iris, Ei_e, is given by the integral of Φ_e over the boundary of the circle divided by its length:

$$Ei_e = -\frac{1}{|\partial R_c|} \int_{\partial R_c} \Phi_e(x)ds$$

where ∂R_c corresponds to the boundary of the circle, while $|\partial R_c|$ represents the length of the boundaries. This term is used for matching the template to the edge of the iris.

5. The peak energy term, E_p, is defined as the integral of Φ_p over the area between the circle and the two parabolae divided by the corresponding areas:

$$E_p = -\frac{1}{|R_w|} \int_{R_w} \Phi_p(x)dA$$

where R_w represents the whites of an eye and its area is given by $|R_w|$.

6. The image energy term E_i is defined as the integral of Φ_i over the circle and over the area between the circle and the two parabolae divided by the corresponding areas:

$$E_i = \frac{1}{|R_c|} \int_{R_c} \Phi_i(x)dA - \frac{1}{|R_w|} \int_{R_w} \Phi_i(x)dA$$

This term attempts to minimise the total brightness inside the circle while maximising it between the circle and the parabolae.

7. The prior energy terms are defined in order to control the shape of the eye model:

$$E_1 = \left\| \mathbf{X}_4 - \frac{\mathbf{X}_0 + \mathbf{X}_1}{2} \right\|$$

$$E_2 = (b - 2r)^2$$

$$E_3 = (b - 2a)^2 + (a - 2c)^2$$

where $\qquad b = \frac{1}{2} \|\mathbf{X}_1 - \mathbf{X}_0\|$

a, c = the distance of points, $\mathbf{X}_2$ and $\mathbf{X}3$, to the line joining $\mathbf{X}_0$ and $\mathbf{X}_1$, respectively.

A greedy algorithm can also be used to minimise the total energy function of the feature model in the course of optimisation. The total potential for the eye is

$$E_t = F(E_c, E_e, Ei_e, E_v, E_p, E_i, E_1, E_2, E_3)$$ (19-21)

$$= k_1 \frac{E_c}{E_{c\_max}} + k_2 \frac{E_e}{E_{e\_max}} + k_3 \frac{Ei_e}{Ei_{e\_max}} + k_4 \frac{E_v}{E_{v\_max}} + k_5 \frac{E_p}{E_{p\_max}} +$$

$$k_6 \frac{E_i}{E_{i\_max}} + k_7 \frac{E_1}{E_{1\_max}} + k_8 \frac{E_2}{E_{2\_max}} + k_9 \frac{E_3}{E_{3\_max}}$$

In each calculation, the maximum value of each component of the potentials is found and used to normalise the corresponding terms. This facilitates the selection of the weighting factors k_1 to k_9. In the course of optimisation, the points and the parameter of the feature models relate to different potentials. For the eye template, the potentials related to each of the points, X_0 to X_4, and the parameter, r, are shown as follows:

$$E_t(X_0) = F(E_c(X_0), E_e(X_2)+E_e(X_3), E_p, E_i, E_1, E_2, E_3)$$ (19-22)
$$E_t(X_1) = F(E_c(X_1), E_e(X_2)+E_e(X_3), E_p, E_i, E_1, E_2, E_3)$$
$$E_t(X_2) = F(E_c(X_0), E_e(X_1)+E_e(X_2), Ei_e, E_v, E_p, E_i, E_3)$$
$$E_t(X_3) = F(E_c(X_0), E_e(X_1)+E_e(X_3), Ei_e, E_v, E_p, E_i, E_3)$$
$$E_t(X_4, r) = F(E_v, Ei_e, E_i, E_p, E_1, E_2)$$

In each iteration, a point moves to a position or a parameter changes resulting in a minimum potential. The iteration stops if none of the points of the feature model moves and the parameter does not change, or if the number of iterations is larger than a pre-specified threshold. This scheme allows the points of the feature model to move to the desired positions very quickly. Figure 19-14 shows the initial templates and final templates based on this scheme.

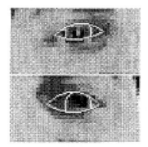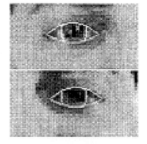

Figure 19-14 The initial and final models of the eye. The initial models of the eyes are shown on the left while the final models are illustrated on the right.

19.5 Searching Human Faces From a Face Database

The problem of human face recognition has been studied for more than thirty years. A practical human face recognition technique should not only be robust to lighting conditions, facial expressions, perspective variations, noise, etc., it should also be efficient and accurate. That means human faces similar to a query face input can be searched from a large face database quickly and accurately. In this section, the eigenface technique and the use of Hausdorff distances will be described for searching face images from a face database.

FACE RECOGNITION BASED ON EIGENFACES
The idea of this technique is to project face images onto a feature space that spans the significant variations among known face images. This is done by projecting a face image onto a set of eigenfaces, e_i. For recognition or identification, a smaller number k' ($< k$) of eigenfaces are sufficient. Each face $\mathbf{F}$ is transformed into its eigenface components by

$$\omega_k = e_i^{\mathrm{T}}(\mathbf{F} - \mathbf{\Psi}), \quad \text{for } i = 1, \ldots, k' \tag{19-23}$$

The weights form a vector $\mathbf{\Omega}^{\mathrm{T}} = [\omega_1, \omega_2, \ldots, \omega_{k'}]$, which is used to represent the face. Each face in a database is represented by the its corresponding weight vector $\mathbf{\Omega}_j$, where j represents its class. With a query image, its corresponding weight vector, $\mathbf{\Omega}_q$, is computed. The following Euclidean distance is then calculated for each class. The input face is classified to be class m if ε_m is a minimum among the classes.

$$\varepsilon_i = \|\mathbf{\Omega} - \mathbf{\Omega}_i\|^2 \tag{19-24}$$

Eigenface is an optimal technique for representing a human face, but not optimal for recognising a face. This is because eigenfaces are the principal components of the face images in a training set. All human faces also have a similar structure, so the eigenfaces are efficient for representing the faces. However, to recognise or identify a face, the fine differences between the faces are of the utmost important.

FACE RECOGNITION USING HAUSDORFF DISTANCE MEASURES
Psychological studies have revealed that humans can categorize a human face at a glance and recognize the line drawings of objects as quickly and almost as accurately as photographs [30]. It is suggested that the edge-like retinal images of faces can provide useful information for face identification at the level of early vision. Furthermore, when the template matching method [31] is used, edge-like maps can achieve a better performance than the feature-based techniques. The advantage of using edges as image features is that they can provide robustness to illumination change and simplicity of presentation.

In this section, we describe the use of the Hausdorff distance for face recognition. For a practical face recognition approach, the computation and the memory capacity required should be small. These are especially important when the size of the face database is very large. We will show shortly that the computational complexity of calculating the Hausdorff distance is simple, and memory required to represent a face image is small.

Hausdorff Distance

Hausdorff distance [32] is a shape comparison method based on measuring the distance between the edge maps of two objects. A major advantage of this distance measure is that the distance can be calculated without the explicit pairing of points in their respective data sets. Given two finite point sets $A=\{a_1,\ldots,a_p\}$ and $B=\{b_1,\ldots,b_q\}$, Hausdorff distance is defined as follows:

$$H(A,B) = \max\{h(A,B), h(B,A)\} \qquad (19\text{-}25)$$

where $h(A,B) = \max_{a \in A} \min_{b \in B} \|a - b\|$

$\|\cdot\|$ is an underlying norm on the point sets A and B. The function $h(A,B)$ is called the directed Hausdorff distance from point set A to B. It identifies the point $a \in A$ that is farthest from any point of B and measures the distance from point a to its nearest neighbor in B. Hence, Hausdorff distance measures the mismatch between the two point sets and can be used as a measure for shape comparsion. This shape comparsion method does not require any explicit correspondence between the model and the image data set, as it does not build a one-to-one pairing between the model and the image feature points.

An efficient way to compute the distance $h(A,B)$ and $h(B,A)$ is by means of distance transform. The distance map generated by the transform represents the distance of a point from its nearest edge point in an image. In the implementation for face recognition, the position of a human face in an input face image is detected and the corresponding position of the two eyes is located. The edge map of an input facial image is generated and then normalized to a specific size based on the inter-distance between the two eyes. This edge map is used as a point set in measuring the hausdorff distance. This normalized edge face image and the face image in the database to be compared are then aligned based on the location of the eyes. Finally, distance transform [33] is applied to the input edge image and different Hausdorff distances are used to measure the similarity between the input human face and the face in the database. The input edge face image is matched against a database of faces and the results are ranked according to the measure of similarity.

The data stored in a database for a face image is its distance map. Suppose that the size of the normalized face is $N \times N$, then the possible longest distance in the distance map is $\sqrt{2}(N-1)$. The total number of bits to be required to represent a distance map is therefore $N^2 \left\lceil \log_2 \left(\sqrt{2}(N-1) \right) \right\rceil$.

Modified Hausdorff distances

A number of modified Hausdorff distance measures [34-35] have been proposed which provide a more reliable and robust measure than the original one. Hausdorff distances designed for face recognition are described in the following.

'Doubly' modified Hausdorff distance [30]

This modified Hausdorff distance introduces the notion of neighborhood function N_B^a and associated penalty (P), which is called 'doubly' modified Hausdorff distance (M2HD). This modified Hausdorff distance is defined as follows:

$$H(A, B) = \max\{h(A, B), h(B, A)\} \tag{19-26}$$

where $h(A, B) = \dfrac{1}{N_a} \sum_{u \in A} d(a, B)$ and $d(a, B) = \max(I \cdot \min_{b \in N_B^u} \|a - b\|, (1 - I) \cdot P)$

In this formulation, N_B^a is the neighborhood of point a in set B. I is an indicator, which is equal to 1 if there exists a point $b \in N_B^a$, and 0 if otherwise. This formulation may have two different values for $h(A,B)$: when all matching pairs fall within a given neighborhood, its value will be the same as the original Hausdorff distance; however, if no matching pair is found, then the penalty value P is considered. It has been shown that the 'doubly' modified Hausdorff distance is suitable for face recognition, because it accounts for small and non-rigid local distortions.

Spatially Weighted Hausdorff Distances [36-37]
In human face recognition, the different facial regions have different degrees of importance; especially the eyes, mouth, face contour, etc. However, traditional Hausdorff distances do not consider the relative importance between the different facial regions, and make no distinction between different parts of the face. The spatially weighted Hausdorff distance (SWHD) assigns different weighting factors for different facial regions in computing the distance. The following is the definition of this kind of Hausdorff distance:

Given two finite point sets $A=\{a_1,...,a_p\}$ and $B=\{b_1,...,b_q\}$,

$$H(A, B) = \max(h_{sw}(A, B), h_{sw}(B, A)) \tag{19-27}$$

The spatially weighted directed Hausdorff distance, $h_{sw}(A,B)$, is defined as follows:

$$h_{sw}(A, B) = \frac{1}{N_a} \cdot \sum_{N_a} w_e(b) \min_{b \in B} \|a - b\| \tag{19-28}$$

where $\|.\|$ is an underlying norm on the points of A and B; and N_a is the number of points in set A. In SWHD, rectangular windows are set for the eyes and mouth, where the weight is set at 1 in computing the Hausdorff distance. For other face regions and the background, the weighting factors are set at 0.5 and 0, respectively. The 3D representation of this spatially weighted function is illustrated in Figure 19-15. The SWHD can also be combined with M2HD to have the spatially weighted 'doubly' Hausdorff distance (SW2HD). Even though both SWHD and SW2HD consider the different importance of facial regions in computing the distances, the setting is rough and cannot fully reflect the exact structure of a human face.

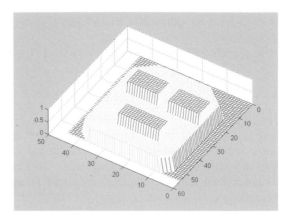

Figure 19-15 The 3D graph of a spatially weighted function $w(x)$.

Spatially Eigen-Weighted Hausdorff distances

Eigenfaces represent the most significant variations in a set of training face images. At those regions where the difference among the training images is large, the corresponding regions at the eigenfaces will have a large magnitude. These regions are usually the eyes, mouth, and face boundary in the images. In order to distinguish two faces, the differences at these important regions should be emphasized. The eigenface is therefore used to serve as a weighting function for this purpose in computing the Hausdorff distances. Therefore, this modified Hausdorff distance measure is called the spatially eigen-weighted Hausdorff distance (SEWHD). In this distance measure, the weight function is denoted as $w_e(x)$, which is the eigen-weighted function emphasizing the measured distances according to the importance of the corresponding point in distinguishing two face images. This weighting function is generated by the first eigenvector which is derived from a set of training face images. Figure 19-16(a) shows a first eigenvector or eigenface, which can be seen to have a significant magnitude at the important facial features and at the face contour.

Figure 19-16 (a) First eigenvector, and (b) the corresponding eigen-weighted function.

Suppose that $e_1(x,y)$ represents the first eigenface. The eigen-weighted function is generated by normalizing the eigenface as follows:

$$n_{mask}(x, y) = (\frac{255}{\max(e) - \min(e)})e(x, y) \qquad (19\text{-}29)$$

where max(*e*) and min(*e*) represent the maximum and minimum magnitudes of the image *e*(*x*, *y*), and *e*(*x*, *y*) = |*e*₁(*x*, *y*)|. The eigen-weighted function, $w_e(b)$, is obtained by normalizing $n_{mask}(x,y)$ to a range between 0 and 1, which is shown in Figure 19-16(b). This eigen-weighted function is incorporated with the spatially weighted Hausdorff distance and the 'doubly' modified Hausdorff distance, which are denoted as SEWHD and SEW2HD, respectively.

Edge Detection and Adaptive Thresholding
In applying Hausdorff distance to face matching, the shape comparison basically operates on edge maps. Consequently, the edge detection algorithm used will have a significant effect on the matching performance based on the Hausdorff distance measure. However, when converting a gray scale edge image to a binary edge image, a fixed threshold usually cannot work well. This is because the contrast in different edge images may vary significantly. In our approach, the threshold is set adaptively for converting an edge image to a binary image, which is then used as the point set for representing the face. The edge image *E*(*x,y*) is obtained by applying morphological operations on the face image *F*(*x,y*). In order to emphasize the important facial features and the face contour which usually show up as valleys and have lower gray level intensities than other parts of a face, the following function is considered:

$$n(x, y) = \frac{E(x, y)}{F(x, y)} \qquad (19\text{-}30)$$

The values of the function *n*(*x,y*) are then sorted in descending order, and the threshold is set so that 18% of the points with the largest magnitudes in *n*(*x,y*) are selected. Figure 19-17 shows some binary images generated by using this adaptive thresholding scheme. The first column shows the original images, the second column displays the gray-level edge images obtained by the morphological edge detection, and the third column shows the thresholded images.

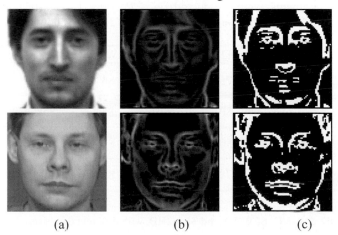

(a)   (b)   (c)

Figure 19-17 (a) Original facial images; (b) gray-level edge images by morphological gradient; and (c) binary images by using the adaptive thresholding method.

Evaluations

To evaluate the performances of the different Hausdorff distance measures for face recognition, the ORL face database (from the Oliver Research Laboratory in Cambridge, U.K), Yale face database (from Yale University), and MIT face database were used in the experiments The MIT database has 144 face images with 16 distinct subjects, while the Yale database has 150 face images with 15 distinct subjects. For the ORL database, there are 400 different face images corresponding to 40 distinct persons, but only 6 images for each of the 40 subjects were included in the testing set. The experimental setup consisted of database composed of upright frontal views of the subjects - at a suitable scale and with a normal facial expression. The performances of the different Hausdorff distance measures and the eigenface technique were then evaluated based on the respective database, as well as the combined one. The Hausdorff distances considered include the conventional Hausdorff distance (HD), the 'doubly' modified Hausdorff distance (M2HD) [30], the spatially weighted Hausdorff distance (SWHD) [36], the spatially weighted 'doubly' Hausdorff distance (SW2HD) [37], the spatially eigen-weighted Hausdorff distance (SEWHD), and the spatially eigen-weighted 'doubly' Hausdorff distance (SEW2HD). A query image is compared to all the faces in a database and the faces are then ranked and arranged in ascending order according to their corresponding measured distances. Table 19-2 tabulates the recognition rates for PCA, HD, M2HD, SWHD, SW2HD, SEWHD, and SEW2HD for each of the three individual databases and a combination of the three databases. Experimental results show that the two eigen-weighted Hausdorff distances, SEWHD and SEW2HD, outperform other methods, while the SEW2HD produces the best performance.

The recognition rates of SEW2HD based on the ORL, Yale, MIT, and combined databases are 91%, 89%, 92%, and 83%, respectively. Figure 19-18 illustrates the cumulative recognition rates of the different methods based on the combined face database. The experiments were conducted on a Pentium II 450MHz computer system. The average runtimes for the SEWHD and SEW2HD are about two seconds.

Method	PCA	HD	M2HD	SWHD	SW2HD	SEWHD	SEW2HD
ORL database	63%	46%	75%	82%	88%	88%	91%
Yale database	50%	66%	80%	82%	83%	85%	89%
MIT database	50%	49%	80%	84%	89%	90%	92%
Combined database	60%	55%	74%	80%	81%	81%	83%

Table 19-2 Recognition rates of the seven face recognition methods: PCA, HD, M2HD, SWHD, SW2HD, SEWHD, and SEW2HD, with different databases.

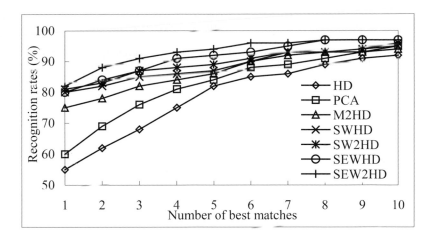

Figure 19-18 Comparison of the overall recognition rates using 534 testing face images and the combined database.

19.6 Conclusion

In this chapter, techniques for the implementation of a face recognition system are described. The techniques include the detection of a human face in an image, extraction of facial features, and computing the difference between two faces. All these techniques should be computationally simple and require small storage capacity, which are especially important when the size of the database under consideration is very large. In case of a large database, besides the accuracy of the face recognition method, the searching time is also an important concern.

References

[1]   R. Chellappa, C.L. Wilson, and S. Sirohey, "Human and machine recognition of faces: a survey," *Proceedings of the IEEE*, Vol. 83, No. 5, pp.705-741, May 1995.

[2]   The Face Recognition Home Page: http://www.cs.rug.nl/~peterkr/FACE/face.html.

[3]   B. Ballard and G.C. Stockman, "Controlling a computer via facial aspect," *IEEE Transactions on Systems, Man, and Cybernetics*, Vol. 25, No. 4, pp.669-677, 1995.

[4]   K. Aizawa, H. Harashima, and T. Saito, "Model-based analysis synthesis image coding (MBASIC) system for a person's face," *Signal Processing: Image Communications*, Vol. 1, No. 2, pp.139-152, 1989.

[5]   M. Kirby and L. Sirovich, "Application of the Karhunen-Loeve procedure for the characterisation of human faces," *IEEE Transactions on Pattern Analysis and Machine Intelligence*, Vol. 12, pp.103-108, 1990.

[6]   M. Turk and A. Pentland, "Eigenfaces for recognition," *Journal of Cognitive Neuroscience*, 3(1), pp.71-86, 1991.

[7]   M.H. Yang, D.J. Kriegman, and N. Ahuja, "Detecting faces in images: a survey," *IEEE Transactions on Pattern Analysis and Machine Intelligence*, Vol. 24, No. 1, pp.34-58, 2002.

[8]   Y. Yokoo and M. Hagiwara, "Human faces detection method using genetic algorithm," *Proceedings of the IEEE International Conference on Evolutionary Computation*, pp.113-1181996.

[9]   Y. Suzuki, H. Saito, and S. Ozawa, "Extraction of the human face from the natural background using GAs," *Proceedings of IEEE TENCON: Digital Signal Processing Applications*, pp.221-226, Vol. 1, 1996.

[10]   J Miao, B. Yin, K.Q. Wang, L.S. Shen and X.C. Chen, "A hierarchical multiscale and multiangle system for human face detection in a complex background using gravity-center template," *Pattern Recognition*, 32(7), pp. 1237-1248, 1999.

[11]   C.H. Lin and K.C. Fan, "Triangle-based approach to the detection of human face," *Pattern Recognition*, 34(6), pp. 1271-1284, 2001.

[12]   G. Yang and T.S. Huang, "Human face detection in a complex background," *Pattern Recognition*, 27(1), pp. 53-63, 1994.

[13]   K.M Lam, "A fast approach for detecting human faces in a complex background," *Proceedings of the IEEE International Symposium on Circuits and Systems*, Vol. 4, pp.85-88, 1998.

[14]   K.W. Wong and K.M. Lam, "A reliable approach for human face detection using genetic algorithm," *Proceedings of the IEEE International Symposium on Circuits and Systems*, Vol. 4, pp. 499-502, 1999.

[15]   Gu Qian and S.Z. Li, "Combining feature optimization into neural network based face detection", *Proceedings of International Conference on Pattern Recognition*, Vol. 2, pp.814-817, 2000.

[16]   K.K. Sung and T. Poggio, "Example-based learning for view-based human face detection," *IEEE Transactions on Pattern Analysis and Machine Intelligence*, Vol. 20, No.1, pp.39-51, 1998.

[17]   Y.J. Wang and B.Z. Yuan, "A novel approach for human face detection from color images under complex background," *Pattern Recognition*, 34, pp.1983-1992, 2001.

[18]   K. Sobottka and I. Pitas, "A novel method for automatic face segmentation, facial feature extraction and tracking," *Signal Processing: Image Communication*, 12 (3, pp. 263-281, 1998.

[19]   D. Chai, and K.N. Ngan, "Face segmentation using skin-color map in videophone applications," *IEEE Transactions on Circuits and System for Video Technology*, Vol. 9, No. 4, pp. 551-564, 1999.

[20]   H.L. Wang and S.F. Chang, "A highly efficient system for automatic face region detection in MPEG video," *IEEE Transactions on Circuits and Systems for Video Technology*, Vol. 7 No. 4, pp. 615-628, 1997.

[21]   Kobus Barnard and Brian Funt, "Camera calibration for colour vision research," *SPIE Conference on Electronic Imaging, Human Vision and Electronic Imaging IV*, SPIE Vol. 3644, pp.576-585, 1999.

[22]   P. Maragos, "Tutorial on advances in morphological image processing and analysis," *Optical Engineering*, 26(7), pp.623-632, 1987.

[23]   K.W. Wong, K.M. Lam and W.C. Siu, "An efficient algorithm for human face detection and facial feature extraction under different conditions", *Pattern Recognition*, 34 (10), pp. 1993-2004, 2001.

[24]   M. Kass, A. Witkin, and D. Terzopoulo, "Snakes, Active contour model," *Proceedings First International Conference on Computer Vision*, pp.259-269, 1987.

[25]   D.J. Williams and M.Shah, "A fast algorithm for active contours and curvature estimation," *CVGIP: Image Understanding*, 55(1), pp.14-26, 1992.

[26]   K.M. Lam and H. Yan, "Fast Greedy Algorithm for Active Contours," *Electronics Letters*, Vol. 30, No. 1, pp. 21-2, 1994.

[27]   Wai-Pak Choi, Kin-Man Lam and Wan-Chi Siu, "An adaptive active contour model for highly irregular boundaries," *Pattern Recognition*, Vol. 34, pp. 323-331, 2001.

[28]   A.L. Yuille, "Deformable templates for face recognition," *Journal of Cognitive Neuroscience*, Vol. 3, pp.59-70, 1991.

[29]   K.M. Lam and H. Yan, "An Improved Method for Locating and Extracting the Eye in Human Face Images," *Proceedings, IEEE International Conference on Pattern Recognition*, pp. 411-5, August 1996.

[30]   B. Takács, "Comparing face images using the modified Hausdorff distance", *Pattern Recognition*, 31(12), pp.1873-1880, 1998.

[31]   R. Brunelli and T. Poggio, "Face recognition: features versus templates," *IEEE Transactions on Pattern Analysis and Machine Intelligence*, Vol. 15, No. 10, pp.1042-1052, 1993.

[32]   D.P. Huttenlocher, G.A. Klanderman, and W.J. Rucklidge, "Comparing images using the Hausdorff distance," *IEEE Transactions on Pattern Analysis and Machine Intelligence*, Vol. 15, No. 9 , pp. 850 -863, 1993.

[33]   P. E. Danielsson, "Euclidean Distance Mapping," *Computer Graphics and Image Processing*, Vol. 14, pp.227-248, 1980.

[34] D. G. Sim, O. K. Kwon and R. H. Park, "Object Matching Algorithms Using Robust Hausdorff Distance Measures," *IEEE Transaction on Image Processing*, Vol. 8, No. 3, pp. 425-429, 1999.

[35] M. P. Dubuisson and A. K Jain, "A Modified Hausdorff distance for Object Matching, " *Proc. 12th International Conference on Pattern Recognition*, pp. 566-568, 1994.

[36] B. Guo, K. M. Lam, W.C. Siu and S. Yang, "Human face recognition using a spatially weighted Hausdorff distance," *Proceedings of the 2001 IEEE International Symposium on Circuits and Systems*, pp. 145 148, 2001.

[37] K. H. Lin, B. Guo, K. M. Lam and W. C. Siu, "Human face recognition using a spatially weighted modified Hausdorff distance," *Proceedings of the International Symposium on Intelligent Multimedia, Video and Speech Processing*, pp. 477-480, 2001.

20 DATA MANAGEMENT FOR LIVE PLANT IDENTIFICATION

Dr. Zheru Chi

Plant identification by computers will surely find a wide range of applications including plant resource survey, plant data management, and education on plant taxonomy. Since computerized plant identification is a very challenging computer vision problem, the research and development in this field is still in its infancy. In this chapter, we first address the necessary and general issues regarding computer-aided plant identification and the management of a living plant database. We then describe a general approach that botanists adopt for plant identification, and typical systems for living plant identification and plant data management. This is followed by a discussion on a sophisticated approach for computer-aided plant identification supported by image processing, intelligent information processing, and plant data management systems. In this chapter, we also discuss various algorithms for processing leaf, flower and plant images, including image segmentation, leaf venation extraction, and flower region localization. In the discussion, we stress the unique features of plant images and special treatments needed for processing these images. Also discussed in this chapter are the feature extraction of leaf and flower images and the use of these features for leaf and flower image retrieval, a very important sub-task for computer-aided living plant identification. Finally, concluding remarks are made at the end of the chapter.

20.1 Introduction

Only about 250,000 species for flowering plants have been named. There are many remaining plant species that have not been classified or named. It is impossible for any one botanist to know more than a tiny fraction of the total number of named species. Moreover, plant taxonomy as a basic research field has not received the attention and funding it deserves. As a result, the number of experienced taxonomists and botanists has dropped significantly. It has been said that the taxonomist is an endangered species in Europe [3]. As in many other fields, plant identification is becoming more dependent on computers. The advanced information technologies provide a potentially very attractive solution for the central management of plant data. One very interesting project is the development of a Chinese medicinal plant identification system that will promote the sharing of knowledge of Chinese medicinal plants. By installing an enquiry system with proper database management, on the Internet, everyone who is interested in Chinese medicinal plants can find reference material for the identification of plants and make contributions to the development of a web-based plant database. A computerized plant identification system will help both botanists and amateurs identify living plants. It will also benefit teaching, learning, and training of plant classification and identification personnel. Computerized plant identification systems may eventually replace human beings in plant identification and classification.

PLANT IDENTIFICATION AND INTELLIGENT INFORMATION PROCESSING

Plant identification is a very demanding and time-consuming task, which has mainly been carried out by taxonomists/botanists. A significant improvement in processing time can be expected if the plant identification can be carried out by a computer automatically or semi-automatically with the aid of image processing and computer vision techniques, and various data management techniques. Automatic (machine) plant identification from color images is and will continue to be one of the most difficult tasks in computer vision due to the lack of proper models or representations, the large number of biological variations that a species of plants can take, and imprecise or ambiguous image pre-processing results.

Developing a fully automatic machine plant identification system seems impossible based on the current technologies. However, in the near future, a computer-aided identification system to help human beings recognize plants from their color images is likely to be developed. Such a system would release botanists from much data searching, routine checking, and typing work. The system would also be useful for field learning and teaching plant recognition and memorization. Machine interpretation of plants images including flowers, leaves, stems, barks, and fruits, becomes more feasible as computer vision and intelligent information techniques advance. (see Figure 20-1)

Figure 20-1 A plant image

PLANT DATA MANAGEMENT

Many applications need a sophisticated plant data management system. A plant management system should provide text, image, and video information about plants of interest. The system should be organized and indexed to enable the efficient retrieval of plant information from both text and image enquiries. Plant identification is one of a number of demanding tasks that need an advanced plant data management system. Other applications include: regional plant resource management for a botany garden, a natural reserve park, or a forest farm; and the management of special groups of plants such as Chinese medicinal plants; education and research on plant taxonomy; and environment protection.

20.2 Botanists' Approaches for Plant Identification

Plant classification and identification is a very old field. In flowering plant taxonomy, one of the first documented efforts to systematize the naming of local flora was made by Theophrastus (370-285 B.C.), a Greek student of Plato and successor to Aristotle as Director of the Lyceum and its botanical garden. A commonly adopted definition for plant classification is "an orderly process resulting in assignment of each individual to a descending series of groups of related plants, as judged by characteristics in common" [2]. The plant kingdom is made up of classically recognized major groups: (1) vascular plants (or tracheophytes) including the flowering plants, gymnosperms, and pteridophytes; (2) mosses and liverworts; and (3) algae and fungi. As shown in Figure 20-2, the seeded plants can be divided into five divisions, each of which can be further divided into a number of classes (subclasses), orders, families, genus, species, and varieties [22].

Unknown plants are identified by using a number of keys. These keys are arranged such that assigning a plant to a category is a process of elimination. A key to the species of conifers is shown in Figure 20-3.

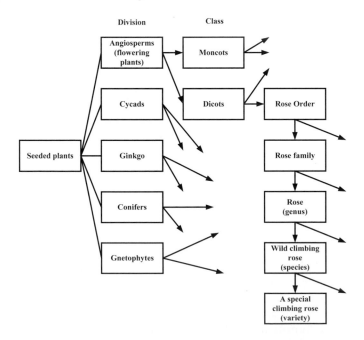

Figure 20-2 The classification of seeded plants

There are keys used to differentiate between the different divisions of vascular plants. Different keys are used to differentiate between the different classes under each division. At each of successive levels, more specific keys are adopted.

The keys to the divisions of vascular plants include the absence or presence of seeds, plant with roots and leaves, plant with no roots or no true leaves, stems largely above ground, stems commonly but not always below ground, stems markedly jointed or not joined, the repeated forking of stems, etc. The keys to the class of the seed plants include the absence or presence of flowers, seeds enclosed or not in an ovary, leaves simple, leaves pinnate or bipinnate, small trees, etc.

For the flowering plants, various features related to roots, stems, buds, leaves (complexity of leaves, characters of the leaf margins, shapes of leaf blades, venation patterns, the petiole or not, relative position of leaves, number of leaves at a node, special kinds of leaves), color and surface characters, flowers (color, structure and arrangement of flowers on the plant), and fruits (color, shape and quality) and seeds are used to identify subclass, order, family, genus, species, and variety.

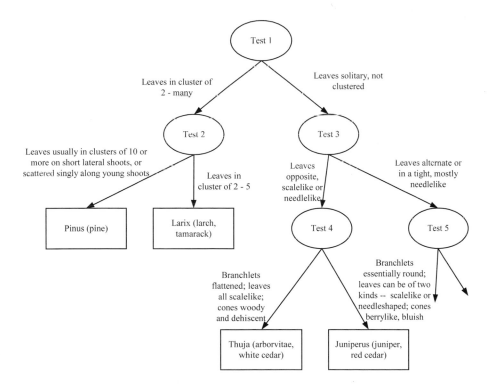

Figure 20-3 A key to conifers

Plants in different regions and locations have different features which should be taken into consideration when identifying plants. Although many manuals are available for identifying the genera, species, and varieties of regional and local floras, there is not a single unified system for classifying all the plants in the world. Much effort has been made in building a unified flora classification system, however a unified classification system is still far from us.

20.3   Plant Identification and Plant Data Management Systems

Several systems have been developed for plant recognition and plant data management. The typical systems include *Lucid* [17], developed by the Centre for Pest Information Technology and Transfer (CPITT) at University of Queenland, *UConn*, the University of Connecticut Plant Database [27], and *CalFlora* [4] hosted by the UC Berkeley Digital Library Project.

Lucid: A Multimedia Knowledge Management Tool
Lucid is a multi-media, knowledge management tool, which helps users make an identification or diagnosis. It is a special type of knowledge system that enables expert knowledge to be integrated and distributed to a wide audience via various media such as CD and the Internet, providing expert guidance to those wishing to identify or diagnose specific objects or situations. As a general-prupose identification tool, it can apply not only to plant identification but also to a variety of other object identification tasks.

Let us use an example to explain the methodology used in *Lucid* for object identification. Suppose you observe a plant in your garden and would like to identify it and obtain more information about it. Developed by taxonomists and botanists, a Lucid key to a group of plants will present you with a series of features or characteristics that you can select as being true or false for your particular plant. This refines the short list of possible matching species. For instance, you could select the shape of leaves, the type of flowers, the geographic location of the garden and so on. As you proceed through the identification, various images and text description, and processing functions can help you refine the short list. Once you have identified the plant species, *Lucid* can provide you with the multimedia information about that species and can link you to other sites that provide further information.

EUCLID [9] is an interactive key and information system for the 324 "eucalypts" of southeastern Australia developed based on *Lucid*. Anyone from professional researchers to the novice with a basic knowledge of plants may use the system to identify eucalypts of southeastern Australia.

UConn Plant Database
The UConn plant database [27] provides a Plant Selector that can search the University of Connecticut Plant Database to find trees, shrubs and vines that meet your needs. You can make an enquiry by giving one or more of following pieces of information:
- Names: genus, species, common, family, etc.;
- Basic traits: plant form and size, foliage character, native/non-native;
- Ornamental traits: flower display, color and fragrance, flowering period, fall foliage color, stem/bark texture and color, fruit quality and color;

- Site characteristics: sun exposure, soil pH, moisture and drainage;
- Cultivar availability: foliage, form and ornamental; and
- Special qualities: invasive tendency, salt/sea spry tolerance, attracts adult butterflies, etc.

For each of these characteristics, there are a number of items that can be selected by the user. For example, under foliage character, there are three items for selection: deciduous, evergreen, and semi-evergreen. For each of the matched plants, the Plant Selector will normally provide descriptions on habitat, habit and form, summer foliage, autumn foliage, flowers, fruit, bark, culture, landscape use, propagation, and cultivars/varieties, together with various photos.

CALFLORA: A Vascular Plant Database

The CALFLORA database [4] provides information about all 8375 currently recognized vascular plants in California, including over 800,000 records of plant observations and 21,000 photographs. Users may browse these collections by scientific or common name, or search by name, location, or other attributes. Information about plant distribution and individual occurrences is presented textually and via interactive map capabilities.

UConn Plant Database and *CalFlora* are information systems with a very limited searching support. *Lucid* is a general-purpose tool that provides a step-by-step procedure to refine the short list and help the user identify an object such as species of plant or animal. Neither of these systems support image and information processing, and intelligent content-based image search techniques.

COMPUTER-AIDED PLANT IDENTIFICATION SYSTEM

With advancing information technology and computer vision techniques, a computer-aided plant identification system with support from well-developed plant data management systems is becoming feasible. The main features of this system include (1) making full use of sophisticated image processing, computer vision, and intelligent information processing techniques to release the user from tasks such as labeling, measuring, classification, and data entry; (2) adopting intelligent search and database management methodologies to aid plant identification by both professionals and amateurs and reduce the amount of time spent searching for information; and (3) supporting full interactive process between the user and the machine. As shown in Figure 20-4, the system would accept both image and text input. For text input, traditional text-based information processing and search techniques can be used. For image input, dedicated image processing algorithms have to be applied to the processing of flower, leaf, and stem images as well as the whole plant images. The functions of plant images processing may include:

- Leaf classification according to its shape, margin, and venation;
- Measurement of degrees of lobing of leaves;
- Counting of the number of leaves at a node;
- Classification of corolla into apetalous (without petals), choripetalous (with separate petals), and sympetalous (with coalescent petals) groups;
- Determination of degrees of lobing and shapes of sympetalous corollas;

- Determination of flower forms (spiral and cyclic) and the symmetry of the flower (radially symmetrical and bilaterally symmetrical);
- Fruit classification according to its shape, color and quality;
- Determination of directions of stems; and
- Color determination of flower, leaf, stem, etc.

   The plant components are classified into botanical categories by locating regions of interest and extracting various features. Together with the text input, the processed results from these images can be used for information retrieval. The information retrieved can provide suggested matches to help the user identify the plant or provide instructions on how to focus the investigation in order to identify a certain plant.

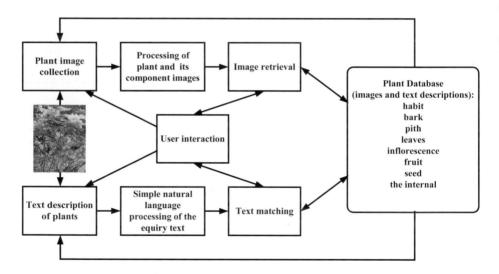

Figure 20-4 Block diagram of a computer-aided living plant identification system

20.4 Plant Image Processing and Feature Extraction

LEAF IMAGE PROCESSING

Since the characteristics of plant leaves are one of the most important set of features for identifying various plants visually, leaf image processing and retrieval is an important step for plant identification. Im et al. [11] and Abbasi et al. [1] have done some preliminary work on plant recognition and classification using the shape features of plant leaves. Im et al. used a hierarchical polygon approximation representation of leaf shape to recognize the Acer family variety. Curvature Scale Space (CSS) was used to represent leaf shapes for Chrysanthemum variety classification by Abbasi et al. [1].

Figure 20-5 A leaf image

Shape Characterization of Leaves

Figure 20-5 shows a leaf image. Many investigations on shape representation such as chain codes, medial axis transform (MAT), Fourier descriptors, wavelet descriptors, moment invariants, deformable templates as well as the centroid-contour distance curve, have been carried out [16, 19]. Rui et al. [23] proposed a Modified Fourier Descriptor (MFD) to decrease the discretization noise and avoid the starting point issue of the traditional Fourier descriptor representation. Intuitively, it is expected that combining several sets of features can better represent a shape, as indicated in some successful applications reported in the literature [8, 14, 29]. A general discussion on shape features and shape-based retrieval can be found in Chapter 1.

Centroid-Contour Distance Curve
Tracing a contour can be considered as circling around its centroid. The tracing path along one direction (clockwise or anti-clockwise) from a fixed starting point represents a shape contour uniquely, that is, a contour point sequence is unique to a shape if its centroid and the starting point are fixed. By using the centroid-contour distance curve [16], a two-dimensional object image can be represented by a one-dimensional curve. Figure 20-6 illustrates the definition of the centroid-contour distance curve.

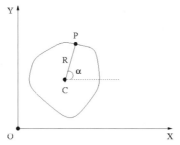

Figure 20-6 Illustration of the centroid-contour distance curve

Generally, the translation, scaling and rotation invariances are expected for shape features. Since the centroid-contour distance is relative to the object centroid, obviously, it is invariant to object translation. The size of plant leaves vary with age and vary at the different parts of the plant at a given time. There are two differences between two scaled versions of a contour: the number of the sample points and the absolute centroid-contour distance of a contour point. To make matching possible, the number of the sample points of two contours should be the same. The problem can be solved by downsampling the contour of the larger object. The second difference can be eliminated with a normalization process. By proper downsampling and normalizing operations, the centroid-contour distance curve is made scaling invariant. However, the contour details such as the sawtooth shape may be partially lost due to the normalization process. The angle code histogram may be adopted to overcome this problem [29].

Rotation may be introduced in data acquisition such as picture grabbing by a camera or scanner. It requires that the shape feature is invariant to rotation. As shown in Figure 20-7, an object contour with m points is rotated one pixel clockwise, so the centroid-contour distance sequences will change from $|OP_1|, |OP_2|, \ldots, |OP_m|$ to $|OP_m|, |OP_1|, |OP_2|, \ldots, |OP_{m-1}|$. It is noticed that the only difference between the two sequences is the rotation of contour points. If the latter sequence is rotated back by one point, the two sequences will be identical. Generally, for an arbitrary rotation, we need to rotate the sequence m times with an exhaustive comparison in order to determine the best alignment.

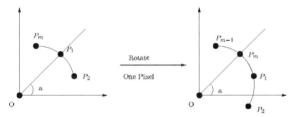

Figure 20-7 Illustration of the rotation of the centroid-contour distance curve

The key for the measure with centroid-contour distance curve to be rotation-invariant is to locate the fixed starting point(s). In order to solve this problem, the possible starting point(s) should be located and the sequence is rotated accordingly. For the worst case, the distance sequence is rotated one by one to locate the starting point(s) in terms of the minimal difference between these two sequences. Fourier transformation and correlation can be used to locate the starting point(s) but they are computationally costly [8, 20]. Rui et al. [23] solved this problem by measuring shape similarity with the magnitude and phase difference respectively with modified Fourier descriptor (MFD). A thinning based starting point locating method was proposed by Wang et al. to reduce the computational cost [29].

Perceptually, the convex corners are the most convenient aligning points. The SUSAN algorithm [25] proposed by Smith and Brady, is one of the most popular corner detection methods, however it occasionally produces false positives, making the computation more expensive. There are more redundant corners for irregular contours. Because the corner detection algorithm only makes use of the local properties of each contour point, occasionally it fails to obtain good corners. Principal axis analysis is another method to align two objects, however, it is very sensitive to discretization error and does not handle circular objects and irregular shapes well. Actually, leaf shape is approximated symmetry and irregular. The thinning based method, which makes use of the global information as well as the local information of the contour, can achieve a better result in locating possible starting points [29]. In this approach, a binarized leaf image is thinned to obtain its skeleton and several end-points are located on the skeleton. For example, in Figure 20-8 (b), there are four end-points on the skeleton. The possible starting points will be located on the contour for each end-point. There are two rules for selecting the possible start points for each end-point:

1. The distance between the end-point and the possible starting point is locally maximal as the skeleton is defined as the set of points whose distance from the nearest contour is locally maximal [13].
2. The angle code of a possible starting point is locally minimal, that is, the locally sharpest contour point will be selected as a possible starting point, which agrees with human perception.

An end-point that satisfies the above two rules on the leaf contour is a possible starting point.

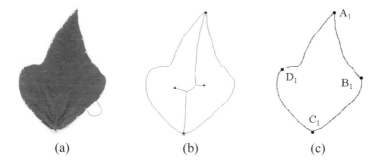

| (a) | (b) | (c) |

Figure 20-8 An example of the thinning based starting point localization. (a) Original leaf image; (b) The skeleton of the leaf and its end points; (c) Possible stating points located

As shown in Fig. 20-8(c), points A_1, B_1, C_1, and D_1 are the identified possible starting points on the contour derived from the corresponding skeleton end point points shown in Fig. 20-8(b). In the matching process, one of the possible starting points of a contour is selected as the starting point and fixed. The other contours are rotated so one of their possible starting points is aligned to the fixed point of the original contour. The matching time is significantly reduced because the number of possible starting points is much smaller the total number of contour points. Because of the quantization error, in practice, five neighboring points for each possible starting point will also be considered.

Angle Code Histogram
Both global and local features play important roles in characterizing a shape. It is observed that the centroid-contour distance cannot represent local features

effectively. However, local features such as the leaf margin (for example, sawteeth) are also very important for characterizing leaves. The leaves with or without a sawtooth margin may belong to different groups. Peng and Chen [18] proposed an angle code to represent a sequence of line segments with two successive line segments forming an angle for trademark and logo retrieval. The angles at contour points on each closed contour were computed and the resulting sequence of successive angles was used to characterize the contour.

Degree	(-5, 5)	(5, 40)	(40, 50)
Code	0	1	2
Degree	(50, 85)	(85, 95)	(95, 130)
Code	3	4	5
Degree	(130, 140)	(140, 175)	(175, 185)
Code	6	7	8

Table 20-1. Angle code

An example of the code assignments is summarized in Table 20-1. Based on the code, an object contour is encoded to generate an angle code string. The angle for each point can be computed based on two approximate lines coming to and leaving the point. The greater the number of points with large angle values the rounder of the leaf shape. It is reasonable to assume that the distribution of angle codes for two similar contours is also similar. The angle code histogram can be used to characterize the local features of a leaf shape.

Eccentricity
The eccentricity of leaves is helpful in the coarse classification of leaf shapes. The eccentricity of an object is defined with moments as:

$$e_I = \frac{(u_{20} - u_{02})^2 + 4u_{11}}{A} \tag{20-1}$$

where $u_{pq} = \iint_R (x - x_c)^p (y - y_c)_q \, dxdy$, R is the whole region that an object occupies, and A is its area, and (x_c, y_c) is the object centroid [13]. It is easy to verify that the eccentricity is translation, scaling and rotation invariant.

Leaf Venation Extraction
Leaf venation is another important feature for characterizing leaves. Soille [26] has done some work on venation network extraction, but his purpose is only to mark and segment the plants. The leaf blade constitutes its venation [21]. There are about three chief types of principle veins, parallel (or nerved), pinnate, and palmate (examples are shown in Figure 20-9). The first type includes paralleled, side, radiate and arc-like veins. The second and third types normally are recticulate (net-veined). Sometimes a combination of these types can be found, for example, ternately pinnate

veins have three principle pinnate veins that form the base. The distribution of leaf veins reflect most of the leaf shape and the configuration of the vein embranchments may give more information which can be used to identify the leaves. So it is important to extract the principle leaf veins and apply the veins' feature to identify different leaves.

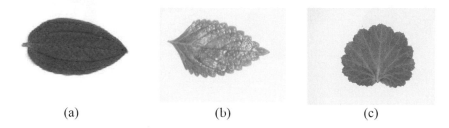

(a)   (b)   (c)

Figure 20-9 Three main types of leaf veins. (a) Parallel (nerved); (b) Pinnate; (c) Palmate.

Leaf Vein Extraction

Leaf veins are thin and long and the intensity of the veins in the grayscale image is either brighter or darker than the adjoining area of leaf veins. The research on vein-like object extraction mostly concentrates on biometric identification using palm vein patterns [12], gestalt identification in pathophysiology to assess the morphology of the retinal vascular tree [15], and classification of granitoid rocks in geography to distinguish different perthite textures [6]. A Gaussian low pass filter and the modified median filter have been used in extracting palm veins. An operator based on vertical-horizontal run length ratios has been utilized to identify and separate vein elements. The retinal vessel was segmented using a model-based top-down scheme where a 2-D second-order polynomial was employed to estimate the gray-value-function of the vessel. To extract leaf venation, Soille [26] applies morphological filters.

Leaf vein extraction can also be considered as a classification problem. A neural network can be trained to classify each individual pixel into a vein pixel or a background pixel. Yan et al. [30] and Chi et al. [5] extracted characters and lines from geographic map images using a neural network approach and fuzzy rule approach, respectively. The background texture of leaf images is more complex than that of map images where the background is quite uniform. A set of nine features including a measure of the local contrast of a darker pixel against its background can be used as the input of the neural network classifier for leaf vein extraction. To train the neural network classifier, a sufficient number of training samples (both vein and background pixels) were manually picked up from training images. A three-layer neural network with two output nodes (for vein and background classes) was employed. Figure 20-10 shows two leaf examples of vein extraction by using the Laplacian and Sobel operators, and a neural network classifier. We can see that a trained neural network classifier can achieve much better results than the Laplacian and Sobel operators.

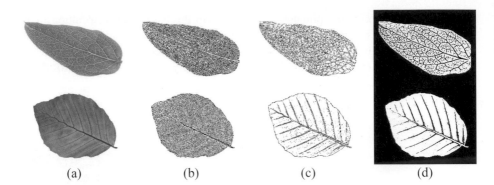

(a)   (b)   (c)   (d)

Figure 20-10 Vein extraction results. (a) Original images; (b) Vein extraction using the Laplacian operator; (b) Vein extraction using the Sobel operator; (d) Vein extraction using a neural network classifier.

Characterization of Leaf Vein
Kaupp [15] used a three-layered neural network and the BP algorithm to classify the segmentation of objects like arteries, vein optic disks, macula and background of retinal vessel images using 30 features as the input and 8 interactively classified images as training samples. Giakoumis [10] separated brush strokes from cracks of painting by implementing an MRBF neural network, which was robust based on median operators.

Leaf venation mainly contains straight lines and curves and may also be depicted by Angle Code Histogram (ACH), as discussed for leaf contour characterization. Since there are only a very limited number of vein types, a model for each type of vein can be developed for vein classification and retrieval.

FLOWER IMAGE PROCESSING
There are two main problems in flower identification: the localization of flowers and the characterization of flower regions in a way similar to human visual perception. The flower characterization should allow flexible querying by example or text description.

 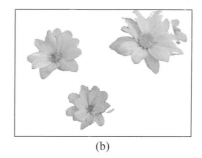

(a) (b)

Figure 20-11 (a) A flower image; (b) Flower region extraction using gradient-based edge detection.

Flower Region Extraction

Figure 20-11(a) shows a flower image. Commonly used edge detection operators such as Laplacian and Sobel operators can be used to extract flower regions if the background is relative simple. Figure 20-11(b) shows the edge extraction result of a flower image from which three flower regions can be clearly seen, although some sort of postprocessing has to be performed in order to remove other non-relevant patterns in the background. If the background of an image is complex, it is difficult to locate flower regions. In this case, a domain knowledge-driven segmentation has to be adopted.

Color is a main feature that can be used to differentiate flowers from the background including leaves, stems, shadows, soils, etc. Flowers are rarely green, black, gray or brown in color. Such color-based domain knowledge can be adopted to delete pixels that do not belong to flower regions [7]. Another observation helpful in identifying background regions is that the background colors are usually visible on the boundary regions of the image. For flower images, it is reasonable to assume that a flower region would occupy a reasonable part of the image. The ratio of a flower region over the whole image can be estimated from database flower images. Usually, only the same type of flowers appears in an image. Segmenting the most likely flower region, which is an easier job than extracting all of the flower regions, is sufficient to characterize the flower. Modeling on flower shapes and types is also helpful in flower region extraction. Das et al. [7] has proposed an iterative segmentation algorithm with knowledge-driven to extract a flower region from the background. As in leaf vein extraction, a neural network can also be trained to detect the flower edge and therefore to extract the flower regions.

Flower Characterization

Flowers can be characterized by color, shape, and texture. Color was used to characterize flowers in [7]. In Das et al.'s approach, a natural language color classification scheme that reflects the human visual perception was adopted to the indexing and retrieval of flower images. Both query by name and query by example were supported. Van der Heijden and Vossepoel proposed a general contour-oriented shape dissimilarity measure for a comparison of flowers of potato species [28]. The basic model of potato flowers was characterized by the five petal tips, which should be at equal intervals.

A feature extraction and learning approach is commonly adopted for the recognition of various objects, such as handwritten characters, human faces, etc. This approach has also been used for automatic recognition of wild flowers [24]. Experimental results reported in [24] showed that four flower features together with two leaf features achieved quite a good performance, recognizing 16 wild flowers. The four flower features were (1) the ratio of the average petal width over the average petal height; (2) the number of petals; (3) the first-order moment; and (4) the color of the second largest cluster in the HSV coordinate. A neural network was trained with training samples to classify 16 wild flowers.

20.5 Leaf and Flower Image Retrieval

LEAF IMAGE RETRIEVAL

The characteristics of plant leaves are very helpful in identifying plants of various species. Leaf image retrieval is a necessary step for computer-aided live plant identification. Based on the discussion in Section 20.4 on leaf image processing and leaf characterization, a leaf image retrieval approach shown in Figure 20-12 can be applied to content-based leaf retrieval [29].

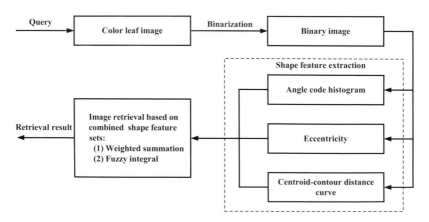

Figure 20-12 Block diagram of a leaf image retrieval approach

Similarity Measure
Similarity between Centroid-Contour Distance (CCD) Curves
The CCD curve can be used to measure the dissimilarity between images. We define the following distance function *D* to measure the dissimilarity between two CCD curves:

$$D = \sqrt{\frac{\sum_{i=1}^{n}\left|f_1(i)-f_2(i)\right|}{n}}$$ (20-2)

where $f_1(i)$ and $f_2(i)$ are the CCD distances of two object contours at the i-th point and n is the total number of the contour points. The distance measure with CCD curves can be made rotation invariant if two curves are properly aligned. In order to find the best matching result, one curve has to be rotated m times, where m is the number of possible starting points. The dissimilarity D_c between two object contours is defined as:

$$D_c = \min\left(D_1, D_2, \ldots, D_j, \ldots, D_m\right)$$ (20-3)

where D_j is the distance between two object contours when one of the contours is rotated by a number of points so that its j-th starting point is aligned to the one of the possible starting points of the other curve.

Angle Code Histogram Similarity

If the distributions of the angle codes of two contours are similar, they will have similar local features. The difference between two angle code histograms is defined as:

$$D_h(I,J) = \sum_{i=1}^{m}\left|H_i(I) - H_i(J)\right|$$ (20-4)

where m is the number of bins in which the angle code histogram is partitioned, and $H_i(I)$ and $H_i(J)$ are the angle code histograms at the i-th bin for image I and image J, respectively.

Eccentricity Similarity

The eccentricity dissimilarity D_e between two leaf images I and J is defined as:

$$D_e(I,J) = \sum_{i=1}^{m}\left|e_I - e_J\right|$$ (20-5)

where e_I and e_J are the eccentricity measures of leaf images I and J, respectively.

Two methods, weight summation and fuzzy integral, can be applied to combining the similarity (dissimilarity) measures from the three sets of features.

Weight Summation Method

The weight summation method is defined as:

$$D_s(I,J) = \frac{w_1 D_e(I,J) + w_2 D_c(I,J) + w_3 D_h(I,J)}{w_1 + w_2 + w_3}$$ (20-6)

where I and J denote two leaf images, the query leaf image and one database leaf image. Weights w_1, w_2 and w_3 are used to weight the relative importance of the three feature sets, which will be determined by simulation or tuned by the user. The weight tuning is time-consuming and of a heuristic nature.

Fuzzy Integral
Fuzzy set theory features handling uncertain problems and has been successfully applied to image processing and pattern recognition [5]. Fuzzy measure and fuzzy integral have been proposed for various applications. Fuzzy measure is used as a scale to express the fuzziness grade. Based on the properties of fuzzy measure, a fuzzy integral is an aggregation operator on multi-attribute fuzzy information. They are used here to combine three distance measures from three different feature sets.

Let $X = \{x_1, x_2, \ldots, x_n\}$ be a finite set with each element being a feature set and $g_i = g(\{x_i\}), i = 1, 2, \ldots, n$, denotes the retrieval rate using feature set i individually. $g_\lambda(X)$ can be expressed as [5]

$$g_\lambda(X) = \sum_{i=1}^{n} g_i + \lambda \sum_{i_1=1}^{n-1} \sum_{i_2=i_1+1}^{n} g_{i_1} g_{i_2} + \cdots + \lambda^{n-1} g_1 \cdots g_n$$

$$= \frac{1}{\lambda} \left[\prod_{x_i \in X} (1 + \lambda g_i) - 1 \right] = 1 \tag{20-7}$$

If fuzzy densities $g_i (i = 1, 2, \ldots, n)$ are known, then the g_λ-measure can be obtained by solving the following equations for λ:

$$\lambda + 1 = \prod_{i=1}^{n} (1 + \lambda g_i) \tag{20-8}$$

Let $f : X \to [0,1]$ be a measurable function. $X = \{x_1, x_2, \ldots, x_n\}$ is so arranged such that $0 \le f(x_1) \le f(x_2) \le \ldots \le f(x_n) \le 1$ (if not, the elements of X are rearranged to satisfy this condition) and $A_i = \{x_i, x_{i+1}, \ldots, x_n\}$ and $X = \{x_1, x_2, \ldots, x_n\}$. In this application, $f(x_i)$ is the membership grade of the normalized similarity measure $s(x_i) (0 \le s(x_i) \le 1)$ on feature set i (centroid-contour distance, angle code histogram, or eccentricity) for an input leaf image. The S-function defined as below is adopted as the membership function:

$$f(s) = \begin{cases} 2s^2 & : & 0 \le s < 0.5 \\ 1 - 2(s-1)^2 & : & 0.5 \le s \le 1 \end{cases} \tag{20-9}$$

The Choquet integral of f with respect to g_λ is given by:

$$\int_A f dg = \sum_{i=1}^{n} \left[\left(f(x_i) - f(x_{i-1}) \right) g_\lambda(A_i) \right] \qquad (20\text{-}10)$$

where $f(x_0) = 0$. A fuzzy integral can be understood as the maximal grade of agreement between the objective evidence and the expectation, herein, the contribution of the three feature sets to the overall similarity between two images are measured. The images with the top integrated values are retrieved.

Experiments with Leaf Image Retrieval

A database containing 440 color leaf images of size 320×240 from 44 Chinese medicinal plants (10 samples from each plant) was used to evaluate leaf image retrieval performance. Figure 20-13 shows some leaf samples in the database.

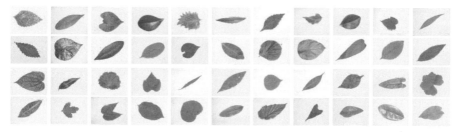

Figure 20-13 Examples of leaf images

The average number of contour points for an image was 939 and the average number of starting points was 4.4. The thinning based starting point locating method can reduce the comparison time significantly.

Figure 20-14(a) illustrates a leaf used as a query image. Figures 20-14(b) and 20-14(c) show the retrieval results for the leaf image shown in Figure 20-14 (a). The retrieved leaf images are perceptually similar to the query image, with the exception of the last one or two images. The weights w_1, w_2 and w_3 are set to 0.25, 0.60 and 0.15, respectively.

The recall rate is used to measure the retrieval performance. It is defined as $(N_{re} / N_t) \times 100\%$ where N_{re} is the number of the returned images that belongs to the same class as that of the query image and N_t is the total number of database images that has the same class. Obviously, N_t is equal to 10 in these experiments.

Several experiments were conducted by using (1) the combination of three feature sets, eccentricity (ECC), CCD and angle code histogram (ACH) based on the weighted summation (1a) and the fuzzy integral (1b); (2) curvature scale space (CSS) features; and (3) modified Fourier descriptor (MFD). For both CSS and MFD, the contour was sampled with 250 equal-distant points. Table 20-2 summarizes the average recall rate for these experiments. The average recall rate for the weighted feature set combination was obtained when $w_1 = 0.15$, $w_2 = 0.60$, and

$w_3 = 0.25$, which is the best result among many trials. When the weights were set to $w_1 = w_2 = w_3 = 1/3$ (equal weights), the result was degraded. The fuzzy integral approach achieved one of the best performances. This may be due to the weighted summation which exhaustively tuned the three weights. The clear advantage of the fuzzy integral approach is that it does not require any time-consuming parameter tuning. In the fuzzy integral approach, the computed parameters $\left(\lambda, g_1, g_2, g_3\right)$ were (0.218, 0.303, 0.523, 0.253), (0.749, 0.433, 0.655, 0.361), and (0.945, 0.618, 0.786, 0.514), respectively when the number of returned images was 20, 40, and 80.

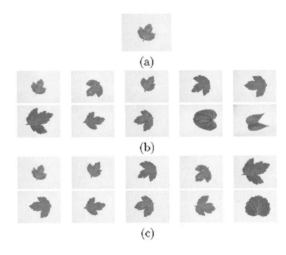

Figure 20-14 (a) Query image; (b) Top 10 retrieved images with the weighted summation approach; (c) Top 10 retrieved images with the fuzzy integral approach.

Since different species of plants may have very similar leaves and the leaf shapes of the same species may be very different, the leaf shape alone is not enough to distinguish different plant species, although it plays an important role in plant identification. The characteristics of flowers, stems, barks, fruits etc. are also important in plant identification.

FLOWER IMAGE RETRIEVAL

The task of flower image retrieval is to search for images of flowers in the database which match the flowers in a query flower image. The main challenge of the task is the localization of flower regions and the characterization of flowers. A number of similarity measures based on color histogram, spatial color distribution, and color clustering have been tested on flower images. A general discussion on color features and histograms, and associated similarity measures can be found in Chapter 1.

Method	The Number of Return Images	Retrieval Rate (%)
Weighted summation (the best weight setting)	20	60.3
	40	72.8
	80	84.8
Weighted summation (equal weights)	20	41.9
	40	67.5
	80	79.7
Fuzzy Integral	20	60.6
	40	70.6
	80	80.4
CSS	20	39.1
	40	51.8
	80	66.8
MFD	20	33.1
	40	45.0
	80	60.3

Table 20-2 Recall rates of different approaches

Feature Extraction and Similarity Measure

Distance between Color Histograms

Assume the color histogram consists of N bins. The color histogram of image I is represented by N bins $(p_1, p_2, \ldots, p_N)$ and that of image J is represented by $(p'_1, p'_2, \ldots, p'_N)$. The distance (dissimilarity) between two images I and J is defined as:

$$d(I, J) = \sum_{i=1}^{N} |p_i - p'_i| \qquad (20\text{-}11)$$

Figure 20-15(b) shows the top 10 retrieved images from a flower image database for the query image shown in Figure 20-15(a) when the above scheme is used with the RGB color system and 4096 ($16 \times 16 \times 16$) histogram bins.

Correlation between Color Histograms

Assume the normalized color histograms for images I and J are $(p_1, p_2, \ldots, p_N)$ and $(p'_1, p'_2, \ldots, p'_N)$, respectively. The correlation is defined as:

$$d(I, J) = \sum_{i=1}^{N} \sum_{j=1}^{N} w_{ij} s_{i,j} \qquad (20\text{-}12)$$

where $s_{i,j}$ is the color grade correlation coefficient of colors i and j, w_{ij} is to measure the color grade difference between two colors i and j, $\sum_{j=1}^{N} S_{i,j} = p_i$, and

$$\sum_{i=1}^{N} S_{i,j} = p'_j \quad (i, j = 1, \cdots, N).$$

The algorithm for obtaining s_{ij} is given below:

1. Set $s_{i,j} = 0$, $(i, j = 1, \cdots, N)$.

2. Set color grade difference $k = 0$.

3. If $\sum_{i=1}^{N} p_i = 0$ or $\sum_{i=1}^{N} p'_i = 0$, stop and obtain $s_{i,j}$; otherwise, set color grade $n = 1$.

4. If $p_n > 0$, $n - k > 0$ and $p'_{n-k} > 0$, then

   $$S_{n,n-k} = \min(p_n, p'_{n-k}), \ P_n = p_n - S_{n,n-k}, \ P'_{n-k} = p'_{n-k} - S_{n,n-k}.$$

   If $p_n > 0$, $n + k < N + 1$ and $p'_{n+k} > 0$, then

   $$S_{n,n+k} = \min(p_n, p'_{n+k}), \ P_n = p_n - S_{n,n+k}, \ P'_{n+k} = p'_{n+k} - S_{n,n+k}.$$

5. $n = n + 1$, if $n < N + 1$, go to Step 4 ; if $n > N$ and $k < N$, then $k = k + 1$ and go to Step 3.

Figure 20-15(c) shows the retrieval results using this scheme when $N = 4 \times 4 \times 4 = 64$ and $w_{ij} = (i - j)^2$.

Spatial Color Histograms

Assume the color histograms of ring-shaped regions of images I and J are A_{ij} and B_{ij} $(i = 1,2,\ldots,N, j = 1,2,\ldots,M)$, where M is the number of ring-shaped regions that the image is partitioned and N is the number of histogram bins in each region. The distance is defined as:

$$d(I,J) = \sum_{i=1}^{N} \sum_{j=1}^{M} |A_{ij} - B_{ij}| \qquad (20\text{-}13)$$

Figure 20-15(d) shows the retrieval results using this scheme when $N = 4 \times 4 \times 4 = 64$ and $M = 5$.

Hybrid Measure on Spatial Color Histogram

Assume the spatial color histogram is considered to be an $N \times M$ (where M is the number of spatial partition such as rings and N is the number of bins in the color histogram of a region) vector and the normalized spatial color histogram for images I

and J are $\left(A_1, A_2, \ldots, A_{N \times M}\right)$ and $\left(B_1, B_2, \ldots, B_{N \times M}\right)$, respectively. The correlation is defined as:

$$d(I,J) = \sum_{i=1}^{N \times M} \sum_{j=1}^{N \times M} w_{ij} s_{i,j} \tag{20-14}$$

where $s_{i,j}$ is similarly defined as in *Correlation between Color Histograms* and

$$w_{ij} = \left(\left[\frac{i}{M}\right] - \left[\frac{j}{M}\right]\right)^2 + \frac{|i \bmod M - j \bmod M|}{M} \tag{20-15}$$

Figure 20-15(e) shows the retrieval results using this scheme when $N = 4 \times 4 \times 4 = 64$ and $M = 5$.

Color Clustering
This scheme is normally implemented in the HSV color space. An image in the RGB space needs to be converted to the HSV color space. A clustering technique is utilized to find out the clusters and their centers. Denote the clustering parameters obtained by $C_i = \{c_i, p_i\}, i = 1, 2, \ldots, m$, where c_i and p_i are the color quantization value of cluster i in the HSV space and the percentage of pixels belonging to cluster i, and m is the number of clusters. Assume the clustering parameters for images I and J are $C_i = \{c_i, p_i\}, i = 1, 2, \ldots, n$ and $C'_i = \{c'_i, p'_i\}, i = 1, 2, \ldots, m$, respectively. Let d_{ij} define the distance between cluster i in image I and cluster j in image J, which is given by $d_{ij} = |c_i - c'_j|$. The distance between these two images is then defined as:

$$d(I,J) = \sum_{i=1}^{n} \sum_{j=1}^{m} v_{ij} d_{i,j} . \tag{20-16}$$

where v_{ij} is the similarity factor between clusters i and j, which is computed based on p_i ($i = 1, 2, \ldots, n$) and p'_i ($i = 1, 2, \ldots, m$). The algorithm is summarized as follows:

1. Set $L = \{\}$ and $M = \{(i, j) : i = 1, \cdots n, j = 1, \cdots m\}$.

2. Choose the minimum $d(r_i, r'_j)$ for $(i, j) \in M - L$. Label the corresponding (i, j) as (i', j'). $\min(p_{i'}, p'_{j'}) \to v_{i',j'}$.

3. If $P_{i'} < p'_{j'}$, set $v_{i,j} = 0, j \neq j'$; otherwise $v_{i,j} = 0, i \neq i'$.

4. $p_{i'} - \min(p_{i'}, p'_{j'}) \to p_{i'}$, $p'_{j'} - \min(p_{i'}, p'_{j'}) \to p'_{j'}$, and $L + \{(i', j')\} \to L$.

5. If $\displaystyle\sum_{i=1}^{m} p_i > 0$ and $\displaystyle\sum_{j=1}^{n} p_j' > 0$, go to Step 2; otherwise, stop.

Figure 20-15(f) shows the retrieval results using this scheme when the HSV color space was quantized into 36 different colors.

Discussion

In the color-based retrieval schemes discussed above, the methods using *Distance between Color Histograms* and *Spatial Color Histograms* achieve the best retrieval results. It is expected that an object-based approach that makes use of extracted flower regions and their characterization would produce better retrieval results.

20.6 Conclusion

In this chapter, a number of issues in developing a computer-aided live plant identification system have been addressed. These issues include: (1) a general procedure for plant identification; (2) the use of computer vision and intelligent information processing techniques for leaf and flower image processing and feature extraction; (3) the utilization of a plant database to assist plant identification; and (4) various techniques for leaf and flower image retrieval. The chapter also discussed various problems in computer-aided live plant identification and proposed some possible solutions to these problems.

Computer-aided plant identification is a very challenging task. The research in this area is still at a very early stage. The main difficulties in developing a sophisticated plant identification system that can achieve human-like performance are: (1) dealing with a huge number of plant species; (2) handling a large number of biological variations for the same species; (3) the segmentation of interesting features such as flowers and stems from an plant image; (4) the modeling of 3-D objects such as flowers with 2-D data.

Leaf modeling and classification is easier than the processing of other parts of a plant such as flowers, stems, and fruits. This is because (1) individual leaf images can be easily acquired, and (2) leaves are normally flat - so 2-D models are good enough to characterize leaves. Leaf types can be successfully classified using a combination of the leaf contour shape, leaf margin, and leaf venation. Leaf retrieval based on these features can also produce good results.

Flower modeling and classification is more difficult than leaf processing. Color is still a main feature used in flower matching and retrieval. There are still many problems in accurately locating flower regions when the background is complex. Due to its complex structure and the nature of 3-D objects, more work needs to be done on flower modeling. One of the most challenging problems is how to restore 3-D information from a 2-D image.

There is not much literature on the processing of stems and fruits, although stems and fruits are also very helpful in identifying plant species. The combining of various pieces of information is also a difficult problem and may require a sophisticated representation scheme and advanced adaptive processing algorithms.

References

[1] S. Abbasi, F. Mokhtarian, and J. Kittler, Reliable classification of chrysanthemum leaves through curvature scale space. *Proc. the International Conference on Scale-Space Theory in Computer Vision*, 284-295, Netherlands, 1997.
[2] L. Benson, *Plant Classification*, D. C. Heath and Company, 1957.
[3] B. Buyck, Taxonomists are an endangered species in Europe, *Nature*, Vol. 401, p. 321, 23 September 1999.
[4] CalFlora, http://www.calflora.org/calflora/botanical.html
[5] Z. Chi, H. Yan, and T. Pham, *Fuzzy Algorithm: with Application to Image Processing and Pattern Recognition*, Word Scientific, 1996.
[6] B. Cohen, I. Dinstein, M. Eyal, Computerized classification of color textured perthite images, *Proc. the 13th International Conference on Pattern Recognition*, Vol.2, pp.601 –605, 1996.
[7] M. Das, R. Manmatha, and E.M. Riseman, Indexing flowers by color names using domain knowledge-driven segmentation, *Proc. the Fourth IEEE Workshop on Applications of Computer Vision*, pp. 94-99, 1998.
[8] X. Ding, W. Kong, C. Hu, and S. Ma, Image retrieval using Schwartz representation of one-dimensional feature, *Proc. the International Conference on Visual Information and Information Systems*, pp.443-450, Amsterdam, The Netherlands, June 1999.
[9] EUCLID, http://www.publish.csiro.au/books/samples/euclid/index.htm
[10] I. Giakoumis and I. Pitas, Digital restoration of painting cracks, *Proc. the 1998 IEEE International Symposium on Circuits and Systems (ISCAS '98)*, Vol.4, pp. 269 –272, 1998.
[11] C. Im, H. Nishda, and T.L. Kunii, Recognizing plant species by leaf shapes-a case study of the acer family, *Proc. the 1998 IEEE international Conference of Pattern Recognition*, pp.1171-1173,1998.
[12] S.-K. Im, H.-M. Park, S.-W. Kim; Improved vein pattern extracting algorithm and its implementation, *Proc. the 2000 International Conference on Consumer Electronics*, pp.2 - 3, 2000.
[13] A. K. Jain, *Fundamentals of Digital Image Processing*, Prentice-Hall, London, UK, 1989.
[14] A.K. Jain and A. Vailaya, Shape-based retrieval: a case study with trademark image databases, *Pattern Recognition*, 31(9): 1369-1390, 1998.
[15] A. Kaupp, A. Dolemeyer, R. Wilzeck, R. Schlosser, S. Wolf and D. Meyer-Ebrecht, Measuring morphologic properties of the human retinal vessel system using a two-stage image processing approach, *Proc. the 1994 IEEE International Conference on Image Processing, Image Processing*, Vol.1, pp.431 –435, 1994.
[16] S. Loncaric, A survey of shape analysis techniques, *Pattern Recognition*, 31(8): 983-1001, 1998.
[17] Lucid, http://www.lucidcentral.com/lucid/about.htm
[18] H.-L. Peng and S.-Y., Chen, Trademark shape recognition using closed contours, *Pattern Recognition Letters*, Vol.18, pp. 791-803, 1997.
[19] B.V.M. Mehtre, M.S. Kankanhalli, and W.F. Lee, Shape measures for content-based image retrieval: a comparison, *Information Processing & Management*, 33(3): 319-337, 1997.
[20] E. Person and K.S. Fu, Shape discrimination using Fourier description, *IEEE Trans. on PAMI*, 3:208-210, 1981.
[21] C.L. Porter, *Taxonomy of flower plants*, 2nd edition, W.H. Freeman and Company, San Francisco, U.S.A, 1967.
[22] P.H. Raven, *Biology of Plants*, Worth Publishers, 1992.
[23] Y. Rui, A.C. She, and T.S. Huang, Modified Fourier descriptors for shape representation – a practical approach, *Proc. the First International Workshop on Image Database and Multi Media Search*, Amsterdam, The Netherlands, August 1996.
[24] T Saitoh and T. Kaneko, Automatic recognition of wild flowers, *Proc. the 15th International Conference on Pattern Recognition*, Vol. 2, pp. 507-510, 2000.
[25] S.M. Smith and J.M. Brady, SUSAN – a new approach to low level image processing, Technical Report, Dept. of Engineering Science, Oxford University, Oxford, UK, 1995.
[26] P. Soille, Morphological image analysis applied to crop field mapping, *Image and Vision Computing*, 18:1025-1032, 2000.
[27] UConn Plant Database of Trees, Shrubs and Vines, http://www.hort.uconn.edu:591/search.html
[28] G.W.A.M. Van-der-Heijden and A.M. Vossepoel, A landmark-based approach of shape dissimilarity," *Proc. the 13th International Conference on Pattern Recognition*, Vol. 1, pp. 120-124, 1996.

[29]   Z. Wang, Z. Chi, and D. Feng, "Leaf image retrieval using a two-step approach with shape features," *Proc. the 2000 IEEE Pacific-Rim Conference on Multimedia (PCM2000)*, pp. 380-383, Sydney, Australia, December 12-15, 2000.

[30]   H. Yan and J. Wu. Character and line extraction from color map images using a multi-layer neural network, *Pattern Recognition Letters*, Vol. 15, pp. 97-103, 1994.

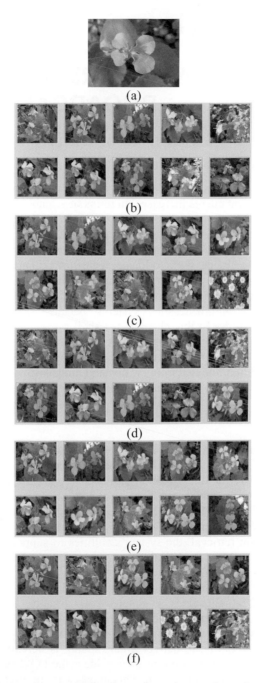

Figure 20-15. Flower image retrieval results using various feature extraction and similarity measures. (a) A query image; (b) – (f) The retrieval results using various similarity/dissimilarity measures.

21 FAST STARTUP AND INTERACTIVE RETRIEVALS OF BROADCAST VIDEOS

Dr. Jimmy To and Dr. C. K. Li

Ideally, the performance of a video delivery system is to be maintained even when the number of concurrent clients increases. In terms of scalability, the startup latency is expected to decrease exponentially with the bandwidth allocated. Broadcast VoD is a cost-effective approach to provide high-demand videos over a wide geographic area. However, in the context of information retrieval, basic VCR interactive functions are often used by the users to locate, review or skip areas of interest within a video. In the presence of reserved spare bandwidth, contingency channels can support VCR functions. While this is feasible in multicast systems, it may not be economically possible in broadcasting systems, because spare channels are too expensive to be reserved. In this chapter, we shall present an overview of some techniques for providing short startup latency and fundamental VCR interactions (play/resume, pause, rewind, fast-forward, skip-forward and backward repositioning) that facilitate information retrieval from on-demand broadcasting video systems.

21.1 Introduction

Technological advancements in information retrieval and communications have made on-line Video-on-Demand (VoD) services feasible. The VoD service enables clients to select and view a video from an on-line video library. Home viewers may not demand interactivity while watching a movie or documentary programme, however it is desirable that their ability to control of the viewing of the video is similar to that exercised over a conventional videocassette recorder (VCR). In the context of information retrieval, the system's abilities to support search, backtrack, freeze and skip can become critical, particularly for applications that require selective viewing of video recordings.

In a VoD system, video data streams are delivered to clients via a high-speed network. Figure 21-1 illustrates a simple VoD system. Since video is an isochronous medium, a video server has to reserve a sufficient amount of network and I/O bandwidth for a new data stream before committing to a client's request. The set of resources required to deliver a data stream is hereafter referred to as a *channel*. The resource bottleneck on the server side is the I/O disk bandwidth, or the aggregated sum of the disks' bandwidth. This imposes a limit on the number of data streams that can be delivered by a server.

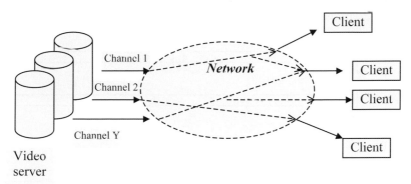

Figure 21-1 A VoD system

Clients' requests for the playback of videos are independent of one another and arrive at random intervals. Intuitively, the simplest scheme for scheduling the delivery of data streams is to dedicate one data stream to each client. However, this results in an enormous demand for data streams and will often result in low availability of service.

To maximize the utilization of the available data streams, requests from multiple clients for the same video, arriving within a certain time span, can be grouped together and served by a single common stream. Such a technique is referred to as *batching* [5]. Batching techniques can be divided into two categories, namely, *multicast* and *broadcast*. If the new data streams are dispatched based on predefined criteria and on the accumulation of user requests, the technique is deemed as multicast. If new data streams are dispatched periodically and independent of the number of pending requests, the technique is referred to as broadcast.

In multicast VoD, when resources needed to start a new data stream become available, the server selects a batch of clients waiting for the same movie according to a predefined policy. For instance, First-Come-First-Served (FCFS) selects the batch with the longest waiting request to be served next. The objective of this approach is to be fair over all videos. The *startup latency*, or the elapsed time between the arrival of a request and the time the request is fulfilled, will then depend on the arrival pattern. Various scheduled multicast schemes have been proposed and reported in the literature [1, 3-7, 10, 12, 18, 24-25, 28].

The basic concept of broadcast VoD can be illustrated by a simple scheme hereafter referred to as *staggered broadcasting*. As shown in Figure 21-2, several channels are used to broadcast a video periodically with staggered start times. For a video of duration L, time units being broadcast over K channels, in a staggered manner, clients can start viewing the video without waiting for more than (L/K) time units. However, the startup latency decreases only linearly as the number of channels K is increased.

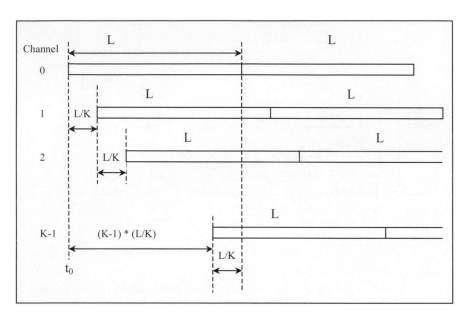

Figure 21-2 Staggered Broadcasting

To improve startup latency, Viswanathan and Imielinski proposed a novel broadcasting scheme called ***Pyramid Broadcasting*** (PB) [27] in which the startup latency decreases exponentially with the increase in the number of allocated channels. However, considerable buffering capacity on the client side is required. Another scheme called the ***Skyscraper Broadcasting*** (SB) [13] modifies the segmentation of video in PB and can keep the channel bandwidth equal to the playout rate. SB also imposes a tighter bound on the size of the largest segment, such that the client-side buffer requirement can be reduced. Other variants of broadcasting techniques are reported in the literature [2, 8-9, 11, 13-17, 19-23, 26-27].

One of the most desirable features of a VoD service is the provision of interactive functions similar to those found on typical VCR machines at home. These interactive functions include play/resume, pause, skip forward/backward, slow forward/backward, fast forward and rewind. Denoting the *point-of-play* of a client as the point in the video which the client is viewing, and, the *end point* of a VCR function as the point in the video where the client stops the function, each VCR function can be described as a pair $(\Delta t, \Delta l)$ [8], where Δt is the time duration of the actions and Δl is the relative position of the end point with respect to the current point-of-play. Assuming that the video normal playout rate is b, the nature of each interaction is defined in the table below.

Interaction	$(\Delta t, \Delta l)$
Play/Resume	$\Delta l / \Delta t = b$
Pause	$\Delta l = 0, \Delta t > 0$
Skip Forward/Backward	$\Delta t = 0, \Delta l \neq 0$
Slow Forward/Backward	$|\Delta l| / \Delta t < b$
Fast Forward/Rewind	$|\Delta l| / \Delta t > b$

In multicast systems, extensive buffering on the client side can support interactive VCR functions. Emergency interactive channels needs to be deployed when the client's buffered contents are not sufficient to satisfy the desired interaction [1, 4, 6, 18, 24, 28]. To a certain extent, the client-side buffering technique can be applied to broadcasting systems, but extra interactive channels are unlikely to be available because reserving these channels is too expensive in broadcasting.

In an on-line video library, there can be thousands of videos of different sizes. The size of the video library expands as new videos are added. The videos can have differing popularity. There may be tens of very popular and hundreds of less popular videos in the library, while the remaining videos will be viewed only rarely. In order to utilize the network bandwidth effectively and efficiently, videos of different degree of popularity should be delivered by different techniques. Since broadcasting excels in supporting an unlimited number of clients without any degradation in performance, it can be used to deliver the most popular videos, as well as the latest videos. For less popular videos, multicasting remains a sound alternative.

In the context of information retrieval, a client may want to browse through the video library visually to find the desired video and the point of interest. Hence a short startup latency is important, as a long startup latency may cause a client to quit with dissatisfaction. During the browsing, the client may also need VCR interactions, such as skip-forward and fast-forward, in order to decrease the time required to locate the right video(s). In attempting to support VCR interactivity, the designer of a VoD system needs to avoid excessive sacrifices in the *startup latency* and *client-side buffer requirements*. The cost of client-side buffer also affects the user-born hardware cost and will affect the competitiveness of the system. In subsequent sections, we examine how critical issues pertaining to the support of short startup latency and VCR interactions arise, and how such issues can be addressed.

21.2 The Generic Broadcasting Schemes

In basic staggered broadcasting, successive instances of the same video are broadcast on different channels with a time difference of (L/K) time units, as shown in Figure 21-2. The entire video is periodically broadcast over the channel at the normal playout rate. Since the consumption rate of the video is the same as the downloading rate of the data, minimal buffer space is required at the client-side. In the absence of a large buffer space, interactive functions can only be performed by switching channels. Without a large buffer and the ability to concurrently download data from multiple channels, the latency for interactions can easily be as large as L/K time units. Thus, the staggered broadcasting scheme suffers from both long startup latency and difficulties in supporting interactive operations.

THE PYRAMID BROADCASTING SCHEME

The Pyramid Broadcasting (PB) scheme [27] was the first broadcasting scheme to achieve a significant reduction in the startup latency, without needing a large number of channels. Figure 21-3 shows details of how PB divides a video into segments of geometrically increasing sizes, where the size of a segment is made α times the size of preceding segment and $\alpha > 1$. Each segment is repeatedly broadcast on a channel that has a bandwidth greater than the playout rate. Clients requesting a video wait for the next instance of the first segment to appear. They then collect the first segment and begin playout immediately. Since the broadcast duration of an instance of a segment is different from other segments, different starting times for viewing may dictate the collection of only certain instances of the subsequent segments, at a *relative time* with respect to the starting time. For different starting times, the relative downloading time of trailing segments will not necessarily be the same, neither will they necessarily be from the same cycle of the broadcast pattern. The complete sequence for downloading all segments from the channels for continuous playout is referred to as a *download sequence*, and is starting time dependent.

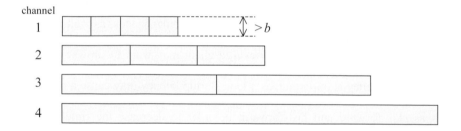

Figure 21-3 Pyramid Broadcasting scheme

Each client collects the next segment of data while playing the current data segment. Since the startup latency is proportional to the length of the first segment, with PB the startup latency improves exponentially with the number of allocated channels. Unfortunately, a large client buffer, of size of up to 70% of the video file, is implied.

THE SKYSCRAPER BROADCASTING SCHEME

The Skyscraper Broadcasting (SB) scheme [13] modifies the segmentation method of PB. In SB, a video is divided into segments of sizes according to the series {1, 2, 2, 5, 5, 12, 12, 25, 25, ...}. The segments are broadcast on channels at the normal playout rate. Similar to PB, a client downloads the data concurrently from a maximum of two channels. To restrict the growth of the largest segment size, an upper bound is imposed on the segment size. SB is able to achieve the same low startup latency as PB with a smaller buffer requirement. However, more channels are needed.

THE GREEDY DISK-CONSERVING BROADCASTING SCHEME

Following the direction on fixed client-I/O bandwidth of PB and SB, another broadcasting scheme called the Greedy Disk-Conserving Broadcasting (GDB) scheme [9] is proposed. GDB exploits the client-I/O bandwidth capacity to improve the system performance. As in SB, each channel with a bandwidth equal to the video playout rate is used by GDB to repeatedly broadcast a distinct video segment. An upper bound on segment size is applied to limit the growth of the segment size.

In GDB, video segment sizes are selected to ensure that the client can receive a video segment before the playback time of the segment. Since the startup latency is proportional to the size of the first video segment, GDB minimizes the size of the first video segment by selecting the largest video segment from amongst all subsequent video segment. Furthermore, a client downloads a video segment as late as possible to reduce the buffer requirement while preserving viewing continuity.

Before partitioning the video into segments, we need to decide on the maximum number of channels, denoted as m, each client can tune into concurrently to download the video segments on those channels. When the value of K is larger than two, the video is partitioned according to the series $\{1, 2, 4, 8, 16, ..., 2^m, 2^{m+1} -2, 2^{m+2} - 8, ...\}$. When the value of m is two, the series is $\{1, 2, 4, 4, 10, 10, 24, 24, 50, ...\}$. With the same client I/O bandwidth, GDB needs only 70% of the client-side buffer requirement required by SB or one less channel than that required by SB.

Figure 21-4 shows the performance in startup latency for a staggered broadcasting scheme, PB (where $\alpha = 3$), SB and GDB (where $m = 3$). SB and GDB are assumed to have the largest segment size, bounded to 120 times the first segment size. The playtime of the video is set at 90 minutes. We can see that the startup latency for broadcasting schemes with a geometrical segment size decreases exponentially with the increase in the network bandwidth allocated to a video and is much shorter than a staggered broadcasting scheme.

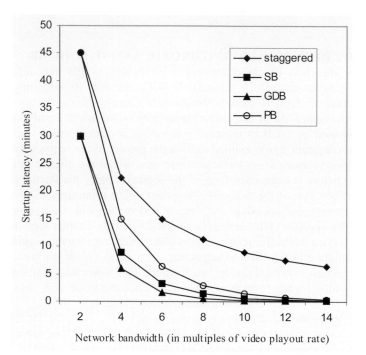

Figure 21-4 Startup latency

Although support for interactive functions is not mentioned in PB, SB and GDB, limited VCR interactions can be provided with these schemes because the data is stored in the buffer before being consumed. The PB has greater potential to provide support for interactions than SB and GDB because of the larger channel bandwidth and a larger buffer. To some extent, interactions can be provided by the data already accumulated in the large buffer of PB. The larger channel bandwidth of PB also implies a shorter access time for a missing segment.

From the above discussions, we can see that there is a trade-off between performance and resources. Staggered broadcasting requires little buffer space, but has a poor performance in the startup latency and the support of interactions. PB can be extended to support interactive functions more easily than SB and GDB, however a larger buffer space is required.

21.3 Prefetching Modes - Aggressive vs. Just-in-Time

By dividing a video into unequal segments of increasing sizes, pyramid-based broadcasting schemes reduce the startup latency more than linearly as the number of channels allocated to a video increases. In these schemes, one channel will be employed to periodically broadcast each segment. Since the segments are of unequal sizes, the broadcast duration for each segment is different. In other words, smaller segments will be broadcast more frequently than the larger segments. The appearance pattern of such segments on the channels is referred to as a *broadcast schedule*. A broadcast pattern will repeatedly appear in the broadcast schedule.

Either of the two modes of downloading can be employed by the client. The first mode is *aggressive* downloading of all segments as soon as they appear. In this mode, segments not immediately needed will nevertheless be collected and kept in the client's buffer space before consumption. We may also call this kind of data storing as *prefetching*. The other mode is *just-in-time* downloading, in which a segment is downloaded only when it is being viewed.

From Section 21.2, we know that prefetching will facilitate the implementation of interactive functions. This property gives the aggressive downloading approach an advantage over just-in-time downloading. Once a client has started downloading the first segment, the remaining segments from all other channels can be downloaded aggressively. However, this approach requires a very large client-side buffer and high I/O bandwidth. If the client's buffer space is to be kept small, the client should download the useful segment(s) as late as possible.

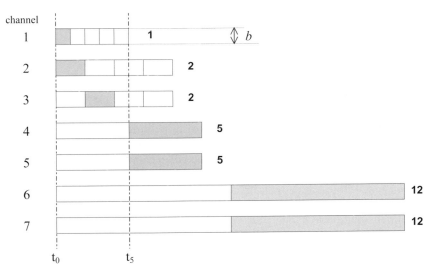

Figure 21-5 A scenario in the Skyscraper Broadcasting scheme

Figure 21-5 shows one of the download sequences of SB (in the shaded blocks). The figure shows that the downloading of segments from channel 3 onwards can be delayed. In practice, the two downloading modes can be mixed in a download sequence. However, the broadcast schedule inherent to a chosen scheme often imposes constraints under which only one of the two download modes can be used. These constraints may originate from the client-side buffer space or the client-side I/O bandwidth. Let us examine Figure 21-5 in detail. For the acclaimed performance, the design of SB restricts concurrent downloading to no more than two channels. If downloading of the first segment from channel 1 begins at t_0, the segment on channel 4 will be downloaded at t_5. Segments downloaded from channels 1, 2 and 3 can only sustain viewing at the normal playout rate for five time units. This implies that the segment chosen on channel 4 will be a just-in-time download. This can be avoided if the client-side I/O bandwidth is large enough for concurrent downloading from four channels. A similar situation can also be found in GDB.

Prefetching can facilitate, but cannot guarantee, the support of interactive functions. Good support of VCR functions depends on *what* data is being prefetched from the buffer. It is possible that a large time difference can exist between prefetched data and the current point-of-play. Data far away from the current point-of-play is not 'useful' for interactive operations.

Even when the prefetched data are close together and are considered 'useful', interaction may still not be feasible. As a VCR interaction can move the point-of-play to a new place in the video, the client may be forced to adopt to a different download sequence for subsequent segmentsThe client may encounter viewing discontinuity when normal playback resumes if they are missing data, that, according to the newly adopted download sequence, should have already been prefetched.

In the following sections, we shall review a number of recent methods to show how interactive functions can be incorporated into a broadcasting system.

21.4 Fast-Forward and Skip-Forward

Fast-forward (FF) interaction is a common function provided by VCRs. It is a visible forward scan of the video at a rate higher than the normal playout rate. Figure 21-6 illustrates the concept of an FF interaction. Let p be the current point-of-play of a video at t_0. After running t units of time, in the normal play mode the new point-of-play will be moved to $(p + a_1)$. But in the FF mode, the new point-of-play will become $(p + a_2)$, where $a_2 > a_1$, and the video data from p to $(p + a_2)$ will be consumed within t units of time. In other words, during FF the buffered video data will be consumed in less time than its normal playback duration. The ratio a_2/a_1 is the FF speed.

An intuitive solution to support FF interaction is to provide an FF version of the video from the video server. When a client requests an FF interaction, the delivery of video data is switched to the FF version. This solution, while feasible in multicast systems, is not applicable to broadcasting systems. Firstly, extra channels are needed for the delivery of such FF version. Secondly, the service policy of the video version will affect the performance and the efficiency of the interaction. If each interaction is to be allocated one independent channel, then scalability will suffer.Another solution is to deliver video data through the channels with a bandwidth larger than the playout rate, as in PB. In this case, the download sequence should be carefully designed such that the buffer space needed to store the prefetched data can be kept within reasonable bounds.

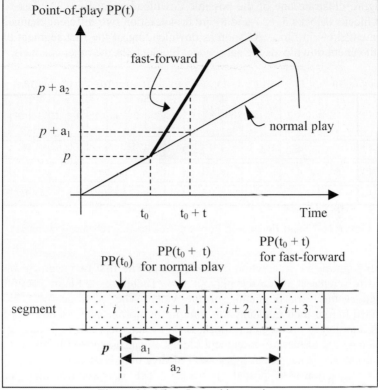

Figure 21-6 Fast-forward interaction

Next we consider three schemes which support FF interaction in broadcasting, namely the Short-Range Fast-Forward Broadcasting (SRFFB) scheme [21] and its variant, the Active Buffer Management (ABM) scheme [8] and the Mirrored-Pyramid Broadcasting (MPB) scheme [19-20]. All these schemes employed geometrically-increasing segment sizes such that the startup latency can be reduced more than linearly with the increase in the network bandwidth.

THE SHORT-RANGE FAST-FORWARD BROADCASTING SCHEME

The objective of the Short-Range Fast-Forward Broadcasting (SRFFB) scheme [21] is to enable clients to perform double-speed fast-forward (2xFF) interactions at any point-of-play during viewing. In SRFFB, a video file is equally divided into 2^K segment blocks where K is the number of channels allocated to the video. The segment blocks are then put onto each channel for periodic broadcast according to the series $\{2, 2, 4, 8, 16, 32, ...\}$. The bandwidth of the first channel is twice that of the normal playout rate while the bandwidth of the other channels is the same as the normal playout rate.

The download sequence of SRFFB employs a principle called "buffer-space-conserving" by which the segment blocks are downloaded as late as possible. Figure 21-7 shows the broadcast schedule of SRFFB with four channels. The shaded blocks in the figure illustrate one of the possible download sequences. The last set of the segment blocks (blocks 5, 6, 7 and 8) are broadcast on two different channels with a half-segment skew in time. As soon as downloading of the first segment block has started, the client downloads the subsequent blocks from the other channels.

timeslot #	T0		T1		T2		T3		T4		T5		T6		T7		T8		T9	
Channel 1	1	2	1	2	1	2	**1**	**2**	1	2	1	2	1	2	1	2	1	2	1	2
Channel 2	3		4		3		4		**3**		**4**		3		4		3		4	
Channel 3	5		6		7		8		5		6		7		8		5		6	
Channel 4	7		8		5		6		7		8		**5**		**6**		**7**		**8**	

Figure 21-7 Short-Range Fast-Forward Broadcasting scheme (4 channels)

SRFFB supports FF interaction by prefetching. The first two segment blocks are delivered to clients at twice the playout rate. Therefore, 2xFF interaction can be performed at the very beginning of the video. The other segment blocks are downloaded before their consumption time in the normal play mode by at least the playback duration of one segment block. If no FF interaction is requested during the downloading time of these two segment blocks, the next segment block, with respect to the current playing segment block, will always be prefetched in the buffer. As a result, FF interaction is supported by SRFFB. SRFFB guarantees that a client can perform a 2xFF action within a range of two segment blocks at any point-of-play of the video.

SRFFB allows the 2xFF for a distance of two segment blocks. As the number of channels allocated to a video is increased by one, the total number of segment blocks in SRFFB is doubled. This means that the playback duration of a segment block will be halved. Thus, SRFFB may be less effective when the number of allocated channels is very large. Moreover, the client-side I/O bandwidth is more demanding because in the worst case clients needs to concurrently download data from $(K - 2)$ channels for a K-channel SRFFB scheme. On the bright side, the broadcast schedule of SRFFB reduces the client-side buffer requirement to only one-quarter of the video file size.

The Medium-Range Fast-Forward Broadcasting (MRFFB) scheme [22] is a variation of SRFFB but with a larger buffer size requirement. MRFFB shares the same fundamental scheme design as SRFFB. The main difference is the way in which the video segments are downloaded. In SRFFB, the video segments are downloaded as late as possible in order to minimize the buffer size requirement. In MRFFB, however, the video segments are prefetched earlier. This allows the average FF range to be longer, though the guaranteed FF range for a 2xFF is still two segment blocks.

timeslot #	T0		T1		T2		T3		T4		T5		T6		T7		T8		T9
Channel 1	1	2	1	2	1	2	1	2	1	2	1	2	1	2	1	2	1	2	1	2
Channel 2	3		4		3		4		3		4		3		4		3		4
Channel 3	5		6		7		8		5		6		7		8		5		6
Channel 4	7		8		5		6		7		8		5		6		7		8

Figure 21-8 Medium-Range Fast-Forward Broadcasting scheme (4 channels)

In MRFFB, a new client has to catch the next appearance of the first segment of the target video. The playback of the video begins as soon as downloading starts. In the mean time, except for the last segment, the client downloads all the subsequent blocks from each channel proactively. The downloading of the first block of the last segment will not start until the last block of the first half of the second-last segment has been downloaded. The other blocks of the last segment will begin to be downloaded soon after. Figure 21-8 shows the download sequence of MRFFB with a starting timeslot the same as that of the download sequence of SRFFB in Figure 21-7.

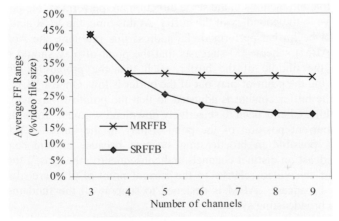

Figure 21-9 Average FF Range of MRFFB and SRFFB

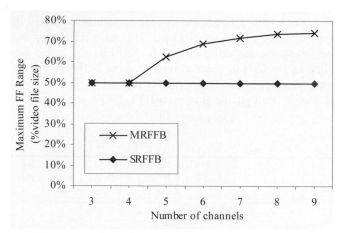

Figure 21-10 Maximum FF Range of MRFFB and SRFFB

Since the video segments are prefetched earlier in MRFFB than in SRFFB, MRFFB has a larger buffer size requirement than SRFFB, which is about 37.5% of video file size. With a larger buffer, MRFFB has a larger FF range than SRFFB, which is shown in Figure 21-9 and Figure 21-10. Figure 21-9 shows the average 2xFF range of both schemes. For each scheme given the same network bandwidth, we measure the feasible FF range in each block over all possible download sequences and take the average value. We can see that the MRFFB has a longer average FF range than SRFFB, up to 50% longer relative to that of SRFFB. Furthermore, the maximum FF range in MRFFB is also much longer than that of SRFFB over a wide range of network bandwidth.

THE ACTIVE BUFFER MANAGEMENT SCHEME

The effects of VCR interactions performed in the same direction are cumulative. When consecutive interactions in the same direction are performed, the point-of-play will finally move to the boundary of the buffer. At this time, the VCR actions in such direction cannot be further performed. To address this problem, the Active Buffer Management (ABM) scheme [8] suggests that the point-of-play should be actively maintained in the middle of the buffer, such that the probability for a VCR interaction moving the point-of-play out of the buffer is low. Once a VCR interaction is performed, the buffer content is adjusted so that the point-of-play will be moved back to the middle. Clients need to selectively download segments from the channels based on the current position of the point-of-play in the buffer. Active buffer management is possible in broadcasting systems because all the segments are repeatedly broadcast on distinct channels and consequently, the 'past', 'present' and 'future' parts of the video relative to the current point-of-play are always being simultaneously broadcast. ABM is designed to support all the fundamental VCR interactions via broadcasting systems.

As an example of the ABM scheme, let us consider three segments, namely y, $(y+1)$ and $(y+2)$, which have already been prefetched in the buffer. The point-of-play will be in segment (y+1). Suppose that an FF interaction moves the point-of-play to a point in segment (y+2). The point-of-play is then no longer in the middle and the client has to start to download segments (y+3) and (y+4) at the same time such that when the point-of-play moves into segment (y+3), segment (y+3) is the 'middle' segment in the buffer.

If the end point of an interaction is projected to cause future viewing discontinuity after normal playback is resumed, ABM will adjust the end point to a nearest possible point-of-play such that the client can be able to play the rest of the video without discontinuity, provided that there are no further VCR interactions. ABM called this kind of interaction the *discontinuous interactive function*.

In ABM, a video file is divided into segments with sizes following the series {1, 2, 4, 4, 8, 16, 16, 32, 64, 64, ...}. Similar to SB, in order to cut down the buffer requirement ABM also imposes a bound on the maximum size of segments . The series is designed based on the following two observations.

1. If the segment containing the end point of the interaction is not in the buffer and the video data being broadcast on this channel is chosen as the end point, the size of this segment must be set equal to the size of the next segment to guarantee viewing continuity.

2. The sizes of all segments beginning from the latter of the two above equal segments must satisfy the viewing continuity condition, assuming that the client can download data from m channels concurrently.

The value of m is set to be 3 in ABM. Therefore, clients are required to have three buffers where each of them is the same size of the largest segment.

Simulation results of ABM [8] show that the percentage of VCR interactions that can be supported by the buffer content varies between 93% and 87% as the probability of issuing an interaction increases from 0.1 to 0.9. As the maximum size of a segment increases, the distance between the requested end point and the implemented end point increases. The increase in the maximum size of a segment also decreases the percentage of successful VCR interactions.

THE MIRRORED-PYRAMID BROADCASTING SCHEME

In general, the bandwidth of the channels of SRFFB, MRFFB and ABM are equal to the playout rate of the video. In the Mirrored-Pyramid Broadcasting (MPB) scheme [19-20], however, all channels have a bandwidth twice the video playout rate to support the fast-forward interaction.

Suppose that the number of channels allocated to a video in MPB is ($2p$ + 1). Each channel will broadcast one distinct video segment periodically, hence there are ($2p$ + 1) video segments. The channels with channel number between 0 and (p −1) are called the *leading channels*, where the channel numbering starts from zero. The channels from (p + 1) to ($2p$) are referred to as *trailing channels* and the channel p is called the *peak channel*.

Starting from the second channel to the peak channel, each video segment is twice as large as the segment on its previous channel. However, instead of monotonically doubling the video segment size, MPB's segmentation algorithm changes to halving once the video segment size reaches a certain peak value. From then on, the size of video segments on the trailing channels will successively shrink by a factor of half. Thus, the segmentation model of the video is a pyramid plus its mirror image. The total number of video segment blocks in MPB is $(3 * 2^p - 2)$.

In such video segment distribution, the network bandwidth requirement is more demanding. To reduce this requirement, the video segments in the trailing set of channels can be packed into one [20]. Thus, the new total number of channels required is reduced to $(p + 2)$, without affecting the startup latency and buffer size requirement.

In MPB, clients need to concurrently download data from a maximum of two channels in order to have a continuous viewing of the video. Once downloading of the first video segment has begun, playback of the video is instantaneous. The downloading of an instance of the subsequent video segment begins whenever playback of the current video segment being downloaded has started.

Now let us discuss the FF interaction support in the scheme. In MPB, the channels deliver data at twice the video playout rate. So, the scheme design can intrinsically support double-rate fast-forward (2xFF) interaction.

Suppose the downloading of the video segment i ($i < (p+2)$), denoted as S_i, starts at timeslot t. Since the size of S_i is 2^i blocks, the latest time of the downloading of S_{i+1} starts at timeslot $(t + 2*2^i)$. Assume the FF period on S_i lasts for y timeslots. In order to prevent viewing discontinuity, the playback (including FF period) of S_i should not be allowed to be finished before $(t + 2*2^i)$. Thus, we have:

$$(t + 1) + y + 2*(2^i - y) \geq t + 2*2^i$$
$$1 + y - 2y \geq 0$$
$$y \leq 1$$

As a result, the guaranteed FF range in MPB is one block.

After the consumption (included FF) of S_i, S_{i+1} starts downloading and plays back instantly. The downloading of S_{i+2} will start either at the same time as S_{i+1} or just after the downloading of S_{i+1}. Since the video data is always delivered at twice the playout rate, the 2xFF interaction can be performed again. For instance, the client performs FF on S_{i+1} immediately after consumption of S_i has been completed. Let us assume that the downloading of S_{i+1} starts at timeslot t_1. The downloading of S_{i+2}, in the worst case, starts at timeslot $(t_1 + 2^{i+1})$. Supposing that the FF on S_{i+1} lasts for y_1 timeslots this time, we have:

$$t_1 + 2*(2^{i+1} - y_1) + y_1 \geq t_1 + 2^{i+1}$$
$$2*2^{i+1} - 2y_1 + y_1 \geq 2^{i+1}$$
$$2^{i+1} - y_1 \geq 0$$
$$y_1 \leq 2^{i+1}$$

The feasible FF range in the second attempt is 2^{i+1} timeslots which is much longer than that of the first attempt and is equal to the delivery time of S_{i+1}. This means that the 2xFF interaction can be performed immediately after the first attempt within the whole S_{i+1} video segment. Therefore, if the first attempt to FF is issued near the end of the consumption of S_i, we may lower the FF rate slightly to extend the consumption time of S_i to close the temporal gap between the FF interactions on S_i and S_{i+1} such that the FF can be continued.

SKIP-FORWARD

Skip-forward (SF) is an instant jump to a future point-of-play of the video. Although intuitively similar to fast-forward in that the point-of-play is moved in the forward direction, SF is quite different from fast-forward. The SF interaction can be implemented by tuning into the channel which is broadcasting the segment containing the end point. However, it is not as trivial as it seems. The instant jump to the other point-of-play may cause viewing discontinuity as the segments on the channel are of varying sizes and the size of a segment determines the frequency of itself appearing on the channel. Based on prefetched data, SRFFB, ABM and MPB can only support a limited range for this kind of interaction. For longer ranges of skip-forward operations, further research is needed.

21.5 Pause-Resume, Rewind and Backward Repositioning

In multicasting, each video is delivered in its entirety through a data stream. When a client pauses the viewing, (s)he is giving up the current stream. Upon resumption, the client is accommodated by one of the streams which started later than the one being given up (if there is a stream delivering a nearby point-of-play), or otherwise by a dedicated contingency stream (if there is one). As the starting interval between successive streams can be irregular in multicasting, it is difficult to estimate the minimal buffer space required if buffer alone is used to handle pause-resume interactions. In broadcasting schemes, however, pause-resume interactions can actually be supported by buffering.

Video data are periodically delivered on the channels in broadcasting systems. The access time for the segment with the largest size is the longest. Thus, if the playback time of the video data in the buffer can cover that access time, instantaneous resumption can be achieved. In other words, to support pause-resume, the buffer space size required is equivalent to the size of the largest segment.

The implementation of rewind is more complicated than that of fast-forward interactions. In FF, most of the difficulties can be resolved by prefetching. In the case of rewind, prefetching will make the implementation more complex. In broadcasting, even though part of video data are read-ahead in the buffer, often parts of the data will not be used in the near future but takes up buffer space. At any time, the buffered data is only piecewise continuous and does not necessarily form a continuous piece.

As the downloading and consumption of data proceeds concurrently in time, consumed data needs to be discarded to free buffer space for new data being downloaded. Data being prefetched is effectively competing with the consumed data for the limited buffer space. Due to the nature of compressed video format, video frames have decoding dependencies. As a result, compared to fast-forward, more consumed data needs to reside in the buffer to ensure a smooth rewind operation.

The range for rewind operations is also difficult to guarantee. During a rewind action, the discarded past-data relative to the current point-of-play may need to be downloaded from the channels again. When normal playback is resumed, the discarded data will become future-data relative to the current point-of-play and will need to be downloaded again. Since the rewind duration is unpredictable, it is difficult to determine whether past or future-data should be removed from the buffer to guarantee viewing continuity .

A simple method for providing the rewind interaction in broadcasting schemes is to double the size of the buffer such that there is enough space to keep the consumed data in the buffer to support rewind interaction. The current point-of-play is then kept in the middle of the data in the buffer, as in ABM. In the proposed broadcasting schemes, the segment size grows geometrically and the buffer size requirement is determined by the largest segment size. By using the same buffer size (as that of holding prefetch data) to keep the consumed data, the time span of the consumed data in the buffer will always be large enough to download an entire instance of its preceding segment and hence allows the client to perform the rewind interaction seamlessly. This solution, however, will inevitably increase the user-born hardware cost and affect the competitiveness of the system.

The problems faced by backward repositioning is similar to those faced by skip-forward. Backward repositioning can be performed by tuning to another channel. For short ranges, it can be satisfied by the buffer content. However, an abrupt change of point-of-play often implies a change in the download sequence. Depending on the buffer content and the new download sequence, viewing discontinuity could still occur and the problem still needs to be addressed.

21.6 Conclusion

We have presented an overview of the issues to be resolved in the short startup latency and the support for VCR interactions which are important for browsing through the contents of broadcasting videos. Broadcasting schemes divide a video into segments, and deliver each segment on one channel in cycles so as to reduce the startup latency. The interactions discussed include pause, play/resume, fast-forward, rewind, skip-forward and backward repositioning. In broadcasting, it is undesirable to reserve spare channels to support the VCR interactions because the solution is not scalable and spare channels are expensive. To support the interactive functions of broadcasting systems, we have considered combinations of techniques such as prefetching video data in the client-side buffer; dividing the video into segments with sizes similar to VCR-oriented series; and delivering the segment on channels at a rate larger than the normal video playout rate. Some of these solutions, however, compromise system performance (such as startup latency) and/or have resources implications (such as client-side buffer requirement and I/O bandwidth requirement). While individual techniques for specific VCR functions have been proposed, most schemes are not designed to support all VCR functions. In particular, better solutions to backward interactions remain to be found.

Acknowledgement

Some of the work conducted by the authors was supported by the Hong Kong Polytechnic University's internal research grant G-T229.

References

[1]   E.L. Abram-Profeta, and K.G. Shin, "Providing Unrestricted VCR Functions in Multicast Video-on-Demand Servers". *Proceedings of the 1998 IEEE International Conference on Multimedia Computing and Systems*, Austin, TX, USA, 28 June-1 July, 1998, pp.66-75 (1998).

[2]   C.C. Aggarwal, J.L. Wolf, and P.S. Yu, "Design and Analysis of Permutation-Based Pyramid Broadcasting". *Multimedia Systems*, Vol. 7, No. 6, pp.439-448 (1999).

[3]   C. C. Aggarwal, J.L. Wolf, and P.S. Yu, "On Optimal Batching Policies for Video-on-Demand Storage Servers". *Proceedings of the Third IEEE International Conference on Multimedia Computing and Systems 1996*, Hiroshima, Japan, 17-23 June, 1996, pp.253-258 (1996).

[4]   K.C. Almeroth, and M.H. Ammar, "The Use of Multicast Delivery to Provide a Scalable and Interactive Video-on-Demand Service". *IEEE Journal on Selected Areas in Communications*, Vol. 14, No. 6, pp.1110-1122 (1996).

[5]   A. Dan, D. Sitaram, and P. Shahabuddin, "Scheduling Policies for an On-Demand Video Server with Batching". *Proceedings of the Second ACM International Conference on Multimedia*, San Francisco, USA, October, 1994, pp.15-24 (1994).

[6]   A.Dan, P. Shahabuddin, D. Sitaram, and D. Towsley, "Channel Allocation under Batching and VCR Control in Video-on-Demand Systems". *Journal of Parallel and Distributed Computing*, Vol. 30, No. 2, pp.168-179 (1995).

[7]   A. Dan, D. Sitaram, and P. Shahabuddin, "Dynamic Batching Policies for an On-Demand Video Server". *Multimedia Systems*, Vol. 4, No. 3, pp.12-121 (1996).

[8]   Z. Fei, M.H.Ammar, I. Kamel, and S. Mukherjee, "Providing Interactive Functions Through Active Client Buffer Management in Partitioned Video Broadcast". *Proc. of First International Workshop on Networked Group Communication*, Pisa, Italy, Novemeber, 1999, pp.152-169 (1999).

[9]   L. Gao, J. Kurose, and Don Towsley, "Efficient Schemes for Broadcasting Popular Videos". *Proceedings of the 8th International Workshop on NOSSDAV*, 8-10 July, 1998 (1998).

[10]   L.Golubchik, J.C.S. Lui, and R.R. Muntz, "Adaptive Piggybacking: A Novel Technique for Data Sharing in Video-on-Demand Storage Servers". *Multimedia Systems*, Vol.4, No.3, pp.140-155 (1996).

[11]   K.A. Hua, Y. Cai, and S. Sheu, "Exploiting Client Bandwidth for More Efficient Video Broadcast". *Proceedings of the 7th International Conference on Computer Communications and Networks*, Lafayette, L.A., 12-15 October, 1998, pp.848-856 (1998).

[12]   K.A. Hua, Y. Cai, and S. Sheu, "Patching: A Multicast Technique for True Video-on-Demand Services". *Proceedings of the Sixth ACM International Conference on Multimedia*, Bristol, United Kingdom, 13-16 September, 1998, pp.191-200 (1998).

[13]   K.A. Hua, and S. Sheu, "Skyscraper Broadcasting: A New Broadcasting Scheme for Metroplitan Video-on-Demand Systems". *Proceedings of the ACM SIGCOMM '97 Conference on Applications, Technologies, Architectures, and Protocols for Computer Communication*, Cannes, France, 14-18 September, 1997, pp.89-100 (1997).

[14]   L.S. Juhn and L.M. Tseng, "Harmonic Broadcasting for Video-on-Demand Service". *IEEE Transactions on Broadcasting*, Vol. 43, No.3, pp.268-271, (1997).

[15]   L.S. Juhn, and L.M. Tseng, "Staircase Data Broadcasting and Receiving Scheme for Hot Video Service". *IEEE Transactions on Consumer Electronics*, Vol. 43, No.4, pp.1110-1117 (1997).

[16]   L.S. Juhn, and L.M. Tseng, "Fast Data Broadcasting and Receiving Scheme for Popular Video Service". *IEEE Transactions on Broadcasting*, Vol 44, No.1, pp.100-105 (1998).

[17]   S.B, Jim, and W.S. Lee, "Video Allocation Methods in a Multi-Level Server for Large-Scale VoD Services". *IEEE Transactions on Consumer Electronics*, Vol. 44, No. 4, pp.1309-1318 (1998).

[18]   W.J. Liao, and V.O.K. Li, "The Split and Merge Protocol for Interactive Video-on-Demand". *IEEE Multimedia*, Vol. 4, No. 4, pp51-62 (1997).

[19]   H.S. Ma, and T.P.J. To, "Mirrored-Pyramid Broadcasting for Metropolitan On-Demand Video Delivery". *Proceedings of the 2000 Workshop on Multimedia Data Storage, Retrieval, Integration and Applications*, Hong Kong, 13-15 January, 2000, pp.104-109 (2000).

[20]   H.S. Ma, T.P.J. To, and P.K. Lun, "Enchanced Mirrored-Pyramid Broadcasting for Video-On-Demand Delivery". *Proceedings of the 2000 IEEE Asian Pacific Conference on CAS*, Tianjin, China, 4-6 December, 2000, pp.875-878 (2000).

[21]   H.S. Ma, T.P.J. To, and C.K. Li, "A New Broadcasting Scheme Supporting Fast-Forward for Video-on-Demand Service", *Proceedings of the Third International Network Conference (INC 2002)*, Plymouth, UK, 16-18 July, 2002, pp.215-222 (2002).

[22]   H.S. Ma, T.P.J. To, and C.K. Leung, "A Fast-Forward-Oriented Broadcasting Scheme for Video-on-Demand Service", *Proc. of the 6th IASTED International Conference in Internet and Multimedia System and Applications*, Kauai, Hawaii, USA, 12-14 August, 2002, pp.366-370 (2002).

[23]   J.F. Pâris, D.D.E. Long, and P.E. Mantey, "Zero-Delay Broadcasting Protocols for Video-on-Demand". *Proceedings of the 1999 International Conference on Multimedia*, Orlando, FL USA, 30 October – 5 November, 1999, pp.189-197 (1999).

[24]   W.F. Poon, K.T. Lo, and J. Feng, "Multicast Video-on-Demand System with VCR Functionality". *Proceedings of the 1998 International Conference on Communication Technology*, Beijing, China, 22-24 October, 1998, pp.S23-10 (1998).

[25]   S. Sheu, Hua, K.A. and W. Tavanapong, "Chaining: A Generalized Batching Technique for Video-on-Demand Systems". *Proceedings of the IEEE International Conference on Multimedia Computing and Systems '97*, Ottawa, Ontario, Canada, 3-6 June, 1997, pp.110-117 (1997).

[26]   Y.C. Tseng, C.M. Hsieh, M.H. Yang, W.H. Liao, and J.P. Sheu, "Data Broadcasting and Seamless Channel Transition for Highly-Demanded Videos". *Proceedings of IEEE INFOCOM 2000*, USA, 26-30 March, 2000, pp.727-736 (2000).

[27]   S. Viswanathan, and T. Imielinski, "Metropolitan Area Video-on-Demand Service Using Pyramid Broadcasting". *Multimedia Systems*, Vol. 4, pp.197-208 (1996).

[28]   P.S. Yu, J.L. Wolf, and H. Shachnai, "Design and Analysis of a Look-Ahead Scheduling Scheme to Support Pause-Resume for Video-on-Demand Applications". *Multimedia Systems*, Vol. 3, pp.137-149 (1995).

Printing (Computer to Plate): Saladruck Berlin
Binding: Stürtz AG, Würzburg